OXFORD MASTER SERIES IN STATISTICAL, COMPUTATIONAL, AND THEORETICAL PHYSICS

OXFORD MASTER SERIES IN PHYSICS

The Oxford Master Series is designed for final year undergraduate and beginning graduate students in physics and related disciplines. It has been driven by a perceived gap in the literature today. While basic undergraduate physics texts often show little or no connection with the huge explosion of research over the last two decades, more advanced and specialized texts tend to be rather daunting for students. In this series, all topics and their consequences are treated at a simple level, while pointers to recent developments are provided at various stages. The emphasis in on clear physical principles like symmetry, quantum mechanics, and electromagnetism which underlie the whole of physics. At the same time, the subjects are related to real measurements and to the experimental techniques and devices currently used by physicists in academe and industry. Books in this series are written as course books, and include ample tutorial material, examples, illustrations, revision points, and problem sets. They can likewise be used as preparation for students starting a doctorate in physics and related fields, or for recent graduates starting research in one of these fields in industry.

CONDENSED MATTER PHYSICS

1. M. T. Dove: *Structure and dynamics: an atomic view of materials*
2. J. Singleton: *Band theory and electronic properties of solids*
3. A. M. Fox: *Optical properties of solids*
4. S. J. Blundell: *Magnetism in condensed matter*
5. J. F. Annett: *Superconductivity*
6. R. A. L. Jones: *Soft condensed matter*

ATOMIC, OPTICAL, AND LASER PHYSICS

7. C. J. Foot: *Atomic physics*
8. G. A. Brooker: *Modern classical optics*
9. S. M. Hooker, C. E. Webb: *Laser physics*

PARTICLE PHYSICS, ASTROPHYSICS, AND COSMOLOGY

10. D. H. Perkins: *Particle astrophysics*
11. Ta-Pei Cheng: *Relativity, gravitation, and cosmology*

STATISTICAL, COMPUTATIONAL, AND THEORETICAL PHYSICS

12. M. Maggiore: *A modern introduction to quantum field theory*
13. W. Krauth: *Statistical mechanics: algorithms and computations*
14. J. P. Sethna: *Entropy, order parameters, and complexity*

A Modern Introduction to Quantum Field Theory

Michele Maggiore

Département de Physique Théorique
Université de Genève

OXFORD
UNIVERSITY PRESS

OXFORD
UNIVERSITY PRESS
Great Clarendon Street, Oxford OX2 6DP
United Kingdom

Oxford University Press is a department of the University of Oxford.
It furthers the University's objective of excellence in research, scholarship,
and education by publishing worldwide. Oxford is a registered trade mark of
Oxford University Press in the UK and in certain other countries

First published 2005
Reprinted 2013

British Library Cataloguing in Publication Data
Data available

Library of Congress Cataloging in Publication Data
Data available

ISBN 978-0-19-852074-0

A Maura, Sara e Ilaria

Contents

Preface

This book grew out of the notes of the course on quantum field theory that I give at the University of Geneva, for students in the fourth year.

Most courses on quantum field theory focus on teaching the student how to compute cross-sections and decay rates in particle physics. This is, and will remain, an important part of the preparation of a high-energy physicist. However, the importance and the beauty of modern quantum field theory resides also in the great power and variety of its methods and ideas. These methods are of great generality and provide a unifying language that one can apply to domains as different as particle physics, cosmology, condensed matter, statistical mechanics and critical phenomena. It is this power and generality that makes quantum field theory a fundamental tool for any theoretical physicist, independently of his/her domain of specialization, as well as, of course, for particle physics experimentalists.

In spite of the existence of many textbooks on quantum field theory, I decided to write these notes because I think that it is difficult to find a book that has a modern approach to quantum field theory, in the sense outlined above, and at the same time is written having in mind the level of fourth year students, which are being exposed for the first time to the subject.

The book is self-contained and can be covered in a two semester course, possibly skipping some of the more advanced topics. Indeed, my aim is to propose a selection of topics that can really be covered in a course, but in which the students are introduced to many modern developments of quantum field theory.

At the end of some chapters there is a Solved Problems section where some especially instructive computations are presented in great detail, in order to give a model of how one really performs non-trivial computations. More exercises, sometimes quite demanding, are provided for Chapters 1 to 8, and their solutions are discussed at the end of the book. Chapters 9, 10 and 11 are meant as a bridge toward more advanced courses at the PhD level.

A few parts which are more technical and can be skipped at a first reading are written in smaller characters.

Acknowledgments. I am very grateful to Stefano Foffa, Florian Dubath, Alice Gasparini, Alberto Nicolis and Riccardo Sturani for their help and for their careful reading of the manuscript. I also thank Jean-Pierre Eckmann for useful comments, and Sonke Adlung, of Oxford University Press, for his friendly and useful advice.

Notation

Our notation is the same as Peskin and Schroeder (1995). We use units $\hbar = c = 1$; their meaning and usefulness is illustrated in Section 1.2. The metric signature is

$$\eta_{\mu\nu} = (+,-,-,-)\,.$$

Indices. Greek indices take values $\mu = 0,\ldots,3$, while spatial indices are denoted by Latin letters, $i,j,\ldots = 1,2,3$. The totally antisymmetric tensor $\epsilon^{\mu\nu\rho\sigma}$ has $\epsilon^{0123} = +1$ (therefore $\epsilon_{0123} = -1$). Observe that, e.g. $\epsilon^{1230} = -1$ since, to recover the reference sequence 0123, the index zero has to jump three positions. Therefore $\epsilon^{\mu\nu\rho\sigma}$ is anticyclic. Repeated upper and lower Lorentz indices are summed over, e.g. $A_\mu B^\mu \equiv \sum_{\mu=0}^{3} A_\mu B^\mu$. When the equations contain only spatial indices, we will keep all indices as upper indices,[1] and we will sum over repeated upper indices; e.g. the angular momentum commutation relations are written as $[J^k, J^l] = i\epsilon^{klm} J^m$, and the totally antisymmetric tensor ϵ^{ijk} is normalized as $\epsilon^{123} = +1$. The notation $\mathbf{A}$ denotes a spatial vector whose components have upper indices, $\mathbf{A} = (A^1, A^2, A^3)$.

[1] We will never use lower spatial indices, to avoid the possible ambiguity due to the fact that in equations with only spatial indices it would be natural to use δ_{ij} to raise and lower them, while with our signature it is rather $\eta_{ij} = -\delta_{ij}$.

The partial derivative is denoted by $\partial_\mu = \partial/\partial x^\mu$ and the (flat space) d'Alambertian is $\Box = \partial_\mu \partial^\mu = \partial_0^2 - \boldsymbol{\nabla}^2$. With our choice of signature the four-momentum operator is represented on functions of the coordinates as $p^\mu = +i\partial^\mu$, so $p^0 = i\partial/\partial x^0 = i\partial/\partial t$ and $p^i = i\partial^i = -i\partial_i = -i\partial/\partial x^i$. Therefore $p^i = -i\nabla^i$ with $\nabla^i = \partial/\partial x^i = \partial_i$ or, in vector notation, $\mathbf{p} = -i\boldsymbol{\nabla}$ and $\boldsymbol{\nabla} = \partial/\partial\mathbf{x}$.

The symbol $\overleftrightarrow{\partial_\mu}$ is defined by $f \overleftrightarrow{\partial_\mu} g = f\partial_\mu g - (\partial_\mu f)g$. We also use the Feynman slash notation: for a four-vector A^μ, we define $\not{A} = A_\mu \gamma^\mu$. In particular, $\not{\partial} = \gamma^\mu \partial_\mu$.

Dirac matrices. Dirac γ matrices satisfy

$$\{\gamma^\mu, \gamma^\nu\} \equiv \gamma^\mu\gamma^\nu + \gamma^\nu\gamma^\mu = 2\eta^{\mu\nu}\,.$$

Therefore $\gamma_0^2 = 1$ and, for each i, $(\gamma^i)^2 = -1$; γ^0 is hermitian while, for each i, γ^i is antihermitian,

$$(\gamma^0)^\dagger = \gamma^0\,, \qquad (\gamma^i)^\dagger = -\gamma^i\,,$$

or, more compactly, $(\gamma^\mu)^\dagger = \gamma^0\gamma^\mu\gamma^0$. The matrix γ^5 is defined as

$$\gamma^5 = +i\gamma^0\gamma^1\gamma^2\gamma^3\,,$$

and satisfies

$$(\gamma^5)^2 = 1\,, \qquad (\gamma^5)^\dagger = \gamma^5\,, \qquad \{\gamma^5, \gamma^\mu\} = 0\,.$$

We also define

$$\sigma^{\mu\nu} = \frac{i}{2}\left[\gamma^\mu, \gamma^\nu\right].$$

Two particularly useful representations of the γ matrix algebra are

$$\gamma^0 = \begin{pmatrix} 0 & 1 \\ 1 & 0 \end{pmatrix}, \quad \gamma^i = \begin{pmatrix} 0 & \sigma^i \\ -\sigma^i & 0 \end{pmatrix}, \quad \gamma^5 = \begin{pmatrix} -1 & 0 \\ 0 & 1 \end{pmatrix}$$

(here 1 denotes the 2×2 identity matrix), which is called the chiral or Weyl representation, and

$$\gamma^0 = \begin{pmatrix} 1 & 0 \\ 0 & -1 \end{pmatrix}, \quad \gamma^i = \begin{pmatrix} 0 & \sigma^i \\ -\sigma^i & 0 \end{pmatrix}, \quad \gamma^5 = \begin{pmatrix} 0 & 1 \\ 1 & 0 \end{pmatrix},$$

which is called the ordinary, or standard, representation.

The Pauli matrices are

$$\sigma^1 = \begin{pmatrix} 0 & 1 \\ 1 & 0 \end{pmatrix}, \quad \sigma^2 = \begin{pmatrix} 0 & -i \\ i & 0 \end{pmatrix}, \quad \sigma^3 = \begin{pmatrix} 1 & 0 \\ 0 & -1 \end{pmatrix},$$

and satisfy

$$\sigma^i \sigma^j = \delta^{ij} + i\epsilon^{ijk}\sigma^k .$$

We also define

$$\sigma^\mu = (1, \sigma^i), \qquad \bar{\sigma}^\mu = (1, -\sigma^i).$$

In the calculation of cross-sections and decay rates we often need the following traces of products of γ matrices,

$$\begin{aligned} \mathrm{Tr}(\gamma^\mu\gamma^\nu) &= 4\,\eta^{\mu\nu}, \\ \mathrm{Tr}(\gamma^\mu\gamma^\nu\gamma^\rho\gamma^\sigma) &= 4\left(\eta^{\mu\nu}\eta^{\rho\sigma} - \eta^{\mu\rho}\eta^{\nu\sigma} + \eta^{\mu\sigma}\eta^{\nu\rho}\right), \\ \mathrm{Tr}(\gamma^5\gamma^\mu\gamma^\nu\gamma^\rho\gamma^\sigma) &= -4i\epsilon^{\mu\nu\rho\sigma} . \end{aligned}$$

Fourier transform. The four-dimensional Fourier transform is

$$\begin{aligned} f(x) &= \int \frac{d^4k}{(2\pi)^4}\, e^{-ikx} \tilde{f}(k), \\ \tilde{f}(k) &= \int d^4x\, e^{ikx} f(x), \end{aligned}$$

and, because of our choice of signature, the three-dimensional Fourier transform is defined as

$$\begin{aligned} f(\mathbf{x}) &= \int \frac{d^3k}{(2\pi)^3}\, e^{+i\mathbf{k}\cdot\mathbf{x}} \tilde{f}(\mathbf{k}), \\ \tilde{f}(\mathbf{k}) &= \int d^3x\, e^{-i\mathbf{k}\cdot\mathbf{x}} f(\mathbf{x}). \end{aligned}$$

For arbitrary n, the n-dimensional Dirac delta satisfies

$$\int d^nx\, e^{ikx} = (2\pi)^n \delta^{(n)}(k).$$

Electromagnetism. The electron charge is denoted by e, and $e < 0$. As is customary in quantum field theory and particle physics, we use the Heaviside–Lorentz system of units for electromagnetism (also called *rationalized* Gaussian c.g.s. units). This means that the fine structure constant $\alpha = 1/137.035\,999\,11(46)$ is related to the electron charge by

$$\alpha = \frac{e^2}{4\pi\hbar c},$$

or simply $\alpha = e^2/(4\pi)$ when we set $\hbar = c = 1$. With this definition of the unit of charge there is no factor of 4π in the Maxwell equations,

$$\boldsymbol{\nabla}\cdot\mathbf{E} = \rho, \qquad \boldsymbol{\nabla}\times\mathbf{B} - \partial_0\mathbf{E} = \mathbf{J},$$

while the Coulomb potential between two static particles of charges $Q_1 = q_1 e$ and $Q_2 = q_2 e$ is

$$V(r) = \frac{Q_1 Q_2}{4\pi r} = q_1 q_2 \frac{\alpha}{r} \tag{1}$$

(where in the last equality we have used $\hbar = c = 1$), and the energy density of the electromagnetic field is

$$\varepsilon = \frac{1}{2}(\mathbf{E}^2 + \mathbf{B}^2).$$

In quantum electrodynamics nowadays these conventions on the electric charge are almost universally used, but it is useful to remark that they differ from the (*unrationalized*) Gaussian units commonly used in classical electrodynamics; see, e.g. Jackson (1975) or Landau and Lifshitz, vol. II (1979), where the electron charge is rather defined so that $\alpha = e^2_{\rm unrat}/(\hbar c) \simeq 1/137$, and therefore $e_{\rm unrat} = e/\sqrt{4\pi}$. The unrationalized electric and magnetic fields, $\mathbf{E}_{\rm unrat}, \mathbf{B}_{\rm unrat}$ by definition are related to the rationalized electric and magnetic fields, $\mathbf{E}, \mathbf{B}$ by $\mathbf{E}_{\rm unrat} = \sqrt{4\pi}\,\mathbf{E}$, $\mathbf{B}_{\rm unrat} = \sqrt{4\pi}\,\mathbf{B}$, i.e. $A^\mu_{\rm unrat} = \sqrt{4\pi}\,A^\mu$. The form of the Lorentz force equation is therefore unchanged, since with these definitions $e\mathbf{E} = e_{\rm unrat}\mathbf{E}_{\rm unrat}$ and $e\mathbf{B} = e_{\rm unrat}\mathbf{B}_{\rm unrat}$. However, a factor 4π appears in the Maxwell equations, $\boldsymbol{\nabla}\cdot\mathbf{E}_{\rm unrat} = 4\pi\rho_{\rm unrat}$ and $\boldsymbol{\nabla}\times\mathbf{B}_{\rm unrat} - \partial_0\mathbf{E}_{\rm unrat} = 4\pi\mathbf{J}_{\rm unrat}$; the Coulomb potential becomes $V(r) = (Q_1Q_2)_{\rm unrat}/r$, and the electromagnetic energy density becomes $\varepsilon = (\mathbf{E}^2_{\rm unrat} + \mathbf{B}^2_{\rm unrat})/(8\pi)$.

In quantum electrodynamics, since $eA^\mu = e_{\rm unrat}A^\mu_{\rm unrat}$, the interaction vertex is $-ie\gamma^\mu$ in rationalized units and $-ie_{\rm unrat}\gamma^\mu$ in unrationalized units. However, in unrationalized units the gauge field is not canonically normalized, as we see for instance from the form of the energy density. Therefore in unrationalized units the factor associated to an incoming photon in a Feynman graph becomes $\sqrt{4\pi}\epsilon^\mu$ rather than just ϵ^μ, to an outgoing photon it is $\sqrt{4\pi}\epsilon^{*\mu}$ rather than just $\epsilon^{*\mu}$, and in the photon propagator the factor $1/k^2$ becomes $4\pi/k^2$. In quantum theory it is more convenient to have a canonically normalized gauge field, which is the reason why, except in Landau and Lifshitz, vol. IV (1982), rationalized units are always used.[2]

[2] Observe that, once the result is written in terms of α, it is independent of the conventions on e, since α is always the same constant $\simeq 1/137$. For instance, the Coulomb potential between two electrons (in units $\hbar = c = 1$) is always $V(r) = \alpha/r$.

Experimental data. Unless explicitly specified otherwise, our experimental data are taken from the 2004 edition of the Review of Particle Physics of the Particle Data Group, S. Eidelman *et al.*, *Phys. Lett.* B592, 1 (2004), also available on-line at http://pdg.lbl.gov.

Introduction

1

1.1 Overview

Quantum field theory is a synthesis of quantum mechanics and special relativity, and it is one of the great achievements of modern physics. Quantum mechanics, as formulated by Bohr, Heisenberg, Schrödinger, Pauli, Dirac, and many others, is an intrinsically non-relativistic theory. To make it consistent with special relativity, the real problem is not to find a relativistic generalization of the Schrödinger equation.[1] Wave equations, relativistic or not, cannot account for processes in which the number and the type of particles changes, as in almost all reactions of nuclear and particle physics. Even the process of an atomic transition from an excited atomic state A^* to a state A with emission of a photon, $A^* \to A + \gamma$, is in principle unaccessible to this treatment (although in this case, describing the electromagnetic field classically and the atom quantum mechanically, one can get some correct results, even if in a not very convincing manner). Furthermore, relativistic wave equations suffer from a number of pathologies, like negative-energy solutions.

A proper resolution of these difficulties implies a change of viewpoint, from wave equations, where one quantizes a single particle in an external classical potential, to quantum field theory, where one identifies the particles with the modes of a field, and quantizes the field itself. The procedure also goes under the name of second quantization.

The methods of quantum field theory (QFT) have great generality and flexibility and are not restricted to the domain of particle physics. In a sense, field theory is a universal language, and it permeates many branches of modern research. In general, field theory is the correct language whenever we face collective phenomena, involving a large number of degrees of freedom, and this is the underlying reason for its unifying power. For example, in condensed matter the excitations in a solid are quanta of fields, and can be studied with field theoretical methods. An especially interesting example of the unifying power of QFT is given by the phenomenon of superconductivity which, expressed in the field theory language, turns out to be conceptually the same as the Higgs mechanism in particle physics. As another example we can mention that the Feynman path integral, which is a basic tool of modern quantum field theory, provides a formal analogy between field theory and statistical mechanics, which has stimulated very important exchanges between these two areas. Beside playing a crucial role for physicists,

[1] Actually, Schrödinger first found a relativistic equation, that today we call the Klein–Gordon equation. He then discarded it because it gave the wrong fine structure for the hydrogen atom, and he retained only the non-relativistic limit. See Weinberg (1995), page 4.

quantum field theory even plays a role in pure mathematics, and in the last 20 years the physicists' intuition stemming in particular from the path integral formulation of QFT has been at the basis of striking and unexpected advances in pure mathematics.

QFT obtains its most spectacular successes when the interaction is small and can be treated perturbatively. In quantum electrodynamics (QED) the theory can be treated order by order in the *fine structure constant* $\alpha = e^2/(4\pi\hbar c) \simeq 1/137$. Given the smallness of this parameter, a perturbative treatment is adequate in almost all situations, and the agreement between theoretical predictions and experiments can be truly spectacular. For example, the electron has a magnetic moment of modulus $g|e|\hbar/(4m_e c)$, where g is called the gyromagnetic ratio. While classical electrodynamics erroneously suggests $g = 1$, the Dirac equation gives $g = 2$, and QED predicts a small deviation from this value; the experimentally measured value is

$$\left(\frac{g-2}{2}\right)\bigg|_{\rm exp} = 0.001\ 159\ 652\ 187(4) \tag{1.1}$$

(the digit in parentheses is the experimental error on the last figure), and the theoretical prediction, computed perturbatively up to order α^4, is

$$\begin{aligned}\left(\frac{g-2}{2}\right)\bigg|_{\rm th} &= \frac{\alpha}{2\pi} - (0.328\,478\,965\ldots)\left(\frac{\alpha}{\pi}\right)^2 + (1.176\,11\ldots)\left(\frac{\alpha}{\pi}\right)^3 \\ &\quad -(1.434\ldots)\left(\frac{\alpha}{\pi}\right)^4 = 0.001\ 159\ 652\ 140(5)(4)(27)\,.\end{aligned}$$

Different sources of errors on the last figures are written separately in parentheses. The theoretical error is due partly to the numerical evaluation of Feynman diagrams (there are 891 of them at order α^4!) and partly to the fact that, at this level of precision, hadronic contributions come into play. We also need to know α with sufficient accuracy; this is provided by the quantum Hall effect.

The gyromagnetic ratio has been measured very precisely also for the muon, and the accuracy of this measurement has been improved recently,[2] with the result $(g-2)/2|_{\rm exp} = 0.001\,165\,9208(6)$, and a theoretical prediction $(g-2)/2|_{\rm th} = 0.001\,165\,9181(7)$. The remaining discrepancy has aroused much interest, in the hope that it might be a signal of new physical effects, but to see whether this is actually the case requires first a better theoretical understanding of hadronic contributions, which are more difficult to compute. In any case, an agreement between theory and experiment at the level of 10 decimal figures for the electron (or eight for the muon) is spectacular, and it is among the most precise in physics.

[2]See http://www.g-2.bnl.gov/. This values updates the value reported in the 2004 edition of the Review of Particle Physics.

As we know today, QED is only a part of a larger theory. As we approach the scales of nuclear physics, i.e. length scales $r \sim 10^{-13}$ cm or energies $E \sim 200$ MeV, the existence of new interactions becomes evident: strong interactions are responsible for instance for binding together neutrons and protons into nuclei, and weak interactions are responsible for a number of decays, like the beta decay of the neutron into the proton, electron and antineutrino, $n \to pe^-\bar{\nu}_e$. A successful theory of beta decay was already proposed by Fermi in 1934. We now understand the Fermi theory as a low energy approximation to a more complete theory, that unifies the weak and electromagnetic interactions into a single conceptual framework, the electroweak theory. This theory, developed in the early 1970s, together with the fundamental theory of strong interactions, quantum chromodynamics (QCD), has such spectacular experimental successes that it now goes under the name of the Standard Model. In the last decade of the 20th century the LEP machine at CERN performed a large number of precision measurements, at the level of one part in 10^4, which are all completely reproduced by the theoretical predictions of the Standard Model. These results show that we do understand the laws of Nature down to the scale of 10^{-17} cm, i.e. four orders of magnitude below the size of a nucleus and nine orders of magnitude below the size of an atom. Part of the activity of high energy physicists nowadays is devoted to the search of physics beyond the Standard Model. The best hint for new physics presently comes from the recent experimental evidence for neutrino oscillations. These oscillations imply that neutrinos have a very small mass, whose deeper origin is suspected to be related to physics beyond the Standard Model.

The Standard Model has a beautiful theoretical structure; its discovery and development, due among others to Glashow, Weinberg, Salam and 't Hooft, requires a number of new concepts compared to QED. A detailed explanation of the Standard Model is beyond the scope of this course, but we will discuss two of its main ingredients: non-abelian gauge fields, or Yang–Mills theories, and spontaneous symmetry breaking through the Higgs mechanism.

In spite of the remarkable successes of the Standard Model, the search for the fundamental laws governing the microscopic world is still very far from being completed. In the Standard Model itself there is still a missing piece, since it predicts a particle, the Higgs boson, which plays a crucial role and which has not yet been observed. LEP, after 11 years of glorious activity, was closed in November 2000, after reaching a maximum center of mass energy of 209 GeV. The new machine, LHC, is now under construction at CERN, and together with the Tevatron collider at Fermilab aims at exploring the TeV ($= 10^3$ GeV $= 10^{12}$ eV) energy range. It is hoped that they will find the Higgs boson and that they will test theoretical ideas like supersymmetry that, if correct, are expected to give observable signals at this energy scale.

Looking much beyond the Standard Model, there is a very substantial reason for believing that we are still far from a true understanding of the fundamental laws of Nature. This is because gravity cannot be included in the conceptual schemes that we have discussed so far. General relativity is incompatible with quantum field theory. From an experimental point of view, at present, this causes no real worry; the energy scale at which quantum gravity effects are expected to become important is so huge (of order 10^{19} GeV) that we can forget them altogether in accelerator experiments.[3] There remains the conceptual need for a new theoretical scheme where these two pillars of modern physics, quantum field theory and general relativity, merge consistently. And, of course, one should also be subtle enough to find situations where this can give testable predictions. A consistent theoretical scheme is perhaps slowly emerging in the form of string theory; but this would lead us very far from the scope of this course.

[3] However, this could change in theories with large extra dimensions. In fact, both in quantum field theory and in string theory, have been devised mechanisms such that some extra dimensions are accessible only to gravitational interactions, and not to electromagnetic, weak or strong interactions. In this case, it turns out that the extra dimensions could even be as large as the millimeter without conflicting with any experimental result, and the huge value 10^{19} GeV of the gravitational scale would emerge from a combination of the large volume of the extra dimensions and a much smaller mass-scale which characterizes the energy where genuine quantum gravity effects set in. This new gravitational mass-scale might even be as low as a few tens of TeV, and in this case it could be within the reach of future particle physics experiments.

1.2 Typical scales in high-energy physics

Before entering into the technical aspects of quantum field theory, it is important to have a physical understanding of the typical scales of atomic and particle physics and to be able to estimate what are the orders of magnitudes involved. Often this can be done just with elementary dimensional considerations, supplemented by some very basic physical inputs. We will therefore devote this section to an overview of order of magnitude estimates in particle physics.

These estimates are much simplified by the use of units $\hbar = c = 1$. To understand the meaning of these units, observe first of all that $\hbar$ and c are universal constants, i.e. they have the same numerical value for all observers. The speed of light has the value $c = 299\,792\,458$ m/s, with no error because, after having defined the unit of time from a particular atomic transition (a hyperfine transition of cesium-133) this value of c is taken as the *definition* of the meter. However, instead of using the meter, we can decide to use a new unit of length (or a new unit of time) defined by the statement that in these units $c = 1$. Then, the velocity v of a particle is measured in units of the speed of light, which is very natural since in particle physics we typically deal with relativistic objects. In these units $0 \leqslant v < 1$ for massive particles, and $v = 1$ for massless particles.

The Planck constant $\hbar$ is another universal constant, and it has dimensions [energy] $\times$ [time] or [length] $\times$ [momentum] as we see for instance from the uncertainty principle. We can therefore choose units of energy such that $\hbar = 1$. Then all multiplicative factors of $\hbar$ and c disappear from our equations and formally, from the point of view of dimensional analysis,

$$[\text{velocity}] = \text{pure number}\,, \tag{1.2}$$

$$[\text{energy}] = [\text{momentum}] = [\text{mass}]\,, \tag{1.3}$$

$$[\text{length}] = [\text{mass}]^{-1}\,. \tag{1.4}$$

The first two equations follow immediately from $c = 1$ while the third follows from the fact that $\hbar/(mc)$ is a length. Thus all physical quantities have dimensions that can be expressed as powers of mass or, equivalently, as powers of length. For instance an energy density, [energy]/[length]^3, becomes a [mass]^4. Units $\hbar = c = 1$ are called *natural units.*

The fine structure constant $\alpha = e^2/(4\pi\hbar c) \simeq 1/137$ is a pure number, and therefore in natural units the electric charge e becomes a pure number.

To make numerical estimates, it is useful to observe that $\hbar c$, in ordinary units, has dimensions [energy$\times$time]$\times$[velocity] = [energy]$\times$[length]. In particle physics a useful unit of energy is the MeV (= 10^6 eV) and a typical length-scale is the fermi: 1 fm = 10^{-13} cm; one fm is the typical size of a proton. Expressing $\hbar c$ in MeV$\times$fm, one gets

$$\boxed{\hbar c \simeq 200 \text{ MeV fm}\,.} \tag{1.5}$$

(The precise value is 197.326 968 (17) MeV fm.) Then, in natural units, $1\,\text{fm} \simeq 1/(200\,\text{MeV})$. The following examples will show that sometimes we can go quite far in the understanding of physics with just very simple dimensional estimates.

If we want to make dimensional estimates in QED the two parameters that enter are the fine structure constant $\alpha \simeq 1/137$ and the electron mass, $m_e \simeq 0.5$ MeV/c^2. Note that in units $c = 1$ masses are expressed simply in MeV, as energies. We now consider a few examples.

The Compton radius. The simplest length-scale associated to a particle of mass m in its rest frame is its Compton radius, $r_C = 1/m$. In particular, for the electron

$$\boxed{r_C = \frac{1}{m_e} \simeq \frac{200 \text{ MeV fm}}{0.5 \text{ MeV}} = 4 \times 10^{-11}\,\text{cm}\,.} \tag{1.6}$$

Since r_C does not depend on α, it is the relevant length-scale in situations in which there is no dependence on the strength of the interaction. Historically, r_C made its first appearance in the Compton scattering of X-rays off electrons. Classically, the wavelength of the scattered X-rays should be the same as the incoming waves, since the process is described in terms of forced oscillations. Quantum mechanically, treating the X-rays as photons, we understand that part of the momentum $h\nu$ of the incoming photon is used to produce the recoil of the electron, so the momentum of the outgoing photon is smaller, and its wavelength is larger. The wavelength of the outgoing photon is fixed by energy–momentum conservation, and therefore is independent of α, so the relevant length-scale must be r_C. Indeed, a simple computation gives

$$\lambda' - \lambda = r_C(1 - \cos\theta)\,, \tag{1.7}$$

where λ, λ' are the initial and final X-ray wavelengths and θ is the scattering angle.

The hydrogen atom. Let us first estimate the Bohr radius r_B. The only mass that enters the problem is the reduced mass of the electron–

proton system; since $m_p \simeq 938$ MeV is much bigger than m_e we can identify the reduced mass with m_e, within a precision of 0.05 per cent. Dimensionally, again $r_B \sim 1/m_e$, but now α enters. Clearly, the radius of the bound state is smaller if the interaction responsible for the binding is stronger, while it must go to infinity in the limit $\alpha \to 0$, so α must be in the denominator and it is very natural to guess that $r_B \sim 1/(m_e\alpha)$. This is indeed the case, as can be seen with the following argument: by the uncertainty principle, an electron confined in a radius r has a momentum $p \sim 1/r$. If the electron in the hydrogen atom is non-relativistic (we will verify the consistency of this hypothesis a posteriori) its kinetic energy is $p^2/(2m_e) \sim 1/(2m_e r^2)$. This kinetic energy must be balanced by the Coulomb potential, so at the equilibrium radius $1/(2m_e r^2) \sim \alpha/r$, which indeed gives $r_B \sim 1/(m_e\alpha)$. In principle factors of 2 are beyond the power of dimensional estimates, but here it is quite tempting to observe that the virial theorem of classical mechanics states that, for a potential proportional to $1/r$, at equilibrium the kinetic energy is one half of the absolute value of the potential energy, so we would guess, more precisely, that $1/(2m_e r_B^2) = \alpha/(2r_B)$, i.e.

$$\boxed{r_B = \frac{1}{m_e\alpha} \simeq 0.5 \times 10^{-8}\,\text{cm}\,,} \tag{1.8}$$

which is indeed the definition of the Bohr radius as found in the quantum mechanical treatment. The typical potential energy of the hydrogen atom is then

$$\langle V \rangle \sim V(r_B) = -\frac{\alpha}{r_B} = -m_e\alpha^2\,, \tag{1.9}$$

and, again using the virial theorem, the kinetic energy is

$$E = -\frac{1}{2}V \sim \frac{1}{2}m_e\alpha^2\,. \tag{1.10}$$

This is the kinetic energy of a non-relativistic electron with typical velocity

$$v \sim \alpha\,. \tag{1.11}$$

Since $\alpha \ll 1$, our approximation of a non-relativistic electron is indeed consistent. This of course was expected, since we know that, in a first approximation, the non-relativistic Schrödinger equation gives a good description of the hydrogen atom.

The sum of the kinetic and potential energy is $-(1/2)m_e\alpha^2$ so the binding energy of the hydrogen atom is

$$\text{binding energy} = \frac{1}{2}m_e\alpha^2 \simeq \frac{1}{2}\,0.5\text{MeV}\left(\frac{1}{137}\right)^2 \simeq 13.6\ \text{eV}\,. \tag{1.12}$$

The Rydberg energy is indeed defined as $(1/2)m_e\alpha^2$, and the Schrödinger equation gives the energy levels

$$E_n = -\frac{m_e\alpha^2}{2n^2}\,. \tag{1.13}$$

In QED this is just the first term of an expansion in α; at next order one finds the *fine structure* of the hydrogen atom,

$$E_{n,j} = m_e \left[-\frac{\alpha^2}{2n^2} - \frac{\alpha^4}{2n^4}\left(\frac{n}{j+\frac{1}{2}} - \frac{3}{4}\right) + \ldots \right], \tag{1.14}$$

where j is the total angular momentum and, to be more accurate, the electron mass should be replaced by the reduced mass $m_e m_p/(m_e+m_p)$. We will derive eq. (1.14) in Solved Problem 3.1. The fine structure constant α gets its name from this formula. From eq. (1.11) we understand that, in the hydrogen atom, the expansion in α is the same as an expansion in powers of v, and the fine structure of the hydrogen atom is just the first relativistic correction.

Electron–photon scattering. We want to estimate the cross-section for the scattering of a photon by an electron, which we take initially at rest, $e^-\gamma \to e^-\gamma$. We denote by ω the initial photon energy (in natural units the energy of the photon $E = \hbar\omega$ becomes simply ω). The energy of the final photon is fixed by the initial energy ω and by the scattering angle θ, so the total cross-section (i.e. the cross-section integrated over the scattering angle) can depend only on two energy scales, m_e and ω, and on the dimensionless coupling α. The dependence on α is determined observing that the scattering process takes place via the absorption of the incoming photon and the emission of the outgoing photon. As we will study in detail in Chapters 5 and 7, this is a process of second order in perturbation theory and its amplitude is $O(e^2)$ so the cross-section, which is proportional to the squared amplitude, is $O(e^4)$, i.e. $O(\alpha^2)$. For a generic incoming photon energy ω, we have two different scales in the problem and we cannot go very far with dimensional considerations. Things simplify in the limit $\omega \ll m_e$. In this limit we can neglect ω compared to m_e and we have basically only one mass-scale, m_e. Since the cross-section has dimensions [length]^2, we can estimate $\sigma \sim \alpha^2/m_e^2$. It is therefore useful to define r_0,

$$\boxed{r_0 = \frac{\alpha}{m_e} \simeq 2.8 \times 10^{-13}\,\mathrm{cm}\,,} \tag{1.15}$$

so that the cross-section is $\sigma \sim r_0^2$. The exact computation gives the result

$$\sigma_T = \frac{8}{3}\pi r_0^2 \tag{1.16}$$

and the factor of π is also easily understood, since a cross-section is an effective area, so it is $\sim \pi r_0^2$. The electron–photon cross-section at $\omega \ll m_e$ is known as the Thomson cross-section and can be computed just with classical electrodynamics, since when $\omega \ll m_e$ the photons are well described by a classical electromagnetic field; r_0 is therefore called the *classical electron radius*, and gives a measure of the size of an electron, as seen using classical electromagnetic fields as a probe.

Consider now the opposite limit $\omega \gg m_e$. In this case the cross-section must have a dependence on the energy of the photon and, because of Lorentz invariance, the cross-section integrated over the angles will depend on the energy of the photon through the energy in the center of mass system. If k is the initial four-momentum of the photon and p_e is the initial four-momentum of the electron, the total initial four-momentum is $p = k + p_e$ and the square of the energy in the center of mass is $s = p^2$. In the rest frame of the electron $p_e = (m_e, 0, 0, 0)$ and $k = (\omega, 0, 0, \omega)$, so $s = (m_e + \omega)^2 - \omega^2 = 2m_e\omega + m_e^2$. In the limit $\omega \gg m_e$ we have $s \gg m_e^2$ and we would expect that we can neglect m_e. Then the only energy scale is provided by $\sqrt{s}$, and we would expect that $\sigma \sim \alpha^2/s$. Here however there is a subtlety. In the previous case, $\omega \ll m_e$, we have implicitly assumed that in the limit $\omega \to 0$ the cross-section is finite. This is indeed the case, since in this limit the electromagnetic field can be treated classically, and the classical computation gives a finite answer.[4] If instead $\omega \gg m_e$, we are effectively taking the limit $m_e \to 0$; it turns out that this limit is problematic in QED, and taking $m_e \to 0$ one finds so-called infrared divergences. In fact, from the explicit computation one finds that the correct high-energy limit of the cross-section is

$$\sigma \simeq \frac{2\pi\alpha^2}{s} \log\left(\frac{s}{m_e^2}\right). \tag{1.17}$$

This is an example of the fact that divergences, which are typical of quantum field theory, can spoil naive dimensional analysis. We will examine this issue in a more general context in Section 5.9.

[4]In general, not every quantum computation has a well-defined classical limit; just think of what happens to the black body spectrum when $\hbar \to 0$ (indeed, this example was just the original motivation of Planck for introducing $\hbar$!). However, reinstating $\hbar$ and c explicitly, the classical electron radius is $r_0 = \alpha(\hbar/m_e c) = (e^2/4\pi\hbar c)(\hbar/m_e c)$ and $\hbar$ cancels, so the limit $\hbar \to 0$ is well defined.

In conclusion, we have found three different scales that can be constructed with m_e and α. The largest is $r_B = 1/(m_e\alpha)$ and gives the characteristic size of an electron bound by the Coulomb potential of a proton; $r_C = 1/m_e$ is the characteristic length-scale associated with a free electron in its rest frame, and the smallest, $r_0 = \alpha/m_e$, is associated with classical $e\gamma$ scattering.

Nucleons and strong interactions. Nuclei are bound states of nucleons, i.e. of protons and neutrons, with a radius $r \sim A^{1/3} \times 1$ fm, where A is the total number of nucleons (so that the volume is proportional to A). From the uncertainty principle, a particle confined within 1 fm has a momentum $p \sim 1/(1\,\text{fm}) \simeq 200$ MeV. If the nucleons in the nucleus are non-relativistic, their kinetic energy is

$$E_N \simeq \frac{p_N^2}{2m_N} \simeq 20\,\text{MeV} \tag{1.18}$$

so this must be the typical scale of nuclear binding energies; the typical velocity is

$$v_N \simeq \frac{p_N}{m_N} \simeq 0.2\,. \tag{1.19}$$

This values of v shows that the non-relativistic approximation is roughly correct, but relativistic corrections in nuclei are numerically more important than in atoms. Since the corrections are proportional to v^2 (compare eqs. (1.11) and (1.14)), in nuclei they are of order 4%.

It is also interesting to estimate the analogue of α for the strong interactions. For this we need to know that the nucleon–nucleon strong potential is not Coulomb-like, but rather decays exponentially at large distances,

$$V \simeq -\frac{\alpha_s}{r} e^{-m_\pi r}, \tag{1.20}$$

where α_s is the coupling constant of strong interactions and $m_\pi \simeq 140$ MeV is the mass of a particle, the pion, that at length-scales $l \gtrsim 1$ fm can be considered the mediator of the strong interaction (we will derive this result in Section 6.6). Consider for instance a proton–neutron system, which makes a bound state (the nucleus of deuterium) of radius $r \sim 1$ fm. At equilibrium, $(-1/2)V$ must be equal to the kinetic energy $p^2/(2m) \sim 1/(2mr^2)$, where $m \simeq m_p/2$ is the reduced mass of the two-nucleon system (and the $-1/2$ comes again from the virial theorem). Since we already know that the equilibrium radius is at $r \simeq 1$ fm, we find $\alpha_s \sim 2(m_p r)^{-1} \exp\{m_\pi r\}|_{r=1\,\text{fm}} \sim 0.8$. The precise numerical value is not of great significance, since we are making order of magnitude estimates, but anyway this shows that the coupling α_s is not a small number, and strong interactions cannot be treated perturbatively in the same way as QED.[5]

[5] We will see in Section 5.9 that the coupling constants actually are not constant at all, but rather depend on the length-scale at which they are measured. We will see that the correct statement is that the theory of strong interactions, QCD, cannot be treated perturbatively at length-scales $l \gtrsim 1$ fm, while α_s becomes small at $l \ll 1$ fm, and there perturbation theory works well.

Lifetime and cross-sections of strong interactions. Hadrons are defined as particles which have strong interactions. If a particle decays by strong interactions it is possible to estimate its lifetime τ as follows. The quantities that can enter the computation of the lifetime are the coupling α_s, the masses of the particles involved, and the typical interaction radius of the strong interactions. However, these particles have typical masses in the GeV range, and the interaction range of the strong interaction $\sim 1\text{fm} \simeq (200\text{MeV})^{-1}$. Then all energy scales in the problem are between a few hundred MeV and a few GeV, so in a first approximation we can say that the only length-scale in the problem is of the order of the fermi. Furthermore, we have seen that $\alpha_s = O(1)$. This means that, in order of magnitude, the lifetimes of particles which decay by strong interactions are in the ballpark of $\tau \sim 1\,\text{fm}/c \sim 3 \times 10^{-24}$ s. Particles with such a small lifetime only show up as peaks in a plot of a scattering cross-section against the energy, and are called resonances, since the mechanism that produces the peak is conceptually the same as the resonance in classical mechanics (we will discuss resonances in detail in Section 6.5). The width Γ of the peak is related to the lifetime by $\Gamma = \hbar/\tau$ or, in natural units,

$$\Gamma = \frac{1}{\tau} \sim \frac{1}{1\,\text{fm}} \simeq 200\text{ MeV}. \tag{1.21}$$

We can estimate similarly the typical cross-sections of processes mediated by strong interactions. Since a cross-section is an effective area, we must typically have $\sigma \sim \pi\,(1\,\text{fm})^2 \sim 3 \times 10^{-26}\text{cm}^2$. A common unit for cross-sections is the barn, 1 barn $= 10^{-24}\,\text{cm}^2$. Therefore a typical strong interactions cross-section, in the absence of dynamical phenomena like resonances, is of the order of 30 millibarns. Here we have implicitly assumed that the particles are relativistic, i.e. their relative speed is close to one. Otherwise we must take into account that the relevant length-scale for a particle of mass m and velocity $v \ll 1$ is given by the De Broglie wavelength $\lambda = 1/(mv) \gg 1/m$, and a typical nuclear cross-section for slow particles, in the absence of resonances, is of the order $\sigma \sim \pi\lambda^2$, see Exercise 1.3.

Electroweak decays. Leptons do not have strong interactions and either are stable or decay through electroweak interactions. Furthermore, strong interactions obey a number of conservation laws, which result in the fact that also many hadrons cannot decay via the strong interaction; in this case they decay through electroweak interactions (except for the proton, which in the Standard Model is stable) and their lifetime is considerably longer than the typical lifetimes $\tau \sim 10^{-24}$ s of strong decays. Weak decays span a broad range of lifetimes because they depend on quite different mass-scales: the electroweak scale, the mass of the decaying particle, and the masses of the decay products. While in the case of hadronic resonances the scales which are involved are all between a few hundred MeV and a few GeV, for weak decays these scales can be very different from each other: the electroweak scale is $O(100)$ GeV, while the masses of the decaying particle or of the decay products can be anywhere between zero (for the photon) or less than a few eV (for the electron neutrino) up to hundreds of GeV. Furthermore the electroweak coupling constants are not of order one. Rather, the electromagnetic coupling is $\alpha \sim 1/137 \simeq 0.007$ while, as we will discuss in Chapter 8, weak interactions are characterized by two coupling constants $g^2/(4\pi)$ and $\bar{g}^2/(4\pi)$ both numerically of order 0.1. For these reasons the electroweak lifetimes, even in order of magnitude, vary from case to case. Some examples are given in Table 1.1.

Table 1.1 Examples of electroweak decays. In the right column we give the lifetime of the decaying particle and in the left column its main decay mode. Observe the broad range of lifetimes. For lifetimes so small as for the Z^0, it is more convenient to give the decay width. For the Z^0, the full width is $\Gamma = 2.4952(23)$ GeV.

main mode			lifetime (sec)
n	$\to$	$pe^-\bar{\nu}_e$	$0.8857(8) \times 10^3$
μ^-	$\to$	$e^-\bar{\nu}_e\nu_\mu$	$2.19703(4) \times 10^{-6}$
π^+	$\to$	$\mu^+\nu_\mu$	$2.6033(5) \times 10^{-8}$
Λ^0	$\to$	$p\pi^-$	$2.632(20) \times 10^{-10}$
K^0_S	$\to$	$\pi^+\pi^-$	$0.8958(6) \times 10^{-10}$
π^0	$\to$	$\gamma\gamma$	$0.84(6) \times 10^{-16}$
Σ^0	$\to$	$\Lambda\gamma$	$0.74(7) \times 10^{-19}$
Z^0	$\to$	hadrons	$2.6379(24) \times 10^{-25}$

The lifetime can be written as

$$\tau = \frac{\hbar}{\Gamma} = \frac{\hbar}{\sum_i \Gamma_i} \tag{1.22}$$

where in the last equality the sum runs over all decay channels. Γ is called the full width, while the Γ_i are the partial widths relative to the decay mode labeled by i. In the first column of Table 1.1 we give the dominant decay mode, i.e. the mode with the largest partial width. In the second column we give the lifetime, i.e. the inverse of the full width. The quantity Γ_i/Γ is called the branching ratio of the mode labeled by i. We will compute explicitly many weak decays in the Solved Problems section of Chapter 8.

The Planck mass. Using simple dimensional estimates we can also understand the statement made at the end of Section 1.1 that, in the realm of particle physics, gravity enters into play only at huge energies. Comparing the Newton potential $V = -G_N m^2/r$ with a Coulomb potential $V = -e^2/4\pi r = -(\alpha\hbar c)/r$, we see that G_N times a mass squared has the dimensions of $\hbar c$. Therefore from the fundamental constants $\hbar, c, G_N$ we can build a mass-scale

$$M_{\rm Pl} = \sqrt{\frac{\hbar c}{G_N}}, \tag{1.23}$$

known as the Planck mass, whose numerical value is $M_{\rm Pl} \simeq 1.2 \times 10^{19}\,{\rm GeV}/c^2$. In natural units, then, $G_N = 1/M_{\rm Pl}^2$ and we see, comparing the Newton and Coulomb laws, that the gravitational analogue of the fine structure constant is $(m/M_{\rm Pl})^2$. More precisely, since in general relativity any form of energy is a source for the gravitational field, particles with an energy E have an effective gravitational coupling

$$\alpha_G = \frac{E^2}{M_{\rm Pl}^2}. \tag{1.24}$$

At the typical energies of particle physics, say $E \sim 1$ GeV, we have $\alpha_G \sim 10^{-38}$ and gravity is completely irrelevant. In the realm of particle physics, gravity becomes important only at energies comparable to the Planck scale. These considerations only apply to the microscopic domain. On the macroscopic scale, gravity can become more important than electric interactions because it is always attractive, so it has a cumulative effect, while on a large scale the electrostatic forces are screened by the formation of electrically neutral objects, and the residual force decreases faster than $1/r^2$.

Since $M_{\rm Pl}$ provides a natural mass-scale, in quantum gravity it is customary to use units in which not only $\hbar$ and c but also $M_{\rm Pl}$ are set equal to one. These are called Planck units, and in these units all physical quantities are dimensionless. We will not use them in this book.

Further reading

- A historical introduction to quantum field theory is given in Weinberg (1995), Chapter 1.
- The standard compilation of experimental data for high-energy physics is the Review of Particle Physics of the Particle Data Group. Unless explicitly stated otherwise, our experimental data are taken from the 2004 edition, S. Eidelman *et al.*, *Phys. Lett.* B592, 1 (2004), also available on-line at http://pdg.lbl.gov.
- Precision measurements are a fascinating field by themselves; the experimentally minded student might enjoy browsing the detailed article by F. J. M. Farley and E. Picasso, *The muon g-2 experiment*, in T. Kinoshita ed., *Quantum Electrodynamics*, World Scientific, Singapore 1990. Recently the measure of the $g-2$ of the muon has been further improved by an experiment in Brookhaven, see the link http://www.g-2.bnl.gov/
- A well-written popular book, which gives a flavor of modern research in quantum gravity and string theory is B. Greene, *The elegant universe: superstrings, hidden dimensions, and the quest for the ultimate theory*, Norton, New York 1999.
- QFT is a domain where there can be an interplay between frontier research in theoretical physics and in pure mathematics, and in the last decades this has generated important advances in both fields. The physicist who wishes an introduction to the ap-

plication to physics of important concepts of geometry and topology (like cohomology groups, complex manifolds, fibre bundles, characteristic classes, etc.) can consult, for instance, Nakahara (1990). These concepts find many applications in the theory of non-abelian gauge fields and in string theory. Conversely, the mathematician interested in the mathematical applications of QFT, supersymmetry and string theory is referred to P. Deligne *et al.* eds., *Quantum Fields and Strings: A Course for Mathematicians*, AMS IAS 1999.

Exercises

(1.1) The Universe is permeated by a thermal background of electromagnetic radiation at a temperature $T = 2.725(1)$ K (the cosmic microwave background radiation, or CMB). Estimate with dimensional arguments the energy density of this gas of photons and compare it with the critical density for closing the Universe, $\rho_c \sim 0.5 \times 10^{-5}\text{GeV/cm}^3$.

[Hint: a useful mnemonic for k_B is given by the fact that, at room temperature $T = 300$ K, $k_B T \simeq (1/40)$ eV. In the energy density, the numerical constant in front of $(k_B T)^4$ turns out to be $(\pi^2/30)g(T)$, where $g(T)$ is of the order of the number of particles which are relativistic at a temperature T, i.e. which have $m \ll T$. With $T \simeq 2.7$ K, only the photon and at most three neutrinos are relativistic and $g(T)$ is between 3 and 4. Then, for the purpose of this exercise, the only thing that matters is that the constant $(\pi^2/30)g(T)$ is of order one.]

(1.2) Model the Sun as an ionized plasma of electrons and protons, with an average temperature $T \simeq 4.5 \times 10^6$ K and an average mass density $\rho \simeq 1.4\,\text{gm/cm}^3$. Estimate the mean free path of photons in the Sun's interior, and compare the contribution to the mean free path coming from the scattering on electrons with that from the scattering on protons. Knowing that the radius of the Sun is $R_\odot \simeq 6.96 \times 10^{10}$ cm, estimate the total time that a photon takes to escape from the Sun.

[Hint: recall that the mean free path l of a particle scattering off an ensemble of targets with number density (i.e. particles per unit volume) n and cross-section σ is

$$l = \frac{1}{n\sigma} \tag{1.25}$$

or, if there are different species of targets, $l = 1/\sum_i n_i\sigma_i$.]

(1.3) Estimate the cross-section for a non-relativistic neutron with kinetic energy $E \sim 1$ MeV, scattering on a proton at rest.

Lorentz and Poincaré symmetries in QFT

2

We mentioned in the Introduction that quantum field theory (QFT) is a synthesis of the principles of quantum mechanics and of special relativity. Our first task will be to understand how Lorentz symmetry is implemented in field theory. We will study the representations of the Lorentz group in terms of fields and we will introduce scalar, spinor, and vector fields. We will then examine the information coming from Poincaré invariance. This chapter is rather mathematical and formal. The effort will pay, however, since an understanding of this group theoretical approach greatly simplifies the construction of the Lagrangians for the various fields in Chapter 3 and gives in general a deeper understanding of various aspects of QFT.

From now on we always use natural units $\hbar = c = 1$.

2.1 Lie groups

Lie groups play a central role in physics, and in this section we recall some of their main properties. In the next sections we will apply these concepts to the study of the Lorentz and Poincaré groups.

A Lie group is a group whose elements g depend in a continuous and differentiable way on a set of real parameters θ^a, $a = 1, \ldots, N$. Therefore a Lie group is at the same time a group and a differentiable manifold. We write a generic element as $g(\theta)$ and without loss of generality we choose the coordinates θ^a such that the identity element e of the group corresponds to $\theta^a = 0$, i.e. $g(0) = e$.

A (linear) *representation* R of a group is an operation that assigns to a generic, abstract element g of a group a linear operator $D_R(g)$ defined on a linear space,

$$g \mapsto D_R(g) \tag{2.1}$$

with the properties that

(i): $D_R(e) = 1$, where 1 is the identity operator, and

(ii): $D_R(g_1)D_R(g_2) = D_R(g_1 g_2)$, so that the mapping preserves the group structure.

The space on which the operators D_R act is called the *basis* for the representation R. A typical example of a representation is a *matrix representation*. In this case the basis is a vector space of finite dimension

n, and an abstract group element g is represented by a $n \times n$ matrix $(D_R(g))^i{}_j$, with $i, j = 1, \ldots, n$. The *dimension* of the representation is defined as the dimension n of the base space. Writing a generic element of the base space as $(\phi^1, \ldots, \phi^n)$, a group element g induces a transformation of the vector space

$$\phi^i \to (D_R(g))^i{}_j \phi^j \,. \tag{2.2}$$

Equation (2.2) allows us to attach a physical meaning to a group element: before introducing the concept of representation, a group element g is just an abstract mathematical object, defined by its composition rules with the other group members. Choosing a specific representation instead allows us to interpret g as a transformation on a certain space; for instance, taking as group $SO(3)$ and as base space the spatial vectors $\mathbf{v}$, an element $g \in SO(3)$ can be interpreted physically as a rotation in three-dimensional space.

A representation R is called reducible if it has an invariant subspace, i.e. if the action of any $D_R(g)$ on the vectors in the subspace gives another vector of the subspace. Conversely, a representation with no invariant subspace is called irreducible. A representation is *completely reducible* if, for all elements g, the matrices $D_R(g)$ can be written, with a suitable choice of basis, in block diagonal form. In other words, in a completely reducible representation the basis vectors ϕ^i can be chosen so that they split into subsets that do not mix with each other under eq. (2.2). This means that a completely reducible representation can be written, with a suitable choice of basis, as the direct sum of irreducible representations.

Two representations R, R' are called *equivalent* if there is a matrix S, independent of g, such that for all g we have $D_R(g) = S^{-1} D_{R'}(g) S$. Comparing with eq. (2.2), we see that equivalent representations correspond to a change of basis in the vector space spanned by the ϕ^i.

When we change the representation, in general the explicit form and even the dimensions of the matrices $D_R(g)$ will change. However, there is an important property of a Lie group that is independent of the representation. This is its *Lie algebra*, which we now introduce.

By the assumption of smoothness, for θ^a infinitesimal, i.e. in the neighborhood of the identity element, we have

$$D_R(\theta) \simeq 1 + i\theta_a T_R^a \,, \tag{2.3}$$

with

$$T_R^a \equiv -i \left. \frac{\partial D_R}{\partial \theta_a} \right|_{\theta=0} \,. \tag{2.4}$$

The T_R^a are called the *generators* of the group in the representation R. It can be shown that, with an appropriate choice of the parametrization far from the identity, the generic group elements $g(\theta)$ can always be represented by[1]

$$\boxed{D_R(g(\theta)) = e^{i\theta_a T_R^a} \,,} \tag{2.5}$$

[1] To be precise, this is only true for the component of the group manifold connected with the identity.

whose infinitesimal form reproduces eq. (2.3). The factor i in the definition (2.4) is chosen so that, if in the representation R the generators are hermitian, then the matrices $D_R(g)$ are unitary. In this case R is a *unitary representation.*

Given two matrices $D_R(g_1) = \exp(i\alpha_a T_R^a)$ and $D_R(g_2) = \exp(i\beta_a T_R^a)$, their product is equal to $D_R(g_1 g_2)$ and therefore must be of the form $\exp(i\delta_a T_R^a)$, for some $\delta_a(\alpha, \beta)$,

$$e^{i\alpha_a T_R^a}\, e^{i\beta_a T_R^a} = e^{i\delta_a T_R^a}\,. \tag{2.6}$$

Observe that T_R^a is a matrix. If A, B are matrices, in general $e^A e^B \neq e^{A+B}$, so in general $\delta_a \neq \alpha_a + \beta_a$. Taking the logarithm and expanding up to second order in α and β we get

$$\begin{aligned} i\delta_a T_R^a &= \log\left\{[1 + i\alpha_a T_R^a + \frac{1}{2}(i\alpha_a T_R^a)^2][1 + i\beta_a T_R^a + \frac{1}{2}(i\beta_a T_R^a)^2]\right\} \quad (2.7)\\ &= \log\left[1 + i(\alpha_a + \beta_a)T_R^a - \frac{1}{2}(\alpha_a T_R^a)^2 - \frac{1}{2}(\beta_a T_R^a)^2 - \alpha_a \beta_b T_R^a T_R^b\right]. \end{aligned}$$

Expanding the logarithm, $\log(1+x) \simeq x - x^2/2$, and paying attention to the fact that the T_R^a do not commute we get

$$\alpha_a \beta_b \left[T_R^a, T_R^b\right] = i\gamma_c(\alpha, \beta) T_R^c\,, \tag{2.8}$$

with $\gamma_c(\alpha, \beta) = -2(\delta_c(\alpha, \beta) - \alpha_c - \beta_c)$. Since this must be true for all α and β, γ_c must be linear in α_a and in β_a, so the relation between γ and α, β must be of the general form $\gamma_c = \alpha_a \beta_b f^{ab}{}_c$ for some constants $f^{ab}{}_c$. Therefore

$$\boxed{[T^a, T^b] = i f^{ab}{}_c T^c\,.} \tag{2.9}$$

This is called the *Lie algebra* of the group under consideration. Two important points must be noted here. The first is that, even if the explicit form of the generators T^a depends on the representation used, the *structure constants* $f^{ab}{}_c$ are independent of the representation. In fact, if $f^{ab}{}_c$ were to depend on the representation, γ^a and therefore δ^a would also depend on R, so it would be of the form $\delta_R^a(\alpha, \beta)$. Then from eq. (2.6) we would conclude that the product of the group elements g_1 and g_2 gives a result which depends on the representation. This is impossible, since the result of the multiplication of two abstract group element $g_1 g_2$ is a property of the group, defined at the abstract group level without any reference to the representations. Therefore, we conclude that $f^{ab}{}_c$ are independent of the representation.[2] The second important point is that this equation has been derived requiring the consistency of eq. (2.6) to second order; however, once this is satisfied, it can be proved that no further requirement comes from the expansion at higher orders.

Thus the structure constants define the Lie algebra, and the problem of finding all matrix representations of a Lie algebra amounts to the algebraic problem of finding all possible matrix solutions T_R^a of eq. (2.9).

[2] Actually, the generators of a Lie group can even be defined without making any reference to a specific representation. One makes use of the fact that a Lie group is also a manifold, parametrized by the coordinates θ^a, and defines the generators as a basis of the tangent space at the origin. One then proves that their commutator (defined as a Lie bracket) is again a tangent vector, and therefore it must be a linear combination of the basis vector. In this approach no specific representation is ever mentioned, so it becomes obvious that the structure constants are independent of the representation. See, e.g., Nakahara (1990), Section 5.6.

A group is called *abelian* if all its elements commute between themselves, otherwise the group is *non-abelian*. For an abelian Lie group the structure constants vanish, since in this case in eq. (2.6) we have $\delta_a = \alpha_a + \beta_a$. The representation theory of abelian Lie algebras is very simple: any d-dimensional abelian Lie algebra is isomorphic to the direct sum of d one-dimensional abelian Lie algebras. In other words, all irreducible representations of abelian groups are one-dimensional. The non-trivial part of the representation theory of Lie algebras is related to the non-abelian structure.

In the study of the representations, an important role is played by the *Casimir operators*. These are operators constructed from the T^a that commute with all the T^a. In each irreducible representation, the Casimir operators are proportional to the identity matrix, and the proportionality constant labels the representation. For example, the angular momentum algebra is $[J^i, J^j] = i\epsilon^{ijk} J^k$ and the Casimir operator is $\mathbf{J}^2$. On an irreducible representation, $\mathbf{J}^2$ is equal to $j(j+1)$ times the identity matrix, with $j = 0, \frac{1}{2}, 1, \ldots$.

A Lie group that, considered as a manifold, is a compact manifold is called a compact group. Spatial rotations are an example of a compact Lie group, while we will see that the Lorentz group is non-compact. A theorem states that non-compact groups have no unitary representations of finite dimension, except for representations in which the non-compact generators are represented trivially, i.e. as zero. The physical relevance of this theorem is due to the fact that in a unitary representation the generators are hermitian operators and, according to the rules of quantum mechanics, only hermitian operators can be identified with observables. If a group is non-compact, in order to identify its generators with physical observables we need an infinite-dimensional representation. We will see in this chapter that the Lorentz and Poincaré groups are non-compact, and that infinite-dimensional representations are obtained introducing the Hilbert space of one-particle states.

2.2 The Lorentz group

The Lorentz group is defined as the group of linear coordinate transformations,

$$x^\mu \to x'^\mu = \Lambda^\mu{}_\nu x^\nu \tag{2.10}$$

which leave invariant the quantity

$$\eta_{\mu\nu} x^\mu x^\nu = t^2 - x^2 - y^2 - z^2 \,. \tag{2.11}$$

The group of transformations of a space with coordinates $(y_1, \ldots y_m, x_1, \ldots x_n)$, which leaves invariant the quadratic form $(y_1^2 + \ldots + y_m^2) - (x_1^2 + \ldots + x_n^2)$ is called the orthogonal group $O(n,m)$, so the Lorentz group is $O(3,1)$. The condition that the matrix Λ must satisfy in order to leave invariant the quadratic form (2.11) is

$$\eta_{\mu\nu} x'^\mu x'^\nu = \eta_{\mu\nu} (\Lambda^\mu{}_\rho x^\rho)(\Lambda^\nu{}_\sigma x^\sigma) = \eta_{\rho\sigma} x^\rho x^\sigma \,. \tag{2.12}$$

Since this must hold for x generic, we must have

$$\eta_{\rho\sigma} = \eta_{\mu\nu}\Lambda^{\mu}{}_{\rho}\Lambda^{\nu}{}_{\sigma}\,. \tag{2.13}$$

In matrix notation, this can be rewritten as $\eta = \Lambda^T\eta\Lambda$. Taking the determinant of both sides, we therefore have $(\det\Lambda)^2 = 1$ or $\det\Lambda = \pm 1$. Transformations with $\det\Lambda = -1$ can always be written as the product of a transformation with $\det\Lambda = 1$ and of a discrete transformation that reverses the sign of an odd number of coordinates, e.g. a parity transformation $(t, x, y, z) \to (t, -x, -y, -z)$, or a reflection around a single spatial axis $(t, x, y, z) \to (t, -x, y, z)$, or a time-reversal transformation, $(t, x, y, z) \to (-t, x, y, z)$. Transformations with $\det\Lambda = +1$ are called *proper Lorentz transformations.* The subgroup of $O(3,1)$ with $\det\Lambda = 1$ is denoted by $SO(3,1)$.

Writing explicitly the 00 component of eq. (2.13) we find

$$1 = (\Lambda^0{}_0)^2 - \sum_{i=1}^{3}(\Lambda^i{}_0)^2 \tag{2.14}$$

which implies that $(\Lambda^0{}_0)^2 \geqslant 1$. Therefore the proper Lorentz group has two disconnected components, one with $\Lambda^0{}_0 \geqslant 1$ and one with $\Lambda^0{}_0 \leqslant -1$, called orthochronous and non-orthochronous, respectively. Any non-orthochronous transformation can be written as the product of an orthochronous transformation and a discrete inversion of the type $(t, x, y, z) \to (-t, -x, -y, -z)$, or $(t, x, y, z) \to (-t, -x, y, z)$, etc. It is convenient to factor out all these discrete transformations, and to redefine the Lorentz group as the component of $SO(3,1)$ for which $\Lambda^0{}_0 \geqslant 1$.

If we consider an infinitesimal transformation

$$\Lambda^{\mu}{}_{\nu} = \delta^{\mu}_{\nu} + \omega^{\mu}{}_{\nu} \tag{2.15}$$

eq. (2.13) gives

$$\omega_{\mu\nu} = -\omega_{\nu\mu}\,. \tag{2.16}$$

An antisymmetric 4×4 matrix has six independent elements, so the Lorentz group has six parameters. These are easily identified: first of all we have the transformations which leave t invariant. This is just the $SO(3)$ rotation group, generated by the three rotations in the $(x, y), (x, z)$ and (y, z) planes. Furthermore, we have three transformations in the (t, x), (t, y) and (t, z) planes that leave invariant $t^2 - x^2$, etc. A transformation that leaves $t^2 - x^2$ invariant is called a *boost* along the x axis, and can be written as

$$t \to \gamma(t + vx)\,, \qquad x \to \gamma(x + vt)\,. \tag{2.17}$$

with $\gamma = (1 - v^2)^{-1/2}$ and $-1 < v < 1$. Its physical meaning is understood looking at the small v limit, where it reduces to the velocity transformation of classical mechanics. It is therefore the relativistic generalization of a velocity transformation. The six independent parameters of the Lorentz group can therefore be taken as the three rotation angles and the three components of the velocity $\mathbf{v}$.

Since $-1 < v < 1$, we can write $v = \tanh\eta$, with $-\infty < \eta < +\infty$. Then $\gamma = \cosh\eta$ and eq. (2.17) can be written as a hyperbolic rotation,

$$\begin{aligned} t &\to (\cosh\eta)t + (\sinh\eta)x \\ x &\to (\sinh\eta)t + (\cosh\eta)x\,. \end{aligned} \tag{2.18}$$

The variable η is called the *rapidity.*

We see that the Lorentz group is parametrized in a continuous and differentiable way by six parameters, and it is therefore a Lie group. However, in the Lorentz group one of the parameters is the modulus of the boost velocity, $|\mathbf{v}|$, which ranges over the non-compact interval $0 \leqslant |\mathbf{v}| < 1$. Therefore the Lorentz group is non-compact.

2.3 The Lorentz algebra

We have seen that the Lorentz group has six parameters, the six independent elements of the antisymmetric matrix $\omega_{\mu\nu}$, to which correspond six generators. It is convenient to label the generators as $J^{\mu\nu}$, with a pair of antisymmetric indices (μ,ν), so that $J^{\mu\nu} = -J^{\nu\mu}$. A generic element Λ of the Lorentz group is therefore written as

$$\Lambda = e^{-\frac{i}{2}\omega_{\mu\nu}J^{\mu\nu}}\,. \tag{2.19}$$

The factor 1/2 in the exponent compensates for the fact that we are summing over all μ,ν rather than over the independent pairs with $\mu < \nu$, and therefore each generator is counted twice.

By definition a set of objects ϕ^i, with $i = 1,\ldots,n$, transforms in a representation R of dimension n of the Lorentz group if, under a Lorentz transformation,

$$\phi^i \to \left[e^{-\frac{i}{2}\omega_{\mu\nu}J_R^{\mu\nu}}\right]^i{}_j\,\phi^j\,, \tag{2.20}$$

where $\exp\{-(i/2)\omega_{\mu\nu}J_R^{\mu\nu}\}$ is a matrix representation of dimension n of the abstract element (2.19) of the Lorentz group; $J_R^{\mu\nu}$ are the Lorentz generators in the representation R, and are $n \times n$ matrices. Under an infinitesimal transformation with infinitesimal parameters $\omega_{\mu\nu}$, the variation of ϕ^i is

$$\delta\phi^i = -\frac{i}{2}\omega_{\mu\nu}(J_R^{\mu\nu})^i{}_j\phi^j\,. \tag{2.21}$$

In $(J_R^{\mu\nu})^i{}_j$ the pair of indices μ,ν identify the generator while the indices i,j are the matrix indices of the representation that we are considering.

All physical quantities can be classified accordingly to their transformation properties under the Lorentz group. A scalar is a quantity that is invariant under the transformation. A typical Lorentz scalar in particle physics is the rest mass of a particle. A *contravariant four-vector* V^μ is defined as an object that satisfies the transformation law

$$V^\mu \to \Lambda^\mu{}_\nu V^\nu\,, \tag{2.22}$$

with $\Lambda^\mu{}_\nu$ defined by the condition (2.13). A covariant four-vector V_μ transforms as $V_\mu \to \Lambda_\mu{}^\nu V_\nu$, with $\Lambda_\mu{}^\nu = \eta_{\mu\rho}\eta^{\nu\sigma}\Lambda^\rho{}_\sigma$. One immediately verifies that, if V^μ is a contravariant four-vector, then $V_\mu \equiv \eta_{\mu\nu} V^\nu$ is a covariant four-vector. We refer generically to covariant and contravariant four-vectors simply as four-vectors. The space-time coordinates x^μ are the simplest example of four-vector. Another particularly important example is given by the four-momentum $p^\mu = (E, \mathbf{p})$.

The explicit form of the generators $(J_R^{\mu\nu})^i{}_j$ as $n \times n$ matrices depends on the particular representation that we are considering. For a scalar ϕ, the index i takes only one value, so it is a one-dimensional representation, and $(J^{\mu\nu})^i{}_j$ is a 1×1 matrix, i.e. a number, for each given pair (μ, ν). But in fact, by definition, on a scalar a Lorentz transformation is the identity transformation, so $\delta\phi = 0$ and $J^{\mu\nu} = 0$. A representation in which all generators are equal to zero is trivially a solution of eq. (2.9), for any Lie group, and so it is called the trivial representation.

The four-vector representation is more interesting. In this case i, j are themselves Lorentz indices, so each generator $J^{\mu\nu}$ is represented by a 4×4 matrix $(J^{\mu\nu})^\rho{}_\sigma$. The explicit form of this matrix is

$$(J^{\mu\nu})^\rho{}_\sigma = i \left(\eta^{\mu\rho} \, \delta^\nu_\sigma - \eta^{\nu\rho} \, \delta^\mu_\sigma \right) . \tag{2.23}$$

This can be shown observing that, from eqs. (2.22) and (2.15), the variation of a four-vector V^μ under an infinitesimal Lorentz transformation is $\delta V^\mu = \omega^\mu{}_\nu V^\nu$, which can be rewritten as

$$\delta V^\rho = -\frac{i}{2} \omega_{\mu\nu} (J^{\mu\nu})^\rho{}_\sigma \, V^\sigma , \tag{2.24}$$

with $(J^{\mu\nu})^\rho{}_\sigma$ given by eq. (2.23) (this solution for $J^{\mu\nu}$ is unique because we require the antisymmetry under $\mu \leftrightarrow \nu$). This representation is irreducible since a generic Lorentz transformation mixes all four components of a four-vector and therefore there is no change of basis that allows us to write $(J^{\mu\nu})^\rho{}_\sigma$ in block diagonal form. We can now use the explicit expression (2.23) to compute the commutators, and we find

$$[J^{\mu\nu}, J^{\rho\sigma}] = i \left(\eta^{\nu\rho} J^{\mu\sigma} - \eta^{\mu\rho} J^{\nu\sigma} - \eta^{\nu\sigma} J^{\mu\rho} + \eta^{\mu\sigma} J^{\nu\rho} \right) . \tag{2.25}$$

This is the Lie algebra of $SO(3, 1)$. It is convenient to rearrange the six components of $J^{\mu\nu}$ into two spatial vectors,

$$J^i = \frac{1}{2} \epsilon^{ijk} J^{jk} , \qquad K^i = J^{i0} . \tag{2.26}$$

In terms of J^i, K^i the Lie algebra of the Lorentz group (2.25) becomes

$$\left[J^i, J^j \right] = i\epsilon^{ijk} J^k , \tag{2.27}$$

$$\left[J^i, K^j \right] = i\epsilon^{ijk} K^k , \tag{2.28}$$

$$\left[K^i, K^j \right] = -i\epsilon^{ijk} J^k . \tag{2.29}$$

Equation (2.27) is the Lie algebra of $SU(2)$ and this shows that J^i, defined in eq. (2.26), is the angular momentum. Instead eq. (2.28) expresses the fact that $\mathbf{K}$ is a spatial vector.

We also introduce the definitions $\theta^i = (1/2)\epsilon^{ijk}\omega^{jk}$ and $\eta^i = \omega^{i0}$. Then

$$\frac{1}{2}\omega_{\mu\nu}J^{\mu\nu} = \omega_{12}J^{12} + \omega_{13}J^{13} + \omega_{23}J^{23} + \sum_{i=1}^{3}\omega_{i0}J^{i0}$$
$$= \boldsymbol{\theta}\cdot\mathbf{J} - \boldsymbol{\eta}\cdot\mathbf{K}\,, \tag{2.30}$$

where we used $\omega_{i0} = -\omega^{i0} = -\eta^i$ while $\omega_{12} = \omega^{12} = \theta^3$, etc. Then a Lorentz transformation can be written as

$$\Lambda = \exp\{-i\boldsymbol{\theta}\cdot\mathbf{J} + i\boldsymbol{\eta}\cdot\mathbf{K}\}\,. \tag{2.31}$$

With our definitions $\theta^i = +(1/2)\epsilon^{ijk}\omega^{jk}$ and $\eta^i = +\omega^{i0}$ a rotation by an angle $\theta > 0$ in the (x, y) plane rotates *counterclockwise* the position of a point P with respect to a fixed reference frame,[3] while performing a boost of velocity $\mathbf{v}$ on a particle at rest we get a particle with velocity $+\mathbf{v}$. To check these signs, we can consider infinitesimal transformations, and use the explicit form (2.23) of the generators. Performing a rotation by an angle θ around the z axis, eqs. (2.31) and (2.23) give

$$\delta x^\mu = -i\theta(J^{12})^\mu{}_\nu x^\nu = \theta\,(\eta^{1\mu}\delta^2_\nu - \eta^{2\mu}\delta^1_\nu)x^\nu \tag{2.32}$$

and therefore $\delta x = -\theta y$ and $\delta y = +\theta x$, corresponding to a counterclockwise rotation. Similarly, performing a boost along the x axis,

$$\delta x^\mu = +i\eta(J^{10})^\mu{}_\nu x^\nu = -\eta\,(\eta^{1\mu}\delta^0_\nu - \eta^{0\mu}\delta^1_\nu)x^\nu \tag{2.33}$$

and therefore $\delta t = +\eta\, x$ and $\delta x = +\eta\, t$, which is the infinitesimal form of eq. (2.18).

[3] This is the "active" point of view. Alternatively, we can say that we keep P fixed and we rotate the reference frame clockwise; this is the "passive" point of view.

2.4 Tensor representations

By definition a tensor $T^{\mu\nu}$ with two contravariant (i.e. upper) indices is an object that transforms as

$$T^{\mu\nu} \to \Lambda^\mu{}_{\mu'}\Lambda^\nu{}_{\nu'}T^{\mu'\nu'}\,. \tag{2.34}$$

In general, a tensor with an arbitrary number of upper and lower indices transforms with a factor $\Lambda^\mu{}_{\mu'}$ for each upper index and a factor $\Lambda_\mu{}^{\mu'}$ for each lower index.

Tensors are examples of representations of the Lorentz group. For instance, a generic tensor $T^{\mu\nu}$ with two indices has 16 components and eq. (2.34) shows that these 16 components transform among themselves, i.e. they are a basis for a representation of dimension 16. However, this representation is reducible. From eq. (2.34) we see that, if $T^{\mu\nu}$ is antisymmetric, after a Lorentz transformation it remains antisymmetric, while if it is symmetric it remains symmetric. So the symmetric and antisymmetric parts of a tensor $T^{\mu\nu}$ do not mix, and the 16-dimensional representation is reducible into a six-dimensional antisymmetric representation $A^{\mu\nu} = (1/2)(T^{\mu\nu} - T^{\nu\mu})$ and a 10-dimensional symmetric representation $S^{\mu\nu} = (1/2)(T^{\mu\nu} + T^{\nu\mu})$. Furthermore, also the trace of a symmetric tensor is invariant,

$$S \equiv \eta_{\mu\nu} S^{\mu\nu} \to \eta_{\mu\nu} \Lambda^{\mu}{}_{\rho} \Lambda^{\nu}{}_{\sigma} S^{\rho\sigma} = S \,, \tag{2.35}$$

where in the last step we used the defining property of the Lorentz group, eq. (2.13). This means, in particular, that a traceless tensor remains traceless after a Lorentz transformation, and thus the 10-dimensional symmetric representation decomposes further into a nine-dimensional irreducible symmetric traceless representation, $S^{\mu\nu} - (1/4)\eta^{\mu\nu} S$, and the one-dimensional scalar representation S.

The following notation is commonly used: an irreducible representation is denoted by its dimensionality, written in boldface. Thus the scalar representation is denoted as **1**, the four-vector representation as **4**, the antisymmetric tensor as **6** and the traceless symmetric tensor as **9**.[4] The tensor representation (2.34) is a tensor product of two four-vector representations, which means that each of the two indices of $T^{\mu\nu}$ transforms separately as a four-vector index, i.e. with the matrix Λ. The tensor product of two representations is denoted by the symbol $\otimes$. We have found above that the tensor product of two four-vector representations decomposes into the direct sum of the **1**, **6**, and **9** representations. Denoting the direct sum by $\oplus$, we have[5]

$$\mathbf{4} \otimes \mathbf{4} = \mathbf{1} \oplus \mathbf{6} \oplus \mathbf{9} \,. \tag{2.36}$$

The decomposition into irreducible representations of tensors with more than two indices can be obtained similarly. The most general irreducible tensor representations of the Lorentz group are found starting from a generic tensor with an arbitrary number of indices, removing first all traces, and then symmetrizing or antisymmetrizing over all pairs of indices. Note that, using $\eta^{\mu\nu}$, we can always restrict to contravariant tensors; for instance V^{μ} and V_{μ} are equivalent representations.

All tensor representations are in a sense derived from the four-vector representation, since the transformation law of a tensor is obtained applying separately on each Lorentz index the matrix $\Lambda^{\mu}{}_{\nu}$ that defines the transformation of four-vectors. This means that (as the name suggests) tensor representations are tensor products of the four-vector representation. For this reason, the four-vector representation plays a distinguished role and is called the *fundamental* representation of $SO(3,1)$.[6]

Another representation of special importance is the *adjoint* representation. It is a representation which has the same dimension as the number of generators. This means that we can use the same type of indices a, b, c for labeling the generator and its matrix elements, and for any Lie group it can be written in full generality in terms of the structure constants, as

$$(T^{a}_{\text{adj}})^{b}{}_{c} = -i f^{ab}{}_{c} \,. \tag{2.37}$$

The Lie algebra (2.9) is automatically satisfied by (2.37). This follows from the fact that, for all matrices A, B, C, there is an algebraic identity

[4] If two inequivalent representations happen to have the same dimensionality one can use a prime or an index to distinguish between them.

[5] In Exercise 2.5 we discuss the separation of the representation **6**, i.e. the antisymmetric tensor, into its self-dual and anti-self-dual parts, both in Minkowski space and in a Euclidean space with metric $\delta^{\mu\nu}$. We will see that in the Euclidean case the antisymmetric tensor $A^{\mu\nu}$ is reducible and decomposes into two three-dimensional representations corresponding to self-dual and anti-self-dual tensors, while in Minkowski space an antisymmetric tensor $A^{\mu\nu}$ with real components is irreducible.

[6] To avoid all misunderstanding, we anticipate that in Section 2.5 we will enlarge the definition of the Lorentz group to include spinorial representations. With this enlarged definition, four-vectors are no longer the fundamental representation of the Lorentz group. Instead, all representations of the Lorentz group will be built from the spinorial representations $(1/2, 0)$ and $(0, 1/2)$ that will be defined in Section 2.5.

known as the *Jacobi identity*,

$$[A,[B,C]]+[B,[C,A]]+[C,[A,B]]=0\,, \tag{2.38}$$

which is easily verified writing the commutators explicitly. Setting in this identity $A=T^a, B=T^b$ and $C=T^c$ we find that the structure constants of any Lie group obey the identity

$$f^{ab}{}_d f^{cd}{}_e + f^{bc}{}_d f^{ad}{}_e + f^{ca}{}_d f^{bd}{}_e = 0\,. \tag{2.39}$$

If we substitute eq. (2.37) into eq. (2.9), we see that the Lie algebra is automatically satisfied because of eq. (2.39).

For the Lorentz group, the adjoint representation has dimension six, so it is given by the antisymmetric tensor $A^{\mu\nu}$. The adjoint representation plays an especially important role in non-abelian gauge theories, as we will see in Chapter 10.

All the representation theory on tensors that we have developed having in mind $SO(3,1)$ goes through for $SO(n)$ or $SO(n,m)$ generic, simply replacing $\eta_{\mu\nu}$ with $\delta_{\mu\nu}$ for $SO(n)$, or with a diagonal matrix with n minus signs and m plus sign for $SO(n,m)$.

2.4.1 Decomposition of Lorentz tensors under $SO(3)$

Since we know how a tensor behaves under a generic Lorentz transformation, we know in particular its transformation properties under the $SO(3)$ rotation subgroup, and we can therefore ask what is the angular momentum j of the various tensor representations. Recall that the representations of $SO(3)$ are labeled by an index j which takes integer values $j=0,1,2,\ldots$, and the dimension of the representation labeled by j is $2j+1$. Within each representation, these $2j+1$ states are labeled by $j_z=-j,\ldots,j$. For $SO(3)$, it is more common to denote the representation as $\mathbf{j}$, i.e. to label it with the angular momentum rather than with the dimension of the representation, $2j+1$. In this notation, $\mathbf{0}$ is the scalar (also called the singlet), $\mathbf{1}$ is a triplet with components $j_z=-1,0,1$, while $\mathbf{2}$ is a representation of dimension 5, etc. (if we rather use the same convention as in the case of the Lorentz group, i.e. we label them by their dimensionality, we should write $\mathbf{1},\mathbf{3},\mathbf{5},\ldots$).

A Lorentz scalar is of course also scalar under rotations, so it has $j=0$. A four-vector $V^\mu=(V^0,\mathbf{V})$ is an irreducible representation of the Lorentz group, since a generic Lorentz transformation mixes all four components, but from the point of view of the $SO(3)$ subgroup it is reducible: spatial rotations do not mix V^0 with $\mathbf{V}$; V^0 is invariant under spatial rotations, so it has $j=0$, while the three spatial components V^i form an irreducible three-dimensional representation of $SO(3)$, so they have $j=1$. In group theory language we say that, from the point of view of spatial rotations, a four-vector decomposes into the direct sum of a scalar and a $j=1$ representation,

$$V^\mu \in \mathbf{0}\oplus\mathbf{1} \tag{2.40}$$

or, if we prefer to label the representations by their dimension, rather than by j, we write $\mathbf{4} = \mathbf{1} \oplus \mathbf{3}$. The former notation indicates more clearly what are the spins involved while the latter makes apparent that the number of degrees of freedom on the left-hand side matches those on the right-hand side.

We now want to understand what angular momenta appear in a generic tensor $T^{\mu\nu}$ with two indices. By definition a tensor $T^{\mu\nu}$ transforms as the tensor product of two four-vector representations. Since, from the point of view of $SO(3)$, a four-vector is $\mathbf{0} \oplus \mathbf{1}$, a generic tensor with two indices has the following decomposition in angular momenta

$$T^{\mu\nu} \in (\mathbf{0} \oplus \mathbf{1}) \otimes (\mathbf{0} \oplus \mathbf{1}) = (\mathbf{0} \otimes \mathbf{0}) \oplus (\mathbf{0} \otimes \mathbf{1}) \oplus (\mathbf{1} \otimes \mathbf{0}) \oplus (\mathbf{1} \otimes \mathbf{1})$$
$$= \mathbf{0} \oplus \mathbf{1} \oplus \mathbf{1} \oplus (\mathbf{0} \oplus \mathbf{1} \oplus \mathbf{2}) . \tag{2.41}$$

In the last step we used the usual rule of composition of angular momenta, which says that composing two angular momenta j_1 and j_2 we get all angular momenta between $|j_1 - j_2|$ and $j_1 + j_2$, so $\mathbf{0} \otimes \mathbf{0} = \mathbf{0}$, $\mathbf{0} \otimes \mathbf{1} = \mathbf{1}$ and $\mathbf{1} \otimes \mathbf{1} = \mathbf{0} \oplus \mathbf{1} \oplus \mathbf{2}$. Thus, in the decomposition of a generic tensor $T^{\mu\nu}$ in representations of the rotation group, the $j = 0$ representation appears twice, the $j = 1$ representation appears three times, and the $j = 2$ once.

It is interesting to see how these representations are shared between the symmetric traceless, the trace and the antisymmetric part of the tensor $T^{\mu\nu}$, since these are the irreducible Lorentz representations. The trace is a Lorentz scalar, so it is in particular scalar under rotations and therefore is a $\mathbf{0}$ representation. An antisymmetric tensor $A^{\mu\nu}$ has six components, which can be written as A^{0i} and $(1/2)\epsilon^{ijk}A^{jk}$. These are two spatial vectors and therefore

$$A^{\mu\nu} \in \mathbf{1} \oplus \mathbf{1} . \tag{2.42}$$

For example, an important antisymmetric tensor in electromagnetism is the field strength tensor $F_{\mu\nu}$, and in this case the two vectors are $E^i = -F^{0i}$ and $B^i = -(1/2)\epsilon^{ijk}F^{jk}$, i.e. the electric and magnetic fields. Another example of an antisymmetric tensor is given by the Lorentz generators $J^{\mu\nu}$ themselves; in this case the two spatial vectors are the angular momentum and the boost generators that have been introduced in eq. (2.26).

Since we have identified the trace S with a $\mathbf{0}$ and $A^{\mu\nu}$ with $\mathbf{1} \oplus \mathbf{1}$, comparison with eq. (2.41) shows that the nine components of a symmetric traceless tensor $S^{\mu\nu}$ decompose, from the point of view of spatial rotations, as

$$S^{\mu\nu} \in \mathbf{0} \oplus \mathbf{1} \oplus \mathbf{2} . \tag{2.43}$$

Observe that, when in eq. (2.41) we write $T^{\mu\nu}$ as $(\mathbf{0} \oplus \mathbf{1}) \otimes (\mathbf{0} \oplus \mathbf{1})$, the first $\mathbf{0}$ corresponds to taking the index $\mu = 0$, the first $\mathbf{1}$ corresponds to taking the index $\mu = i$, and similarly for the second factor $(\mathbf{0} \oplus \mathbf{1})$ and the index ν. Therefore the term $(\mathbf{0} \otimes \mathbf{0})$ in eq. (2.41) corresponds to T^{00}, $(\mathbf{0} \otimes \mathbf{1})$ is T^{0i}, $(\mathbf{1} \otimes \mathbf{0})$ is T^{i0} and $(\mathbf{1} \otimes \mathbf{1})$ is T^{ij}. It is clear that T^{00} is a scalar under spatial rotations, while T^{0i} and T^{i0} are spatial vectors. As for T^{ij}, the antisymmetric part $A^{ij} = T^{ij} - T^{ji}$ is a vector, as can be seen considering $\epsilon^{ijk}A^{jk}$; this gives the third **1** representation. The symmetric part $S^{ij} = T^{ij} + T^{ji}$ can be separated into its trace, which gives the second **0** representation, and the traceless symmetric part, which therefore must have $j = 2$. For example, gravitational waves can be described by a traceless symmetric spatial tensor (transverse to the propagation direction) and therefore have spin 2, see Exercise 2.6.

In general, a symmetric tensor with N indices contains angular momenta up to $j = N$. In four dimensions, higher antisymmetric tensors are instead less interesting, because the index μ takes only four values $0, \ldots, 3$ and therefore we cannot antisymmetrize over more than four indices, otherwise we get zero. Furthermore, a totally antisymmetric tensor with four indices, $A^{\mu\nu\rho\sigma}$, has only one independent component A^{0123}, so it must be a Lorentz scalar. An antisymmetric tensor with three indices, $A^{\mu\nu\rho}$, has $4 \cdot 3 \cdot 2/3! = 4$ components and it has the same transformation properties of a four-vector.

The last point can be better understood introducing the totally antisymmetric tensor defined as follows. In a given reference frame $\epsilon^{\mu\nu\rho\sigma}$ is defined by $\epsilon^{0123} = +1$ and by the condition of total antisymmetry, so it vanishes if any two indices are equal and it changes sign for any exchange of indices, e.g. $\epsilon^{1023} = -1$, etc. Normally, if one gives the numerical value of the components of a tensor in a given frame, in another frame they will be different. The ϵ tensor is however special, because under (proper) Lorentz transformations

$$\epsilon^{\mu\nu\rho\sigma} \rightarrow \Lambda^{\mu}{}_{\mu'}\Lambda^{\nu}{}_{\nu'}\Lambda^{\rho}{}_{\rho'}\Lambda^{\sigma}{}_{\sigma'}\epsilon^{\mu'\nu'\rho'\sigma'} = (\det\Lambda)\epsilon^{\mu\nu\rho\sigma} = \epsilon^{\mu\nu\rho\sigma} \,. \tag{2.44}$$

So, the components of the ϵ tensor have the same numerical value in all Lorentz frames. In terms of this tensor, it is immediate to understand that the four independent components of $A^{\mu\nu\rho}$ can be rearranged in a four-vector $A_\mu = \epsilon_{\mu\nu\rho\sigma}A^{\nu\rho\sigma}$, and that $A^{0123} = (1/4!)\epsilon_{\mu\nu\rho\sigma}A^{\mu\nu\rho\sigma}$ is a scalar.

A tensor which is invariant under all group transformations (i.e. for the Lorentz group, a tensor which has the same form in all Lorentz frames) is called an *invariant tensor.* The only other invariant tensor of the Lorentz group is $\eta_{\mu\nu}$; its invariance follows from the defining property of the Lorentz group, eq. (2.13).

2.5 Spinorial representations

2.5.1 Spinors in non-relativistic quantum mechanics

Tensor representations do not exhaust all physically interesting finite-dimensional representations of the Lorentz group. We can understand the issue considering spatial rotations, i.e. the $SO(3)$ subgroup of the Lorentz group. The tensor representations of $SO(3)$ are constructed exactly as before, with scalars ϕ, spatial vectors v^i, tensors T^{ij}, etc. with

$i = 1, 2, 3$. However we know from non-relativistic quantum mechanics that, beside the tensor representations, there are other representations of great physical interest. These are the spinorial representations. Strictly speaking, these are not $SO(3)$ representations, because under a rotation of 2π a spinor changes sign, while an $SO(3)$ rotation by 2π is the same as the identity transformation. However, since the observables are quadratic in the wave function, this sign ambiguity is perfectly acceptable physically, and these representations must be included. In more formal terms, this means that, for spatial rotations, the physically relevant group is not $SO(3)$ but rather $SU(2)$.

We recall some facts about $SU(2)$ representations, well known from non-relativistic quantum mechanics. The Lie algebras of $SU(2)$ and of $SO(3)$ are the same, and are given by the angular momentum algebra

$$[J^i, J^j] = i\epsilon^{ijk} J^k \,. \tag{2.45}$$

From the discussion in Section 2.1, we see that the Lie algebra knows only about the properties of a group near the identity element, and the fact that $SU(2)$ and $SO(3)$ have the same Lie algebra means that they are indistinguishable at the level of infinitesimal transformations. However, $SU(2)$ and $SO(3)$ differ at the global level, i.e. far from the identity. In $SO(3)$ a rotation by 2π is the same as the identity. Instead, it can be shown that $SU(2)$ is periodic only under rotations by 4π. This means that an object that picks a minus sign under a rotation by 2π is an acceptable representation of $SU(2)$, while it is not an acceptable representation of $SO(3)$. Therefore when we consider $SU(2)$ we include the solutions of eq. (2.45) that correspond to half-integer spin, while for $SO(3)$ we only retain representations with integer spin. Thus, the representations of $SU(2)$ are labeled by an index j which takes values $0, \frac{1}{2}, 1, \frac{3}{2}, \ldots$ and gives the spin of the state, in units of $\hbar$. The spin-j representation has dimension $2j + 1$, and the various states within it are labeled by j_z, which takes the values $-j, \ldots, j$ in integer steps. The representation $j = 1/2$ is called the spinorial representation, and has dimension 2: on it the J^i are represented as

$$J^i = \frac{\sigma^i}{2} \,, \tag{2.46}$$

where σ^i are the Pauli matrices,

$$\sigma^1 = \begin{pmatrix} 0 & 1 \\ 1 & 0 \end{pmatrix} \quad \sigma^2 = \begin{pmatrix} 0 & -i \\ i & 0 \end{pmatrix} \quad \sigma^3 = \begin{pmatrix} 1 & 0 \\ 0 & -1 \end{pmatrix} . \tag{2.47}$$

They satisfy the algebraic identity

$$\sigma^i \sigma^j = \delta^{ij} + i\epsilon^{ijk}\sigma^k \,, \tag{2.48}$$

from which it follows immediately that $\sigma^i/2$ obey the commutation relations (2.45).

The spinorial is the fundamental representation of $SU(2)$ since all representations can be constructed with tensor products of spinors. In physical terms, this means that with spin 1/2 particles we can construct composite systems with all possible integer or half-integer spin. For instance, the composition of two spin 1/2 states gives spin zero and spin 1,

$$\frac{\mathbf{1}}{\mathbf{2}} \otimes \frac{\mathbf{1}}{\mathbf{2}} = \mathbf{0} \oplus \mathbf{1}\,. \tag{2.49}$$

If we denote by $\uparrow$ and $\downarrow$ the $j = 1/2$ states with $j_z = +1/2$ and $j_z = -1/2$, respectively, then the three states with $j = 1$ are given by

$$(\uparrow\uparrow)\,, \quad \frac{1}{\sqrt{2}}(\uparrow\downarrow + \downarrow\uparrow)\,, \quad (\downarrow\downarrow) \tag{2.50}$$

while the singlet (i.e. the scalar state) is

$$\frac{1}{\sqrt{2}}(\uparrow\downarrow - \downarrow\uparrow)\,. \tag{2.51}$$

2.5.2 Spinors in the relativistic theory

We certainly want to keep spinors in the relativistic theory. This means that we must enlarge the set of representations of the Lorentz group, compared to the tensor representations discussed above. This is most easily done starting from the Lorentz algebra in the form given by eqs. (2.27)–(2.29), and defining

$$\mathbf{J}^{\pm} = \frac{\mathbf{J} \pm i\mathbf{K}}{2}\,. \tag{2.52}$$

The Lie algebra becomes

$$\left[J^{+,i}, J^{+,j}\right] = i\epsilon^{ijk} J^{+,k} \tag{2.53}$$

$$\left[J^{-,i}, J^{-,j}\right] = i\epsilon^{ijk} J^{-,k} \tag{2.54}$$

$$\left[J^{+,i}, J^{-,j}\right] = 0\,. \tag{2.55}$$

Therefore we have two copies of the angular momentum algebra, which commute between themselves.[7]

Having written the Lorentz group in this form, it is now easy to include spinorial representations: we simply take all solutions of the algebra (2.53)–(2.55), including spinor representations.

Since we know the representations of $SU(2)$, and here we have two commuting $SU(2)$ factors, we find that:

- The representations of the Lorentz algebra can be labeled by two half-integers: (j_-, j_+).
- The dimension of the representation (j_-, j_+) is $(2j_- + 1)(2j_+ + 1)$.
- The generator of rotations $\mathbf{J}$ is related to $\mathbf{J}^+$ and $\mathbf{J}^-$ by $\mathbf{J} = \mathbf{J}^+ + \mathbf{J}^-$; therefore, by the usual addition of angular momenta in quantum mechanics, in the representation (j_-, j_+) we have states with all possible spin j in integer steps between the values $|j_+ - j_-|$ and $j_+ + j_-$.

[7] The fact that the Lorentz algebra can be written as the algebra of $SU(2) \times SU(2)$ does *not* mean that the Lorentz group $SO(3,1)$ is the same as $SU(2) \times SU(2)$. First of all, the Lie algebra only reflects the properties of the group close to the identity. Furthermore, $\mathbf{J}^{\pm}$ are complex combinations of $\mathbf{J}$ and $\mathbf{K}$. Observe that, because of the factor i in eq. (2.52), a representation of $SU(2) \times SU(2)$ with $\mathbf{J}^{\pm}$ hermitian induces a representation of $SO(3,1)$ with $\mathbf{J}$ hermitian but $\mathbf{K}$ antihermitian. For the more mathematical reader: $SU(2) \times SU(2)$ is the universal covering group of $SO(4)$ (similarly to the fact that $SU(2)$ is the universal covering group of $SO(3)$) and $SO(4)$ is the Euclidean version of the Lorentz group, i.e. it is obtained taking the time variable t purely imaginary. The universal covering group of $SO(3,1)$ is $SL(2, C)$.

The representations are in general complex and the dimension of the representation is the number of independent complex components. In some cases we can impose a reality condition and $(2j_-+1)(2j_++1)$ becomes the number of independent real components. The representations (j_-,j_+) must include all tensor representations discussed in the previous section, plus spinorial representations. We examine the simplest cases.

$(\mathbf{0},\mathbf{0})$. This representation has dimension one. On it, $\mathbf{J}^\pm=0$ so also $\mathbf{J},\mathbf{K}$ are zero. Therefore it is the scalar representation.

$(\frac{\mathbf{1}}{\mathbf{2}},\mathbf{0})$ and $(\mathbf{0},\frac{\mathbf{1}}{\mathbf{2}})$. These representations have both dimension two and spin 1/2, so they are spinorial representations. We denote by $(\psi_L)_\alpha$, with $\alpha=1,2$, a spinor in $(1/2,0)$ and by $(\psi_R)_\alpha$ a spinor in $(0,1/2)$ (sometimes in the literature the index of ψ_L is instead denoted by $\dot\alpha$ to stress that it is an index in a different representation compared to the index of ψ_R). ψ_L is called a *left-handed Weyl spinor* and ψ_R is called a *right-handed Weyl spinor*:

$$\boxed{\text{Weyl spinors:}\quad \psi_L\in\left(\frac{1}{2},0\right),\qquad \psi_R\in\left(0,\frac{1}{2}\right).} \tag{2.56}$$

We want to determine the explicit form of the generators $\mathbf{J},\mathbf{K}$ on Weyl spinors. Consider first the representation $(1/2,0)$. By definition, on this representation $\mathbf{J}^-$ is represented by a 2×2 matrix, while $\mathbf{J}^+=0$. The solution of (2.54) in terms of 2×2 matrices is of course $\mathbf{J}^-=\boldsymbol{\sigma}/2$, and therefore

$$\mathbf{J}=\mathbf{J}^++\mathbf{J}^-=\frac{\boldsymbol{\sigma}}{2} \tag{2.57}$$

$$\mathbf{K}=-i(\mathbf{J}^+-\mathbf{J}^-)=i\,\frac{\boldsymbol{\sigma}}{2}\,. \tag{2.58}$$

Observe that in this representation the generators K^i are not hermitian, in agreement with the comment in note 7. This is a consequence of the fact that the Lorentz group is non-compact and of the theorem mentioned on page 16, which states that non-compact groups have no unitary representations of finite dimension, except for representations in which the non-compact generators (in this case the K^i) are represented trivially, i.e. $K^i=0$. We can now write explicitly how a Weyl spinor transforms under Lorentz transformations, using eq. (2.31),

$$\boxed{\psi_L\to\Lambda_L\psi_L=\exp\left\{(-i\boldsymbol{\theta}-\boldsymbol{\eta})\cdot\frac{\boldsymbol{\sigma}}{2}\right\}\psi_L\,.} \tag{2.59}$$

Repeating the argument for the $(0,1/2)$ representation, we find again $\mathbf{J}=\boldsymbol{\sigma}/2$ but $\mathbf{K}=-i\boldsymbol{\sigma}/2$ and

$$\psi_R\to\Lambda_R\psi_R=\exp\left\{(-i\boldsymbol{\theta}+\boldsymbol{\eta})\cdot\frac{\boldsymbol{\sigma}}{2}\right\}\psi_R\,. \tag{2.60}$$

Note that $\Lambda_{L,R}$ are complex matrices, and therefore necessarily the two components of a Weyl spinor are complex numbers.

Using the property of the Pauli matrices $\sigma^2\sigma^i\sigma^2 = -\sigma^{i*}$ and the explicit form of $\Lambda_{L,R}$ it is easy to show that

$$\sigma^2\Lambda_L^*\sigma^2 = \Lambda_R\,. \tag{2.61}$$

From this it follows that

$$\sigma^2\psi_L^* \to \sigma^2(\Lambda_L\psi_L)^* = (\sigma^2\Lambda_L^*\sigma^2)\sigma^2\psi_L^* = \Lambda_R(\sigma^2\psi_L^*) \tag{2.62}$$

where we used the fact that $\sigma^2\sigma^2 = 1$. Therefore , if $\psi_L \in (1/2, 0)$, then $\sigma^2\psi_L^*$ is a right-handed Weyl spinor,

$$\boxed{\sigma^2\psi_L^* \in \left(0, \frac{1}{2}\right).} \tag{2.63}$$

We define the operation of charge conjugation on Weyl spinors as an operation that transforms ψ_L into a new spinor ψ_L^c defined as

$$\psi_L^c = i\sigma^2\psi_L^*\,. \tag{2.64}$$

Then charge conjugation transforms a left-handed Weyl spinor into a right-handed one. Taking the complex conjugate of eq. (2.64) and denoting the right-handed spinor ψ_L^c by ψ_R, we have $\psi_L = -i\sigma^2\psi_R^*$ (having used the fact that σ^2 is purely imaginary and $\sigma^2\sigma^2 = 1$). Therefore we define charge conjugation on a right-handed spinor ψ_R as

$$\psi_R^c = -i\sigma^2\psi_R^*\,, \tag{2.65}$$

so that charge conjugation transforms a right-handed Weyl spinor into a left-handed one. The factor i in eq. (2.64) is chosen so that, iterating the transformation twice, we get the identity operation,

$$(\psi_L^c)^c = (i\sigma^2\psi_L^*)^c = -i\sigma^2(i\sigma^2\psi_L^*)^* = \psi_L\,. \tag{2.66}$$

We will understand the physical meaning of charge conjugation in Chapter 4.

$(\frac{1}{2}, \frac{1}{2})$. This representation has (complex) dimension four and $|1/2 - 1/2| \leqslant j \leqslant 1/2 + 1/2$, i.e. $j = 0, 1$. Comparing with eq. (2.40) we see that it is a complex four-vector representation. A generic element of the $(1/2, 1/2)$ representation can be written as a pair $((\psi_L)_\alpha, (\xi_R)_\beta)$, where ψ_L and ξ_R are two independent Weyl spinors, left-handed and right-handed, respectively, and α, β take the values $1, 2$. We want to make explicit the relation between these four (complex) quantities and the four components of a (complex) four-vector.

First of all, we have seen above that, given a right-handed spinor ξ_R, we can form a left-handed spinor $\xi_L \equiv -i\sigma^2\xi_R^*$, and similarly from ψ_L we can build $\psi_R \equiv i\sigma^2\psi_L^*$. We define the matrices σ^μ and $\bar{\sigma}^\mu$ as

$$\sigma^\mu = (1, \sigma^i)\,, \qquad \bar{\sigma}^\mu = (1, -\sigma^i)\,, \tag{2.67}$$

where σ^i are the Pauli matrices and 1 is the 2×2 identity matrix. Then, it is easy to show (see Exercise 2.3) that

$$\xi_R^\dagger \sigma^\mu \psi_R \tag{2.68}$$

and

$$\xi_L^\dagger \bar{\sigma}^\mu \psi_L \,. \tag{2.69}$$

are contravariant four-vectors. These four vectors are by construction complex. Since the matrix $\Lambda^\mu{}_\nu$ that represents the Lorentz transformation of a four-vector is real, given a complex four-vector V^μ it is consistent with Lorentz invariance to impose on it a reality condition, $V_\mu = V_\mu^*$ because, if we impose it in a given frame, it will remain true in all Lorentz frames. Therefore we obtain the real four-vector representation.

$(\mathbf{1}, \mathbf{0})$ and $(\mathbf{0}, \mathbf{1})$. These correspond to self-dual and anti-self-dual antisymmetric tensors $A^{\mu\nu}$, and each have complex dimension three, i.e. real dimension six. We discuss them in Exercise 2.5.

2.6 Field representations

Our main motivation for studying Lorentz symmetry is to construct a Lorentz-invariant field theory. A field $\phi(x)$ is a function of the coordinates with some definite transformation properties under the Lorentz group. In general, if

$$x^\mu \to x'^\mu = \Lambda^\mu{}_\nu x^\nu \tag{2.70}$$

the field $\phi(x)$ will transform into a new function of the new coordinates,

$$\phi(x) \to \phi'(x') \,. \tag{2.71}$$

To define how a field transforms means to state how $\phi'(x')$ is related to $\phi(x)$.

2.6.1 Scalar fields

The simplest possible transformation is that of a *scalar field*,

$$\boxed{\phi'(x') = \phi(x) \,.} \tag{2.72}$$

In other words, the numerical value of a scalar field at a point is Lorentz invariant: a point P has coordinates x in a reference frame and x' in the transformed frame, and the functional form of the field changes so that its numerical value in P is the same, independently of how P is labeled.

Consider now an infinitesimal Lorentz transformation

$$x^\rho \to x'^\rho = x^\rho + \delta x^\rho \tag{2.73}$$

with

$$\delta x^\rho = \omega^\rho{}_\sigma x^\sigma = -\frac{i}{2}\omega_{\mu\nu}(J^{\mu\nu})^\rho{}_\sigma\, x^\sigma\,, \tag{2.74}$$

and $(J^{\mu\nu})^\rho{}_\sigma = i(\eta^{\mu\rho}\,\delta^\nu_\sigma - \eta^{\nu\rho}\,\delta^\mu_\sigma)$, as in eqs. (2.23) and (2.24). Under this transformation $\delta\phi \equiv \phi'(x') - \phi(x) = 0$ by definition. This corresponds to the fact that the scalar representation gives a trivial representation of the generators, $J^{\mu\nu} = 0$. However, in the case of fields we have a more interesting possibility, namely we can consider an infinitesimal variation *at fixed coordinate* x (rather than at a given point P),

$$\delta_0\phi \equiv \phi'(x) - \phi(x)\,. \tag{2.75}$$

To understand the difference between $\delta\phi$ and $\delta_0\phi$ we observe that, when we compute $\delta\phi = \phi'(x') - \phi(x)$, we are studying how a single degree of freedom (the field evaluated at the point P) changes when we change the label of the point P from x to x'. However the point P is kept fixed, so the base space is made by the single degree of freedom $\phi(P)$ and therefore is one-dimensional. More generally, when in the next subsections we consider spinor or vector fields, we will see that $\delta\psi$ or δA_μ always provides a finite-dimensional representation of the generators. For instance the four degrees of freedom $A_\mu(P)$ provide a four-dimensional base space. Instead, when we compute $\delta_0\phi$, we are comparing the fields at two different space-time points P and P', so we are comparing different degrees of freedom. The base space now becomes the set of $\phi(P)$ with P varying over all of space-time, or in other words is a space of functions, and therefore it is an infinite-dimensional base-space. We then obtain an infinite-dimensional representation of the generators.

To find the generators in this representation, we expand eq. (2.75) to first order in δx,

$$\delta_0\phi = \phi'(x' - \delta x) - \phi(x) = -\delta x^\rho \partial_\rho \phi(x)\,. \tag{2.76}$$

Using eq. (2.74) for δx^ρ, this can be rewritten as

$$\delta_0\phi = \frac{i}{2}\omega_{\mu\nu}(J^{\mu\nu})^\rho{}_\sigma\, x^\sigma \partial_\rho \phi \equiv -\frac{i}{2}\omega_{\mu\nu} L^{\mu\nu}\phi\,, \tag{2.77}$$

where we defined

$$L^{\mu\nu} = -(J^{\mu\nu})^\rho{}_\sigma\, x^\sigma \partial_\rho = i(x^\mu\partial^\nu - x^\nu\partial^\mu)\,. \tag{2.78}$$

We can easily check that the operators $L^{\mu\nu}$ satisfy the Lie algebra (2.25) and therefore give a representation of the generators of the Lorentz group. As discussed above, the basis for the representation is the space of scalar fields. This is a space of functions, so it is infinite-dimensional, and therefore this is an *infinite-dimensional representation* of the Lorentz algebra. We have not yet specified what is the scalar product in the field space, so we cannot yet ask whether this representation is unitary. We postpone the issue to the next chapter.

Recalling that with our metric signature $p^\mu = +i\partial^\mu$ (see the Notation), we find $L^{\mu\nu} = x^\mu p^\nu - x^\nu p^\mu$. In particular, for spatial rotations we have $L^{ij} = x^i p^j - x^j p^i$ and $L^i = (1/2)\epsilon^{ijk} L^{jk} = \epsilon^{ijk} x^j p^k$, and we recognize that L^i is the orbital angular momentum.

2.6.2 Weyl fields

A left-handed Weyl field $\psi_L(x)$ is defined as a field that, under $x^\mu \to x'^\mu = \Lambda^\mu{}_\nu x^\nu$, transforms as

$$\boxed{\psi_L(x) \to \psi'_L(x') = \Lambda_L \psi_L(x)\,,} \tag{2.79}$$

with Λ_L given by eq. (2.59). Similarly a right-handed Weyl field ψ_R transforms with Λ_R given in eq. (2.60). In the classical theory we will consider ψ_L, ψ_R as ordinary, commuting, c-numbers.

The representation of the Lorentz generators on ψ_L can be found computing

$$\begin{aligned}\delta_0\psi_L &\equiv \psi'_L(x) - \psi_L(x) = \psi'_L(x' - \delta x) - \psi_L(x)\\ &= \psi'_L(x') - \delta x^\rho \partial_\rho \psi_L(x) - \psi_L(x)\\ &= (\Lambda_L - 1)\psi_L(x) - \delta x^\rho \partial_\rho \psi_L(x)\,.\end{aligned} \tag{2.80}$$

We see that $\delta_0\psi_L$ is made of two parts; one comes from the variation of the coordinate δx^ρ and is the same as for scalar fields. Exactly as in eqs. (2.76) and (2.77), we have

$$-\delta x^\rho \partial_\rho \psi_L = -\frac{i}{2}\omega_{\mu\nu} L^{\mu\nu} \psi_L\,, \tag{2.81}$$

with $L^{\mu\nu}$ given in eq. (2.78). We write Λ_L in the form

$$\Lambda_L = e^{-\frac{i}{2}\omega_{\mu\nu} S^{\mu\nu}}\,. \tag{2.82}$$

Then eq. (2.80) becomes

$$\delta_0\psi_L = -\frac{i}{2}\omega_{\mu\nu} J^{\mu\nu} \psi_L \tag{2.83}$$

with

$$J^{\mu\nu} = L^{\mu\nu} + S^{\mu\nu}\,. \tag{2.84}$$

Comparing eq. (2.82) with eq. (2.59) we see that

$$S^i = \frac{1}{2}\epsilon^{ijk} S^{jk} = \frac{\sigma^i}{2}\,, \tag{2.85}$$

while

$$S^{i0} = i\frac{\sigma^i}{2}\,. \tag{2.86}$$

We recognize in eq. (2.84) the separation of the angular momentum into the orbital and the spin contributions. It is clear that this separation is completely general, and holds for any representation. The orbital part $L^{\mu\nu}$ always has the form (2.78) independently of the representation, while $S^{\mu\nu}$ depends on the specific representation used. For instance, for right-handed Weyl fields S^i are still given by eq. (2.85) while $S^{i0} = -i\sigma^i/2$, as we see from eq. (2.60).

2.6.3 Dirac fields

Consider a parity transformation $(t, \mathbf{x}) \to (t, -\mathbf{x})$. Under this operation the boost generators behave as true vectors and change sign, $\mathbf{K} \to -\mathbf{K}$, since the parity transformation reverses the velocity $\mathbf{v}$ of the boost. The angular momentum generator is instead a pseudovector, $\mathbf{J} \to \mathbf{J}$. Therefore a parity operation exchanges the $J^i_\pm$ generators, $J^i_+ \leftrightarrow J^i_-$. This means that under a parity transformation an object in the (j_-, j_+) representation is transformed into an object in the (j_+, j_-) representation. Therefore the representation (j_-, j_+) of the Lorentz group is not at the same time a basis for a representation of the parity transformation, unless $j_- = j_+$. In particular, ψ_L and ψ_R, separately, are *not* a basis for a representation of the parity transformation.

In Nature, we know experimentally that parity is violated by weak interactions. At the theoretical level, this is reflected in the fact that in the Standard Model the left and right-handed components of the spin 1/2 particles enter the theory in a very different way, as we will see in Chapter 8. However, we saw in Section 1.2 that the typical scale of weak interactions is $O(100)$ GeV, much higher than the scale of strong and of electromagnetic interactions. At sufficiently low energies, therefore, the effect of weak interactions is small, and the dominant contributions come from the electromagnetic and the strong interactions, which both conserve parity. In this case, it is convenient to work with fields which provide a representation of Lorentz *and* parity transformations. We then define a *Dirac field* as[8]

[8] More precisely, this is the Dirac field written in the chiral basis, see Section 3.4.2.

$$\boxed{\Psi = \begin{pmatrix} \psi_L \\ \psi_R \end{pmatrix} .} \tag{2.87}$$

A Dirac field therefore has four complex components, and it provides a basis for a representation of both Lorentz and parity transformations. In fact, under a Lorentz transformation, $\Psi \to \Lambda_D \Psi$ with

$$\Lambda_D = \begin{pmatrix} \Lambda_L & 0 \\ 0 & \Lambda_R \end{pmatrix} , \tag{2.88}$$

and Λ_L, Λ_R given in eqs. (2.59) and (2.60).[9] Under a parity transformation P the coordinates change as $x^\mu \to x'^\mu = (t, -\mathbf{x})$ while

[9] In Section 3.4.2, after introducing the Dirac matrices, we will see how to write Λ_D in terms of the commutator of Dirac matrices, and the result will be independent of the chiral basis that we have used here.

$$\begin{pmatrix} \psi_L(x) \\ \psi_R(x) \end{pmatrix} \to \begin{pmatrix} \psi_R(x') \\ \psi_L(x') \end{pmatrix} \tag{2.89}$$

and therefore

$$\Psi(x) \to \begin{pmatrix} 0 & 1 \\ 1 & 0 \end{pmatrix} \Psi(x') . \tag{2.90}$$

When we study the *quantized* Dirac field we will examine the possibility and the meaning of an overall phase $\eta = \pm 1$ in the transformation law (2.90), see Section 4.2.3.

In eqs. (2.64) and (2.65) we defined the operation of charge conjugation on Weyl spinors. Given a Dirac spinor Ψ as in eq. (2.87), charge conjugation allows us to define a new Dirac spinor

$$\Psi^c = \begin{pmatrix} -i\sigma^2\psi_R^* \\ i\sigma^2\psi_L^* \end{pmatrix} = -i\begin{pmatrix} 0 & \sigma^2 \\ -\sigma^2 & 0 \end{pmatrix}\Psi^* . \tag{2.91}$$

and, as for Weyl spinors, iterating charge conjugation twice one finds the identity transformation,

$$(\Psi^c)^c = \Psi . \tag{2.92}$$

Note that the coordinates x^μ are unchanged under charge conjugation. We will understand the importance of charge conjugation when we quantize the theory and we will find particles and antiparticles.

Dirac spinors are the basic objects in quantum electrodynamics (QED). Since QED preserves parity and charge conjugation, the Weyl spinors always appear in the combination Ψ. On Ψ parity is a well-defined operation, and we can use it to construct a parity-invariant theory while, having for instance only ψ_L at our disposal, it is impossible to build a theory invariant under parity. We will see that in the Standard Model, parity and charge conjugation are not symmetries and ψ_L, ψ_R appear separately, in a non-symmetric way. Therefore, Weyl spinors are more fundamental objects than Dirac spinors.

2.6.4 Majorana fields

A Majorana spinor is a Dirac spinor in which ψ_L and ψ_R are not independent, but rather $\psi_R = i\sigma^2\psi_L^*$,

$$\Psi_M = \begin{pmatrix} \psi_L \\ i\sigma^2\psi_L^* \end{pmatrix} . \tag{2.93}$$

So, it has the same number of degrees of freedom as a Weyl spinor, although it is written in the form of a Dirac spinor. From this definition it follows that a Majorana spinor is invariant under charge conjugation

$$\Psi_M^c = \Psi_M . \tag{2.94}$$

Observe that, if we have a complex scalar field $\phi(x)$, we can impose on it a reality condition $\phi(x) = \phi^*(x)$, and this is a Lorentz-invariant condition: since ϕ and ϕ^* are both Lorentz invariant, if we impose $\phi = \phi^*$ in a frame, we will have $\phi = \phi^*$ in any other frame. The same is true for the four-vector representation, as we already discussed on page 29. For a Dirac spinor Ψ the situation is different; Ψ is a complex field, and the condition $\Psi = \Psi^*$ is not Lorentz invariant, since the matrix Λ_D in eq. (2.88) is not real. Therefore, if we impose the relation $\Psi = \Psi^*$ in a frame, it will not hold in general in another Lorentz frame. Instead, the condition (2.94) is by construction Lorentz invariant, since it is a consequence of the definition (2.93), which in turns expresses the

Lorentz-invariant statement that $i\sigma^2\psi_L^*$ is a right-handed spinor. Since Ψ_M^c involves complex conjugation, see eq. (2.91), the condition (2.94) is a Lorentz-invariant relation between Ψ and Ψ^*, and in this sense it is called a reality condition.

So we can see Majorana fields as "real" Dirac fields, with respect to the only possible Lorentz-invariant reality condition, eq. (2.94).

It is possible that Majorana spinors play an important role in the description of the neutrino. We will come to this issue later.

2.6.5 Vector fields

The definition of vector fields at this point is obvious. A (contravariant) vector field $V^\mu(x)$ is defined as a field that, under $x^\mu \to x'^\mu = \Lambda^\mu{}_\nu x^\nu$, transforms as

$$V^\mu(x) \to V'^\mu(x') = \Lambda^\mu{}_\nu V^\nu(x)\,. \tag{2.95}$$

From the discussion in Section 2.4.1 we see that a general vector field has a spin-0 and a spin-1 component. An example of a vector field that will be important for us is the gauge field $A^\mu(x)$ in electromagnetism. We will see in Section 4.3.1 that $A^\mu(x)$ is subject to some conditions, stemming from gauge invariance, that eliminate the spin-0 component and the state with $(j=1, j_z=0)$, where z is the propagation direction.

Since a vector field belongs to the $(1/2, 1/2)$ representation, it has $j_- = j_+$ and therefore it is a basis for the representation of parity. A true vector transform as $(V^0, \mathbf{V}) \to (V^0, -\mathbf{V})$ while a pseudovector (or axial vector) transforms as $(V^0, \mathbf{V}) \to (-V^0, \mathbf{V})$.

Tensor fields are defined similarly.

2.7 The Poincaré group

Beside invariance under Lorentz transformations, we require also invariance under space-time translations. A generic element of the translation group is written as

$$\exp\{-iP^\mu a_\mu\} \tag{2.96}$$

where a_μ are the parameters of the translation, $x^\mu \to x^\mu + a^\mu$, and the components of the four-momentum operator P^μ are the generators. Translations plus Lorentz transformations form a group, called the *Poincaré group*, or the inhomogeneous Lorentz group (it is sometimes denoted as $ISO(3,1)$, where "I" stands for inhomogeneous).

Since the translations commute, we have

$$[P^\mu, P^\nu] = 0\,. \tag{2.97}$$

To find the commutator between P^μ and $J^{\rho\sigma}$ we can start from the commutators

$$[J^i, P^j] = i\epsilon^{ijk}P^k\,, \tag{2.98}$$

$$[J^i, P^0] = 0\,, \tag{2.99}$$

which express the facts that P^i is a vector under rotations and that the energy is a scalar under rotations. The unique Lorentz-covariant generalization of eqs. (2.98) and (2.99) is

$$[P^\mu, J^{\rho\sigma}] = i\left(\eta^{\mu\rho}P^\sigma - \eta^{\mu\sigma}P^\rho\right). \tag{2.100}$$

Together with the Lorentz algebra (2.25), eqs. (2.97) and (2.100) define the Poincaré algebra. In terms of $J^i, K^i, P^0 = H$ and P^i it reads

$$\left[J^i, J^j\right] = i\epsilon^{ijk}J^k\,, \quad \left[J^i, K^j\right] = i\epsilon^{ijk}K^k\,, \quad \left[J^i, P^j\right] = i\epsilon^{ijk}P^k\,, \tag{2.101}$$

$$\left[K^i, K^j\right] = -i\epsilon^{ijk}J^k\,, \quad \left[P^i, P^j\right] = 0\,, \quad \left[K^i, P^j\right] = iH\delta^{ij}\,, \tag{2.102}$$

$$\left[J^i, H\right] = 0\,, \qquad \left[P^i, H\right] = 0\,, \qquad \left[K^i, H\right] = iP^i\,. \tag{2.103}$$

Equations (2.101) express the fact that the J^i generate spatial rotations and K^i, P^i are vectors under rotations. Equations (2.103) state that J^i and P^i commute with the generator of time translations and therefore are conserved quantities; the K^i instead are not conserved, and this is the reason why the eigenvalues of $\mathbf{K}$ are not used for labeling physical states.

2.7.1 Representation on fields

We saw in Section 2.6 that fields provide an infinite-dimensional representation of the Lorentz group, and that on fields the generators $J^{\mu\nu}$ are represented as

$$J^{\mu\nu} = L^{\mu\nu} + S^{\mu\nu} \tag{2.104}$$

where

$$L^{\mu\nu} = i(x^\mu\partial^\nu - x^\nu\partial^\mu) \tag{2.105}$$

and $S^{\mu\nu}$ depends on the spin of the field in question, but not on the coordinates x^μ. To obtain a representation of the full Poincaré group on fields, we must now find how to represent the four-momentum operator P^μ, i.e. we have to specify the transformation law of fields under translations.

We require that all fields, independently of their transformation property under the Lorentz group, behave as *scalars* under space-time translation. Let us label by ϕ a generic field, either a Lorentz scalar field, or a component of a spinor field ξ_α with α given, or a given component V^μ of a vector field, etc. Then, under a translation $x \to x' = x + a$, all fields, independently of their Lorentz properties, transform as

$$\phi'(x') = \phi(x)\,. \tag{2.106}$$

Under an infinitesimal translation $x^\mu \to x'^\mu = x^\mu + \epsilon^\mu$ we have, to first order in ϵ,

$$\delta_0\phi \equiv \phi'(x) - \phi(x) = \phi'(x' - \epsilon) - \phi(x) = -\epsilon^\mu\partial_\mu\phi(x)\,. \tag{2.107}$$

On the other hand, from the form (2.96) of the translation operator, it follows that

$$\phi'(x'-\epsilon) = e^{-iP^\mu(-\epsilon_\mu)}\phi'(x') = e^{i\epsilon_\mu P^\mu}\phi(x) \tag{2.108}$$

and therefore to first order in ϵ

$$\delta_0\phi = i\epsilon_\mu P^\mu\phi(x)\,. \tag{2.109}$$

Comparing eqs. (2.107) and (2.109) we see that the momentum operator is represented as

$$P^\mu = +i\partial^\mu\,. \tag{2.110}$$

Therefore

$$H = i\frac{\partial}{\partial x^0} = i\frac{\partial}{\partial t}\,, \qquad P^i = i\partial^i = -i\partial_i = -i\frac{\partial}{\partial x^i}\,. \tag{2.111}$$

The explicit form of $J^{\mu\nu}$ and of P^μ has been found requiring that the fields have well-defined transformation properties under the Poincaré group; therefore these explicit expressions must automatically satisfy the Poincaré algebra. We can check this easily observing that $S^{\mu\nu}$ does not depend on the coordinates and therefore commutes with ∂^μ, while $[\partial^\mu, x^\nu] = \eta^{\mu\nu}$. Therefore

$$\begin{aligned}[P^\mu, J^{\rho\sigma}] &= [i\partial^\mu, i(x^\rho\partial^\sigma - x^\sigma\partial^\rho)] = -\eta^{\mu\rho}\partial^\sigma + \eta^{\mu\sigma}\partial^\rho \\ &= i\left(\eta^{\mu\rho}P^\sigma - \eta^{\mu\sigma}P^\rho\right)\,,\end{aligned} \tag{2.112}$$

in agreement with eq. (2.100). The commutator $[P^\mu, P^\nu] = 0$ is also satisfied by $P^\mu = i\partial^\mu$ and we already know that the commutator $[J^{\mu\nu}, J^{\rho\sigma}]$ is correctly reproduced, so the full Poincaré algebra is obeyed.

2.7.2 Representation on one-particle states

The representation of the Poincaré group on fields allows us to construct Poincaré invariant Lagrangians, as we will study in the next chapter. At the classical level, a Lagrangian description is all that we need in order to specify the dynamics of the system. At the quantum level, however, one of our aims will be to understand how the concept of particle emerges from field quantization. It is therefore useful to see how the Poincaré group can be represented using as a basis the Hilbert space of a free particle. We will denote the states of a free particle with momentum $\mathbf{p}$ as $|\mathbf{p}\,, s\rangle$, where s labels collectively all other quantum numbers. Since $\mathbf{p}$ is a continuous and unbounded variable, this base space is infinite-dimensional. A theorem by Wigner (see Weinberg (1995), Chapter 2) states that on this Hilbert space any symmetry transformation can be represented by a unitary operator.[10] Therefore in this base space a Poincaré transformation is represented by a unitary matrix, and the generators J^i, K^i, P^i and H by hermitian operators.

The representations are labeled by the Casimir operators. One is easily found, and is $P_\mu P^\mu$. On a one-particle state it has the value m^2,

[10] Actually there is also the possibility of an anti-unitary operator; the only symmetry transformation where this happens is time-reversal, and we postpone the definition of anti-unitary operators to Chapter 4.

where m is the mass of the particle. Using the commutation relations of the Poincaré group one can verify that there is a second Casimir operator given by $W_\mu W^\mu$, where

$$W^\mu = -\frac{1}{2}\epsilon^{\mu\nu\rho\sigma} J_{\nu\rho} P_\sigma \tag{2.113}$$

is called the Pauli–Lubanski four-vector. To prove that $W_\mu W^\mu$ is a Casimir operator is straightforward. First of all, W^μ is clearly a four-vector, so $W_\mu W^\mu$ is Lorentz-invariant and therefore commutes with $J^{\mu\nu}$. From the explicit form it also follows that

$$[W^\mu, P^\nu] = 0\,, \tag{2.114}$$

(using eq. (2.100) and the antisymmetry of $\epsilon^{\mu\nu\rho\sigma}$), and then $W_\mu W^\mu$ commutes also with P^ν.

Since $W_\mu W^\mu$ is Lorentz-invariant, we can compute it in the frame that we prefer. If $m \neq 0$, it is convenient to choose the rest frame of the particle; in this frame $W^\mu = (-m/2)\epsilon^{\mu\nu\rho 0} J_{\nu\rho} = (m/2)\epsilon^{0\mu\nu\rho} J_{\nu\rho}$, so $W^0 = 0$ while

$$W^i = \frac{m}{2}\epsilon^{0ijk} J^{jk} = \frac{m}{2}\epsilon^{ijk} J^{jk} = mJ^i\,. \tag{2.115}$$

Therefore on a one-particle state with mass m and spin j we have

$$-W_\mu W^\mu = m^2 j(j+1)\,, \qquad (m \neq 0)\,. \tag{2.116}$$

If instead $m = 0$ the rest frame does not exist, but we can choose a frame where $P^\mu = (\omega, 0, 0, \omega)$; in this frame a straightforward computation gives $W^0 = W^3 = \omega J^3$, $W^1 = \omega(J^1 - K^2)$ and $W^2 = \omega(J^2 + K^1)$. Therefore

$$-W_\mu W^\mu = \omega^2[(K^2 - J^1)^2 + (K^1 + J^2)^2]\,, \qquad (m = 0)\,. \tag{2.117}$$

Comparing eqs. (2.116) and (2.117) we see that the limit $m \to 0$ is quite subtle, and we must study separately the massive and massless representations.

Massive representations: In this case on the one-particle states we have $P^\mu P_\mu = m^2$ while $W_\mu W^\mu = -m^2 j(j+1)$. We will restrict to m real[11] and positive. Therefore the representations are labeled by the mass m and by the spin j. We can understand this better observing that, if $m \neq 0$, with a Lorentz transformation we can bring P^μ into the form $P^\mu = (m, 0, 0, 0)$. This choice of P^μ still leaves us with the freedom of performing spatial rotations. In other words, the space of one-particle states with momentum $P^\mu = (m, 0, 0, 0)$ is still a basis for the representation of spatial rotations. The group of transformations which leaves invariant a given choice of P^μ is called the *little group*. In this case, since we want to include spinor representations, the little group is $SU(2)$. The massive representations are therefore labeled by the mass m and by the spin $j = 0, 1/2, 1, \ldots$, and states within each

[11] In principle there is also the possibility of representations with $m^2 < 0$, known as tachyons. In field theory the emergence of a tachyonic mode is the signal of an instability, and reflects the fact that we have expanded around the wrong vacuum, e.g. around a maximum rather than a minimum of a potential.

representation are labeled by $j_z = -j, -j+1, \ldots, j$. This means that *massive particles of spin j have $2j+1$ degrees of freedom.*

Massless representations: When $P^2 = 0$ the rest frame does not exist, but we can reduce P^μ to the form $P^\mu = (\omega, 0, 0, \omega)$. The little group is the set of Poincaré transformations that leaves this vector unchanged. One sees immediately that the rotations in the (x, y) plane leave this P^μ invariant; this is an $SO(2)$ group, generated by J^3.

This part is more technical and can be omitted at a first reading. Just assume that the little group is $SO(2)$ and skip the part written in smaller characters.

Furthermore there are two less evident Lorentz transformations that do not change P^μ; to find the most general solution, it is sufficient to restrict to infinitesimal Lorentz transformations $\Lambda^\mu{}_\nu = \delta^\mu_\nu + \omega^\mu{}_\nu$, and to look for the most general matrix $\omega^{\mu\nu}$ which satisfies $\omega^{\mu\nu} = -\omega^{\nu\mu}$ (in order to have a Lorentz transformation) and

$$\omega^{\mu\nu} P_\nu = 0\,, \tag{2.118}$$

for $P_\nu = (\omega, 0, 0, -\omega)$. Therefore

$$\begin{pmatrix} 0 & \omega^{01} & \omega^{02} & \omega^{03} \\ -\omega^{01} & 0 & \omega^{12} & \omega^{13} \\ -\omega^{02} & -\omega^{12} & 0 & \omega^{23} \\ -\omega^{03} & -\omega^{13} & -\omega^{23} & 0 \end{pmatrix} \begin{pmatrix} 1 \\ 0 \\ 0 \\ -1 \end{pmatrix} = 0\,, \tag{2.119}$$

which gives $\omega^{03} = 0$, $\omega^{01} + \omega^{13} = 0$ and $\omega^{02} + \omega^{23} = 0$. Denoting $\omega^{01} = \alpha$, $\omega^{02} = \beta$ and $\omega^{12} = \theta$ we see that the most general Lorentz transformation that leaves P^μ invariant can be written as

$$\Lambda = e^{-i(\alpha A + \beta B + \theta C)} \tag{2.120}$$

where (lowering the second Lorentz index)

$$A^\mu{}_\nu = i \begin{pmatrix} 0 & -1 & 0 & 0 \\ -1 & 0 & 0 & 1 \\ 0 & 0 & 0 & 0 \\ 0 & -1 & 0 & 0 \end{pmatrix}, \quad B^\mu{}_\nu = i \begin{pmatrix} 0 & 0 & -1 & 0 \\ 0 & 0 & 0 & 0 \\ -1 & 0 & 0 & 1 \\ 0 & 0 & -1 & 0 \end{pmatrix} \tag{2.121}$$

and

$$C^\mu{}_\nu = i \begin{pmatrix} 0 & 0 & 0 & 0 \\ 0 & 0 & -1 & 0 \\ 0 & 1 & 0 & 0 \\ 0 & 0 & 0 & 0 \end{pmatrix}. \tag{2.122}$$

Comparison with eq. (2.23) shows that $C^\mu{}_\nu$ is nothing but $(J^3)^\mu{}_\nu$, i.e. the explicit expression of J^3 as a 4×4 matrix in the four-vector representation. Similarly we find that $A^\mu{}_\nu = (K^1 + J^2)^\mu{}_\nu$ and $B^\mu{}_\nu = (K^2 - J^1)^\mu{}_\nu$. These are just the combinations that appear in eq. (2.117), so in the massless case

$$-W_\mu W^\mu = \omega^2 (A^2 + B^2)\,. \tag{2.123}$$

Using the commutation rules of the Lorentz group, or directly the explicit expressions given above, one finds that the operators J^3, A and B close an algebra:

$$[J^3, A] = +iB\,, \qquad [J^3, B] = -iA\,, \qquad [A, B] = 0\,. \tag{2.124}$$

Formally this is the same algebra generated by the operators p^x, p^y and $L^z = xp^y - yp^x$, which describe the translations and rotations of a Euclidean plane, with A and B playing the role of the translation operators. This algebra is denoted by $ISO(2)$. The matrices $A^\mu{}_\nu$ and $B^\mu{}_\nu$ given in eq. (2.121) are not hermitian.[12] This is as it should be, since they are 4×4 matrices, and therefore are a finite-dimensional representation of non-compact Lorentz generators.

[12] They would be hermitian if we write them as $A^{\mu\nu}, B^{\mu\nu}$ and $C^{\mu\nu}$. However, δx^ρ is proportional to $\omega_{\mu\nu}(J^{\mu\nu})^\rho{}_\sigma x^\sigma$, so the representation is provided by the matrices with one upper and one lower index, and it is for these matrices that the algebra (2.124) holds.

We can however represent the algebra (2.124) taking as the base space the one-particle states with momentum $\mathbf{p}$. In this representation A and B are hermitian operators because of Wigner's theorem and, since they are commuting, they can be diagonalized simultaneously. We denote by a, b the respective eigenvalues. Then

$$A|\mathbf{p}\,;a,b\rangle = a|\mathbf{p}\,;a,b\rangle\,, \qquad B|\mathbf{p}\,;a,b\rangle = b|\mathbf{p}\,;a,b\rangle\,. \tag{2.125}$$

However, if a and b are non-zero, we can now find a continuous set of eigenvalues! Consider in fact the state

$$|\mathbf{p}\,;a,b,\theta\rangle \equiv e^{-i\theta J^3}|\mathbf{p}\,;a,b\rangle\,, \tag{2.126}$$

with θ an arbitrary angle. We have

$$A\,e^{-i\theta J^3}|\mathbf{p}\,;a,b\rangle = e^{-i\theta J^3}\left(e^{i\theta J^3}Ae^{-i\theta J^3}\right)|\mathbf{p}\,;a,b\rangle\,. \tag{2.127}$$

Using the commutation rules (2.124) we find that

$$e^{i\theta J^3}Ae^{-i\theta J^3} = A\cos\theta - B\sin\theta \tag{2.128}$$

(this can be proved expanding the exponentials in power series) and therefore

$$A|\mathbf{p}\,;a,b,\theta\rangle = (a\cos\theta - b\sin\theta)|\mathbf{p}\,;a,b,\theta\rangle\,, \tag{2.129}$$

and similarly

$$B|\mathbf{p}\,;a,b,\theta\rangle = (a\sin\theta + b\cos\theta)|\mathbf{p}\,;a,b,\theta\rangle\,. \tag{2.130}$$

This means that, unless $a = b = 0$, we find representations corresponding to massless particles with a continuous internal degree of freedom θ. These representations do not so far find physical applications, and we therefore restrict to states with $a = b = 0$. Since for massless particles we found $-W_\mu W^\mu = \omega^2(A^2 + B^2)$, on these states (and only on these states) we have $-W_\mu W^\mu = 0$, which agrees with the $m \to 0$ limit of eq. (2.116). On the states with $a = b = 0$ the little group is simply $SO(2)$ or, equivalently, $U(1)$.

As for any abelian group, the irreducible representations of $SO(2)$ are one-dimensional. The generator of the group $SO(2)$ of rotations in the (x, y) plane is the angular momentum J^3 and therefore the one-dimensional representations are labeled by the eigenvalue h of J^3; it represents the angular momentum in the direction of propagation of the particle (in this case, the z axis), and is called the *helicity*.

It can be shown that h is quantized, $h = 0, \pm 1/2, \pm 1, \ldots$. Actually, there is a subtle technical point in the quantization of h: the elementary proof that, for $SU(2)$, j_z is quantized is of an algebraic nature. One defines $\lambda_m = \langle j, m+1|(J_x + iJ_y)|jm\rangle$ and, using the commutation relations between the three J_i, one finds a recursion relation $|\lambda_{m-1}|^2 - |\lambda_m|^2 = 2m$. The condition that this recursion relation does not produce a negative $|\lambda_m|^2$ provides the quantization of $m = j_z$.[13] In the case of the little group of massless particles we do not have J_x, J_y at our disposal, but only the single $SO(2)$ generator J_z and therefore this algebraic proof does not go through. There is however a topological proof, based on the fact that the universal covering of the Lorentz group is $SL(2, C)$; this is a double covering, therefore any Lorentz rotation by 4π is the same

[13] See any book on quantum mechanics, e.g. L. Schiff, Quantum Mechanics, third edition, McGraw-Hill, New York 1968, eq. (27.23).

as the identity matrix. A detailed discussion can be found in Weinberg (1995), pages 86–90.

This analysis shows that *massless particles have only one degree of freedom, and are characterized by the value h of their helicity.* On a state of helicity h, a $U(1)$ rotation of the little group is represented by

$$U(\theta) = \exp\{-ih\theta\}\,. \tag{2.131}$$

From the point of view of the representations of the Poincaré group, a massless particle with helicity $+h$ and a massless particle with helicity $-h$ are logically two different species of particles, since they belong to two different representations of the Poincaré group. However, the helicity is the projection of the angular momentum along the direction of motion, so it can be written as

$$h = \hat{\mathbf{p}} \cdot \mathbf{J} \tag{2.132}$$

where $\hat{\mathbf{p}}$ is the unit vector in the direction of propagation. We see from eq. (2.132) that the helicity is a pseudoscalar, i.e. it changes sign under parity. If the interaction conserves parity, to each particle of helicity h there must correspond another particle with helicity $-h$, and these two helicity states must enter into the theory in a symmetric way. Since the electromagnetic interaction conserves parity, it is more natural to define the photon as a representation of the Poincaré group *and* of parity, i.e. to assemble together the two states of helicity $h = \pm 1$. The two states $h = \pm 1$ are then referred to as left-handed ($h = -1$) and right-handed ($h = +1$) photons.

Similarly the two states with helicity $h = \pm 2$ that mediate the gravitational interaction are better considered as two polarization states of the same particle, the graviton:

Photon:	$m^2 = 0$, two polarization states $h = \pm 1$.
Graviton:	$m^2 = 0$, two polarization states $h = \pm 2$.

On the contrary, neutrinos have only weak interactions (apart from the much smaller gravitational interaction), which do not conserve parity, and the two states with helicity $h = \pm 1/2$ are given different names: neutrino is reserved for $h = -1/2$, and antineutrino for $h = +1/2$.

Summary of chapter

In this chapter we have introduced a number of mathematical tools that will greatly simplify our construction of classical and quantum field theories in the next chapters. We recall some important points.

- Lie group, Lie algebras and their representations have been discussed in Section 2.1. They are central concepts in modern theoretical physics, independently of our applications to the Lorentz and Poincaré group. Basically, Lie groups are the correct language for describing continuous symmetries.

- The Lorentz group is generated by rotations and boosts, and its algebra is given in eqs. (2.27)–(2.29). We have discussed its tensor representations in Section 2.4 and its spinorial representations in Section 2.5. This leads in particular to the introduction of Weyl spinors, eq. (2.56); Dirac spinors are obtained assembling a left-handed and a right-handed Weyl spinor, and are a representation of Lorentz and of parity transformations.
- Fields are functions of the coordinates with well-defined transformation properties under Poincaré transformations. Depending on their transformation properties under the Lorentz group, we have scalar fields, Weyl fields, Dirac fields, vector fields, etc.
- The study of the representations of the Poincaré group using as base space the Hilbert space of one-particle states leads to massive particles, characterized by the spin j and having $2j+1$ degrees of freedom, and massless particles, which have one degree of freedom and a definite helicity h. For the photon and for the graviton, parity considerations suggest assembling the two states with helicity $h = \pm 1$ (for the photon) and $h = \pm 2$ (for the graviton) into a single particle.

Further reading

- For Weyl and Dirac spinors see Ramond (1990), Chapter 1 and Peskin and Schroeder (1995), Chapter 3. Observe that our definitions of ψ_L and ψ_R are inverted compared to Ramond (but agree with Peskin and Schroeder). In particular, for us the boost generator on ψ_L is $+i\boldsymbol{\sigma}/2$ while for Ramond is $-i\boldsymbol{\sigma}/2$ and as a consequence for us the four-vector made with left-handed spinors is $\xi_L^\dagger \bar{\sigma}^\mu \psi_L$, see eq. (2.69), while for Ramond it is $\xi_L^\dagger \sigma^\mu \psi_L$ (the fact that we both say that ψ_L belongs to the $(1/2, 0)$ representation is due to the fact that we write (j_-, j_+) while Ramond writes (j_+, j_-)). In the next chapter we will see that with our definition ψ_L has helicity $-1/2$ (and therefore with the definition of Ramond it has $h = +1/2$).
- A very clear book on Lie groups for physicists is Georgi (1999). The second edition contains many improvements of the already 'classical' first edition. For a more geometrical approach to Lie groups, see, e.g., Nakahara (1990), Section 5.6. An advanced book is J. Fuchs and C. Schweigert, *Symmetries, Lie Algebras and Representations*, Cambridge University Press 1997.
- For the representations of the Poincaré group see Sections 2.4 and 2.5 of Weinberg (1995).

Exercises

(2.1) Consider a massive particle moving with velocity $v = \tanh \eta$. Show that, if E is the energy of the particle and p its momentum along the propagation direction, then

$$\eta = \frac{1}{2} \log \frac{E+p}{E-p}, \tag{2.133}$$

and verify that, under a boost in the direction of motion with velocity v', η transforms additively: $\eta \to \eta + \text{arctanh}(v')$. Therefore $d\eta$ is invariant under longitudinal boosts.

(2.2) Show that a totally symmetric and traceless spatial tensor $T^{i_1 \cdots i_N}$ with N spatial indices has angular momentum $j = N$. Discuss some physically interesting examples.

(2.3) Prove that, if ψ_R and ξ_R are right-handed Weyl spinors, $\xi_R^\dagger \sigma^\mu \psi_R$ is a four-vector, and similarly for $\xi_L^\dagger \bar{\sigma}^\mu \psi_L$, where ξ_L and ψ_L are left-handed Weyl spinors.

(2.4) Find the explicit form of the variation of an antisymmetric tensor $F^{\mu\nu}$ under an infinitesimal Lorentz transformation. Writing $F^{0i} = -E^i$ and $F^{ij} = -\epsilon^{ijk} B^k$, find the infinitesimal transformation of E^i and B^i.

(2.5) Consider an antisymmetric tensor $A^{\mu\nu}$. (i) Prove that, with Minkowski signature $\eta^{\mu\nu}$, if we try to impose the condition $A^{\mu\nu} = (1/2)\epsilon^{\mu\nu\rho\sigma} A_{\rho\sigma}$, or the condition $A^{\mu\nu} = -(1/2)\epsilon^{\mu\nu\rho\sigma} A_{\rho\sigma}$, then the only solution is $A^{\mu\nu} = 0$.

(ii) Repeat the same exercise in a Euclidean space with metric $\delta^{\mu\nu}$. A Euclidean tensor is called self-dual if $A^{\mu\nu} = (1/2)\epsilon^{\mu\nu\rho\sigma} A_{\rho\sigma}$, or anti-self-dual if $A^{\mu\nu} = -(1/2)\epsilon^{\mu\nu\rho\sigma} A_{\rho\sigma}$. Verify that self-dual and anti-self-dual tensors are irreducible representations of (real) dimension three of the Euclidean group $SO(4)$, and verify that the six-dimensional representation $A^{\mu\nu}$ of $SO(4)$ decomposes into its self-dual and anti-self-dual parts.

(iii) With Minkowski signature, an antisymmetric tensor $A^{\mu\nu}$ with *complex* components is called self-dual if $A^{\mu\nu} = (i/2)\epsilon^{\mu\nu\rho\sigma} A_{\rho\sigma}$ and anti-self-dual if $A^{\mu\nu} = -(i/2)\epsilon^{\mu\nu\rho\sigma} A_{\rho\sigma}$. Prove that the self-dual and anti-self-dual parts are irreducible representations of the Lorentz group of complex dimension three (i.e. real dimension six) and identify these representations with the appropriate (j_-, j_+) representations.

(iv) Write the Maxwell tensor $F^{\mu\nu}$ as a sum of a self-dual and anti-self-dual part. Realize that a *real* antisymmetric tensor, such as $F^{\mu\nu}$, is an irreducible representation of $SO(3,1)$.

(2.6) (i) A classical electromagnetic wave propagating in the direction $\hat{\mathbf{n}} = (0,0,1)$ is described by the linear polarization vectors $\mathbf{e}^1 = (1,0,0)$ and $\mathbf{e}^2 = (0,1,0)$. Define the circular polarizations as $\mathbf{e}^\pm = \mathbf{e}^1 \pm i\mathbf{e}^2$. Compute the transformation of $\mathbf{e}^\pm$ under a rotation in the (x,y) plane and conclude that electromagnetic waves are made of massless spin-1 particles.

(ii) A classical gravitational wave propagating in the direction $\hat{\mathbf{n}}$ is described by a polarization tensor h^{ij} symmetric, traceless, and transverse to the propagation direction, $\hat{n}^i h^{ij} = 0$, i.e. (setting $\hat{\mathbf{n}} = \hat{\mathbf{z}}$) by a matrix of the general form

$$h^{ij} = \begin{pmatrix} h_+ & h_\times & 0 \\ h_\times & -h_+ & 0 \\ 0 & 0 & 0 \end{pmatrix}, \tag{2.134}$$

and $h_{+,\times}$ are called the plus and cross polarizations, respectively. Compute the transformation properties of $h_{+,\times}$ under a rotation in the (x,y) plane, and the transformation properties of the circular polarization tensors $h_\times \pm i h_+$. Conclude that gravitational waves are made of massless spin-2 particles.

Classical field theory

3

In the previous chapter we defined our basic objects, the fields. We now introduce their dynamics, first of all at the classical level. We will discuss the Lagrangian and the Hamiltonian formalism and we will present the Noether theorem, which provides the relation between symmetries and conservation laws. We will see that at the classical level the fields obey relativistic wave equations, such as the Klein–Gordon, Dirac, and Maxwell equations. Finally, even if it is a subject logically distinct from classical field theory, we will include in this chapter a discussion of the first quantization of the relativistic wave equations and we will see that, despite the intrinsic limitations of the method, this can nevertheless be useful to compute the lowest-order relativistic corrections to the Schrödinger equation. In the Solved Problems section we will present some explicit computations using first quantized relativistic wave equations, and in particular we will compute the fine structure of the hydrogen atom using the Dirac equation.

3.1 The action principle

We first briefly recall the basic principles of classical mechanics in the Lagrangian formalism. A classical system with N degrees of freedom is described by a set of coordinates $q_i(t)$, with $i = 1, \ldots, N$, which we often denote collectively simply as q. The Lagrangian L is a function of the q_i's and of their first time derivatives $\dot{q}_i$, i.e. $L = L(q, \dot{q})$; in the simplest case it is given by a kinetic term minus a potential term, $L(q, \dot{q}) = \sum_i (m_i/2)\dot{q}_i^2 - V(q)$. The action S is

$$S = \int dt\, L(q, \dot{q})\,. \tag{3.1}$$

The action principle states that, if we fix the values of the coordinates at the initial time t_{in} and at the final time t_{f}, so that $q(t_{\text{in}}) = q_{\text{in}}, q(t_{\text{f}}) = q_{\text{f}}$, then the classical trajectory which satisfies these boundary conditions is an extremum of the action,

$$\delta \int_{t_{\text{in}}}^{t_{\text{f}}} dt\, L(q, \dot{q}) = 0\,. \tag{3.2}$$

The variation is performed holding fixed the boundary conditions, i.e. one must find the function $q(t)$ which extremizes the action, within the space of functions that satisfy $q(t_{\text{in}}) = q_{\text{in}}, q(t_{\text{f}}) = q_{\text{f}}$. The variation of the Lagrangian is

$$\delta L = \sum_i \frac{\partial L}{\partial q_i}\delta q_i + \frac{\partial L}{\partial \dot{q}_i}\delta \dot{q}_i\,. \tag{3.3}$$

$\delta q_i, \delta \dot{q}_i$ are variations in the space of functions of t (with the given boundary conditions), and therefore in an infinite-dimensional space. The time derivative commutes[1] with the variation operator δ,

$$\delta \dot{q}_i = \frac{\partial}{\partial t}\delta q_i\,. \tag{3.4}$$

The variation of the action then becomes

$$\delta S = \int_{t_{\rm in}}^{t_{\rm f}} dt \sum_i \left[\frac{\partial L}{\partial q_i}\delta q_i + \frac{\partial L}{\partial \dot{q}_i}\frac{\partial}{\partial t}\delta q_i\right] = 0\,. \tag{3.5}$$

Integrating the second term by parts, the boundary term does not contribute because the boundary conditions are held fixed: $\delta q_i(t_{\rm in}) = \delta q_i(t_{\rm f}) = 0$. Therefore we get

$$\int_{t_{\rm in}}^{t_{\rm f}} dt \sum_i \left[\frac{\partial L}{\partial q_i} - \frac{\partial}{\partial t}\frac{\partial L}{\partial \dot{q}_i}\right]\delta q_i = 0\,. \tag{3.6}$$

This must be true for any functional form of the variations δq_i (we are considering systems which are not subject to constraints, therefore all q_i are independent), so we obtain the *Euler–Lagrange* equations,

$$\boxed{\frac{\partial L}{\partial q_i} - \frac{\partial}{\partial t}\frac{\partial L}{\partial \dot{q}_i} = 0\,,} \tag{3.7}$$

with $i = 1, \ldots, N$. These are the equations of motion in the Lagrangian formulation.

In order to pass to the Hamiltonian formalism, one defines the *conjugate momenta*

$$p_i = \frac{\partial L}{\partial \dot{q}_i}\,, \tag{3.8}$$

and the Hamiltonian

$$H(p, q) = \sum_i p_i \dot{q}_i - L\,, \tag{3.9}$$

where $\dot{q}_i$ is expressed in terms of p_i and possibly q_i using eq. (3.8).

In the previous chapter we introduced the fields, defined as functions of the space-time point x with given transformation properties under the Poincaré group. The classical dynamics of a field can be defined extending the Lagrangian formalism from the case of functions of time, $q_i(t)$, to the case of functions of space-time $\phi_i(x)$. We will only be interested in *local* field theories, in which case the Lagrangian has the general form

$$L = \int d^3x\, \mathcal{L}(\phi, \partial_\mu \phi)\,, \tag{3.10}$$

[1] This can be understood discretizing this functional space (for instance taking t discrete rather than continuous), so that it becomes a finite-dimensional space, described by a finite number of parameters that we denote collectively by α. Then $\delta q_i = (\partial q_i/\partial \alpha)\delta\alpha$ becomes an ordinary variation of a function of a finite number of variables α and of time, and similarly $\delta \dot{q}_i = (\partial \dot{q}_i/\partial \alpha)\delta\alpha$; on a sufficiently regular function $q(\alpha, t)$ we can interchange the derivatives with respect to t and to α, $\delta \dot{q}_i = \frac{\partial}{\partial \alpha}(\frac{\partial q_i}{\partial t})\delta\alpha = \frac{\partial}{\partial t}(\frac{\partial q_i}{\partial \alpha})\delta\alpha = \frac{\partial}{\partial t}\delta q_i$.

where $\mathcal{L}$ is called the Lagrangian density (however, following standard use in field theory, we will often refer to $\mathcal{L}$ simply as the Lagrangian) and depends only on a finite number of derivatives. We will often denote collectively the fields ϕ_i simply as ϕ. To make contact with the variables $q_i(t)$ of classical mechanics, we can think of $\phi_i(t, \mathbf{x})$ as functions of time labeled both by the index i and by the continuous label $\mathbf{x}$ (the analogy becomes exact if we discretize space and we put the system in a finite box). In most cases of interest, $\mathcal{L}$ depends only on the first derivative.[2] In a Lorentz-covariant theory $\mathcal{L}$ depends on the time and space derivatives of ϕ only through the four-vector $\partial_\mu\phi$. For the moment we do not assume anything about the transformation properties of the field, so for instance ϕ_i could denote a set of scalars, or the four components of a vector field, etc. The action has the general form

$$S = \int dt\, L = \int d^4x\, \mathcal{L}(\phi, \partial_\mu\phi)\,. \tag{3.11}$$

While for point particles we considered the time integral between two values $t_{\text{in}}, t_{\text{f}}$, in classical field theory we will rather be interested in the situation where the integral extends over all of space-time, and the boundary conditions are that all fields decrease sufficiently fast at infinity so that, in particular, all boundary terms can be neglected.

The classical dynamics is again defined by the principle that the action is stationary. The same manipulations performed above in the case of a function $q(t)$ are immediately generalized to the case of a function $\phi(x)$, and

$$\begin{aligned} \delta S &= \int d^4x \sum_i \left[\frac{\partial \mathcal{L}}{\partial \phi_i}\delta\phi_i + \frac{\partial \mathcal{L}}{\partial(\partial_\mu\phi_i)}\delta(\partial_\mu\phi_i)\right] \\ &= \int d^4x \sum_i \left[\frac{\partial \mathcal{L}}{\partial \phi_i} - \partial_\mu \frac{\partial \mathcal{L}}{\partial(\partial_\mu\phi_i)}\right]\delta\phi_i = 0\,. \end{aligned} \tag{3.12}$$

Therefore the equations of motion, or Euler–Lagrange equations, are

$$\boxed{\frac{\partial \mathcal{L}}{\partial \phi_i} - \partial_\mu \frac{\partial \mathcal{L}}{\partial(\partial_\mu\phi_i)} = 0\,,} \tag{3.13}$$

with $i = 1, \ldots, N$. Consider now the theory obtained replacing the original Lagrangian density $\mathcal{L}$ by a new Lagrangian density $\mathcal{L}'$ which differs from $\mathcal{L}$ only by a four-divergence,

$$\mathcal{L}' = \mathcal{L} + \partial_\mu K^\mu\,, \tag{3.14}$$

with $K^\mu = K^\mu(\phi)$. In a finite volume V bounded by a surface Σ we have, by Stokes theorem,

$$\int_V d^4x\, \partial_\mu K^\mu = \int_\Sigma dA\; n_\mu K^\mu\,, \tag{3.15}$$

where dA is the surface element and n^μ is the outward normal to the surface. This is a boundary term and therefore vanishes on field configurations that go to zero sufficiently fast at infinity. More generally,

[2] The variational principle can be generalized to Lagrangians containing higher derivatives. For instance, if $L = L(q, \dot{q}, \ddot{q})$, then $\delta L = (\delta L/\delta q)\delta q + (\delta L/\delta\dot{q})\delta\dot{q} + (\delta L/\delta\ddot{q})\delta\ddot{q}$. At the boundaries we must hold fixed both q and $\dot{q}$ and, after integrations by parts, the equation of motion is $(\delta L/\delta q) - d/dt(\delta L/\delta\dot{q}) + d^2/dt^2(\delta L/\delta\ddot{q}) = 0$.

we can consider the variational principle in a finite four-dimensional volume V but in this case, similarly to the situation discussed in classical mechanics, in order to have a well-defined variational principle we must impose the boundary condition that the fields are kept constant on Σ. In any case, the term $\int_V d^4x\,\partial_\mu K^\mu$ either vanishes or is anyway a constant, and therefore the condition that S' is stationary is equivalent to the condition that S is stationary. This means that two Lagrangian densities which differ by a total derivative give rise to the same equations of motion and therefore are classically equivalent.

In the Hamiltonian formalism one defines the conjugate momenta as

$$\boxed{\Pi_i(x) = \frac{\partial \mathcal{L}}{\partial\left(\partial_0\phi_i(x)\right)}\,.} \tag{3.16}$$

The *Hamiltonian density* $\mathcal{H}$ is then defined as

$$\boxed{\mathcal{H}(x) = \sum_i \Pi_i(x)\partial_0\phi_i(x) - \mathcal{L}\,,} \tag{3.17}$$

and the total Hamiltonian is

$$H = \int d^3x\,\mathcal{H}\,. \tag{3.18}$$

The Lagrangian formalism has the advantage of keeping explicit at each stage the Lorentz covariance. Instead, in the Hamiltonian formalism Lorentz invariance is less explicit, since the time variable plays a special role in defining the conjugate momenta.

3.2 Noether's theorem

The relation between symmetries and conservation laws is extremely important in field theory and, at the classical level, it is expressed by Noether's theorem. We consider a field theory with fields ϕ_i and action S, and we perform an infinitesimal transformation of the coordinates and of the fields, parametrized by a set of infinitesimal parameters ϵ^a, with $a = 1, \ldots, N$, of the general form

$$x^\mu \to x'^\mu = x^\mu + \epsilon^a A^\mu_a(x) \tag{3.19}$$

$$\phi_i(x) \to \phi'_i(x') = \phi_i(x) + \epsilon^a F_{i,a}(\phi, \partial\phi)\,, \tag{3.20}$$

with $A^\mu_a(x)$ and $F_{i,a}(\phi, \partial\phi)$ given. Equations (3.19, 3.20) define a symmetry transformation if they leave the action $S(\phi)$ invariant, for any ϕ. Note that we are not assuming that the fields ϕ_i satisfy the classical equations of motion. A symmetry by definition leaves the action invariant for every field configuration, solution or not of the equations of motion.

There are two important distinctions to be drawn. The first is between *local* and *global* symmetries. If in eqs. (3.19) and (3.20) the parameters

ϵ^a are constants, we have a global symmetry. Instead, the above transformation is a local symmetry if it leaves invariant the action even when ϵ is allowed to be an arbitrary function of x. Of course, a local symmetry gives rise also to a global symmetry.[3]

The second important distinction is between *internal* and *space-time* symmetries. Internal symmetries do not change the coordinates, so they have $A^\mu_a(x) = 0$, while space-time symmetries involve also a change in the coordinates. For internal symmetries (and also for Lorentz transformations and for translations) d^4x is invariant and the condition of invariance of the action is equivalent to the condition of invariance of $\mathcal{L}$, but in the general case the symmetries are given by invariances of S, not of $\mathcal{L}$.

Now, suppose that eqs. (3.19) and (3.20) are a global, but not a local, symmetry of our theory. Consider what happens if, starting from a given field configuration ϕ_i, we perform the above transformation, but with the ϵ^a slowly varying functions of x, i.e. $|\epsilon^a| \ll 1$ and $l|\partial_\mu\epsilon^a| \ll |\epsilon^a|$, where l is the characteristic scale of variation of ϕ_i. Since the ϵ^a depend on x, and we are assuming that eqs. (3.19) and (3.20) are not a local symmetry, this transformation will not leave the action invariant, and δS will have a non-vanishing term at $O(\epsilon)$. However, since the ϵ^a are slowly varying, we can expand this $O(\epsilon)$ term in powers of the derivatives,

$$S(\phi') = S(\phi) + \int d^4x \, [\epsilon^a(x)K_a(\phi) - (\partial_\mu\epsilon^a)j^\mu_a(\phi) + O(\partial\partial\epsilon)] + O(\epsilon^2)\,, \tag{3.21}$$

where we have denoted by K_a the coefficient of ϵ^a and by $-j^\mu_a$ the coefficient of $\partial_\mu\epsilon^a$. This equation holds for any slowly varying ϵ.[4] We can then apply it also to the case where the ϵ^a are all constants. However we know that, if the ϵ^a are constants, the variation of the action must be zero because in this case ϵ parametrizes a global symmetry. Therefore we learn that, in eq. (3.21), the function $K_a(\phi)$ is actually zero, for any ϕ. Observe that $K_a(\phi)$ is by definition a function of ϕ but is independent of ϵ; all the ϵ dependence is written explicitly in eq. (3.21). Therefore, even if to show that $K_a(\phi) = 0$ we have looked at the limiting case of constant ϵ, once we have shown that K_a vanishes, it vanishes independently of ϵ. Then, for *any* slowly varying function $\epsilon(x)$, we have the expansion

$$S(\phi') = S(\phi) - \int d^4x \, [(\partial_\mu\epsilon^a)j^\mu_a(\phi) + O(\partial\partial\epsilon)] + O(\epsilon^2)\,. \tag{3.22}$$

We now take ϵ to be a function which goes to zero sufficiently fast at infinity. This allows us to integrate the above equation by parts (without making any assumptions on ϕ) and the boundary term vanishes, so we get

$$S(\phi') = S(\phi) + \int d^4x \, \epsilon^a(x)\partial_\mu j^\mu_a(\phi) + O(\partial\partial\epsilon) + O(\epsilon^2)\,. \tag{3.23}$$

We have derived the above result independently of our choice of ϕ, and for ϵ slowly varying and vanishing at infinity.

[3] However, it can happen that the corresponding global symmetry is trivial, i.e. it is just the identity transformation. For example, we will see in Section 3.5 that the free electromagnetic field is invariant under the gauge transformation $A_\mu \to A_\mu - \partial_\mu\theta$. For θ constant the corresponding transformation is just the identity and does not give rise to conserved charges (as long as we do not include matter fields).

[4] We generically denote by ϵ the set of ϵ^a with $a = 1, \ldots N$. The statement "ϵ slowly varying" means that all ϵ^a are slowly varying.

Suppose now that, for ϕ, we choose a classical solution of the equations of motion, ϕ^{cl}. Observe, first of all, that in the action the x are integration variables so, after performing the transformation (3.19, 3.20), we can rename x' as x. Then, as long as we are only interested in studying the variation of the action, the infinitesimal transformation in eqs. (3.19) and (3.20) is equivalent to a transformation in which the coordinates are unchanged, $x^\mu \to x^\mu$, while

$$\phi_i(x) \to \phi_i(x - \epsilon^a A^\mu_a) + \epsilon^a F_{i,a} = \phi_i(x) + \epsilon^a F_{i,a} - \epsilon^a A^\mu_a \partial_\mu \phi_i \,. \quad (3.24)$$

Thus we have rewritten the transformations (3.19, 3.20) in a form that does not change the coordinates, and with $\phi_i(x) \to \phi_i(x) + \delta\phi_i(x)$ with $\delta\phi_i(x)$ some variation that goes to zero at infinity. This is the kind of variation that is used in the derivation of the equations of motion. By definition, a classical solution is an extremum of the action and therefore if we now take $\phi_i = \phi_i^{\text{cl}}$ the linear term in the variation of S in eq. (3.23) must be identically zero for any ϵ that vanishes at infinity, independently of whether or not ϵ depends on x, and of whether or not it is a symmetry transformation (the condition that ϵ has to vanish at infinity follows from the fact that the classical equations of motion are obtained from a variation with fixed boundary conditions, i.e. keeping the fields fixed at infinity). Therefore we arrive at the important conclusion that, *on a classical solution of the equations of motion*, the N currents j^μ_a are conserved,

$$\boxed{\partial_\mu j^\mu_a(\phi^{\text{cl}}) = 0 \,,} \quad (3.25)$$

which is the content of Noether's theorem. In other words, there is a conserved current j^μ_a (with $a = 1, \ldots, N$) for each generator of a symmetry transformation. The fact that the symmetry was global but not local means that $S(\phi') - S(\phi)$ in eq. (3.22) must be non-vanishing, and therefore j^μ_a themselves are non-vanishing.[5] We now define the charges Q_a,

$$\boxed{Q_a \equiv \int d^3x \, j^0_a(\mathbf{x}, t) \,.} \quad (3.26)$$

The current conservation (in the sense of eq. (3.25)) implies that Q_a is conserved (in the sense that it is time-independent). In fact

$$\partial_0 Q_a = \int d^3x \, \partial_0 j^0_a(\mathbf{x}, t) = -\int d^3x \, \partial_i j^i_a(\mathbf{x}, t) \,. \quad (3.27)$$

This is the integral of a total divergence, and vanishes since we assume a sufficiently fast decrease of the fields at infinity. More generally, in a finite volume the variation of the charge is given by a boundary term representing the incoming or outgoing flux.

The explicit form of the current can be obtained performing the variation of the action with ϵ slowly varying, collecting the terms proportional to $\partial_\mu \epsilon$, and comparing with eq. (3.22). This can be done in full generality. Denoting by δ_ϵ the variations induced by the transformation (3.19,

[5] The transformations that we consider depend on the ϵ^a in a continuous and differentiable way, therefore they form a Lie group. If the linear term j^μ_a in eq. (3.22) vanished, integrating the infinitesimal transformation we would find that the finite transformation is the identity; compare with eqs. (2.3) and (2.5).

3.20), we have

$$\delta_\epsilon S = \delta_\epsilon \int d^4x\, \mathcal{L} = \int \left[\delta_\epsilon(d^4x)\, \mathcal{L} + d^4x \left(\frac{\partial \mathcal{L}}{\partial \phi_i} \delta_\epsilon \phi_i + \frac{\partial \mathcal{L}}{\partial(\partial_\mu \phi_i)} \delta_\epsilon(\partial_\mu \phi_i) \right) \right]. \tag{3.28}$$

Computing the Jacobian of the transformation (3.19, 3.20) to linear order, we find $d^4x \to d^4x(1 + A^\mu_a \partial_\mu \epsilon^a)$ plus a term $\sim \epsilon$; $\delta_\epsilon \phi_i$ does not produce terms $\sim \partial\epsilon$ while

$$\delta_\epsilon(\partial_\mu \phi_i) = \frac{\partial \phi'_i}{\partial x'^\mu} - \frac{\partial \phi_i}{\partial x^\mu} = \frac{\partial x^\nu}{\partial x'^\mu} \frac{\partial}{\partial x^\nu}(\phi_i + \epsilon^a F_{i,a}) - \frac{\partial \phi_i}{\partial x^\mu}. \tag{3.29}$$

This produces a term proportional to $\partial_\mu \epsilon$ and equal to

$$-(\partial_\mu \epsilon^a)\, (A^\nu_a \partial_\nu \phi_i - F_{i,a}). \tag{3.30}$$

(Observe that, since δ_ϵ is a variation that also changes the coordinates, $\delta_\epsilon(\partial_\mu \phi_i) \neq \partial_\mu(\delta_\epsilon \phi_i)$.) Putting together all terms $\sim \partial_\mu \epsilon^a$ gives

$$j^\mu_a = \frac{\partial \mathcal{L}}{\partial(\partial_\mu \phi_i)} \left[A^\nu_a(x) \partial_\nu \phi_i - F_{i,a}(\phi, \partial\phi) \right] - A^\mu_a(x) \mathcal{L}. \tag{3.31}$$

For internal symmetries $A^\mu_a = 0$ and the above expression simplifies to

$$j^\mu_a = -\frac{\partial \mathcal{L}}{\partial(\partial_\mu \phi_i)} F_{i,a} \qquad \text{(internal symmetries)}. \tag{3.32}$$

Quite often one is interested in linear transformations of the fields, in which case $F_{i,a}(\phi, \partial\phi) = (M_a)_i{}^j \phi_j$, where $(M_a)_i{}^j$ are N constant matrices.

Finally, let us see what happens if the transformation (3.19, 3.20) is *not* a global symmetry. In this case eq. (3.21) still holds, since it is the most general expansion of the variation of the action when ϵ is slowly varying; however, now $K_a(\phi)$ does not vanish, and indeed it represents the variation of the Lagrangian under the global transformation, $K_a \equiv (\delta_a \mathcal{L})_{\rm global}$, where $(\delta_a \mathcal{L})_{\rm global}$ is defined by $(\delta \mathcal{L})_{\rm global} = \epsilon^a (\delta_a \mathcal{L})_{\rm global}$. Following the same steps that lead to eq. (3.25) we now find that, on the classical solutions,

$$\partial_\mu j^\mu_a = -(\delta_a \mathcal{L})_{\rm global}, \tag{3.33}$$

where j^μ_a is still given by eq. (3.31). This expression is particularly useful when $(\delta_a \mathcal{L})_{\rm global}$ is small compared to the relevant scales, so that the current is approximately conserved.

3.2.1 The energy–momentum tensor

Space-time translations are a symmetry which is present in all the theories that we will consider. In this case the index a appearing in ϵ^a is a Lorentz index and, as explained in Section 2.7, all fields are scalars under translations. Therefore

$$\begin{aligned} x^\mu \to x'^\mu &= x^\mu + \epsilon^\mu = x^\mu + \epsilon^\nu \delta^\mu_\nu \\ \phi_i(x) \to \phi'_i(x') &= \phi_i(x) \end{aligned} \tag{3.34}$$

and in eqs. (3.19) and (3.20) we have $A^\mu_\nu = \delta^\mu_\nu$ and $F_{i,a} = 0$. The four conserved currents $j^\mu_{(\nu)} \equiv \theta^\mu{}_\nu$ therefore form a Lorentz tensor, known as the *energy–momentum* tensor. Using eq. (3.31) and raising the ν index, $\theta^{\mu\nu} = \eta^{\nu\rho}\theta^\mu{}_\rho$ we get

$$\boxed{\theta^{\mu\nu} = \frac{\partial \mathcal{L}}{\partial(\partial_\mu \phi_i)}\partial^\nu \phi_i - \eta^{\mu\nu}\mathcal{L}\,.} \tag{3.35}$$

Equation (3.25) becomes

$$\boxed{\partial_\mu \theta^{\mu\nu} = 0} \tag{3.36}$$

on the solutions of the classical equations of motion. The conserved charge associated to the energy–momentum tensor is the four-momentum,

$$P^\nu \equiv \int d^3x\, \theta^{0\nu}\,. \tag{3.37}$$

This is the definition of four-momentum in field theory. A field configuration, solution of the equations of motion, carries an energy $E = P^0$ and a spatial momentum P^i which can be calculated using eqs. (3.35) and (3.37).

The energy–momentum tensor defined from eq. (3.35) in general is not automatically symmetric in the two indices μ, ν, so for instance in eq. (3.36) one should be careful to contract ∂_μ with the first index of $\theta^{\mu\nu}$. However, consider the "improved" energy–momentum tensor

$$T^{\mu\nu} = \theta^{\mu\nu} + \partial_\rho A^{\rho\mu\nu} \tag{3.38}$$

where $A^{\rho\mu\nu}$ is an arbitrary tensor antisymmetric in the indices ρ, μ. This new tensor is still conserved: $\partial_\mu\partial_\rho A^{\rho\mu\nu} = 0$ because of the antisymmetry in ρ, μ. Furthermore, for $\mu = 0$, $\partial_\rho A^{\rho 0\nu} = \partial_i A^{i0\nu}$ is a spatial divergence, and therefore this term does not contribute to the four-momentum (3.37), if the fields vanish sufficiently fast at infinity. Therefore $T^{\mu\nu}$ and $\theta^{\mu\nu}$ are physically equivalent, and one can choose $A^{\rho\mu\nu}$ such that $T^{\mu\nu}$ is symmetric.

The reader with some knowledge of general relativity may compare this definition with the definition of energy–momentum tensor in general relativity, which is given by the variation of the action with respect to the metric. Since the metric is symmetric, this definition automatically gives the symmetric form of $T^{\mu\nu}$.

3.3 Scalar fields

3.3.1 Real scalar fields; Klein–Gordon equation

We now have all the elements for writing down Poincaré invariant actions. We start with the theory of a single, *real* scalar field ϕ. An action describing a non-trivial dynamics must contain $\partial_\mu\phi$. In order to have a Lorentz invariant action the index μ must be saturated and, for a scalar field, the only possibility is to contract it with another factor $\partial^\mu\phi$. Therefore the kinetic term must be proportional to $\partial_\mu\phi\partial^\mu\phi$. We first consider the action

$$S = \frac{1}{2}\int d^4x \, \left(\partial_\mu\phi\partial^\mu\phi - m^2\phi^2\right) . \tag{3.39}$$

The Euler–Lagrange equation gives

$$\boxed{(\Box + m^2)\phi = 0 ,} \tag{3.40}$$

with $\Box = \partial_\mu\partial^\mu$. This is the free *Klein–Gordon* (KG) equation. A plane wave $e^{\pm ipx}$ is a solution of eq. (3.40) if $p^2 = m^2$, i.e. if

$$(p^0)^2 - \mathbf{p}^2 = m^2 . \tag{3.41}$$

Therefore, the classical KG equation imposes the relativistic dispersion relation, and the parameter m appearing in the action is the mass (we take by definition $m > 0$). Taking into account that ϕ must be real, the most general solution is a real superposition of plane waves,

$$\phi(x) = \int \frac{d^3p}{(2\pi)^3\sqrt{2E_{\mathbf{p}}}} \left(a_{\mathbf{p}}e^{-ipx} + a^*_{\mathbf{p}}e^{ipx}\right)|_{p^0=E_{\mathbf{p}}} , \tag{3.42}$$

where $E_{\mathbf{p}} \equiv +\sqrt{\mathbf{p}^2 + m^2}$. The factor $\sqrt{2E_{\mathbf{p}}}$ is a convenient choice of normalization of the coefficients $a_{\mathbf{p}}$. The solution is evaluated on $p^0 = +E_{\mathbf{p}}$, i.e. on the positive solution of eq. (3.41). Note however that in eq. (3.42) we have both solutions that oscillate as $e^{-iE_{\mathbf{p}}t}$ (*positive frequency modes*) and as $e^{+iE_{\mathbf{p}}t}$ (*negative frequency modes*). The proper interpretation of the latter modes will only come after quantization of the theory.[6]

The overall sign (and normalization) of the action (3.39) is irrelevant as long as we are interested in the equations of motion, but is important for obtaining a positive definite Hamiltonian (and the correct choice depends on our convention for the metric). The momentum conjugate to ϕ is

$$\Pi_\phi = \frac{\partial\mathcal{L}}{\partial(\partial_0\phi)} = \partial_0\phi , \tag{3.43}$$

and the Hamiltonian density is

$$\mathcal{H} = \Pi_\phi\partial_0\phi - \mathcal{L} = \frac{1}{2}\left[\Pi_\phi^2 + (\nabla\phi)^2 + m^2\phi^2\right] . \tag{3.44}$$

[6] Of course, after one has included both the solutions e^{ipx} and e^{-ipx} with $p^0 = +E_{\mathbf{p}}$ one does not get anything new including solutions with $p^0 = -E_{\mathbf{p}}$. For instance, a term e^{-ipx}, when $p^0 = -E_{\mathbf{p}}$, is equal to $\exp\{-ip^0t + i\mathbf{p}\cdot\mathbf{x}\} = \exp\{iE_{\mathbf{p}}t + i\mathbf{p}\cdot\mathbf{x}\}$. After changing the dummy integration variable $\mathbf{p}$ to $-\mathbf{p}$ we get back to $\exp\{iE_{\mathbf{p}}t - i\mathbf{p}\cdot\mathbf{x}\}$ which is just e^{ipx} with $p^0 = +E_{\mathbf{p}}$.

The energy momentum tensor is found from eq. (3.35),

$$\theta^{\mu\nu} = \partial^\mu\phi\partial^\nu\phi - \eta^{\mu\nu}\mathcal{L}\,, \tag{3.45}$$

so that

$$\theta^{00} = (\partial_0\phi)^2 - \mathcal{L} = \frac{1}{2}\left[(\partial_0\phi)^2 + (\nabla\phi)^2 + m^2\phi^2\right]. \tag{3.46}$$

We see that $\theta^{00} = \mathcal{H}$, as expected, and $H = \int d^3x\,\theta^{00}$. The Hamiltonian is the conserved charge related to the invariance under time translations.

We now compute the conserved currents associated to Lorentz invariance. In this case the parameters ϵ^a which appear in eqs. (3.19) and (3.20) are better labeled by an antisymmetric pair of Lorentz indices; we will denote them by $\omega^{\rho\sigma}$, and $\delta x^\mu = \omega^\mu{}_\nu x^\nu$ can be rewritten as

$$\delta x^\mu = \sum_{\rho<\sigma} A^\mu_{(\rho\sigma)}\omega^{\rho\sigma}\,, \tag{3.47}$$

with

$$A^\mu_{(\rho\sigma)} = \delta^\mu_\rho x_\sigma - \delta^\mu_\sigma x_\rho\,. \tag{3.48}$$

This gives the term A^μ_a which appears in eq. (3.31), while for a scalar field $\phi'(x') = \phi(x)$ and therefore $F_{i,a} = 0$. We can then compute the conserved currents $j^\mu_{(\rho\sigma)}$, and we find that they can be written in terms of the energy–momentum tensor as[7]

$$j^{(\rho\sigma)\mu} = x^\rho\theta^{\mu\sigma} - x^\sigma\theta^{\mu\rho}\,. \tag{3.49}$$

[7] We have reabsorbed a minus sign in the definition of $j^{(\rho\sigma)\mu}$, compared to the sign which comes from eq. (3.31). Of course, if a current j^μ is conserved, also $-j^\mu$ is conserved.

In particular, the conserved charge associated to spatial rotations is

$$M^{ij} = \int d^3x\,(x^i\theta^{0j} - x^j\theta^{0i}) = \int d^3x\,\partial_0\phi\,(x^i\partial^j - x^j\partial^i)\phi \tag{3.50}$$

and we recognize the operator $L^{ij} = i(x^i\partial^j - x^j\partial^i)$ found in eq. (2.78) as a representation of the Lorentz generator on the scalar field. Integrating by parts the spatial derivatives we can also rewrite eq. (3.50) as

$$M^{ij} = \frac{i}{2}\int d^3x\,\left[\phi L^{ij}(\partial_0\phi) - (\partial_0\phi)L^{ij}\phi\right]. \tag{3.51}$$

We now define the scalar product

$$\boxed{\langle\phi_1|\phi_2\rangle = \frac{i}{2}\int d^3x\,\phi_1\,\overleftrightarrow{\partial_0}\,\phi_2\,,} \tag{3.52}$$

where $f\overleftrightarrow{\partial_\mu}g \equiv f\partial_\mu g - (\partial_\mu f)g$. This scalar product is time-independent if ϕ_1, ϕ_2 are solutions of the KG equation. In fact, using the KG equation,

$$\partial_0[\phi_1\partial_0\phi_2 - (\partial_0\phi_1)\phi_2] = \phi_1\partial_0^2\phi_2 - (\partial_0^2\phi_1)\phi_2 = \phi_1\nabla^2\phi_2 - (\nabla^2\phi_1)\phi_2 \tag{3.53}$$

and, inserted into eq. (3.52), this expression gives zero after integration by parts. Observe that this scalar product is not positive definite. Equation (3.51) can be written as the expectation value of the operator L^{ij} with this scalar product,

$$M^{ij} = \langle \phi | L^{ij} | \phi \rangle \, . \tag{3.54}$$

This is completely general and elucidates the relation between the representation of the generators as operators acting on fields (here L^{ij}) and the value of the corresponding charges on a given solution of the equations of motion (here M^{ij}) . For instance, for the four-momentum

$$P^{\mu} \equiv \int d^3x \, \theta^{0\mu} \tag{3.55}$$

we have the equality

$$P^{\mu} = \langle \phi | i\partial^{\mu} | \phi \rangle \, . \tag{3.56}$$

We check it for $\mu = 0$,

$$\begin{aligned} \langle \phi | i\partial^0 | \phi \rangle = \langle \phi | i\partial_0 | \phi \rangle &= \frac{i}{2} \int d^3x \, [\phi (i\partial_0)\partial_0 \phi - (\partial_0 \phi) i\partial_0 \phi] \\ &= \frac{1}{2} \int d^3x \, \left[-\phi \partial_0^2 \phi + (\partial_0 \phi)^2 \right] \\ &= \frac{1}{2} \int d^3x \, \left[-\phi(\nabla^2 - m^2)\phi + (\partial_0 \phi)^2 \right] \, . \end{aligned} \tag{3.57}$$

Integrating by parts, $-\phi\nabla^2\phi$ becomes $(\nabla\phi)^2$ and we see that we have reproduced the integral of θ^{00}, eq. (3.46). Note that $i\partial^{\mu}$ and $i(x^{\mu}\partial^{\nu} - x^{\nu}\partial^{\mu})$ are hermitian operators with respect to this scalar product.

Finally, the free KG action can be generalized to a self-interacting scalar field introducing a scalar potential $V(\phi)$,

$$S = \int d^4x \left[\frac{1}{2} \partial_{\mu}\phi \partial^{\mu}\phi - V(\phi) \right] \, . \tag{3.58}$$

The quadratic part of the potential gives the mass term while higher powers, like ϕ^3, ϕ^4, etc. give non-linear contributions to the equations of motion and therefore correspond to self-interactions.

3.3.2 Complex scalar field; $U(1)$ charge

We can assemble two real scalar fields ϕ_1 and ϕ_2, with the same mass m, into a single complex scalar $\phi = (\phi_1 + i\phi_2)/\sqrt{2}$. The KG action is the sum of the actions of ϕ_1 and ϕ_2 and, written in terms of ϕ, reads

$$S = \int d^4x \left(\partial_{\mu}\phi^* \partial^{\mu}\phi - m^2 \phi^* \phi \right) \, . \tag{3.59}$$

The complex field ϕ still satisfies the KG equation, as we can see from the fact that by definition its real and imaginary parts ϕ_1 and ϕ_2 separately satisfy the KG equation. Alternatively, we can obtain the same result considering as independent variables ϕ and ϕ^*, rather than ϕ_1 and ϕ_2, and performing the variation of the action (3.59) with respect to ϕ^*, holding ϕ fixed. The mode expansion of ϕ is

$$\phi(x) = \int \frac{d^3p}{(2\pi)^3\sqrt{2E_{\mathbf{p}}}} \left(a_{\mathbf{p}}e^{-ipx} + b_{\mathbf{p}}^* e^{ipx}\right)|_{p^0=E_{\mathbf{p}}}\,, \tag{3.60}$$

with $a_{\mathbf{p}}, b_{\mathbf{p}}$ independent, since the reality condition on ϕ is now absent. The interesting new aspect of the complex KG field is the existence of a *global* $U(1)$ *symmetry* of the action,

$$\phi(x) \to e^{i\theta}\phi(x)\,, \qquad \phi^*(x) \to e^{-i\theta}\phi^*(x)\,, \tag{3.61}$$

which of course makes sense only on complex fields. The Noether current can be found from eq. (3.32) with $\phi_i = (\phi, \phi^*)$ and $F_{i,a} = (M_a)_i{}^j\phi_j$ with $M_\phi{}^\phi = i, M_{\phi^*}{}^{\phi^*} = -i$,

$$j_\mu = -i(\phi\partial_\mu\phi^* - \phi^*\partial_\mu\phi) = i\phi^* \overleftrightarrow{\partial_\mu}\, \phi\,. \tag{3.62}$$

The conserved $U(1)$ charge is

$$Q_{U(1)} = \int d^3x\, j^0 = i\int d^3x\, \phi^* \overleftrightarrow{\partial_0}\, \phi\,. \tag{3.63}$$

In Section 3.5.4 we will study the interaction of the complex scalar field with the electromagnetic field, and we will see that the $U(1)$ current is coupled to the gauge field, and that the $U(1)$ charge is the electric charge.

The scalar product for the complex scalar field is defined as

$$\langle\phi|\phi'\rangle = i\int d^3x\, \phi^* \overleftrightarrow{\partial_0}\, \phi'\,, \tag{3.64}$$

and it is conserved on the equations of motion. We see that the value of the conserved charge (3.63) on a classical field configuration ϕ is equal to the expectation value of the identity operator with respect to this scalar product,

$$Q_{U(1)} = \langle\phi|\phi\rangle\,. \tag{3.65}$$

This is completely analogous to eqs. (3.54) and (3.56) since the generator of the $U(1)$ transformation is the identity operator, as we can see observing that eq. (3.61) can be written as $\phi(x) \to e^{i\theta T}\phi(x)$, with the generator T equal to the identity.

3.4 Spinor fields

3.4.1 The Weyl equation; helicity

We now consider the theory of a single left-handed Weyl field ψ_L. We found in Section 2.5 that $\psi_L^\dagger\bar{\sigma}^\mu\psi_L$ is a four-vector, where $\bar{\sigma}^\mu = (1, -\sigma^i)$.

It is therefore possible to write a Lorentz-invariant kinetic term which is first-order in the derivative,

$$\mathcal{L}_L = i\psi_L^\dagger \bar{\sigma}^\mu \partial_\mu \psi_L \,. \tag{3.66}$$

The factor i in front is fixed by the condition that the action $\int d^4x\, \mathcal{L}_L$ is real, as we verify immediately using the fact that the matrices $\bar{\sigma}^\mu$ are hermitian. The equation of motion is obtained varying with respect to ψ_L^*, considering ψ_L^* and ψ_L as two independent fields. Since $\partial_\mu \psi_L^\dagger$ does not appear in $\mathcal{L}_L$, the Euler–Lagrange equation is simply $\partial\mathcal{L}/\partial\psi_L^* = 0$, which gives $\bar{\sigma}^\mu\partial_\mu\psi_L = 0$, or, more explicitly

$$\boxed{(\partial_0 - \sigma^i\partial_i)\psi_L = 0 \,.} \tag{3.67}$$

As a consequence (using $\sigma^i\sigma^j = \delta^{ij} + i\epsilon^{ijk}\sigma^k$ and the fact that, on a regular function, $\partial_i\partial_j$ is symmetric in i,j), $\partial_0^2\psi_L = (\sigma^i\partial_i)(\sigma^j\partial_j)\psi_L = \partial_i^2\psi_L$, or

$$\Box\psi_L = 0 \,. \tag{3.68}$$

Then eq. (3.67) implies the massless KG equation. However, eq. (3.67) is a first-order differential equation, and gives further information. Consider for instance a plane wave solution of positive energy,

$$\psi_L(x) = u_L e^{-ipx} \tag{3.69}$$

where u_L is a constant spinor, and all the x-dependence is in the plane wave $\exp\{-ipx\} = \exp\{-iEt + i\mathbf{p}\cdot\mathbf{x}\}$. Then eq. (3.67) gives

$$\frac{\mathbf{p}\cdot\boldsymbol{\sigma}}{E} u_L = -u_L \tag{3.70}$$

and eq. (3.68) gives $E = |\mathbf{p}|$. Since for a spin 1/2 field the angular momentum is $\mathbf{J} = \boldsymbol{\sigma}/2$, eq. (3.70) can be rewritten as

$$(\hat{\mathbf{p}}\cdot\mathbf{J})\, u_L = -\frac{1}{2} u_L \,, \tag{3.71}$$

where $\hat{\mathbf{p}} = \mathbf{p}/|\mathbf{p}|$. This shows that *a left-handed massless Weyl spinor has helicity* $h = -1/2$. This result is consistent with our discussion of the representations of the Poincaré group in Chapter 2, where we found that massless particles are helicity eigenstates.[8]

The energy–momentum tensor is obtained from the general formula (3.35). Observe that on a classical solution $\bar{\sigma}^\mu\partial_\mu\psi_L = 0$, so the Lagrangian (3.66) vanishes. The energy–momentum tensor is therefore

$$\theta^{\mu\nu} = i\psi_L^\dagger\bar{\sigma}^\mu\partial^\nu\psi_L \,, \tag{3.72}$$

and in particular

$$\theta^{00} = i\psi_L^\dagger\partial_0\psi_L \,. \tag{3.73}$$

The Lagrangian (3.66) is invariant under a global $U(1)$ internal transformation,

$$\psi_L \to e^{i\theta}\psi_L \,, \tag{3.74}$$

[8] Let us anticipate that, when we quantize the theory, this result will translate into the existence of massless quanta of the field ψ_L with helicity $h = -1/2$, while the negative energy solutions will correspond to antiparticles with $h = +1/2$; see Section 4.2.2.

with Noether's current $j^\mu_{U(1)} = \psi_L^\dagger \bar\sigma^\mu \psi_L$ and conserved charge

$$Q_{U(1)} = \int d^3x\, \psi_L^\dagger \psi_L\,. \tag{3.75}$$

Of course, the Weyl Lagrangian (3.66) is not invariant under parity. Under parity ψ_L is sent into a right-handed spinor which is not even present in eq. (3.66).

Observe also that the Lagrangian

$$\mathcal{L}'_L = i\psi_L^\dagger \bar\sigma^\mu \partial_\mu \psi_L - \frac{i}{2}\partial_\mu\left(\psi_L^\dagger \bar\sigma^\mu \psi_L\right) = \frac{i}{2}\psi_L^\dagger \bar\sigma^\mu \overleftrightarrow{\partial_\mu}\, \psi_L \tag{3.76}$$

differs from eq. (3.66) only by a total derivative, and it therefore gives the same equations of motion and conserved currents, so it is classically equivalent. We will verify in Exercise 3.5 that the conserved currents computed with $\mathcal{L}'_L$ are different from those computed with $\mathcal{L}_L$, but the conserved charges are the same.

For a right-handed Weyl spinor ψ_R we saw in Section 2.5.2 and in Exercise 2.3 that $\psi_R^\dagger \sigma^\mu \psi_R$ is a four-vector, and therefore we can construct the Lagrangian

$$\mathcal{L}_R = i\psi_R^\dagger \sigma^\mu \partial_\mu \psi_R\,, \tag{3.77}$$

with $\sigma^\mu = (1, \sigma^i)$ in place of $\bar\sigma^\mu = (1, -\sigma^i)$. We then find that the equation of motion is $\sigma^\mu \partial_\mu \psi_R = (\partial_0 + \sigma^i \partial_i)\psi_R = 0$ and the positive energy solution has helicity $h = +1/2$. The energy–momentum tensor of the right-handed Weyl field is $\theta^{\mu\nu} = i\psi_R^\dagger \sigma^\mu \partial^\nu \psi_R$.

The neutrinos come into three species (families) ν_e, ν_μ, ν_τ and are spin 1/2 particles. Until recently, the experimental values of their masses were compatible with zero. Recent results on neutrino oscillations however provide strong evidence that they have small masses. More precisely, these experiments are sensitive to the difference in mass squared between different families, rather than to the values of the masses themselves, and presently suggest a mass difference between different families Δm^2 between 10^{-5} and 10^{-3} eV^2. In most situations the neutrino masses can be considered as extremely small and, if we neglect them, neutrinos are described by massless left-handed Weyl spinors.

One could ask whether it is possible to describe a massive particle with a single Weyl spinor. This is indeed possible, and the resulting mass term is known as a Majorana mass. We will discuss it in Section 3.4.4. We first turn our attention to Dirac spinors.

3.4.2 The Dirac equation

We now consider the theory that we can build having at our disposal both a left-handed Weyl spinor ψ_L and a right-handed Weyl spinor ψ_R. The crucial point is that we can construct two new Lorentz scalars, $\psi_L^\dagger \psi_R$ and $\psi_R^\dagger \psi_L$. In fact, from the explicit form (2.59, 2.60) we see that $\Lambda_L^\dagger \Lambda_R = 1 = \Lambda_R^\dagger \Lambda_L$, and therefore $\psi_L^\dagger \psi_R \to \psi_L^\dagger \Lambda_L^\dagger \Lambda_R \psi_R$ is invariant, and similarly $\psi_R^\dagger \psi_L$.

In particular, we have the two real combinations $\psi_L^\dagger\psi_R+\psi_R^\dagger\psi_L$ and $i(\psi_L^\dagger\psi_R-\psi_R^\dagger\psi_L)$. Under a parity transformation, $\psi_L\leftrightarrow\psi_R$ and therefore the first combination is a scalar and the second is a pseudoscalar.

The Dirac Lagrangian is defined as

$$\boxed{\mathcal{L}_D = i\psi_L^\dagger\bar{\sigma}^\mu\partial_\mu\psi_L + i\psi_R^\dagger\sigma^\mu\partial_\mu\psi_R - m(\psi_L^\dagger\psi_R+\psi_R^\dagger\psi_L)\,.} \tag{3.78}$$

Contrary to the Weyl Lagrangian, the Dirac Lagrangian is invariant under parity transformation, since under parity $\psi_L\leftrightarrow\psi_R$ and $\partial_i\to-\partial_i$ so that $\bar{\sigma}^\mu\partial_\mu\leftrightarrow\sigma^\mu\partial_\mu$.

The Euler–Lagrange equations are obtained considering ψ_L and ψ_L^* independent (since they are complex fields) and similarly for ψ_R,ψ_R^*. Then, performing the variations with respect to ψ_L^* and ψ_R^*, we get

$$\bar{\sigma}^\mu i\partial_\mu\psi_L = m\psi_R\,, \tag{3.79}$$

$$\sigma^\mu i\partial_\mu\psi_R = m\psi_L\,. \tag{3.80}$$

This is the Dirac equation written in terms of Weyl spinors. Note that, because of the mass term, ψ_L and ψ_R are no longer helicity eigenstates. Applying the operator $\sigma^\mu i\partial_\mu$ on both sides of eq. (3.79) and using eq. (3.80) we get $-\sigma^\mu\bar{\sigma}^\nu\partial_\mu\partial_\nu\psi_L=m^2\psi_L$. Since $\partial_\mu\partial_\nu$ is symmetric we can replace $\sigma^\mu\bar{\sigma}^\nu$ with $(1/2)(\sigma^\mu\bar{\sigma}^\nu+\sigma^\nu\bar{\sigma}^\mu)$ and use the identity

$$\sigma^\mu\bar{\sigma}^\nu+\sigma^\nu\bar{\sigma}^\mu = 2\eta^{\mu\nu} \tag{3.81}$$

which follows immediately from the definition of σ^μ and $\bar{\sigma}^\mu$. We then find

$$(\Box+m^2)\psi_L = 0\,, \tag{3.82}$$

and similarly for ψ_R. Therefore the Dirac equation implies a massive KG equation for ψ_L,ψ_R, and the parameter m introduced in the Dirac Lagrangian is indeed a mass term.

It is convenient to write everything in terms of the Dirac spinor. We write the Dirac spinor as in Chapter 2,

$$\Psi = \begin{pmatrix}\psi_L\\ \psi_R\end{pmatrix}. \tag{3.83}$$

It is clear that we might as well define the Dirac spinor taking different combinations of ψ_L,ψ_R. For instance we might take $(\psi_R+\psi_L)/\sqrt{2}$ as the upper component of Ψ and $(\psi_R-\psi_L)/\sqrt{2}$ as the lower component. The choice (3.83) defines the so-called chiral representation. We then define the 4×4 γ matrices in the chiral representation,

$$\gamma^0 = \begin{pmatrix}0 & 1\\ 1 & 0\end{pmatrix}, \qquad \gamma^i = \begin{pmatrix}0 & \sigma^i\\ -\sigma^i & 0\end{pmatrix}, \tag{3.84}$$

or, more compactly,

$$\gamma^\mu = \begin{pmatrix}0 & \sigma^\mu\\ \bar{\sigma}^\mu & 0\end{pmatrix}. \tag{3.85}$$

Using eq. (3.81) we see that the γ matrices satisfy the *Clifford algebra*,

$$\{\gamma^\mu, \gamma^\nu\} = 2\eta^{\mu\nu} \,. \tag{3.86}$$

In terms of Dirac spinors, the Dirac equation becomes

$$\boxed{(i\partial\!\!\!/ - m)\,\Psi = 0 \,.} \tag{3.87}$$

Here we have introduced the Feynman slash notation: for a generic four-vector A^μ, we denote $\gamma^\mu A_\mu$ by $A\!\!\!/$; then $\partial\!\!\!/$ is the notation for $\gamma^\mu \partial_\mu$.

In order to write the Lagrangian in a compact form it is convenient to define the Dirac adjoint,

$$\bar{\Psi} = \Psi^\dagger \gamma^0 \,. \tag{3.88}$$

In the chiral representation $\bar{\Psi} = (\psi_R^\dagger, \psi_L^\dagger)$ and the Dirac Lagrangian can be written as

$$\boxed{\mathcal{L}_D = \bar{\Psi}\,(i\partial\!\!\!/ - m)\,\Psi \,.} \tag{3.89}$$

We also define $\gamma^5 = i\gamma^0\gamma^1\gamma^2\gamma^3$, so in the chiral representation

$$\gamma^5 = \begin{pmatrix} -1 & 0 \\ 0 & 1 \end{pmatrix} . \tag{3.90}$$

Therefore $(1 \pm \gamma^5)/2$ is a projector on the Weyl spinors,

$$\frac{1-\gamma^5}{2}\Psi = \begin{pmatrix} \psi_L \\ 0 \end{pmatrix} \qquad \frac{1+\gamma^5}{2}\Psi = \begin{pmatrix} 0 \\ \psi_R \end{pmatrix} . \tag{3.91}$$

If we take the neutrinos to be massless, a single left-handed Weyl spinor ν_L suffices for their description. Even in this case, however, it can be convenient to use a Dirac notation, i.e. to describe the neutrino with a Dirac spinor ν which, in the chiral representation, has the form

$$\nu = \begin{pmatrix} \nu_L \\ 0 \end{pmatrix} , \tag{3.92}$$

and therefore satisfies

$$\frac{1-\gamma^5}{2}\nu = \nu \,. \tag{3.93}$$

As we already remarked, the form (3.83) for the Dirac spinor is a possible choice but, depending on the problem, other choices might be more convenient. For instance, we can define a new Dirac spinor $\Psi' = U\Psi$, with U a constant unitary matrix. Then the Dirac Lagrangian becomes

$$\mathcal{L}_D = (\Psi')^\dagger U\gamma^0\,(i\gamma^\mu\partial_\mu - m)\,U^\dagger\Psi' = \bar{\Psi}'\,(i\gamma^{\mu\prime}\partial_\mu - m)\,\Psi' \tag{3.94}$$

with $\gamma^{\mu\prime} = U\gamma^\mu U^\dagger$ and $\bar{\Psi}' = (\Psi')^\dagger\gamma^{0\prime}$. So the explicit form of the γ matrices changes, as well as the relation between Ψ and the Weyl spinors ψ_L, ψ_R. However, the action and therefore the Dirac equation maintains the same form in terms of the redefined Dirac field and γ matrices. Furthermore, the algebra (3.86) is invariant under $\gamma^\mu \to U\gamma^\mu U^\dagger$ with U unitary. These different explicit expressions for the γ matrices correspond therefore to equivalent representations of the Clifford algebra. Together with the chiral representation which has been our starting point, another particularly useful representation is obtained acting with the unitary matrix $U = \frac{1}{\sqrt{2}}\begin{pmatrix} 1 & 1 \\ -1 & 1 \end{pmatrix}$, which gives

$$\Psi = \frac{1}{\sqrt{2}}\begin{pmatrix} \psi_R + \psi_L \\ \psi_R - \psi_L \end{pmatrix} \equiv \begin{pmatrix} \phi \\ \chi \end{pmatrix}, \tag{3.95}$$

and the γ matrices become

$$\gamma^0 = \begin{pmatrix} 1 & 0 \\ 0 & -1 \end{pmatrix}, \qquad \gamma^i = \begin{pmatrix} 0 & \sigma^i \\ -\sigma^i & 0 \end{pmatrix}, \tag{3.96}$$

$$\gamma^5 = i\gamma^0\gamma^1\gamma^2\gamma^3 = \begin{pmatrix} 0 & 1 \\ 1 & 0 \end{pmatrix}. \tag{3.97}$$

This representation, known as the ordinary, or standard representation, is useful in the non-relativistic limit (see Section 3.6). The chiral representation is more useful in the ultra-relativistic limit, when the mass can be neglected and ψ_L, ψ_R are helicity eigenstates, and also displays more clearly the group theoretical structure since $\psi_{L,R}$ are the irreducible representations of the Lorentz group $(\frac{1}{2}, 0)$ and $(0, \frac{1}{2})$, respectively.

The general solution of the massive Dirac equation is a superposition of plane waves of the form

$$\Psi(x) = u(p)e^{-ipx} \tag{3.98}$$

and of the form

$$\Psi(x) = v(p)e^{ipx}\,, \tag{3.99}$$

where $u(p), v(p)$ are four-component spinors. The former, in a classical theory, is a positive energy solution and the latter is a negative energy solution. The proper interpretation of the negative energy solutions will come after quantization, and will be discussed in Chapter 4.

The Dirac equation applied to the plane waves (3.98) and (3.99) gives

$$(\not{p} - m)\, u(p) = 0\,, \tag{3.100}$$

$$(\not{p} + m)\, v(p) = 0\,, \tag{3.101}$$

using the Feynman slash notation, $\not{p} = \gamma^\mu p_\mu$. We use the chiral representation, and we write

$$u(p) = \begin{pmatrix} u_L(p) \\ u_R(p) \end{pmatrix}. \tag{3.102}$$

We consider the case $m \neq 0$. Then we can solve the equation in the rest frame, where $p^\mu = (m, 0, 0, 0)$ and eq. (3.100) reads $(\gamma^0 - 1)u(p) = 0$. The solution, using the expression (3.84) for γ^0, is $u_L = u_R$. We see that, while the KG equation imposes only the mass shell condition $p^2 = m^2$, the Dirac equation, being first-order in the derivatives, has also the effect of reducing by a factor of two the number of independent degrees of freedom, since it relates u_L to u_R. A convenient choice of normalization is $u_L = u_R = \sqrt{m}\,\xi$ with ξ a two-component spinor satisfying $\xi^\dagger\xi = 1$.

The solution for generic p can be found either directly from the Dirac equation or performing a boost of the rest frame solution, using the transformation properties of $\psi_{L,R}$ discussed in Chapter 2. In any case, setting $\mathbf{p}$ along the $+z$ direction, the result is

$$u^s(p) = \begin{pmatrix} \left[\sqrt{E+p^3}\,\frac{1-\sigma^3}{2} + \sqrt{E-p^3}\,\frac{1+\sigma^3}{2}\right]\xi^s \\ \\ \left[\sqrt{E+p^3}\,\frac{1+\sigma^3}{2} + \sqrt{E-p^3}\,\frac{1-\sigma^3}{2}\right]\xi^s \end{pmatrix}. \tag{3.103}$$

The index $s = 1, 2$ labels the two independent solutions,

$$\xi^1 = \begin{pmatrix} 1 \\ 0 \end{pmatrix} \qquad \xi^2 = \begin{pmatrix} 0 \\ 1 \end{pmatrix}. \tag{3.104}$$

In the ultra-relativistic limit we have $p^\mu \to (E, 0, 0, E)$. Then, when $s = 1$, eq. (3.103) becomes

$$u^1(p) \to \sqrt{2E}\begin{pmatrix} 0 \\ \xi^1 \end{pmatrix}. \tag{3.105}$$

If instead $s = 2$

$$u^2(p) \to \sqrt{2E}\begin{pmatrix} \xi^2 \\ 0 \end{pmatrix}. \tag{3.106}$$

Thus, in the ultra-relativistic limit (or, equivalently, in the massless limit), u^1 has only the right-handed component while u^2 has only the left-handed component. When we quantize the theory, in Section 4.2, we will see what this means in terms of the helicities of massless particles.

We can study similarly the equation for $v(p)$; the result differs from $u(p)$ in the sign of the lower component,

$$v^s(p) = \begin{pmatrix} \left[\sqrt{E+p^3}\,\frac{1-\sigma^3}{2} + \sqrt{E-p^3}\,\frac{1+\sigma^3}{2}\right]\eta^s \\ \\ -\left[\sqrt{E+p^3}\,\frac{1+\sigma^3}{2} + \sqrt{E-p^3}\,\frac{1-\sigma^3}{2}\right]\eta^s \end{pmatrix}, \tag{3.107}$$

where again we wrote the result in the frame where $\mathbf{p} = (0, 0, p^3)$ and in the chiral representation; η^s is the two-dimensional spinor describing the two distinct solutions for $v^s(p)$,

$$\eta^1 = \begin{pmatrix} 1 \\ 0 \end{pmatrix} \qquad \eta^2 = \begin{pmatrix} 0 \\ 1 \end{pmatrix}. \tag{3.108}$$

We define $\bar{u}^s(p) = u^{s\dagger}\gamma^0$ and $\bar{v}^s(p) = v^{s\dagger}\gamma^0$. From the explicit form of $u(p), v(p)$ one finds a number of useful properties. First of all, from our normalization choice $\xi^{r\dagger}\xi^s = \delta^{rs}, \eta^{r\dagger}\eta^s = \delta^{rs}$, it follows that

$$\bar{u}^r(p)u^s(p) = 2m\,\delta^{rs}, \qquad \bar{v}^r(p)v^s(p) = -2m\,\delta^{rs} \tag{3.109}$$

and

$$u^{r\dagger}(p)u^s(p) = 2E_{\mathbf{p}}\,\delta^{rs}\,, \qquad v^{r\dagger}(p)v^s(p) = 2E_{\mathbf{p}}\,\delta^{rs}\,, \tag{3.110}$$

$$\bar{u}^r(p)v^s(p) = 0\,, \qquad \bar{v}^r(p)u^s(p) = 0\,. \tag{3.111}$$

When computing scattering cross-sections, one is often interested in the situation in which we sum over all possible final spin states and average over all possible initial spin states. In this case one finds a sum over the spins, that is performed with the help of the formulas

$$\sum_{s=1,2} u^s(p)\bar{u}^s(p) = \not{p} + m\,, \tag{3.112}$$

$$\sum_{s=1,2} v^s(p)\bar{v}^s(p) = \not{p} - m\,, \tag{3.113}$$

which again can be found from the explicit expressions for u, v. Note of course that, while $\bar{u}u$ is a number, $u\bar{u}$ is a 4×4 matrix.

We will also need the Dirac equation for $\bar{u}, \bar{v}$. Taking the hermitian conjugate of eqs. (3.100) and (3.101) we find

$$\bar{u}(p)\,(\not{p} - m) = 0\,, \tag{3.114}$$

$$\bar{v}(p)\,(\not{p} + m) = 0\,. \tag{3.115}$$

Another useful identity is obtained multiplying eq. (3.114) (with spin state r) by $u^s(p)$ from the right and using eq. (3.109), which gives $p_\mu \bar{u}^r(p)\gamma^\mu u^s(p) = 2m^2\delta^{rs}$. Since this must be true for generic p^μ, we find that

$$\bar{u}^r(p)\gamma^\mu u^s(p) = 2p^\mu\delta^{rs}\,, \tag{3.116}$$

and similarly

$$\bar{v}^r(p)\gamma^\mu v^s(p) = 2p^\mu\delta^{rs}\,. \tag{3.117}$$

Finally, we observe that the 16 matrices $1, \gamma^\mu, \gamma^5, \gamma^\mu\gamma^5$ and

$$\sigma^{\mu\nu} \equiv \frac{i}{2}\,[\gamma^\mu, \gamma^\nu] \tag{3.118}$$

are linearly independent and therefore the most general 4×4 matrix can be expressed in terms of them. The most general fermion bilinear is therefore a combination of

$$\bar{\Psi}\Psi, \quad \bar{\Psi}\gamma^\mu\Psi, \quad \bar{\Psi}\gamma^5\Psi, \quad \bar{\Psi}\gamma^5\gamma^\mu\Psi, \quad \bar{\Psi}\sigma^{\mu\nu}\Psi\,. \tag{3.119}$$

Their Lorentz and parity transformation properties are most easily understood writing them in the chiral representation, in terms of Weyl spinors. We have already seen that $\bar{\Psi}\Psi = (\psi_L^\dagger\psi_R + \psi_R^\dagger\psi_L)$ is a true scalar while $\bar{\Psi}\gamma^5\Psi = \psi_L^\dagger\psi_R - \psi_R^\dagger\psi_L$ is a pseudoscalar. Consider now

$$j_V^\mu \equiv \bar{\Psi}\gamma^\mu\Psi = \psi_L^\dagger\bar{\sigma}^\mu\psi_L + \psi_R^\dagger\sigma^\mu\psi_R\,. \tag{3.120}$$

We already know that $\psi_L^\dagger\bar{\sigma}^\mu\psi_L$ and $\psi_R^\dagger\sigma^\mu\psi_R$ are four vectors, so $\bar{\Psi}\gamma^\mu\Psi$ is a four-vector. Under parity, $\psi_L \leftrightarrow \psi_R$ and, since $\sigma^0 = \bar{\sigma}^0$ and $\sigma^i = -\bar{\sigma}^i$, we have $j_V^0 \to j_V^0, j_V^i \to -j_V^i$, so j_V^μ is a true four-vector.

Similarly one finds that

$$j_A^\mu \equiv \bar{\Psi}\gamma^5\gamma^\mu\Psi = \psi_L^\dagger\bar{\sigma}^\mu\psi_L - \psi_R^\dagger\sigma^\mu\psi_R \tag{3.121}$$

is a pseudovector and $\bar{\Psi}\sigma^{\mu\nu}\Psi$ is a true tensor.

In eq. (2.88) we have written explicitly the Lorentz transformation of Dirac spinors in the chiral representation. It is straightforward to check that, in terms of γ matrices, this transformation reads

$$\Psi \to \exp\{-\frac{i}{4}\omega_{\mu\nu}\sigma^{\mu\nu}\}\Psi\,. \tag{3.122}$$

We have derived this transformation law working in the chiral representation. However, once a transformation has been written in terms of γ matrices, it holds in any representation: if $\Psi \to \Lambda_D\Psi$, we have $U\Psi \to U(\Lambda_D\Psi) = (U\Lambda_D U^{-1})U\Psi$. However $\Psi' = U\Psi$ is the spinor in the new representation and, expanding in power series the exponential, we see that

$$U\Lambda_D U^{-1} = \exp\{-\frac{i}{4}\omega_{\mu\nu}U\sigma^{\mu\nu}U^{-1}\} = \exp\{-\frac{i}{4}\omega_{\mu\nu}\sigma'^{\mu\nu}\} \tag{3.123}$$

where $\sigma'^{\mu\nu} = (i/2)[\gamma'^\mu, \gamma'^\nu]$ and γ'^μ are the Dirac matrices in the new representation.

Equation (3.122) states that $J^{\mu\nu} = \sigma^{\mu\nu}/2$ provides a representation of complex dimension four of the Lorentz algebra (2.25). This can also be checked directly, from the definition of $\sigma^{\mu\nu}$ in terms of the γ matrices and using $\{\gamma^\mu, \gamma^\nu\} = 2\eta^{\mu\nu}$. This representation is however reducible, since we know that $\Psi \in (\mathbf{\frac{1}{2}}, \mathbf{0}) \oplus (\mathbf{0}, \mathbf{\frac{1}{2}})$.

3.4.3 Chiral symmetry

Let us consider the Dirac Lagrangian (3.78) with the mass term set to zero. In this case the action has a global internal symmetry,

$$\psi_L \to e^{i\theta_L}\psi_L\,, \qquad \psi_R \to e^{i\theta_R}\psi_R\,, \tag{3.124}$$

in which ψ_L and ψ_R are rotated by two independent angles θ_L and θ_R. The above transformation therefore belongs to the group $U(1) \times U(1)$. The transformation with $\theta_L = \theta_R \equiv \alpha$ can be written in terms of the Dirac spinor as

$$\Psi \to e^{i\alpha}\Psi\,, \tag{3.125}$$

while the transformation with $\theta_R = -\theta_L \equiv \beta$ can be written as

$$\Psi \to e^{i\beta\gamma^5}\Psi\,. \tag{3.126}$$

We can also verify that eqs. (3.125) and (3.126) are symmetry transformations directly on the Dirac action in the form (3.89), with $m = 0$. For the transformation (3.125) it is evident. For the transformation (3.126), using $\{\gamma^5, \gamma^\mu\} = 0$ we can show that

$$\gamma^\mu e^{i\beta\gamma^5} = e^{-i\beta\gamma^5}\gamma^\mu\,, \tag{3.127}$$

simply expanding the exponential in power series. Then, using the fact that γ^5 is hermitian,

$$\bar{\Psi}\gamma^\mu\partial_\mu\Psi = \Psi^\dagger\gamma^0\gamma^\mu\partial_\mu\Psi \to \Psi^\dagger e^{-i\beta\gamma^5}\gamma^0\gamma^\mu\partial_\mu e^{i\beta\gamma^5}\Psi\,. \tag{3.128}$$

The factor $\exp\{i\beta\gamma^5\}$ commutes with ∂_μ since β is independent of x and, from eq. (3.127),

$$\gamma^0\gamma^\mu e^{i\beta\gamma^5} = \gamma^0 e^{-i\beta\gamma^5}\gamma^\mu = e^{i\beta\gamma^5}\gamma^0\gamma^\mu\,, \tag{3.129}$$

so in eq. (3.128) $\exp\{i\beta\gamma^5\}$ cancels with $\exp\{-i\beta\gamma^5\}$ and $\bar{\Psi}\gamma^\mu\partial_\mu\Psi$ is invariant.

The transformation (3.125) is a global $U(1)$ symmetry, which will be promoted to a local gauge symmetry in Section 3.5.4. It is also called the vector $U(1)$ since, using the Noether theorem, its conserved current is the vector current

$$j^\mu_V = \bar{\Psi}\gamma^\mu\Psi\,. \tag{3.130}$$

The transformation (3.126) is instead called a *chiral transformation*, or the axial $U(1)$ (often denoted as $U_A(1)$). Its conserved current is

$$j^\mu_A = \bar{\Psi}\gamma^\mu\gamma^5\Psi\,, \tag{3.131}$$

which, as we have seen above, is a pseudovector, and is called the axial current.

If we now switch on the mass term, we see that eq. (3.124) is a symmetry transformation only if $\theta_L = \theta_R$, since the mass term couples ψ_L and ψ_R. Therefore the mass term breaks the axial $U(1)$, while the vector $U(1)$ is preserved. Indeed, if we compute the divergence of the axial current (3.131) using the equations of motion of the massive theory, we find

$$\partial_\mu j^\mu_A = 2im\bar{\Psi}\gamma^5\Psi\,. \tag{3.132}$$

The same result can be obtained using the general formula (3.33).

3.4.4 Majorana mass

In Section 3.4.1 we discussed how to describe a spin 1/2 massless particle using a single Weyl field, e.g. left-handed, ψ_L. In Section 3.4.2 we have shown that we can introduce a mass term using two spinor fields, one left-handed and one right-handed, and this has given rise to a Dirac mass term.

We are now in the position to reexamine a question that we posed at the end of Section 3.4.1. Is it possible to describe a massive particle with a single Weyl field, e.g. left-handed? The answer is yes, because we have seen in eq. (2.64) that, given a left-handed Weyl spinor ψ_L, we can construct a right-handed Weyl spinor $\psi_R = i\sigma^2\psi_L^*$. We can therefore write the Dirac equation (3.79) using $i\sigma^2\psi_L^*$ as the right-handed Weyl field,

$$\bar{\sigma}^\mu i\partial_\mu\psi_L = im\sigma^2\psi_L^*\,, \tag{3.133}$$

and eq. (3.80) gives simply the complex conjugate of this equation. The algebraic manipulations performed for the Dirac equation to show that it implies a massive Klein–Gordon equation can be repeated here without any change (since they are valid independently of whether ψ_L and ψ_R are independent or not), and therefore eq. (3.133) implies $(\Box + m^2)\psi_L = 0$. We have therefore constructed a mass term using only ψ_L. Such a mass term is known as a Majorana mass.

The formal similarity with the Dirac mass term can be seen writing eq. (3.133) in terms of the four-component Majorana spinor Ψ_M, which in the chiral representation is

$$\Psi_M = \begin{pmatrix} \psi_L \\ i\sigma^2\psi_L^* \end{pmatrix} . \tag{3.134}$$

Then eq. (3.133) becomes formally the same as the Dirac equation,

$$(i\partial\!\!\!/ - m)\Psi_M = 0 . \tag{3.135}$$

Naively, one would guess that this can be derived from the variation of a Dirac action, with Ψ replaced by Ψ_M. However, this is not true, because for a Majorana spinor

$$\bar{\Psi}_M\Psi_M = -i\psi_L^T\sigma^2\psi_L + h.c. , \tag{3.136}$$

where ψ_L^T denotes the transpose. In components, $\psi_L^T\sigma^2\psi_L = \psi_{L,a}\sigma^2_{ab}\psi_{L,b}$. In the classical theory, we defined the Weyl fields as ordinary commuting numbers (often called *c*-numbers when one want to stress that they are numbers, rather than operators), therefore $\psi_{L,a}\psi_{L,b}$ is symmetric in the indices (a, b). Instead the Pauli matrix σ^2_{ab} is antisymmetric in (a, b), and therefore $\bar{\Psi}_M\Psi_M$ vanishes identically. If we wish to write a classical action for the Majorana mass term we can do it, but we must state that, already at the classical level, ψ_L is an anticommuting field. Otherwise, we can be satisfied with eq. (3.135), without deriving it from a classical Lagrangian.

Apart from this technical aspect, the Majorana mass has a very important physical difference compared to the Dirac mass. As we saw in Section 3.4.3, the Dirac action with a mass term is invariant under a global $U(1)$ transformations of ψ_L and of ψ_R, $\psi_L \to e^{i\alpha}\psi_L$, $\psi_R \to e^{i\alpha}\psi_R$, see eq. (3.125). Rather than on the action, we can see this invariance directly on the massive Dirac equation in Weyl form, eqs. (3.79) and (3.80). For Majorana spinors, however, ψ_L and ψ_R are not independent but are related by complex conjugation. Therefore, if ψ_L transforms as $\psi_L \to e^{i\alpha}\psi_L$, automatically $\psi_R = i\sigma^2\psi_L^*$ transforms as $\psi_R \to e^{-i\alpha}\psi_R$. It is impossible to define on a Majorana spinor a $U(1)$ transformation under which $\psi_L \to e^{i\alpha}\psi_L$ and at the same time $\psi_R \to e^{i\alpha}\psi_R$.

In other words, the Majorana equation (3.133) is not invariant under global $U(1)$ symmetries of the type that in Section 3.4.3 we called vector $U(1)$ (or simply $U(1)$, while axial $U(1)$ symmetries are denoted by $U_A(1)$). This means that a spin 1/2 particle which carries a $U(1)$ conserved charge cannot have a Majorana mass. For instance, all spin

1/2 particles which have an electric charge, like the electron, cannot have a Majorana mass. Similarly, processes involving a lepton with a Majorana mass violate lepton number (i.e. the number of leptons minus the number of antileptons) since this is again a $U(1)$ symmetry.

A possible candidate for a particle which could have a Majorana mass is the neutrino. As we mentioned already, at present there are indications that the neutrinos have tiny masses, but it is not known if these masses are Dirac or Majorana masses.

A Dirac mass for the neutrino would imply that, together with the left-handed neutrino, there exists also a right-handed neutrino, which combines with the left-handed one to produce the Dirac mass. However, these hypothetical right-handed neutrinos are not seen in weak interactions, and therefore, if they exist, they must be sterile, which means that they do not participate in weak interactions, or at least they participate much more weakly than the left-handed neutrinos. The other possibility is that neutrinos are described by purely left-handed fields and have Majorana masses. In this case the lepton number symmetry is violated. Experiments on neutrino-less double beta decay aim at detecting these violations.

3.5 The electromagnetic field

3.5.1 Covariant form of the free Maxwell equations

The electromagnetic field is described by a four-vector A_μ, the gauge potential. The field strength tensor is defined as

$$F_{\mu\nu} = \partial_\mu A_\nu - \partial_\nu A_\mu \,, \tag{3.137}$$

and it is related to the electric and magnetic fields as

$$F^{0i} = \partial^0 A^i - \partial^i A^0 = \partial_0 A^i + \nabla^i A^0 = -E^i \,, \tag{3.138}$$

$$F^{ij} = -\epsilon^{ijk} B^k \,. \tag{3.139}$$

The Lagrangian of the free electromagnetic field is

$$\boxed{L = -\frac{1}{4} F_{\mu\nu} F^{\mu\nu} = \frac{1}{2}(\mathbf{E}^2 - \mathbf{B}^2) \,.} \tag{3.140}$$

The equations of motion derived from this Lagrangian are

$$\boxed{\partial_\mu F^{\mu\nu} = 0 \,.} \tag{3.141}$$

In terms of the electric and magnetic fields these are the first pair of Maxwell equations in the absence of sources,

$$\nabla \cdot \mathbf{E} = 0 \,, \qquad \nabla \times \mathbf{B} = \partial_0 \mathbf{E} \,, \tag{3.142}$$

written in a Lorentz covariant form. Furthermore, defining

$$\tilde{F}^{\mu\nu} = \frac{1}{2}\epsilon^{\mu\nu\rho\sigma}F_{\rho\sigma}\,, \tag{3.143}$$

we see that (if the gauge field is a regular function) $\partial_\mu \tilde{F}^{\mu\nu} = \epsilon^{\mu\nu\rho\sigma}\partial_\mu\partial_\rho A_\sigma$ vanishes identically, since the antisymmetric tensor $\epsilon^{\mu\nu\rho\sigma}$ is contracted with the symmetric tensor $\partial_\mu\partial_\rho$.[9] Therefore we have the identity

[9] The tensor $\partial_\mu\partial_\rho$ is symmetric only if it acts on a regular function A_σ, so that the derivatives can be interchanged. The important case where this does not happen is the Dirac monopole.

$$\boxed{\partial_\mu \tilde{F}^{\mu\nu} = 0\,,} \tag{3.144}$$

which is the second pair of Maxwell equations, $\nabla\cdot\mathbf{B}=0$ and $\nabla\times\mathbf{E} = -\partial_0\mathbf{B}$. Equations (3.141) and (3.144) are the Maxwell equations written in an explicitly covariant form.

3.5.2 Gauge invariance; radiation and Lorentz gauges

A crucial *local* symmetry of the Maxwell Lagrangian is the symmetry under *gauge transformations*

$$\boxed{A_\mu \to A_\mu - \partial_\mu\theta} \tag{3.145}$$

with $\theta(x)$ an arbitrary (regular) function. One verifies immediately that $F_{\mu\nu}$ is gauge invariant. As long as we consider only the free electromagnetic field without interaction, the global version of eq. (3.145) (i.e. θ independent of x) is trivial and therefore there is no associated conserved current (this is the situation, in the demonstration of the Noether theorem, where j^μ in eq. (3.21) vanishes identically). The situation changes when we switch on the interaction, as we will see in Section 3.5.4.

The existence of the local symmetry (3.145) introduces new problems, since it means that the variables A_μ give a redundant description of the electromagnetic field. We can use the gauge freedom to constraint A_μ, and a convenient choice is the following. First of all, it is easy to find a gauge transformation that sets $A_0 = 0$. It is given simply by

$$A_\mu \to A'_\mu = A_\mu - \partial_\mu \int^t dt'\, A_0(\mathbf{x},t')\,. \tag{3.146}$$

After that, we still have the freedom of performing a gauge transformation with θ independent of t, because this does not modify the condition $A'_0 = 0$. Therefore we perform a further gauge transformation which sends A'_μ into a new field $A''_\mu = A'_\mu - \partial_\mu\theta$, choosing

$$\theta(\mathbf{x}) = -\int \frac{d^3y}{4\pi|\mathbf{x}-\mathbf{y}|}\frac{\partial A'^i(\mathbf{y},t)}{\partial y^i}\,. \tag{3.147}$$

Despite its appearance, this function θ is actually independent of t. In fact, in this gauge $E^i = -\partial_0 A'^i$ since $A'_0 = 0$. Then $\partial_i E^i = 0$ implies

$\partial_0\partial_i A'^i = 0$ and we can see from eq. (3.147) that $\partial_0\theta = 0$. Furthermore, using the identity

$$\nabla_x^2 \left(\frac{1}{4\pi|\mathbf{x}-\mathbf{y}|} \right) = -\delta^3(\mathbf{x}-\mathbf{y}) \tag{3.148}$$

we see that

$$\nabla \cdot \mathbf{A}'' = \nabla \cdot \mathbf{A}' - \nabla^2\theta = 0\,. \tag{3.149}$$

We have therefore used the gauge freedom to set

$$\boxed{A_0 = 0 \qquad\qquad \nabla \cdot \mathbf{A} = 0\,.} \tag{3.150}$$

This gauge is called the *radiation gauge.* Note that it implies the *Lorentz gauge,*

$$\boxed{\partial_\mu A^\mu = 0\,.} \tag{3.151}$$

The equation of motion (3.141) in this gauge becomes

$$0 = \partial_\mu(\partial^\mu A^\nu - \partial^\nu A^\mu) = \partial_\mu\partial^\mu A^\nu - \partial^\nu(\partial_\mu A^\mu) = \Box A^\nu\,, \tag{3.152}$$

and we recognize a massless KG equation for each of the components of A^ν. After quantization, this will translate into the fact that the electromagnetic field describes massless particles. We can look for plane wave solutions,

$$A_\mu(x) = \epsilon_\mu(k)e^{-ikx} + c.c. \tag{3.153}$$

where ϵ_μ is called the polarization vector. Equation (3.152) gives $k^2 = 0$ while our gauge choice implies $\epsilon_0 = 0$ and $\boldsymbol{\epsilon}\cdot\mathbf{k} = 0$. Choosing for instance $\mathbf{k} = (0,0,k)$, i.e. considering an electromagnetic wave traveling in the z-direction, $\boldsymbol{\epsilon}\cdot\mathbf{k} = 0$ becomes $\epsilon^3 = 0$. Therefore *an electromagnetic wave has two degrees of freedom, represented by a vector $\boldsymbol{\epsilon}$ in the plane transverse to the direction of propagation.* The circular polarizations $\epsilon^\pm = \epsilon^1 \pm i\epsilon^2$ are the helicity eigenstates (see Exercise 2.4).

The advantage of the radiation gauge is to expose clearly the physical degrees of freedom of the electromagnetic field. The disadvantage, however, is that Lorentz covariance is no longer explicit, since eq. (3.150) is not Lorentz covariant. The Lorentz gauge (3.151) is instead Lorentz covariant, but alone it is insufficient to eliminate all spurious degrees of freedom. Therefore we must choose between a redundant, but Lorentz covariant description, and a description which breaks the covariance but exposes clearly the physical degrees of freedom. This will be a recurrent theme in the quantization of gauge fields.

3.5.3 The energy–momentum tensor

The energy–momentum tensor of the electromagnetic field can be computed from eq. (3.35),

$$\theta^{\mu\nu} = -F^{\mu\rho}\partial^\nu A_\rho + \frac{1}{4}\eta^{\mu\nu}F^2\,, \tag{3.154}$$

with $F^2 \equiv F^{\mu\nu}F_{\mu\nu}$. This form is not explicitly gauge invariant, since it depends on A_μ and not only on $F_{\mu\nu}$. However, using the equation of motion $\partial_\rho F^{\mu\rho} = 0$, we can see that under a gauge transformation it changes as

$$\theta^{\mu\nu} \to \theta^{\mu\nu} + F^{\mu\rho}\partial^\nu\partial_\rho\theta = \theta^{\mu\nu} + \partial_\rho\left(F^{\mu\rho}\partial^\nu\theta\right), \tag{3.155}$$

and the conserved charges change as

$$P^\nu \to P^\nu + \int d^3x\, \partial_\rho\left(F^{0\rho}\partial^\nu\theta\right) = P^\nu + \int d^3x\, \partial_i\left(F^{0i}\partial^\nu\theta\right). \tag{3.156}$$

The additional term is a total spatial derivative which integrates to zero assuming, as always, that the field decreases sufficiently fast at infinity. Therefore the charges are gauge invariant. If we want to have directly a gauge-invariant form for the energy–momentum tensor then, following the discussion in Section 3.2.1, we can add a term $\partial_\rho(F^{\mu\rho}A^\nu)$ to $\theta^{\mu\nu}$, which is by itself conserved, and its $\mu = 0$ component is a total spatial derivative. We then obtain the "improved" energy–momentum tensor

$$T^{\mu\nu} = F^{\mu\rho}F_\rho{}^\nu + \frac{1}{4}\eta^{\mu\nu}F^2, \tag{3.157}$$

which is explicitly gauge invariant. The energy of the electromagnetic field is

$$E = \int d^3x\, T^{00} = \frac{1}{2}\int d^3x\, (\mathbf{E}^2 + \mathbf{B}^2) \tag{3.158}$$

and the spatial momentum is

$$P^i = \int d^3x\, T^{0i} = \int d^3x\, (\mathbf{E}\times\mathbf{B})^i. \tag{3.159}$$

This is known as the Poynting vector.[10] The spatial components T^{ij} form a spatial tensor, whose negative is known as the Maxwell stress tensor T_M^{ij},

$$T_M^{ij} \equiv -T^{ij} = E^iE^j + B^iB^j - \frac{1}{2}\delta^{ij}(\mathbf{E}^2 + \mathbf{B}^2). \tag{3.160}$$

Observe that, integrating the equation $\partial_\mu T^{\mu i} = 0$ over a volume V bounded by a surface S, one finds

$$\int_V d^3x\, \left[\partial_0 T^{0i} + \partial_j T^{ji}\right] = 0, \tag{3.161}$$

and therefore[11]

$$\frac{d}{dt}P^i = +\int_V d^3x\, \partial_j T_M^{ji} = -\int_S dS\, \hat{n}^j T_M^{ji}, \tag{3.162}$$

where P^i is the momentum contained in the volume V and $\hat{\mathbf{n}}$ is the outward normal to the surface. Therefore the Maxwell stress tensor gives the flow of momentum across a surface.

[10] The factors 1/2 in front of the energy and the factor 1 in front of the Poynting vector depend on our choice of rationalized units for the electric charge, and in unrationalized units they would become $1/(8\pi)$ and $1/(4\pi)$, respectively; see the Notation.

[11] Recall that with our signature we have $\hat{n}_j T_M^{ji} = -\hat{n}^j T_M^{ji}$.

From this example we also learn that in field theory the observable quantities are the charges, rather than the currents. The currents (as in this case the energy–momentum tensor) are not uniquely defined. For instance, we can add a total four-divergence to the Lagrangian density. We saw in Section 3.1 that this does not modify the classical equations of motion, so two Lagrangian densities which differ by a total four-divergence define the same classical theory. However, the currents that one obtains through the Noether theorem using eq. (3.31) are different for the two theories, while the charges are the same (see Exercise 3.5 for an example).

This means also that in general we cannot localize the energy density with arbitrary precision. What is well defined is not the energy density, but rather its integral over a volume V such that the fields go to zero sufficiently fast at the boundaries, so that the ambiguities due to total divergences become irrelevant. For instance, if we have an electromagnetic wave packet centered around a frequency ω, corresponding to a wavelength $\lambda = 2\pi/\omega$, its energy in a volume V is well defined only if V is a box with sides equal to at least several wavelengths. Correspondingly, energy density is well defined only in a smeared sense, i.e. as an average over a box of side equal to a few times λ.[12]

[12] This lack of localizability is also present in the quantum theory, where it can be understood in terms of the uncertainty principle. In fact, in order to know the energy of the electromagnetic field in a box of side L, we must known how many photons there are at a given time and what is the energy of each. However, in order to know if a photon is inside the box, we must know its position with an error $\Delta x < L$, and correspondingly we have $\Delta p > 1/L$. If we take $L \ll \lambda$, the uncertainty on the momentum becomes much bigger than the momentum itself.

3.5.4 Minimal and non-minimal coupling to matter

In classical electrodynamics, when we add an external current j^μ, the second pair of Maxwell equations (3.144) are not modified, since they are just a consequence of the definition of $F^{\mu\nu}$ (assuming again that we have regular gauge fields, and excluding therefore the case of a Dirac monopole), while the first pair (3.141) becomes

$$\partial_\mu F^{\mu\nu} = j^\nu \,. \tag{3.163}$$

Since $F^{\mu\nu}$ is antisymmetric and we are assuming that it is a regular function, it automatically satisfies $\partial_\nu \partial_\mu F^{\mu\nu} = 0$, and therefore eq. (3.163) is consistent only if the current j^ν is conserved, $\partial_\nu j^\nu = 0$. We can understand this requirement as a consequence of gauge invariance. Equation (3.163) is the equation of motion derived from the action

$$S = -\int d^4x \left(\frac{1}{4} F^2 + j^\mu A_\mu \right) . \tag{3.164}$$

The term F^2 is explicitly gauge invariant. The term $j^\mu A_\mu$, under a gauge transformation $A_\mu \to A_\mu - \partial_\mu \theta$, changes by $-j^\mu \partial_\mu \theta$. If we integrate by parts and we disregard the boundary term at infinity (which means that we restrict to gauge transformation or to currents that vanish sufficiently fast at infinity), the variation of the action becomes equal to $-\int d^4x \, \theta \, \partial_\mu j^\mu$. The action is therefore gauge invariant if and only if $\partial_\mu j^\mu = 0$.

Beside being respected by classical electrodynamics, we will see that gauge invariance is also a crucial ingredient for quantizing theories with a massless vector field A_μ. Gauge invariance is a guiding principle in building the theory of fundamental interactions, and the corresponding theories are known as *gauge theories.*

A very general method for writing a gauge invariant action is the following. We start from a theory with a *global* $U(1)$ invariance. Let us consider for definiteness the Dirac action (3.89), but the procedure is completely general. We consider the transformation

$$\Psi \to e^{iq\theta}\Psi\,. \tag{3.165}$$

This transformation is a symmetry of the free Dirac action if θ is a constant, but we want to consider a generic function $\theta(x)$. The free Dirac action is not invariant if θ depends on x, because then the factor $e^{iq\theta}$ coming from Ψ does not commute with ∂_μ and so it cannot be canceled by the factor $e^{-iq\theta}$ coming from $\bar{\Psi}$. We have assigned to the field Ψ a parameter q that defines its transformation properties. We will see that it has the meaning of the electric charge of the particles described by the field, in units of e.

At the same time, the action of the free electromagnetic field is invariant under

$$A_\mu \to A_\mu - \partial_\mu\theta\,. \tag{3.166}$$

We then define the *covariant derivative* of Ψ as

$$\boxed{D_\mu\Psi = (\partial_\mu + iqA_\mu)\Psi} \tag{3.167}$$

and we immediately verify that, under the combined transformations (3.165) and (3.166), with $\theta = \theta(x)$,

$$\boxed{D_\mu\Psi \to e^{iq\theta}D_\mu\Psi\,,} \tag{3.168}$$

i.e. $D_\mu\Psi$ transforms in the same way as Ψ, even when θ is a function of x. It is now easy to construct a Lagrangian with a *local* $U(1)$ invariance. It suffices to replace all derivatives ∂_μ with covariant derivatives D_μ. This procedure is expressed by saying that we have gauged the global $U(1)$ symmetry, promoting it to a local symmetry. The resulting theory is called a *gauge theory* and A_μ is called a *gauge field.* More precisely, it is a $U(1)$, or abelian gauge field, since we have gauged a $U(1)$ symmetry. In Chapter 10 we will study how to gauge non-abelian groups, like $SU(N)$, and this will lead to non-abelian gauge theories and to the Standard Model.

It is important to note that the form of the covariant derivative depends on the transformation properties of the field on which it acts. For instance, for fields transforming as in eq. (3.165), D_μ depends on the parameter q. One can consider more general transformation laws, however. As an example, a gauge field transforms as in eq. (3.166) rather than being multiplied by a phase, and on a gauge field we simply define $D_\mu A_\nu = \partial_\mu A_\nu$ since $F_{\mu\nu}$ is already gauge invariant. We will find more general transformation properties, and more general definitions of the covariant derivative, when we study non-abelian gauge theories.

It is now straightforward to couple a Dirac field of charge q to the electromagnetic field: we just replace ∂_μ by D_μ in the Dirac Lagrangian and we have

$$\mathcal{L}_D = \bar{\Psi}\left(i\gamma^\mu D_\mu - m\right)\Psi \tag{3.169}$$

or, more explicitly,

$$\boxed{\mathcal{L}_D = \bar{\Psi}\left(i\gamma^\mu \partial_\mu - m\right)\Psi - qA_\mu \bar{\Psi}\gamma^\mu\Psi\,.} \tag{3.170}$$

Thus, the electrodynamics of a spinor field is obtained coupling A_μ to the current $\bar{\Psi}\gamma^\mu\Psi$. The resulting theory has by construction the local $U(1)$ symmetry defined by eqs. (3.165) and (3.166), with θ an arbitrary function of x, and therefore it obviously also has the global symmetry $\Psi \to e^{iq\theta}\Psi, A_\mu \to A_\mu$ with θ a constant. Applying the Noether theorem to this global symmetry we see that (modulo of course an arbitrary normalization) the conserved current is the vector current that we already met in Section 3.4.3,

$$j_V^\mu = \bar{\Psi}\gamma^\mu\Psi\,. \tag{3.171}$$

Therefore the electromagnetic field is coupled to a conserved current. The conserved charge is

$$Q = \int d^3x\, \bar{\Psi}\gamma^0\Psi = \int d^3x\, \Psi^\dagger\Psi \tag{3.172}$$

and, as we will see after quantization of the theory, it has the meaning of the electric charge, in units of e.

The equation of motion is

$$\boxed{\left(i\gamma^\mu D_\mu - m\right)\Psi = 0\,.} \tag{3.173}$$

This is the Dirac equation describing a spin 1/2 charged particle interacting with an electromagnetic field. We will discuss some of its consequences in the next section.

Consider now a complex scalar field ϕ transforming under gauge transformations as $\phi \to e^{iq\theta}\phi$. Again $D_\mu\phi$ is defined as $(\partial_\mu + iqA_\mu)\phi$ and the complex KG Lagrangian becomes

$$\begin{aligned}\mathcal{L} &= (D_\mu\phi)^* D^\mu\phi - m^2\phi^*\phi \qquad (3.174)\\ &= \partial_\mu\phi\partial^\mu\phi^* + iqA^\mu(\phi\partial_\mu\phi^* - \phi^*\partial_\mu\phi) + q^2|\phi|^2 A_\mu A^\mu - m^2\phi^*\phi\,.\end{aligned}$$

This is the Lagrangian of scalar electrodynamics. As we discussed in Section 3.3.2, the complex Klein–Gordon theory has a $U(1)$ symmetry, whose conserved current is given in eq. (3.62). We see from eq. (3.175) that A_μ couples to this current, and there is also a term in the Lagrangian proportional to $A_\mu A^\mu|\phi|^2$. The latter term plays an important role in the Higgs mechanism and in superconductivity, as we will see in

Section 11.3. Note that it is not possible to couple in this way a *real* scalar field to the electromagnetic field. For a real field necessarily $q = 0$ otherwise $e^{iq\theta}\phi$ becomes complex. After quantization, a real scalar field describes particles which are neutral under electromagnetism.

However, one can have a neutral scalar particle formed by charged constituents, and this particle will interact with the electromagnetic field not through its electric charge, which is zero, but through its higher electric and magnetic multipoles, exactly as the hydrogen atom is neutral but interacts with the electromagnetic fields through its electric and magnetic dipole moments, quadrupole moments, etc. This means that it must be possible to write a gauge-invariant coupling to the electromagnetic field also for a real scalar field. For example, a possible interaction term is

$$\mathcal{L}_S = a_S\, \phi F_{\mu\nu} F^{\mu\nu}\,, \tag{3.175}$$

where a_S is a coupling constant. Another possibility is

$$\mathcal{L}_{PS} = a_{PS}\, \phi\, \epsilon_{\mu\nu\rho\sigma} F^{\mu\nu} F^{\rho\sigma}\,, \tag{3.176}$$

with another coupling constant a_{PS}. Observe that under parity $F_{\mu\nu}F^{\mu\nu}$ is invariant while $\epsilon_{\mu\nu\rho\sigma}F^{\mu\nu}F^{\rho\sigma}$ is a pseudoscalar. Therefore the interaction Lagrangian $\mathcal{L}_S$ preserves parity only if ϕ is a scalar field, while $\mathcal{L}_{PS}$ is invariant only if ϕ is a pseudoscalar field. For example the neutral pion π^0 is a pseudoscalar, and it decays into two photons; the Lagrangian $\mathcal{L}_{PS}$ gives a good phenomenological description of its interaction with the electromagnetic field.

The coupling to the electromagnetic field which is obtained performing in the free Lagrangian the replacement $\partial_\mu \to D_\mu$ is called the *minimal coupling*. Otherwise the coupling is called non-minimal.

Similarly, for a Dirac fermion we can in principle write non-minimal couplings. For example we can add to the Lagrangian an interaction term

$$\mathcal{L}_{\text{int}} = a\, \bar{\Psi}\sigma^{\mu\nu}\Psi\, F_{\mu\nu}, \tag{3.177}$$

with a coupling constant a. After quantization of the theory we will find that the interaction term $qA_\mu\bar{\Psi}\gamma^\mu\Psi$ describes indeed the coupling of the gauge field with the electric charge of the particle. We will instead show in Solved Problem 7.2 that the coupling (3.177) corresponds to a magnetic dipole interaction.

We leave as an exercise to verify that the non-minimal coupling constants a_S, a_{PS} and a are not dimensionless, but rather have the dimension of the inverse of a mass. We will understand in Section 5.6 why interaction terms of this sort have a less fundamental significance than interactions terms in which the coupling constant is dimensionless, as in the minimal coupling.

3.6 First quantization of relativistic wave equations

As we already discussed in the Introduction, a first quantization of relativistic wave equations cannot be performed consistently. In particular, we have seen that both the free Klein–Gordon and the free Dirac equation have solutions proportional to e^{-ipx} and solutions proportional to e^{+ipx}. The former oscillate in time as e^{-iEt} and therefore in a first quantized formalism are eigenfunctions of the Hamiltonian $H = i\partial/\partial t$ with eigenvalue E, while the latter have eigenvalue $-E$ and therefore correspond to negative energy solutions. The proper interpretation of these solutions comes only after the field quantization that we will discuss in the next chapter.[13] However, as we will show below and in Solved Problem 3.1, in the non-relativistic limit the Dirac equation in an external electromagnetic field reduces to a Schrödinger equation, $i\partial\psi/\partial t = H\psi$, with a Hamiltonian H which contains an expansion in powers of the velocity of the particle, i.e. relativistic corrections. *A posteriori*, the field theoretical treatment shows that the first-order correction produced by the relativistic wave equations is correct, i.e. it coincides with the field theory result.[14] It is therefore useful to examine the non-relativistic limit of the Dirac equation, and to treat it in first quantization, promoting the classical field to a wave function. One should be aware, however, that higher-order corrections are not correctly given by the relativistic wave equations, and the full QFT treatment is needed.

The Dirac equation (3.173) for an electron of charge $q = e$ (with $e < 0$ in our notation) in an external electromagnetic field A_μ is

$$[\gamma^\mu(i\partial_\mu - eA_\mu) - m]\Psi = 0\,. \tag{3.178}$$

To study the non-relativistic limit, it is convenient to use the standard representation, eqs. (3.95) and (3.96), and to define

$$\chi'(\mathbf{x},t) = e^{imt}\chi(\mathbf{x},t)\,, \qquad \phi'(\mathbf{x},t) = e^{imt}\phi(\mathbf{x},t) \tag{3.179}$$

so that, if ϕ,χ have a time-dependence e^{-iEt} with E the relativistic energy, then ϕ' and χ' oscillate as $e^{-iE_{\rm NR}t}$ with $E_{\rm NR} = E - m$. Then the Dirac equation reads

$$(i\partial_0 - eA_0)\phi' = -\boldsymbol{\sigma}\cdot(i\nabla + e\mathbf{A})\chi'\,, \tag{3.180}$$

$$(i\partial_0 - eA_0 + 2m)\chi' = -\boldsymbol{\sigma}\cdot(i\nabla + e\mathbf{A})\phi'\,. \tag{3.181}$$

Observe that in eq. (3.180) the mass term obtained acting with ∂_0 on e^{imt} cancels with the mass term originally present in the Dirac equation, while in eq. (3.181) they add up (recall also that $\partial_i = \nabla^i$, see the Notation). In the non-relativistic limit we have

$$i\partial_0\chi' \ll m\chi'\,, \qquad eA_0 \ll m \tag{3.182}$$

and to lowest order the equation for χ' is easily solved,

$$\chi' \simeq -\frac{1}{2m}\boldsymbol{\sigma}\cdot(i\nabla + e\mathbf{A})\phi'\,. \tag{3.183}$$

[13] For fermions, Dirac found an ingenious solution to the negative energy problem using the Pauli principle and assuming that all states with negative energies are filled. However, this solution does not work for bosons, and today in high energy physics this "filled Dirac sea" has only historical interest. In condensed matter, however, it leads to an interpretation in terms of electrons and holes which is still useful, see Exercise 4.6.

[14] In the language of Feynman graphs that we will discuss in Chapter 5, the relativistic wave equations reproduce the result of tree level graphs.

This solution is the lowest order in a relativistic expansion, and further corrections will be computed in Solved Problem 3.1. We now insert this expression for χ' into eq. (3.180) and use

$$\sigma^i\sigma^j(i\nabla^i + eA^i)(i\nabla^j + eA^j)\phi' = \left[(i\nabla + e\mathbf{A})^2 + i\epsilon^{ijk}\sigma^k ie(\nabla^i A^j)\right]\phi' \tag{3.184}$$

which follows from $\sigma^i\sigma^j = \delta^{ij} + i\epsilon^{ijk}\sigma^k$. Finally, $\epsilon^{ijk}\nabla^i A^j = (\nabla \times \mathbf{A})^k = B^k$ is the magnetic field, and therefore (writing $\mathbf{p} = -i\nabla$) eq. (3.180) becomes

$$i\partial_0\phi' \simeq \left[\frac{(\mathbf{p} - e\mathbf{A})^2}{2m} - \frac{e}{2m}\boldsymbol{\sigma}\cdot\mathbf{B} + eA_0\right]\phi'\,. \tag{3.185}$$

Therefore in the non-relativistic limit the Dirac equation reduces to a Schrödinger equation for the two-component spinor ϕ', with a minimal coupling to the gauge field A_μ, plus an interaction term with a magnetic field. We see that the contribution to the energy due to the term $\boldsymbol{\sigma}\cdot\mathbf{B}$ can be written as $-\boldsymbol{\mu}\cdot\mathbf{B}$ with a magnetic moment $\boldsymbol{\mu}$ given by

$$\boldsymbol{\mu} = \frac{e}{2m}\boldsymbol{\sigma} = \frac{e}{m}\mathbf{S} \tag{3.186}$$

where $\mathbf{S} = \boldsymbol{\sigma}/2$ is the spin of the electron. In non-relativistic mechanics, a charged particle with charge e and angular momentum $\mathbf{L}$ has a magnetic moment

$$\boldsymbol{\mu} = \frac{e}{2m}\mathbf{L}\,. \tag{3.187}$$

It is then customary to write the magnetic moment due to the spin as

$$\boldsymbol{\mu} = \frac{ge}{2m}\mathbf{S}\,, \tag{3.188}$$

where g is called the gyromagnetic ratio, and we see that the Dirac equation predicts $g = 2$ while non-relativistic physics erroneously suggests $g = 1$. The present experimental value is (see the Introduction) $(g-2)/2 = 0.001\,159\,652\,187(4)$ and the deviations from $g = 2$ come from loop corrections that we will discuss in Chapter 7 and, in detail, in Solved Problem 7.2.

3.7 Solved problems

Problem 3.1. The fine structure of the hydrogen atom

In this problem we use the Dirac equation to compute the fine structure of the hydrogen atom. In this case the external potential A_μ is just the Coulomb potential of the nucleus of charge $-Ze$ (with $Z = 1$ for hydrogen, but it takes no effort to keep Z generic; recall also that $e < 0$), therefore $\mathbf{A} = 0$ and

$A_0 = -Ze/(4\pi r)$. The Coulomb potential is $V(r) = eA_0 = -Z\alpha/r$. The Dirac equation in the standard representation becomes

$$(i\partial_0 - V - m)\phi = -i\boldsymbol{\sigma}\cdot\nabla\chi\,, \tag{3.189}$$

$$(i\partial_0 - V + m)\chi = -i\boldsymbol{\sigma}\cdot\nabla\phi\,. \tag{3.190}$$

We look for a solution $\phi(\mathbf{x},t) = e^{-iEt}\phi(\mathbf{x})$, $\chi(\mathbf{x},t) = e^{-iEt}\chi(\mathbf{x})$ and we define $\varepsilon = E - m$. Then

$$(\varepsilon - V)\phi = -i\boldsymbol{\sigma}\cdot\nabla\chi\,, \tag{3.191}$$

$$(2m + \varepsilon - V)\chi = -i\boldsymbol{\sigma}\cdot\nabla\phi\,. \tag{3.192}$$

We now want to perform an expansion in powers of p^2/m^2 of the Dirac equation, keeping corrections $O(p^2/m^2)$ to the kinetic term $p^2/(2m)$ and to the potential V, i.e. we want to keep terms up to $O(p^4/m^3)$ and $O(Vp^2/m^2)$. Equation (3.192) allows us to eliminate χ using

$$\chi = \frac{-i}{2m+\varepsilon-V}\boldsymbol{\sigma}\cdot\nabla\phi = \frac{1}{2m}\left(1+\frac{\varepsilon-V}{2m}\right)^{-1}\boldsymbol{\sigma}\cdot\mathbf{p}\phi \simeq \frac{1}{2m}\left(1-\frac{\varepsilon-V}{2m}\right)\boldsymbol{\sigma}\cdot\mathbf{p}\phi\,. \tag{3.193}$$

We can then obtain an equation of the Schrödinger type for the two-component spinor ϕ. In order to make contact with a Schrödinger equation, however, we must also ensure that the wave function that appears in the Schrödinger equation is properly normalized. To this purpose, we observe that the total charge, in units of e, is given by eq. (3.172),

$$Q = \int d^3x\,\Psi^\dagger\Psi = \int d^3x\,\left[|\phi|^2 + |\chi|^2\right]\,. \tag{3.194}$$

Observe that, in first quantization, Q is positive definite. This will not be true in second quantization, where Q will be the number of electrons minus the number of positrons, as we will see in eq. (4.43). In the first quantized formalism, we require that the wave function expresses the condition that there is one electron in a volume V (with $V \to \infty$). We therefore define a Schrödinger wave function ϕ_S (again a two-components spinor) requiring that

$$\int_V d^3x\,|\phi_S|^2 = \int_V d^3x\,\left[|\phi|^2 + |\chi|^2\right]\,. \tag{3.195}$$

Since $\chi = O(\frac{p}{m}\phi)$, at zeroth order $\phi_S = \phi$. However, we want to substitute eq. (3.193) into eq. (3.191) keeping the first-order correction, so for consistency we must use $\chi \simeq (1/2m)(-i\boldsymbol{\sigma}\cdot\nabla)\phi$ in eq. (3.195). Then to this order

$$\begin{aligned}\int_V d^3x\,|\phi_S|^2 &= \int_V d^3x\,\left[|\phi|^2 + \frac{1}{4m^2}(\boldsymbol{\sigma}\cdot\nabla\phi^*)(\boldsymbol{\sigma}\cdot\nabla\phi)\right]\\ &= \int_V d^3x\,\left[|\phi|^2 - \frac{1}{4m^2}\phi^*(\boldsymbol{\sigma}\cdot\nabla)(\boldsymbol{\sigma}\cdot\nabla)\phi\right]\\ &= \int_V d^3x\,\left[|\phi|^2 - \frac{1}{4m^2}\phi^*\nabla^2\phi\right]\\ &= \int_V d^3x\,\phi^*\left(1+\frac{p^2}{4m^2}\right)\phi\,.\end{aligned} \tag{3.196}$$

Therefore

$$\phi_S = \left(1 + \frac{p^2}{8m^2} + O(\frac{p^4}{m^4})\right)\phi \tag{3.197}$$

and

$$\phi = \left(1 - \frac{p^2}{8m^2} + O(\frac{p^4}{m^4})\right) \phi_S\,. \tag{3.198}$$

In terms of ϕ_S, keeping only the first-order corrections, eq. (3.193) reads

$$\begin{aligned}\chi &\simeq \frac{1}{2m}\left(1 - \frac{\varepsilon - V}{2m}\right)\boldsymbol{\sigma}\cdot\mathbf{p}\left(1 - \frac{p^2}{8m^2}\right)\phi_S \\ &\simeq \frac{1}{2m}\left[\boldsymbol{\sigma}\cdot\mathbf{p}\left(1 - \frac{p^2}{8m^2}\right) - \frac{\varepsilon - V}{2m}\boldsymbol{\sigma}\cdot\mathbf{p}\right]\phi_S\,. \end{aligned}\tag{3.199}$$

We now substitute eqs. (3.199) and (3.198) into eq. (3.191). Performing some simple algebra (and paying attention to the fact that $\mathbf{p} = -i\nabla$ does not commute with the potential $V(r)$!) we get

$$\left[\varepsilon - \frac{p^2}{2m} - V + \varepsilon\frac{p^2}{8m^2} + \frac{p^4}{16m^3} + \frac{Vp^2}{8m^2} - \frac{1}{4m^2}(\boldsymbol{\sigma}\cdot\mathbf{p})V(\boldsymbol{\sigma}\cdot\mathbf{p})\right]\phi_S = 0\,. \tag{3.200}$$

At lowest order, we have of course $\varepsilon\phi_S \simeq (\frac{p^2}{2m} + V)\phi_S$. Therefore the term $\varepsilon\frac{p^2}{8m^2}$ in the above equation can be rewritten as

$$\varepsilon\frac{p^2}{8m^2} = \frac{p^2}{8m^2}\varepsilon \simeq \frac{p^2}{8m^2}\left(\frac{p^2}{2m} + V\right)\,. \tag{3.201}$$

Of course ε is a c-number and we can write it both to the left or to the right of p^2. When we substitute it with $p^2/(2m) + V$ the difference between writing it to the left or to the right is $O(p^6/m^4)$ and therefore can be neglected, at the order at which we are working. Equation (3.200) then becomes

$$\varepsilon\phi_S = \left[\frac{p^2}{2m} + V - \frac{p^4}{8m^3} + \frac{1}{4m^2}\left((\boldsymbol{\sigma}\cdot\mathbf{p})V(\boldsymbol{\sigma}\cdot\mathbf{p}) - \frac{1}{2}(p^2V + Vp^2)\right)\right]\phi_S\,. \tag{3.202}$$

The correction term involving the potential can be rewritten in a more transparent form using the identity $\sigma^i\sigma^j = \delta^{ij} + i\epsilon^{ijk}\sigma^k$, together with

$$[p^i, V] = -i(\nabla^i V) = ieE^i \tag{3.203}$$

(where $\mathbf{E}$ is the electric field) and

$$p^iVp^i = (Vp^i + [p^i, V])p^i = Vp^2 + ie\mathbf{E}\cdot\mathbf{p}\,. \tag{3.204}$$

Then

$$\begin{aligned}\sigma^i\sigma^jp^iVp^j - \frac{1}{2}(p^2V + Vp^2) &= p^iVp^i + i\epsilon^{ijk}\sigma^kp^iVp^j - \frac{1}{2}(p^2V + Vp^2) \\ &= Vp^2 + ie\mathbf{E}\cdot\mathbf{p} - \frac{1}{2}p^2V - \frac{1}{2}Vp^2 + i\epsilon^{ijk}\sigma^k([p^i, V] + Vp^i)p^j \\ &= ie\mathbf{E}\cdot\mathbf{p} + \frac{1}{2}(Vp^2 - p^2V) - e\,\epsilon^{ijk}E^ip^j\sigma^k\,. \end{aligned}\tag{3.205}$$

(In the last line we used the fact that $\epsilon^{ijk}Vp^ip^j = 0$ because p^ip^j is a symmetric tensor; note that this could not be used directly on $\epsilon^{ijk}p^iVp^j$ because p^i and V do not commute). Using eq. (3.203) it is easy to see that

$$Vp^2 - p^2V = -ie(\mathbf{E}\cdot\mathbf{p} + \mathbf{p}\cdot\mathbf{E}) \tag{3.206}$$

(again, one has to be careful since $\mathbf{E}$ and $\mathbf{p}$ do not commute!). Inserting this into eq. (3.205) we find

$$\begin{aligned}\sigma^i\sigma^j p^i V p^j - \frac{1}{2}(p^2V + Vp^2) &= ie\mathbf{E}\cdot\mathbf{p} - \frac{ie}{2}(\mathbf{E}\cdot\mathbf{p} + \mathbf{p}\cdot\mathbf{E}) - e\,\epsilon^{ijk}E^ip^j\sigma^k \\ &= \frac{ie}{2}(\mathbf{E}\cdot\mathbf{p} - \mathbf{p}\cdot\mathbf{E}) - e\,\epsilon^{ijk}E^ip^j\sigma^k \\ &= -\frac{e}{2}\,(\nabla\cdot\mathbf{E}) - e(\mathbf{E}\times\mathbf{p})\cdot\boldsymbol{\sigma}\,. \end{aligned} \tag{3.207}$$

Plugging this into eq. (3.202) we get

$$\varepsilon\phi_S = \left[\frac{p^2}{2m} + V - \frac{p^4}{8m^3} - \frac{e}{4m^2}\boldsymbol{\sigma}\cdot(\mathbf{E}\times\mathbf{p}) - \frac{e}{8m^2}\,(\nabla\cdot\mathbf{E})\right]\phi_S\,. \tag{3.208}$$

All the manipulations that we have performed until now hold for a generic potential $V(\mathbf{x})$. We now use the fact that in the hydrogen atom $V = V(r)$ and therefore

$$e\mathbf{E} = -\nabla V = -\mathbf{r}\left(\frac{1}{r}\frac{dV}{dr}\right), \tag{3.209}$$

so that

$$-\frac{e}{4m^2}\boldsymbol{\sigma}\cdot(\mathbf{E}\times\mathbf{p}) = \frac{1}{2m^2}\frac{1}{r}\frac{dV}{dr}\mathbf{S}\cdot(\mathbf{r}\times\mathbf{p}) = \frac{1}{2m^2}\frac{1}{r}\frac{dV}{dr}\mathbf{S}\cdot\mathbf{L} \tag{3.210}$$

where $\mathbf{S} = \boldsymbol{\sigma}/2$ is the spin of the electron and $\mathbf{L}$ is the orbital angular momentum. Therefore, in a radial potential $V(r)$, the first relativistic correction to the Schrödinger equation is given by

$$\boxed{\varepsilon\phi_S = \left[\frac{p^2}{2m} + V - \frac{p^4}{8m^3} + \frac{1}{2m^2}\frac{1}{r}\frac{dV}{dr}\mathbf{S}\cdot\mathbf{L} - \frac{e}{8m^2}\,(\nabla\cdot\mathbf{E})\right]\phi_S\,.} \tag{3.211}$$

The correction term $-p^4/(8m^3)$ is easily understood, since it comes from the expansion of the relativistic expression $\varepsilon = (p^2 + m^2)^{1/2}$. The term $\sim \mathbf{S}\cdot\mathbf{L}$ is the spin–orbit coupling and the term $\sim \nabla\cdot\mathbf{E}$ is known as the Darwin term. Restricting now to the Coulomb potential

$$V(r) = -\frac{Z\alpha}{r} \tag{3.212}$$

we have

$$\frac{1}{r}\frac{dV}{dr} = \frac{Z\alpha}{r^3}\,. \tag{3.213}$$

Using

$$\nabla^2\frac{1}{r} = -4\pi\delta^{(3)}(\mathbf{x}) \tag{3.214}$$

(see, e.g. Jackson (1975), Section 1.7 for the proof) we find

$$-e\nabla\cdot\mathbf{E} = +\nabla^2 V = -Z\alpha\nabla^2\frac{1}{r} = 4\pi Z\alpha\,\delta^{(3)}(\mathbf{x})\,. \tag{3.215}$$

We can therefore write

$$\varepsilon\phi_S = (H_0 + H_{\rm pert})\phi_s \tag{3.216}$$

where $H_0 = p^2/(2m) + V$ is the unperturbed Hamiltonian of the hydrogen atom and

$$\boxed{H_{\rm pert} = -\frac{p^4}{8m^3} + \frac{Z\alpha}{2m^2r^3}\,\mathbf{S}\cdot\mathbf{L} + \frac{\pi Z\alpha}{2m^2}\,\delta^{(3)}(\mathbf{x})\,.} \tag{3.217}$$

Denoting by $|njl\rangle$ the unperturbed states of the hydrogen atom, to first order in perturbation theory the correction to the energy levels is given by

$$(\Delta E)_{njl} = \langle njl|H_{\rm pert}|njl\rangle \,. \tag{3.218}$$

We must therefore compute the following expectation values:

(1) $\langle njl|p^4|njl\rangle$: if ψ_{njl} is a solution of the unperturbed Schrödinger equation, then by definition $(p^2/(2m)+V)\psi_{njl} = \epsilon_n \psi_{njl}$, or

$$\frac{p^2}{2m}\psi_{njl} = \left(\epsilon_n + \frac{Z\alpha}{r}\right)\psi_{njl} \tag{3.219}$$

where

$$\epsilon_n = -\frac{mZ^2\alpha^2}{2n^2} \tag{3.220}$$

are the unperturbed energy levels. Therefore

$$\int d^3x\, \psi^*_{njl}\, p^4 \psi_{njl} = 4m^2 \langle njl| \left(\epsilon_n + \frac{Z\alpha}{r}\right)^2 |njl\rangle \,. \tag{3.221}$$

For a Coulomb potential $V = -Z\alpha/r$ one has

$$\langle njl|\,\frac{1}{r}\,|njl\rangle = \frac{m\alpha Z}{n^2}\,, \qquad \langle njl|\,\frac{1}{r^2}\,|njl\rangle = \frac{(m\alpha Z)^2}{n^3(l+\frac{1}{2})} \tag{3.222}$$

and therefore

$$\langle njl|\,p^4\,|njl\rangle = 4(mZ\alpha)^4\left(-\frac{3}{4n^4} + \frac{1}{n^3(l+\frac{1}{2})}\right) \,. \tag{3.223}$$

(2) $\langle njl|\mathbf{S}\cdot\mathbf{L}/r^3|njl\rangle$: from $\mathbf{J}=\mathbf{L}+\mathbf{S}$ it follows that

$$j(j+1) = l(l+1) + s(s+1) + 2\mathbf{S}\cdot\mathbf{L} \tag{3.224}$$

with $s = 1/2$, and using the wave function of the hydrogen atoms one has

$$\langle njl|\,\frac{1}{r^3}\,|njl\rangle = \frac{(m\alpha Z)^3}{n^3 l(l+\frac{1}{2})(l+1)}\,, \qquad \text{if } l\neq 0 \tag{3.225}$$

and $\langle njl|1/r^3|njl\rangle = 0$ if $l=0$. Therefore

$$\langle njl|\frac{1}{r^3}\mathbf{S}\cdot\mathbf{L}|njl\rangle = (1-\delta_{l,0})\frac{(m\alpha Z)^3}{2n^3 l(l+\frac{1}{2})(l+1)}[j(j+1)-l(l+1)-\frac{3}{4}]\,. \tag{3.226}$$

(3) $\langle njl|\delta^3(\mathbf{x})|njl\rangle$: this is easily computed:

$$\langle njl|\delta^3(\mathbf{x})|njl\rangle = \int d^3x |\psi_{njl}(\mathbf{x})|^2\delta^3(\mathbf{x}) = |\psi_{njl}(0)|^2 = \frac{(m\alpha Z)^3}{\pi n^3}\delta_{l,0}\,. \tag{3.227}$$

Putting all contributions together and considering the two cases $j = l\pm 1/2$ when $l\neq 0$, and $j=1/2$ when $l=0$, we find that the result can always be expressed only in terms of n, j, and there is no separate dependence on l. The final result is

$$\boxed{(\Delta E)_{njl} = -\frac{m(Z\alpha)^4}{2n^3}\left[\frac{1}{j+\frac{1}{2}} - \frac{3}{4n}\right]\,.} \tag{3.228}$$

Therefore the fine structure removes the degeneracy between states with the same principal quantum number n but different values of j. However, states with the same n, j and different l, as the states $2S_{1/2}$ and $2P_{1/2}$, are still degenerate at the level of the Dirac equation, i.e. at the level of the first relativistic correction. In principle one might look for higher-order corrections coming from the Dirac equation, using perturbation theory with respect to H_{pert} at higher orders (indeed, it is even possible to find a closed form for the energy levels predicted by the Dirac equation to all orders in α), but physically this is not meaningful since, starting from the next order, the corrections due to the quantum nature of the electromagnetic field come into play, and the correct framework for computing these corrections is quantum electrodynamics, rather than the Dirac equation where A_μ is treated as an external, given, classical field.

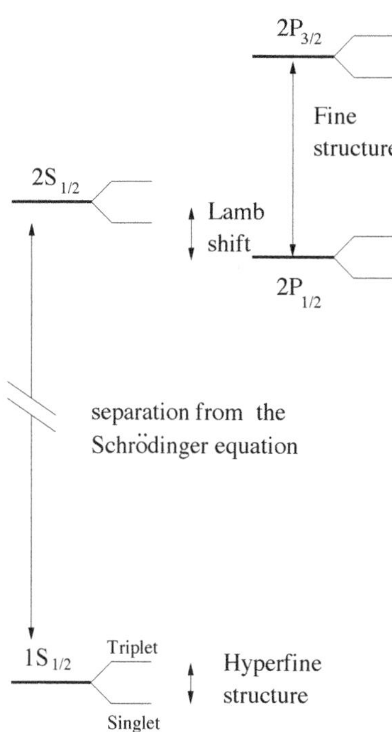

Fig. 3.1 The lowest lying energy levels of the hydrogen atom. Note that the figure is *not* to scale. In reality, the fine structure splittings are smaller by a factor $\sim 10^{-5}$ compared to the separation between the levels $2S_{1/2}$ and $1S_{1/2}$, and the Lamb shift and the hyperfine structure are smaller by a factor ~ 10 compared to the fine structure.

The structure of the energy levels of the hydrogen atom, including the fine structure correction (3.228) is shown in Fig. 3.1. For instance, the separation between the states $2P_{3/2}$ and $2P_{1/2}$ of the hydrogen is, from eq. (3.228)

$$E_{2P_{3/2}} - E_{2P_{1/2}} = -\frac{m\alpha^4}{16}\frac{1}{8} + \frac{m\alpha^4}{16}\frac{5}{8} = \frac{m\alpha^4}{32} \simeq 4.53 \times 10^{-5}\,\text{eV}\,, \tag{3.229}$$

corresponding to a frequency $f = \omega/(2\pi) \simeq 10.9$ GHz, in the domain of microwaves. Actually, the levels $2S_{1/2}$ and $2P_{1/2}$ are not exactly degenerate, as predicted by eq. (3.228), but rather have a splitting

$$E_{2S_{1/2}} - E_{2P_{1/2}} \simeq 1057\,\text{MHz} \tag{3.230}$$

know as the Lamb shift. The explanation for this splitting was, historically, one of the first successes of QED. At a comparable level, we find the hyperfine structure, due to the interaction between the spin of the nucleus and the spin of the electron. Each level then splits into a triplet and a singlet and, for instance,

$$E_{1S_{1/2},\text{triplet}} - E_{1S_{1/2},\text{singlet}} \simeq 1420.4\,\text{MHz}\,. \tag{3.231}$$

The corresponding wavelength is $\lambda = c/f \simeq 21.105$ cm, in the radio waves. This line is of great importance in astrophysics for investigating the presence of neutral hydrogen in our and in other galaxies because radio waves, compared to most other wavelengths, are much less affected by absorption in the interstellar medium, and propagate to a very large distance.

Problem 3.2. Relativistic energy levels in a magnetic field

We consider now an electron in a magnetic field $\mathbf{B} = B\mathbf{z}$ (we take $B > 0$). For the gauge field we can take $A_0 = A_x = A_z = 0$ and $A_y = Bx$. It is even technically simpler to solve the Dirac equation in this external magnetic field exactly, rather than performing a non-relativistic expansion. However, one should recall that only the first-order non-relativistic correction is really correct, and at higher orders effects from quantum field theory come into play.

We write the Dirac equation in the standard representation as

$$(i\partial_0 - m)\phi = \boldsymbol{\sigma}\cdot(\mathbf{p} - e\mathbf{A})\chi \tag{3.232}$$

$$(i\partial_0 + m)\chi = \boldsymbol{\sigma}\cdot(\mathbf{p} - e\mathbf{A})\phi \tag{3.233}$$

where, as usual, $\mathbf{p} = -i\nabla$. We look for a solution of the form

$$\phi(x) = \phi(\mathbf{x})e^{-iEt}\,, \qquad \chi(x) = \chi(\mathbf{x})e^{-iEt}\,. \tag{3.234}$$

Then eqs. (3.232) and (3.233) become

$$(E-m)\phi(\mathbf{x}) = \boldsymbol{\sigma}\cdot(\mathbf{p}-e\mathbf{A})\chi(\mathbf{x}) \tag{3.235}$$

$$(E+m)\chi(\mathbf{x}) = \boldsymbol{\sigma}\cdot(\mathbf{p}-e\mathbf{A})\phi(\mathbf{x}) \tag{3.236}$$

Substituting $\chi(\mathbf{x})$ from eq. (3.236) into eq. (3.235) and performing basically the same manipulations as in Section 3.6, we find[15]

$$\begin{aligned}(E^2-m^2)\phi(\mathbf{x}) &= \left[(\mathbf{p}-e\mathbf{A})^2 - e\boldsymbol{\sigma}\cdot\mathbf{B}\right]\phi(\mathbf{x})\\ &= \left[\mathbf{p}^2 + e^2B^2x^2 - 2ep_yBx - e\sigma_zB\right]\phi(\mathbf{x})\,.\end{aligned} \tag{3.237}$$

[15] Observe that with our choice of $\mathbf{A}$ we have $[p^i, A^j] = 0$ and therefore in $(\mathbf{p}-e\mathbf{A})^2$ we do not have to be careful about ordering. Otherwise the mixed term in $(\mathbf{p}-e\mathbf{A})^2$ is $-e(\mathbf{p}\cdot\mathbf{A}+\mathbf{A}\cdot\mathbf{p})$.

Since p_y, p_z commute with x, we can search for a solution of the form

$$\phi(\mathbf{x}) = e^{i(p_yy+p_zz)}f(x)\,, \tag{3.238}$$

where p_y and p_z are c-numbers and $f(x)$, as $\phi(\mathbf{x})$, is a two-component spinor. The equation for $f(x)$ becomes

$$\left[-\frac{d^2}{dx^2} + (p_y - eBx)^2 - eB\sigma_z\right]f(x) = (E^2 - m^2 - p_z^2)f(x)\,. \tag{3.239}$$

We take $f(x)$ to be an eigenfunction of σ_z with eigenvalue $\sigma = \pm 1$, $\sigma_z f = \sigma f$. Then

$$\left[-\frac{d^2}{dx^2} + \frac{1}{2}(2e^2B^2)(x - \frac{p_y}{eB})^2\right]f(x) = (E^2 - m^2 - p_z^2 + eB\sigma)f(x)\,. \tag{3.240}$$

This is formally identical to the Schrödinger equation of a harmonic oscillator with frequency $2|e|B$. The energy levels therefore are given by

$$E^2 - m^2 - p_z^2 + eB\sigma = (n+\frac{1}{2})2|e|B\,, \tag{3.241}$$

or (using $e = -|e|$)

$$\boxed{E(n,p_z,\sigma) = \left[m^2 + p_z^2 + (2n+1+\sigma)|e|B\right]^{1/2}\,.} \tag{3.242}$$

We observe that there is a continuous degeneracy in p_x and p_y, as well as a discrete degeneracy $E(n,p_z,\sigma=+1) = E(n+1,p_z,\sigma=-1)$. In the non-relativistic limit $p_z^2 \ll m^2, (2n+1)|e|B \ll m^2$, the expansion of eq. (3.242) gives

$$E(n,p_z,\sigma) \simeq m + \frac{p_z^2}{2m} + \left(n + \frac{1+\sigma}{2}\right)\omega_B \tag{3.243}$$

with $\omega_B = |e|B/m$, and we recover the Landau levels of non-relativistic quantum mechanics.

Summary of chapter

- The classical dynamics of a field theory is given by the Euler–Lagrange equation (3.13).
- Noether's theorem states that for any continuous global symmetry there is a current j^μ which is conserved, i.e. $\partial_\mu j^\mu = 0$. Given the symmetry transformation in the form (3.19, 3.20), the current can be computed using eq. (3.31). Given a conserved current, the charge given in eq. (3.26) is time-independent.

- Invariance under space-time translations leads, via Noether's theorem, to the conservation of the energy–momentum tensor, given in eq. (3.35). The corresponding conserved charges are energy and momentum, see eq. (3.37).
- The kinetic term of the actions for the scalar or spinor fields are derived from the requirement of Poincaré invariance. This leads to the free Klein–Gordon equation for scalar particles (eq. (3.40)) and to the free Weyl or Dirac equations for spin 1/2 particles, eqs. (3.67) and (3.87). For the vector fields, there is also an issue of gauge invariance; the equations of motion give a pair of Maxwell equations, while the second pair is a consequence of the definition of $F^{\mu\nu}$, see eqs. (3.141) and (3.144).
- There is large freedom in the choice of interaction terms. For instance, we can add a generic potential $V(\phi)$ to the action of a scalar field, eq. (3.58). For the electromagnetic field, we have seen minimal and non-minimal couplings in Section 3.5.4. When we study the quantum theory, we will see that some choices of interaction terms can be of more fundamental significance than others.
- Relativistic wave equations reduce, in the non-relativistic limit, to a Schrödinger equation plus corrections. We can then treat them in first quantization. We have seen that in this way two remarkable predictions are obtained from the Dirac equation: the gyromagnetic factor of the electron is predicted to be $g = 2$, in contrast with $g = 1$ from classical physics, and the fine structure of the hydrogen atom can be correctly computed. However, higher-order relativistic corrections can only be computed in the framework of second quantization.

Exercises

(3.1) Actions have the same dimensions as $\hbar$, so they are dimensionless in our $\hbar = 1$ units. Find the dimensions of scalar, spinor, and gauge fields in $d = 4$ dimensions. Repeat the analysis in d dimensions, with $S = \int d^dx\,\mathcal{L}$ and $\mathcal{L}$ the same as in the $d = 4$ case.

(3.2) We saw on page 59 that the solution of the massive Dirac equation in the rest frame is $u_L = u_R \equiv \sqrt{m}\,\xi$. Perform a boost on this solution along the z axis and verify that the result is given by eq. (3.103).

(3.3) Consider the KG action with the mass term set to zero,

$$S = \frac{1}{2}\int d^4x\,\eta^{\mu\nu}\partial_\mu\phi\partial_\nu\phi\,, \qquad (3.244)$$

and a *dilatation* transformation with parameter α,

$$\begin{aligned} x^\mu &\to x'^\mu = e^\alpha\, x^\mu \\ \phi(x) &\to \phi'(x') = \phi(x)\exp\{-d_\phi\alpha\}\,. \end{aligned} \qquad (3.245)$$

(i) Show that this transformation is a global symmetry, for an appropriate choice of the parameter d_ϕ. Find the Noether current associated to this symmetry and verify that it is conserved on the equations of motions.

(ii) Show that, if in the KG action we have also a non-vanishing mass term, then the above transformation is not a symmetry. Show that instead a term $V(\phi) = \lambda\phi^4$ does not spoil the dilatation symmetry. What are the dimensions of λ ?

(3.4) (i) Consider the Lagrangian of QED with the electron mass m set to zero, and consider the dilatation $x^\mu \to e^\alpha x^\mu$, $A_\mu \to e^{-d_A \alpha} A_\mu$, $\psi \to e^{-d_\psi \alpha} \psi$. Find the values of d_A, d_ψ for which this transformation is a symmetry.

(ii) Compute the conserved current and express it in terms of the energy–momentum tensor of the theory. Verify that the conservation of the dilatation current follows from the fact that the trace of the energy–momentum tensor vanishes in the massless theory.

(iii) Include the electron mass term in the Lagrangian. Compute the dilatation current using eq. (3.31) with the new Lagrangian. Verify that it is not conserved and relate its divergence to the trace of the energy–momentum tensor.

(3.5) Consider the two Dirac Lagrangians

$$\mathcal{L} = \bar{\psi}(i\partial\!\!\!/ - m)\psi\,, \quad \mathcal{L}' = \bar{\psi}(\frac{i}{2}\overleftrightarrow{\partial\!\!\!/} - m)\psi\,. \quad (3.246)$$

Verify that they are classically equivalent. Compute the energy–momentum tensor in the two cases. Verify that they are different, but they give rise to the same conserved charges.

(3.6) Consider a five-dimensional space-time labeled by $(t, \mathbf{x}, y)$ where $(t, \mathbf{x})$ are the usual coordinates of four-dimensional space-time and y is a compact coordinate which parametrizes the extra dimension, with $-R/2 \leqslant y \leqslant R/2$. Consider the free KG equation in this space, $(\Box_5 + m^2)\phi = 0$, where $\Box_5 \equiv \Box - \partial^2/\partial y^2$ and $\Box = \partial_0^2 - \boldsymbol{\nabla}^2$ is the usual four-dimensional d'Alambertian. Show that, from the point of view of a four-dimensional observer, this equation describes an infinite set of massive particles, and compute their masses. These particles are known as Kaluza–Klein modes. What do you think is the experimental bound on the size R of the extra dimension?

Quantization of free fields

4

4.1 Scalar fields

4.1.1 Real scalar fields. Fock space

From the basic principles of quantum mechanics we know that, to quantize a classical system with coordinates q^i and momenta p^i, in the Schrödinger picture, we promote q^i, p^i to operators and we impose the commutation relation $[q^i, p^j] = i\delta^{ij}$. In the Heisenberg picture, where the operators depend on time, the commutation relation is imposed at equal time. The same principle can be applied to a scalar field theory, where the coordinates $q^i(t)$ are replaced by the fields $\phi(t, \mathbf{x})$ while $p^i(t)$ are replaced by the conjugate momenta $\Pi(t, \mathbf{x})$, and we interpret $\mathbf{x}$ as a label that distinguishes the "coordinates" $\phi(t, \mathbf{x})$ of our system. Since $\mathbf{x}$ is a continuous variable, δ_{ij} must be replaced by a Dirac delta. Thus, the basic principle of canonical quantization is to promote the field ϕ and its conjugate momentum to operators, and to impose the equal time commutation relation

$$\boxed{[\phi(t, \mathbf{x}), \Pi(t, \mathbf{y})] = i\delta^{(3)}(\mathbf{x} - \mathbf{y}),} \tag{4.1}$$

while, at equal time, we impose $[\phi(t, \mathbf{x}), \phi(t, \mathbf{y})] = [\Pi(t, \mathbf{x}), \Pi(t, \mathbf{y})] = 0$. Furthermore, a real field is promoted to a hermitian operator. Let us first apply this procedure to a free real scalar field. The mode expansion of a free real scalar field is given in eq. (3.42). Promoting the real field ϕ to a hermitian operator means to promote $a_{\mathbf{p}}$ to an operator while $a^*_{\mathbf{p}}$ becomes the hermitian conjugate operator $a^\dagger_{\mathbf{p}}$; thus

$$\phi(x) = \int \frac{d^3p}{(2\pi)^3 \sqrt{2E_{\mathbf{p}}}} \left(a_{\mathbf{p}} e^{-ipx} + a^\dagger_{\mathbf{p}} e^{ipx} \right), \tag{4.2}$$

with $p^0 = E_{\mathbf{p}}$. The conjugate momentum is given by eq. (3.43). Using these expressions it is easy to verify that, in terms of $a_{\mathbf{p}}, a^\dagger_{\mathbf{p}}$, the commutation relation (4.1) reads

$$\boxed{[a_{\mathbf{p}}, a^\dagger_{\mathbf{q}}] = (2\pi)^3 \delta^{(3)}(\mathbf{p} - \mathbf{q}),} \tag{4.3}$$

while

$$[a_{\mathbf{p}}, a_{\mathbf{q}}] = 0, \qquad [a^\dagger_{\mathbf{p}}, a^\dagger_{\mathbf{q}}] = 0. \tag{4.4}$$

It is sometimes convenient to put the system into a box of size L, so that the total volume $V = L^3$ is finite. This procedure regularizes divergences coming from the infinite-volume limit or, equivalently, from the small momentum region, and is an example of an *infrared cutoff.* In a finite box of size L, imposing periodic boundary conditions on the fields, the momenta take the discrete values $p^i = (2\pi/L)n^i$ with $n^i = 0, \pm 1, \pm 2, \ldots$, and therefore

$$\int d^3p \to \left(\frac{2\pi}{L}\right)^3 \sum_{\mathbf{n}} . \tag{4.5}$$

The condition $\int d^3p\, \delta^{(3)}(\mathbf{p}-\mathbf{q}) = 1$ then gives

$$\delta^{(3)}(\mathbf{p}-\mathbf{q}) \to \left(\frac{L}{2\pi}\right)^3 \delta_{\mathbf{p},\mathbf{q}} . \tag{4.6}$$

In particular, this implies that

$$(2\pi)^3 \delta^{(3)}(\mathbf{p}=0) \to V . \tag{4.7}$$

Recalling the standard commutation relation of the creation and annihilation operators of a harmonic oscillator, $[a, a^\dagger] = 1$, we see from eq. (4.3) that the commutation relations of the real scalar field are equivalent to that of a collection of harmonic oscillators, with one oscillator for each value of the momentum $\mathbf{p}$ (apart from a normalization factor $1/\sqrt{V}$ in $a_{\mathbf{p}}$ and $a_{\mathbf{p}}^\dagger$).

We can now construct the *Fock space* following the standard procedure for the harmonic oscillator: we interpret $a_{\mathbf{p}}$ as destruction operators and $a_{\mathbf{p}}^\dagger$ as creation operators, and we define a *vacuum state* $|0\rangle$ as the state annihilated by all destruction operators, so for all $\mathbf{p}$

$$\boxed{a_{\mathbf{p}}|0\rangle = 0 .} \tag{4.8}$$

We normalize the vacuum with $\langle 0|0\rangle = 1$. The generic state of the Fock space is obtained acting on the vacuum with the creation operators,

$$|\mathbf{p}_1, \ldots \mathbf{p}_n\rangle \equiv (2E_{\mathbf{p}_1})^{1/2} \ldots (2E_{\mathbf{p}_n})^{1/2} a_{\mathbf{p}_1}^\dagger \ldots a_{\mathbf{p}_n}^\dagger |0\rangle . \tag{4.9}$$

The factors $(2E_{\mathbf{p}})^{1/2}$ are a convenient choice of normalization. In particular, the one-particle states are

$$|\mathbf{p}\rangle = (2E_{\mathbf{p}})^{1/2} a_{\mathbf{p}}^\dagger |0\rangle . \tag{4.10}$$

From the commutation relations and eq. (4.8) we find that

$$\begin{aligned}
\langle \mathbf{p}_1|\mathbf{p}_2\rangle &= (2E_{\mathbf{p}_1})^{1/2}(2E_{\mathbf{p}_2})^{1/2}\langle 0|a_{\mathbf{p}_1} a_{\mathbf{p}_2}^\dagger|0\rangle \\
&= (2E_{\mathbf{p}_1})^{1/2}(2E_{\mathbf{p}_2})^{1/2}\langle 0|[a_{\mathbf{p}_1}, a_{\mathbf{p}_2}^\dagger]|0\rangle \\
&= 2E_{\mathbf{p}_1}(2\pi)^3 \delta^{(3)}(\mathbf{p}_1 - \mathbf{p}_2) .
\end{aligned} \tag{4.11}$$

The factors $(2E_{\mathbf{p}})^{1/2}$ in eq. (4.10) have been chosen so that in the above scalar product the combination $E_{\mathbf{p}}\delta^{(3)}(\mathbf{p}-\mathbf{q})$ appears, which is Lorentz invariant (see Exercise 4.2).

We now compute the energy of these states. The Hamiltonian can be written in terms of $a_{\mathbf{p}}, a_{\mathbf{p}}^\dagger$ substituting eq. (4.2) and the corresponding expression for Π into the Hamiltonian density (3.44), and integrating over d^3x; one finds

$$H = \int \frac{d^3p}{(2\pi)^3} E_p \frac{1}{2} \left(a_{\mathbf{p}}^\dagger a_{\mathbf{p}} + a_{\mathbf{p}} a_{\mathbf{p}}^\dagger \right) = \int \frac{d^3p}{(2\pi)^3} E_p \left(a_{\mathbf{p}}^\dagger a_{\mathbf{p}} + \frac{1}{2} [a_{\mathbf{p}}, a_{\mathbf{p}}^\dagger] \right) . \tag{4.12}$$

The second term is the sum of the zero-point energy of all oscillators, and it is proportional to $(2\pi)^3 \delta^{(3)}(0)$. In a finite volume we see from eq. (4.7) that $(2\pi)^3 \delta^{(3)}(0) \to V$. The zero-point energy is therefore

$$E_{\text{vac}} = \frac{1}{2} V \int \frac{d^3p}{(2\pi)^3} E_p , \tag{4.13}$$

and the energy density of the vacuum is

$$\rho_{\text{vac}} \equiv \frac{1}{V} E_{\text{vac}} = \frac{1}{2} \int \frac{d^3p}{(2\pi)^3} E_p . \tag{4.14}$$

For large $|\mathbf{p}|$, $E_p = \sqrt{\mathbf{p}^2 + m^2} \simeq |\mathbf{p}|$ and the integral diverges. We can regulate the divergence putting a cutoff Λ in the integration over large momenta, so that we integrate only over $|\mathbf{p}| < \Lambda$. This is an example of an *ultraviolet cutoff*. The vacuum energy density then diverges as

$$\rho_{\text{vac}} \sim \int^{\Lambda} p^3 dp \sim \Lambda^4 . \tag{4.15}$$

This is our first encounter with an ultraviolet divergence. In quantum field theory we will get used to divergences and we will see under what conditions these can be cured. In this case, however, the divergence apparently is relatively harmless. Since what we measure are energy differences, we can simply discard this zero-point energy[1] and declare that our Hamiltonian is

$$H = \int \frac{d^3p}{(2\pi)^3} E_{\mathbf{p}} a_{\mathbf{p}}^\dagger a_{\mathbf{p}} . \tag{4.16}$$

[1] This is not true when we include gravity, since any form of energy contributes to the gravitational interaction. In Section 5.7, after having studied the renormalization of field theory, we will come back to this zero-point energy and we will discuss its relation with the cosmological constant problem.

We can formalize this statement introducing the concept of *normal ordering*: given an operator $\mathcal{O}$ its normal ordered form, denoted by $:\mathcal{O}:$, is obtained writing by hand all creation operators to the left of all destruction operators. Thus, for instance, $: a_{\mathbf{p}} a_{\mathbf{p}}^\dagger : = a_{\mathbf{p}}^\dagger a_{\mathbf{p}}$ and we can say that the quantum Hamiltonian (4.16) is obtained from the classical expression (3.44) promoting ϕ to an operator and performing the normal ordering,

$$H = \frac{1}{2} \int d^3x \, : \Pi^2 + (\nabla\phi)^2 + m^2\phi^2 : \; . \tag{4.17}$$

We can now compute the energy of the various states of the Fock space. The vacuum state $|0\rangle$ now, by definition, has zero energy. The operator $a_{\mathbf{p}}^\dagger a_{\mathbf{p}}$ is just the number operator of the oscillator labeled by $\mathbf{p}$ and therefore the energy of the generic state (4.9) is given by the sum of the energies $E_{\mathbf{p}_i}$ of the various particles,

$$H|\mathbf{p}_1, \dots \mathbf{p}_n\rangle = (E_{\mathbf{p}_1} + \dots + E_{\mathbf{p}_n})\,|\mathbf{p}_1, \dots \mathbf{p}_n\rangle\,. \tag{4.18}$$

Similarly we can compute the spatial momentum of these states. From the Noether theorem, we know how to write the spatial momentum as the conserved charge associated to spatial translations. For the real scalar field we found it in Section 3.3.1. Performing the normal ordering we have the quantum expression

$$P^i = \int d^3x \, :\theta^{0i}: \; = \int d^3x \, :\partial_0\phi\partial^i\phi: \,. \tag{4.19}$$

Substituting ϕ from eq. (4.2) we see that the terms quadratic in the destruction operators vanish because they are given by an integral over d^3p of the function $p^i a_{-\mathbf{p}}\, a_{\mathbf{p}}$, which is odd under $\mathbf{p} \to -\mathbf{p}$. Similarly for the terms quadratic in the creation operators, and we are left with[2]

$$P^i = \int \frac{d^3p}{(2\pi)^3}\, p^i a^\dagger_{\mathbf{p}} a_{\mathbf{p}}\,. \tag{4.20}$$

[2] Actually, for the momentum operator it is not really necessary to perform the normal ordering, since the terms that come out from the commutators are odd under $\mathbf{p} \to -\mathbf{p}$ and cancel when we integrate over d^3p.

Therefore *the states* $a^\dagger_{\mathbf{p}}|0\rangle$ *are one-particle states with momentum* $\mathbf{p}$, *energy* $E_{\mathbf{p}} = \sqrt{\mathbf{p}^2 + m^2}$ *and mass* m. The generic state of the Fock space (4.9) is a multiparticle state, and its energy and momentum are the sum of the individual energies and momenta. From the fact that the creation operators commute between themselves we see that the multiparticle states (4.9) are symmetric under the exchange of any two particles, and therefore obey Bose–Einstein statistics. This is an example of the spin-statistics theorem, which states that particles with integer spin are bosons and particles with half-integer spin are fermions.

Finally, we can examine the angular momentum of these states. From the Noether theorem we found that for scalar fields the angular momentum operator has a part interpreted as orbital angular momentum, eq. (3.50), and that there is no intrinsic spin part. Therefore the quanta of the scalar field are spin-0 particles.

4.1.2 Complex scalar field; antiparticles

We now consider a free complex scalar field. Eq. (3.60) becomes

$$\phi(x) = \int \frac{d^3p}{(2\pi)^3\sqrt{2E_{\mathbf{p}}}} \left(a_{\mathbf{p}} e^{-ipx} + b^\dagger_{\mathbf{p}} e^{ipx}\right) \tag{4.21}$$

and the complex conjugate field ϕ^* becomes the hermitian conjugate operator,

$$\phi^\dagger(x) = \int \frac{d^3p}{(2\pi)^3\sqrt{2E_{\mathbf{p}}}} \left(a^\dagger_{\mathbf{p}} e^{ipx} + b_{\mathbf{p}} e^{-ipx}\right)\,. \tag{4.22}$$

Imposing the canonical commutation relation (4.1) gives

$$[a_{\mathbf{p}}, a^\dagger_{\mathbf{q}}] = [b_{\mathbf{p}}, b^\dagger_{\mathbf{q}}] = (2\pi)^3\delta^{(3)}(\mathbf{p} - \mathbf{q})\,, \tag{4.23}$$

while all other commutators $[a,a], [a^\dagger,a^\dagger], [b,b], [b^\dagger,b^\dagger]$ and all commutators between the $a, a^\dagger$ and $b, b^\dagger$ are equal to zero. The Fock space is constructed defining the vacuum state as the state annihilated by all $a_{\mathbf{p}}$ and $b_{\mathbf{p}}$, for each $\mathbf{p}$,

$$a_{\mathbf{p}}|0\rangle = b_{\mathbf{p}}|0\rangle = 0\,. \tag{4.24}$$

Acting with $a^\dagger_{\mathbf{p}}, b^\dagger_{\mathbf{p}}$ we generate the Fock space. After normal ordering, the Hamiltonian and spatial momentum are given by

$$H = \int \frac{d^3p}{(2\pi)^3}\, E_{\mathbf{p}}(a^\dagger_{\mathbf{p}}a_{\mathbf{p}} + b^\dagger_{\mathbf{p}}b_{\mathbf{p}})\,, \tag{4.25}$$

$$P^i = \int \frac{d^3p}{(2\pi)^3}\, p^i(a^\dagger_{\mathbf{p}}a_{\mathbf{p}} + b^\dagger_{\mathbf{p}}b_{\mathbf{p}})\,. \tag{4.26}$$

We see that the quanta of a complex scalar are given by two different particle species with the same mass, created by the $a^\dagger_{\mathbf{p}}$ and $b^\dagger_{\mathbf{p}}$ operators respectively.

The $U(1)$ charge is given in eq. (3.63). We compute it explicitly as a prototype of many similar calculations in this section,

$$\begin{aligned}
Q_{U(1)} &= i\int d^3x\, \phi^\dagger \overleftrightarrow{\partial_0}\, \phi = i\int d^3x \frac{d^3q}{(2\pi)^3\sqrt{2E_{\mathbf{q}}}}\frac{d^3p}{(2\pi)^3\sqrt{2E_{\mathbf{p}}}} \\
&\quad\times\Big[\left(a^\dagger_{\mathbf{q}}e^{iqx} + b_{\mathbf{q}}e^{-iqx}\right)\partial_0\left(a_{\mathbf{p}}e^{-ipx} + b^\dagger_{\mathbf{p}}e^{ipx}\right) \\
&\qquad - \left(\partial_0(a^\dagger_{\mathbf{q}}e^{iqx} + b_{\mathbf{q}}e^{-iqx})\right)\left(a_{\mathbf{p}}e^{-ipx} + b^\dagger_{\mathbf{p}}e^{ipx}\right)\Big] \\
&= \int d^3x \frac{d^3q}{(2\pi)^3\sqrt{2E_{\mathbf{q}}}}\frac{d^3p}{(2\pi)^3\sqrt{2E_{\mathbf{p}}}} \\
&\quad\times\Big[\left(a^\dagger_{\mathbf{q}}e^{iqx} + b_{\mathbf{q}}e^{-iqx}\right)E_{\mathbf{p}}\left(a_{\mathbf{p}}e^{-ipx} - b^\dagger_{\mathbf{p}}e^{ipx}\right) \\
&\qquad + E_{\mathbf{q}}\left(a^\dagger_{\mathbf{q}}e^{iqx} - b_{\mathbf{q}}e^{-iqx}\right)\left(a_{\mathbf{p}}e^{-ipx} + b^\dagger_{\mathbf{p}}e^{ipx}\right)\Big]\,.
\end{aligned} \tag{4.27}$$

The integration over d^3x produces $(2\pi)^3\delta^{(3)}(\mathbf{p}-\mathbf{q})$ on the terms $\sim a^\dagger_{\mathbf{q}}a_{\mathbf{p}}$ and $b_{\mathbf{q}}b^\dagger_{\mathbf{p}}$, and $(2\pi)^3\delta^{(3)}(\mathbf{p}+\mathbf{q})$ on the terms $\sim b_{\mathbf{q}}a_{\mathbf{p}}$ and $a^\dagger_{\mathbf{q}}b^\dagger_{\mathbf{p}}$; in both cases $|\mathbf{p}| = |\mathbf{q}|$ and therefore $E_{\mathbf{q}} = E_{\mathbf{p}}$. Using this fact it is straightforward to find that the terms $\sim b_{\mathbf{q}}a_{\mathbf{p}}$ and $a^\dagger_{\mathbf{q}}b^\dagger_{\mathbf{p}}$ cancel and we are left with the terms $a^\dagger_{\mathbf{q}}a_{\mathbf{p}}$ and $b_{\mathbf{p}}b^\dagger_{\mathbf{p}}$. In these terms the exponentials are of the type $\exp\{\pm i(q-p)x\} = \exp\{\pm i[(E_{\mathbf{q}} - E_{\mathbf{p}})t - (\mathbf{q}-\mathbf{p})\cdot\mathbf{x}]\}$, and since $E_{\mathbf{q}} = E_{\mathbf{p}}$ the time dependence cancels, as it should for a conserved charge. Therefore

$$\begin{aligned}
Q_{U(1)} &= \int \frac{d^3p}{(2\pi)^3\sqrt{2E_{\mathbf{p}}}}\frac{d^3q}{(2\pi)^3\sqrt{2E_{\mathbf{q}}}}(2\pi)^3\delta^{(3)}(\mathbf{p}-\mathbf{q})2E_{\mathbf{p}}(a^\dagger_{\mathbf{q}}a_{\mathbf{p}} - b_{\mathbf{q}}b^\dagger_{\mathbf{p}}) \\
&= \int \frac{d^3p}{(2\pi)^3}(a^\dagger_{\mathbf{p}}a_{\mathbf{p}} - b_{\mathbf{p}}b^\dagger_{\mathbf{p}})\,.
\end{aligned} \tag{4.28}$$

Again we normal order this expression,[3] and we obtain

$$\boxed{Q_{U(1)} = \int \frac{d^3p}{(2\pi)^3}(a^\dagger_{\mathbf{p}}a_{\mathbf{p}} - b^\dagger_{\mathbf{p}}b_{\mathbf{p}})\,.} \tag{4.29}$$

Since $a^\dagger a$ is the number operator of a harmonic oscillator, we see that the $U(1)$ charge is equal to the number of quanta created by the operators $a^\dagger_{\mathbf{p}}$ minus the number of quanta created by $b^\dagger_{\mathbf{p}}$, integrated over

[3] Even if one accepts the normal ordering of the Hamiltonian on the grounds that a vacuum energy is unobservable, one cannot accept the normal ordering of the charge on the same grounds, since a charged vacuum would have observable effects. Rather, in this case one can understand the need for normal ordering observing that the classical expression for the charge involves the product of ϕ^* with $\partial_0\phi$, eq. (3.63). When promoted to quantum operators, $\phi^\dagger$ and $\partial_0\phi$ do not commute, and therefore there is an ordering ambiguity already in the starting expression (4.27); for instance, we could write $\phi^\dagger\partial_0\phi$, or $(\partial_0\phi)\phi^\dagger$, or we could take the symmetric combination. The ambiguity is removed requiring that the charge of the vacuum vanishes.

all momenta. Therefore the states $a_{\mathbf{p}}^\dagger|0\rangle$ and $b_{\mathbf{p}}^\dagger|0\rangle$ represent particles with momentum $\mathbf{p}$, mass m, spin zero and opposite charge; $a_{\mathbf{p}}^\dagger|0\rangle$ has $Q_{U(1)} = +1$ while $b_{\mathbf{p}}^\dagger|0\rangle$ has $Q_{U(1)} = -1$ and is called the *antiparticle* of $a_{\mathbf{p}}^\dagger|0\rangle$.[4] We now understand what is the proper interpretation of the negative energy solutions of the KG equation. The coefficient of the positive energy solution e^{-ipx} after quantization becomes the destruction operator of a particle and the coefficient of e^{ipx} becomes the creation operator of its antiparticle. In the case of the real scalar field the reality condition requires $a_{\mathbf{p}} = b_{\mathbf{p}}$ and therefore the particle is its own antiparticle, and it is neutral under any $U(1)$ symmetry.

[4]Of course the overall sign (and normalization) of the Noether charge are arbitrary, since if a current j^μ is conserved also $-j^\mu$ is conserved, and it is also an arbitrary convention what state we call a particle and what an antiparticle.

4.2 Spin 1/2 fields

4.2.1 Dirac field

We start from the Lagrangian (3.89), $\mathcal{L} = \bar\Psi(i\gamma^\mu\partial_\mu - m)\Psi$. The conjugate momentum is

$$\Pi_\Psi = \frac{\delta\mathcal{L}}{\delta(\partial_0\Psi)} = i\bar\Psi\gamma^0 = i\Psi^\dagger\,. \tag{4.30}$$

A basic principle of quantum field theory is the spin-statistic theorem, that requires that fields with half-integer spin are quantized imposing equal time anticommutation relation, while spin with integer spin with equal time commutation relation. We will not discuss this theorem in full generality, but will see below how the need for anticommutators arises in the case of Dirac fields. So we impose

$$\boxed{\{\Psi_a(\mathbf{x},t), \Psi_b^\dagger(\mathbf{y},t)\} = \delta^{(3)}(\mathbf{x}-\mathbf{y})\delta_{ab}\,,} \tag{4.31}$$

where $\{\,,\}$ is the anticommutator and $a, b = 1, \ldots 4$ are the Dirac indices. The expansion of the free Dirac field in plane waves is written

$$\Psi(x) = \int \frac{d^3p}{(2\pi)^3\sqrt{2E_{\mathbf{p}}}} \sum_{s=1,2} \left(a_{\mathbf{p},s}u^s(p)e^{-ipx} + b_{\mathbf{p},s}^\dagger v^s(p)e^{ipx}\right)\,, \tag{4.32}$$

and therefore

$$\bar\Psi(x) = \int \frac{d^3p}{(2\pi)^3\sqrt{2E_{\mathbf{p}}}} \sum_{s=1,2} \left(b_{\mathbf{p},s}\bar v^s(p)e^{-ipx} + a_{\mathbf{p},s}^\dagger \bar u^s(p)e^{ipx}\right)\,. \tag{4.33}$$

The wave functions $u^s(p), v^s(p)$ are given in eqs. (3.103) and (3.107). Writing the anticommutation relations (4.31) in terms of the a, b operators we find

$$\boxed{\{a_{\mathbf{p}}^r, a_{\mathbf{q}}^{s\dagger}\} = \{b_{\mathbf{p}}^r, b_{\mathbf{q}}^{s\dagger}\} = (2\pi)^3\delta^{(3)}(\mathbf{p}-\mathbf{q})\delta^{rs}\,,} \tag{4.34}$$

with all other anticommutators equal to zero. The Fock space is constructed defining first a vacuum state annihilated by all destruction operators

$$a_{\mathbf{p},s}|0\rangle = b_{\mathbf{p},s}|0\rangle = 0\,. \tag{4.35}$$

Then multiparticle states are obtained acting on the vacuum with $a^\dagger_{\mathbf{p},s}$ or $b^\dagger_{\mathbf{p},s}$. Since these operators anticommute between themselves, the resulting multiparticle state is antisymmetric under the exchange of two particles, so spin 1/2 particles (as in general all half-integer spin particles) obey Fermi–Dirac statistics. The one-particle states are normalized as in the case of the scalar field,

$$(2E_{\mathbf{p}})^{1/2}\, a^\dagger_{\mathbf{p},s}|0\rangle\,, \qquad (2E_{\mathbf{p}})^{1/2}\, b^\dagger_{\mathbf{p},s}|0\rangle \tag{4.36}$$

and depend on the momentum as well as on the spin degree of freedom s, which takes the values $s = 1, 2$.

The Hamiltonian density is obtained computing first the classical expression,

$$\begin{aligned}\mathcal{H} &= \Pi_\Psi \partial_0 \Psi - \mathcal{L} = i\Psi^\dagger \partial_0 \Psi - \bar{\Psi}\left(i\gamma^0\partial_0 + i\gamma^i\partial_i - m\right)\Psi \\ &= \bar{\Psi}\left(-i\gamma^i\partial_i + m\right)\Psi\,,\end{aligned} \tag{4.37}$$

and therefore we get the Dirac Hamiltonian

$$H = \int d^3x\, \bar{\Psi}\left(-i\gamma^i\partial_i + m\right)\Psi = \int d^3x\, \bar{\Psi}\left(-i\boldsymbol{\gamma}\cdot\boldsymbol{\nabla} + m\right)\Psi\,. \tag{4.38}$$

We then substitute the mode expansion (4.32) and we perform the normal ordering, which in this case means that we put all $a^\dagger_{\mathbf{p},s}$ to the left of all $a_{\mathbf{p},s}$ and all $b^\dagger_{\mathbf{p},s}$ to the left of all $b_{\mathbf{p},s}$, adding a minus sign each time we exchange the position of any destruction or creation operator, but without paying the price of the Dirac delta; e.g. $: a_{\mathbf{p},s}a^\dagger_{\mathbf{p},s} : = -a^\dagger_{\mathbf{p},s}a_{\mathbf{p},s}$. The final result is

$$H = \int \frac{d^3p}{(2\pi)^3} \sum_{s=1,2} E_{\mathbf{p}}\left(a^\dagger_{\mathbf{p},s}a_{\mathbf{p},s} + b^\dagger_{\mathbf{p},s}b_{\mathbf{p},s}\right)\,. \tag{4.39}$$

If we were to quantize the Dirac field in terms of commutators, at this point we would have found a minus sign in front of the term $b^\dagger_{\mathbf{p},s}b_{\mathbf{p},s}$, and therefore the energy would have been unbounded from below. In this way, instead, we see that the situation becomes completely analogous to the complex scalar field, and the coefficients of the negative energy solutions e^{ipx} become the creation operators of another type of particle. Let us now study the momentum, spin and charge of these particles.

The momentum operator is again obtained from the Noether theorem,

$$\mathbf{P} = \int \frac{d^3p}{(2\pi)^3} \sum_{s=1,2} \mathbf{p}\left(a^\dagger_{\mathbf{p},s}a_{\mathbf{p},s} + b^\dagger_{\mathbf{p},s}b_{\mathbf{p},s}\right)\,. \tag{4.40}$$

The new aspect compared to the complex scalar field is the spin degree of freedom. The angular momentum is the Noether charge associated to spatial rotations and, as discussed in Section 2.6.2, it is made up of the orbital contribution plus a spin term. Using the expressions for the Noether current given in Section 2.6.2, the spin part is

$$\mathbf{S} = \frac{1}{2}\int d^3x\, \Psi^\dagger \mathbf{\Sigma}\, \Psi\,, \tag{4.41}$$

where

$$\Sigma^i = \begin{pmatrix} \sigma^i & 0 \\ 0 & \sigma^i \end{pmatrix}. \tag{4.42}$$

Substituting again the mode expansion one finds that, in the rest frame (i.e. when $\mathbf{p} = 0$), the state created by $a^\dagger_{\mathbf{p},s}$ with $s = 1$ has $J_z = +1/2$ and that with $s = 2$ has $J_z = -1/2$. For the state created by $b^\dagger_{\mathbf{p},s}$ the situation is reversed and $s = 1$ has $J_z = -1/2$ while $s = 2$ has $J_z = +1/2$. Performing a boost in the z direction, J_z is unchanged and therefore a state created by $a^\dagger_{\mathbf{p},s}$ with $\mathbf{p} = (0, 0, p_z)$ and $s = 1$ has helicity $h = +1/2$, etc.

Finally, we saw in Section 3.4.3 that the Dirac action has a $U(1)$ global symmetry $\Psi \to e^{i\alpha}\Psi$. The conserved charge is

$$Q_{U(1)} = \int \frac{d^3p}{(2\pi)^3} \sum_{s=1,2} \left(a^\dagger_{\mathbf{p},s} a_{\mathbf{p},s} - b^\dagger_{\mathbf{p},s} b_{\mathbf{p},s}\right)\,, \tag{4.43}$$

Table 4.1 The quantum numbers of the one-particles states created by $a^\dagger_{\mathbf{p},s}, b^\dagger_{\mathbf{p},s}$. The momentum $\mathbf{p}$ is directed by definition along the z direction. The electric charge is in units of e, with $e < 0$.

state	J_z	$U(1)$ charge
$a^\dagger_{\mathbf{p},1}\|0\rangle$	$+\frac{1}{2}$	$+1$
$a^\dagger_{\mathbf{p},2}\|0\rangle$	$-\frac{1}{2}$	$+1$
$b^\dagger_{\mathbf{p},1}\|0\rangle$	$-\frac{1}{2}$	-1
$b^\dagger_{\mathbf{p},2}\|0\rangle$	$+\frac{1}{2}$	-1

The one-particle states and their quantum numbers are summarized in Table 4.1. The states created by $a^\dagger_{\mathbf{p},s}$ are called particles and the states created by $b^\dagger_{\mathbf{p},s}$ are called antiparticles. In particular, in electrodynamics, we identify $a^\dagger_{\mathbf{p},s}|0\rangle$ with the electron and $b^\dagger_{\mathbf{p},s}|0\rangle$ with the positron. The $U(1)$ charge is equal to the number of particles minus the number of antiparticles.

4.2.2 Massless Weyl field

In this section we consider a massless Weyl field. Its quantization follows immediately from the quantization of the Dirac field. It is convenient to use a Dirac notation for the Weyl fields so, if ψ_L is a left-handed two-component Weyl field and ψ_R a right-handed Weyl field, we write, in the chiral representation

$$\Psi_L = \begin{pmatrix} \psi_L \\ 0 \end{pmatrix}, \qquad \Psi_R = \begin{pmatrix} 0 \\ \psi_R \end{pmatrix}. \tag{4.44}$$

We consider first Ψ_L. As with any Dirac field, we can expand it as in eq. (4.32),

$$\Psi_L(x) = \int \frac{d^3p}{(2\pi)^3\sqrt{2E_{\mathbf{p}}}} \sum_{s=1,2} \left(a_{\mathbf{p},s} u^s_L(p) e^{-ipx} + b^\dagger_{\mathbf{p},s} v^s_L(p) e^{ipx}\right). \tag{4.45}$$

However, by definition $u^s_L(p)$ and $v^s_L(p)$ are Dirac spinors which, in the chiral representation, have the two lowest components vanishing,

$$u^s_L = \begin{pmatrix} U^s_L \\ 0 \end{pmatrix}, \qquad v^s_L = \begin{pmatrix} V^s_L \\ 0 \end{pmatrix}. \tag{4.46}$$

Comparing with eqs. (3.105), (3.106) and (3.107) we see that when $s = 1$ we have $u_L^s = v_L^s = 0$, so only the term with $s = 2$ contributes in eq. (4.45).

Therefore

$$\Psi_L(x) = \int \frac{d^3p}{(2\pi)^3\sqrt{2E_{\mathbf{p}}}} \left(a_{\mathbf{p},2} u_L^2(p) e^{-ipx} + b_{\mathbf{p},2}^\dagger v_L^2(p) e^{ipx} \right), \qquad (4.47)$$

and

$$\bar{\Psi}_L(x) = \int \frac{d^3p}{(2\pi)^3\sqrt{2E_{\mathbf{p}}}} \left(a_{\mathbf{p},2}^\dagger \bar{u}_L^2(p) e^{ipx} + b_{\mathbf{p},2} \bar{v}_L^2(p) e^{-ipx} \right), \qquad (4.48)$$

We see from Table 4.1 that $b_{\mathbf{p},2}^\dagger$ creates an antiparticle with $h = +1/2$ while $a_{\mathbf{p},2}^\dagger$ creates a particle with $h = -1/2$. Therefore:

In the massless case, the operators $\Psi_L, \bar{\Psi}_L$ create or destroy a particle with $h = -1/2$ and its antiparticle with $h = +1/2$.

If we neglect the small masses indicated by the oscillations experiments, the neutrinos in the Standard Model are described by massless left-handed Weyl fields. The neutrino has $h = -1/2$ and the antineutrino has $h = +1/2$.

Repeating the analysis for Ψ_R we see that now only the $s = 1$ term survives and therefore the situation is reversed. A right-handed massless Weyl field describes a particle with $h = +1/2$ and its antiparticle with $h = -1/2$.

4.2.3 C, P, T

In Section 2.6.3 we studied how parity and charge conjugation act on a classical Dirac field. We now want to understand how they act on one-particle states or, equivalently, on the operator Ψ that represents a quantized Dirac field. Let us start with the parity operator P. Under parity the momentum $\mathbf{p} \to -\mathbf{p}$ while the spin s is unchanged since the angular momentum is a pseudovector. Then for a particle of type a we must have

Parity

$$P|\mathbf{p}, s; a\rangle = \eta_a |-\mathbf{p}, s; a\rangle . \qquad (4.49)$$

We have inserted an index a to label the type of particle and we have included the possibility of a constant phase factor η_a, since vectors in the Fock space which differ by a phase still represent the same physical state. We will call η_a the *intrinsic parity* of the particle a. Performing twice the parity transformation on a physical observable gives the identity operation; this is not yet sufficient to conclude that $\eta_a^2 = 1$, since the observables are built from an even number of fermionic operators. Therefore, the condition that P^2 is the identity on the physical observable only implies that either $\eta_a^2 = +1$ or $\eta_a^2 = -1$. However, it can be shown (see Weinberg (1995), page 125) that for all spin 1/2 particles except Majorana fermions it is possible to redefine the parity operation so that $\eta_a^2 = +1$, and therefore $\eta_a = \pm 1$. In the following we will restrict to this case. We will come back to Majorana fermions on page 94, and we will see that the intrinsic parity of Majorana fermions satisfies instead $\eta_a^2 = -1$, i.e. $\eta_a = \pm i$.

In order to implement (4.49) on a generic multiparticle state, the operators $a^\dagger_{\mathbf{p},s}, b^\dagger_{\mathbf{p},s}$ must satisfy $Pa^\dagger_{\mathbf{p},s} = \eta_a a^\dagger_{-\mathbf{p},s}P$ and $Pb^\dagger_{\mathbf{p},s} = \eta_b b^\dagger_{-\mathbf{p},s}P$, so that, for instance,

$$Pa^\dagger_{\mathbf{p},s}b^\dagger_{\mathbf{q},s'}|0\rangle = \eta_a a^\dagger_{-\mathbf{p},s}Pb^\dagger_{\mathbf{q},s'}|0\rangle = \eta_a\eta_b a^\dagger_{-\mathbf{p},s}b^\dagger_{-\mathbf{q},s'}|0\rangle \tag{4.50}$$

and similarly on a generic multiparticle state.[5] Using $P^2 = 1$, this means that

$$\boxed{Pa^\dagger_{\mathbf{p},s}P = \eta_a a^\dagger_{-\mathbf{p},s}\,, \qquad Pb^\dagger_{\mathbf{p},s}P = \eta_b b^\dagger_{-\mathbf{p},s}\,.} \tag{4.51}$$

[5] We are assuming that the vacuum is non-degenerate and invariant under parity, and therefore $P|0\rangle = \eta|0\rangle$. In this case, as part of the definition of the operator P, we can choose $\eta = +1$, so $P|0\rangle = |0\rangle$. The situation in which the vacuum state is degenerate and the parity operation sends a vacuum state into a different vacuum state is an example of spontaneous symmetry breaking, and will be treated in Chapter 11.

As mentioned in Section 2.7.2, the Wigner theorem states that we can implement this symmetry transformation by means of a unitary operator P. Then the conditions $PP^\dagger = P^\dagger P = 1$, together with $PP = 1$, give $P^\dagger = P$. Taking the hermitian conjugate of (4.51) and taking into account that we are restricting to $\eta_{a,b}$ real, we find

$$Pa_{\mathbf{p},s}P = \eta_a a_{-\mathbf{p},s}\,, \qquad Pb_{\mathbf{p},s}P = \eta_b b_{-\mathbf{p},s}\,. \tag{4.52}$$

Therefore

$$\Psi(x) \to \Psi'(x') = P\Psi(x)P\,, \tag{4.53}$$

with

$$P\Psi(x)P = \int\frac{d^3p}{(2\pi)^3\sqrt{2E_{\mathbf{p}}}}\sum_{s=1,2}\left(\eta_a a_{-\mathbf{p},s}u^s(p)e^{-ipx} + \eta_b b^\dagger_{-\mathbf{p},s}v^s(p)e^{ipx}\right). \tag{4.54}$$

We now change the integration variable $\mathbf{p}$ to $\mathbf{p}' = -\mathbf{p}$. This transformation does not change p^0, which is quadratic in $\mathbf{p}$, so $p^{0'} = p^0$. Then $\exp(ipx) = \exp(ip^0t - i\mathbf{px}) = \exp(ip^0t + i\mathbf{p}'\mathbf{x}) = \exp(ip'x')$ with $x' = (t, -\mathbf{x})$, and similarly $\exp(-ipx) = \exp(-ip'x')$. To understand how $u^s(p)$ and $v^s(p)$ transform if we change their argument from p to p' we can, without loss of generality, choose $\mathbf{p}$ along the z axis and use the explicit form given in eqs. (3.103) and (3.107). We see that the transformation $p^3 \to -p^3$ exchanges the upper and lower components in the chiral representation, and therefore can be written in terms of γ^0,

$$u^s(p) = \gamma^0 u^s(p')\,, \qquad v^s(p) = -\gamma^0 v^s(p')\,. \tag{4.55}$$

Therefore, renaming the integration variable $p' = p$,

$$P\Psi(x)P = \gamma^0\int\frac{d^3p}{(2\pi)^3\sqrt{2E_{\mathbf{p}}}}\sum_{s=1,2}\left(\eta_a a_{\mathbf{p},s}u^s(p)e^{-ipx'} - \eta_b b^\dagger_{\mathbf{p},s}v^s(p)e^{ipx'}\right). \tag{4.56}$$

We now require that the quantum operator Ψ is a representation of parity, up to a phase. From the above equation, we see that this is possible if and only if

$$\boxed{\eta_a = -\eta_b\,.} \tag{4.57}$$

This shows that *the intrinsic parity of a spin 1/2 particle and of its antiparticle are opposite.*[6]

The transformation law of the operator Ψ then becomes

$$\Psi(x) \to \Psi'(x') = \eta_a \gamma^0 \Psi(x') \,, \tag{4.58}$$

which, once we recall the form of γ^0 in the chiral representation, is in agreement with the classical result (2.90), plus the novel quantum effect of the intrinsic parity factor η_a.[7] Of course η_a cancels in any fermion bilinear involving only particles of one type. However, the relative phase factors of different particles can be observables,[8] and in particular the opposite sign of the parity of the particle and its antiparticle is observable. An interesting application is to the case of positronium, which is the bound state of an electron and positron, and is discussed in Exercise 4.1.

The situation should be compared with what happens to a complex scalar field. In this case repeating the same arguments we have again $Pa_{\mathbf{p}}P = \eta_a a_{-\mathbf{p}}$ and $Pb_{\mathbf{p}}P = \eta_b b_{-\mathbf{p}}$. We can go through the same steps, with the only difference that in $\phi(x)$ the annihilation and creation operators are multiplied simply by e^{-ipx} and e^{ipx}, respectively, while in $\Psi(x)$ they were multiplied by $u(p)e^{-ipx}$ and $v(p)e^{-ipx}$, respectively. Therefore for scalar fields we do not get the relative minus sign between η_a and η_b, which for Dirac fields originated in the different transformation properties of $u(p)$ and $v(p)$, see eq. (4.55). Then we find that the quantized complex scalar field ϕ gives a representation of parity if $\eta_a = +\eta_b$, so that *the intrinsic parity of a spin-0 particle and of its antiparticle are equal.*

[6] If we consider also Majorana fermions the phase η is no longer restricted to be real (see page 92), and repeating the above steps one finds $\eta_a = -\eta_b^*$.

[7] We have derived the transformation of Ψ working in the chiral representation. As already remarked below eq. (3.122), once the transformation has been written in terms of γ matrices, it holds in any representation; in this case, if $\Psi \to \gamma^0\Psi$, we have $U\Psi \to U\gamma^0\Psi = (U\gamma^0U^{-1})U\Psi$. However, $\Psi' = U\Psi$ is the spinor in the new representation and $(U\gamma^0U^{-1})$ is γ^0 in the new representation.

[8] More precisely, redefining a new parity operator $P' = P\exp\{i\alpha B + i\beta L + i\gamma Q\}$, where B, L, Q are the baryon number, lepton number and electric charge, respectively, we can always set the intrinsic parity of the neutron, proton and electron to the value $+1$. The intrinsic parities of other particles are then fixed, see Weinberg (1995), page 125.

Charge conjugation

The effect of charge conjugation on the classical Dirac field has been obtained in eq. (2.91), working in the chiral representation. In terms of γ matrices, eq. (2.91) reads $\Psi \to -i\gamma^2\Psi^*$. We now study how the charge conjugation C acts on one-particle states. Let us consider the following transformation of the operators $a^\dagger_{\mathbf{p},s}, b^\dagger_{\mathbf{p},s}$

$$\boxed{Ca_{\mathbf{p},s}C = \eta_C b_{\mathbf{p},s} \,, \qquad Cb_{\mathbf{p},s}C = \eta_C a_{\mathbf{p},s} \,.} \tag{4.59}$$

We limit for simplicity to $\eta_C = \pm 1$. As we saw in eq. (2.91), charge conjugation relates Ψ and Ψ^*. Therefore we need to know how $u(p), v(p)$ transform under complex conjugation. The result is[9]

$$u^s(p) = -i\gamma^2 (v^s(p))^* \tag{4.60}$$

and therefore (since $-i\gamma^2$ is real and $(\gamma^2)^2 = -1$) we also have

$$v^s(p) = -i\gamma^2 (u^s(p))^* \,. \tag{4.61}$$

We can now write

$$C\Psi(x)C = \eta_C \int \frac{d^3p}{(2\pi)^3\sqrt{2E_{\mathbf{p}}}} \sum_{s=1,2} \left(b_{\mathbf{p},s} u^s(p) e^{-ipx} + a^\dagger_{\mathbf{p},s} v^s(p) e^{ipx} \right)$$

[9] One can check this result on the explicit expressions (3.103, 3.107) with $\mathbf{p} = (0, 0, p^3)$. However in this frame only σ^3 appears, which is real, so $u(p), v(p)$ are real. To check that $u^s(p)$ is indeed equal to $-i\gamma^2(v^s(p))^*$ rather than to $-i\gamma^2 v^s(p)$ it suffices to consider also the case where $\mathbf{p} = (0, p^2, 0)$.

$$= -i\eta_C \gamma^2 \int \frac{d^3p}{(2\pi)^3 \sqrt{2E_{\mathbf{p}}}} \sum_{s=1,2} \left(b_{\mathbf{p},s} (v^s(p))^* e^{-ipx} + a^\dagger_{\mathbf{p},s} (u^s(p))^* e^{ipx} \right)$$

$$= -i\eta_C \gamma^2 \Psi^* \,. \tag{4.62}$$

We see that this transformation is just the charge conjugation operation defined on the classical field, apart from the quantum phase $\eta_C = \pm 1$ which depends on the particle type.

From eq. (4.59) we see that *charge conjugation exchanges the particle with the antiparticle.* The momentum $\mathbf{p}$ is unchanged by C and also the index s. Recall however from the previous section that the state created by $a^\dagger_{\mathbf{p},s}$ describes a particle with $J_z = +1/2$ when $s = 1$ and $J_z = -1/2$ when $s = 2$, while the state created by $b^\dagger_{\mathbf{p},s}$ describes a particle with $J_z = -1/2$ when $s = 1$ and $J_z = +1/2$ when $s = 2$ (see Table 4.1); J_z was defined as the spin in the rest frame, but if a particle has spin J_z in its rest frame, it also has the same value of J_z if we make a boost along the z direction, since the generators J_z and K_z commute. Then we see that charge conjugation transforms a fermion with momentum $\mathbf{p} = (0, 0, p^3)$ and $J_z = +1/2$ (hence helicity $h = 1/2$) into an antifermion with the same momentum but $J_z = -1/2$, which means $h = -1/2$. Therefore *charge conjugation reverses the helicity.*

Using the definition (4.62), one can verify (see Exercise 4.3) that the current changes sign under charge conjugation,

$$C \left(\bar{\Psi} \gamma^\mu \Psi \right) C = -\bar{\Psi} \gamma^\mu \Psi \,. \tag{4.63}$$

For Majorana spinors we found in eqs. (2.91) and (2.94) that $\Psi^c_M = \Psi_M$, i.e. $-i\gamma^2 \Psi^*_M = \Psi_M$. Then eq. (4.62) becomes $C\Psi_M C = \eta_C \Psi_M$. Expanding Ψ_M in terms of creation and annihilation operators and using eq. (4.59) we find $a_{\mathbf{p},s} = \eta_C b_{\mathbf{p},s}$. Therefore for a Majorana spinor the particle and the antiparticle are identical. As we already remarked in Section 2.6.4, the relation between Majorana spinors and Dirac spinors is similar to the relation between real scalar fields and complex scalar fields. In both cases we have a reality condition ($\phi = \phi^*$ for a scalar field and $\Psi = -i\gamma^2 \Psi^*$ for a Dirac field) which eliminates one half of the degrees of freedom, and identifies the particle with the antiparticle.[10]

[10] As already remarked in Section 2.6.4, a reality condition on a Dirac spinor cannot be imposed in the form $\Psi = \Psi^*$. In terms of Weyl spinors such a condition would imply $\psi_L = \psi^*_L$ and $\psi_R = \psi^*_R$; however, these conditions are not Lorentz invariant, as we see from eqs. (2.59) and (2.60).

As we mentioned in note 6, in the general case where the intrinsic parity η is not assumed to be real, the parity of a fermion and an antifermion are related by $\eta_a = -\eta^*_b$. Since for Majorana fermions the particle is the same as the antiparticle, we have $\eta_a = -\eta^*_a$ and therefore $\eta_a = \pm i$.

Time-reversal

Finally, we consider the time-reversal transformation T. The implementation of time reversal in quantum field theory is somewhat peculiar. In fact, the Wigner theorem states that a symmetry transformation can be implemented either by a linear unitary operator, which is the case that we have met until now, or by an anti-unitary and antilinear operator, i.e. by an operator U that, given two states $|a\rangle$ and $|b\rangle$ with scalar product $\langle a|b\rangle$, satisfies $\langle Ua|Ub\rangle = \langle a|b\rangle^*$ (instead of being equal to $\langle a|b\rangle$ as for a unitary operator) and, for c a complex constant, $Uc|a\rangle = c^* U|a\rangle$. Time reversal is indeed the case of a symmetry that can be implemented only by an anti-unitary and antilinear operator (see Peskin and Schroeder (1995), page 67).

We want to define T in such a way that $T\Psi T$ satisfies the time-reversed Dirac equation. Using the antilinearity of T, it can be shown that this can be obtained defining

$$Ta_{\mathbf{p},s}T = a_{-\mathbf{p},-s}\,, \qquad Tb_{\mathbf{p},s}T = b_{-\mathbf{p},-s} \tag{4.64}$$

where $a_{\mathbf{p},-s} \equiv (a_{\mathbf{p},2}, -a_{\mathbf{p},1})$ and $b_{\mathbf{p},-s} \equiv (b_{\mathbf{p},2}, -b_{\mathbf{p},1})$. Therefore T changes the sign of the momentum and flips the spin, as we expect for time reversal. On the Dirac field this gives

$$T\Psi(t,\mathbf{x})T = -\gamma^1\gamma^3\Psi(-t,\mathbf{x})\,. \tag{4.65}$$

We leave it as an exercise to the reader to show that $-\gamma^1\gamma^3\Psi(-t,\mathbf{x})$ indeed verifies the Dirac equation with $t \to -t$.

Now that we have defined C, P, and T on the field Ψ, we can ask whether the Lagrangian governing the dynamics of Ψ is invariant under these transformations. For the free Dirac action, one immediately sees that C, P and T are indeed symmetry operations, but it is easy to construct interaction terms with fermion bilinears and possibly with derivatives that violate C, P or T separately. However, it is impossible to write a Lorentz-invariant term that violates CPT. In fact, under the combined action of C, P and T, the fermion bilinears $\bar{\Psi}\Psi$, $i\bar{\Psi}\gamma^5\Psi$, and $\bar{\Psi}\sigma^{\mu\nu}\Psi$ are invariant while $\bar{\Psi}\gamma^\mu\Psi$ and $\bar{\Psi}\gamma^5\gamma^\mu\Psi$ change sign. To construct a quadratic Lorentz-invariant term, the free Lorentz indices in $\bar{\Psi}\sigma^{\mu\nu}\Psi$, $\bar{\Psi}\gamma^\mu\Psi$ and $\bar{\Psi}\gamma^5\gamma^\mu\Psi$ must be contracted with a derivative ∂_μ, while in quartic and higher-order terms the indices can also be contracted between the various fermion bilinears. Of course ∂_μ is invariant under C while under the combined action of P, T we have $\partial_\mu \to -\partial_\mu$. We see that each free Lorentz index in a fermion bilinear constructed with $\bar{\Psi}$ and Ψ carries a minus sign under CPT, and the same is true for the Lorentz index in ∂_μ. Therefore all possible Lorentz invariant terms, where all indices are contracted and therefore are even in number, are invariant under CPT. The fact that CPT is conserved, therefore, follows from the fact that it is impossible even to write down a Lorentz-invariant term that violates CPT.

This is an example of the CPT theorem, which states that, independently of the spin of the particle, a (local) Lorentz-invariant field theory with a hermitian Hamiltonian cannot violate CPT.

Since CPT exchanges a particle with the antiparticle, and is an exact symmetry, i.e. it commutes with the Hamiltonian, it implies that the mass of a particle and of its antiparticle must be exactly equal. Experimentally, this is verified to an extraordinary accuracy in the $K^0\bar{K}^0$ system, where the bound on the mass difference is

$$\frac{|m_{K^0} - m_{\bar{K}^0}|}{m_{K^0}} < 10^{-18}\,. \tag{4.66}$$

4.3 Electromagnetic field

4.3.1 Quantization in the radiation gauge

The quantization of the electromagnetic field presents new aspects. The core of the problem is that, because of gauge invariance, the field A_μ gives a redundant description. We have therefore two choices in the quantization procedure. The first possibility is to choose from the beginning a gauge such as the radiation gauge (3.150), which fixes completely the gauge freedom; in this case we work directly with the physical degrees of freedom, but the price that we have to pay is a loss of explicit Lorentz covariance and at the end of the quantization procedure we must verify that we have not really lost Lorentz symmetry. This quantization scheme will be discussed in this section. The second possibility is to work with the full gauge field A_μ. This will introduce some spurious degrees of freedom, which we will have to get rid of. This second quantization procedure will be discussed in the next section.

Thus in this section we choose the radiation gauge (3.150), that we recall here

$$A_0 = 0\,, \qquad \nabla\cdot\mathbf{A} = 0\,. \tag{4.67}$$

We have seen that in this gauge the equation of motion for the three residual components A^i is simply $\Box A^i = 0$, so the most general classical solution is

$$\mathbf{A} = \int \frac{d^3p}{(2\pi)^3\sqrt{2\omega_\mathbf{p}}} \sum_{\lambda=1,2} \left[\boldsymbol{\epsilon}(\mathbf{p},\lambda) a_{\mathbf{p},\lambda} e^{-ipx} + \boldsymbol{\epsilon}^*(\mathbf{p},\lambda) a^*_{\mathbf{p},\lambda} e^{ipx}\right]\,, \tag{4.68}$$

where, for the electromagnetic field, we use the notation $\omega_\mathbf{p} \equiv p^0$. Inserting this expansion in the equation of motion $\Box A^i = 0$ we get $p^2 = 0$ and therefore $\omega_\mathbf{p} = |\mathbf{p}|$; the gauge fixing condition $\nabla\cdot\mathbf{A} = 0$ requires instead $\boldsymbol{\epsilon}\cdot\mathbf{p} = 0$; this equation, for each fixed $\mathbf{p}$, has of course two independent solutions, the two orthogonal vectors, which we label by an index $\lambda = 1, 2$. The physical degrees of freedom of the electromagnetic field are described by two independent polarization vectors $\boldsymbol{\epsilon}(\mathbf{p}, 1)$ and $\boldsymbol{\epsilon}(\mathbf{p}, 2)$.

We now promote $\mathbf{A}$ to a hermitian operator, and we write

$$\mathbf{A}(x) = \int \frac{d^3p}{(2\pi)^3\sqrt{2\omega_\mathbf{p}}} \sum_{\lambda=1,2} \left[\boldsymbol{\epsilon}(\mathbf{p},\lambda) a_{\mathbf{p},\lambda} e^{-ipx} + \boldsymbol{\epsilon}^*(\mathbf{p},\lambda) a^\dagger_{\mathbf{p},\lambda} e^{ipx}\right]\,, \tag{4.69}$$

where now $a_{\mathbf{p},\lambda}, a^\dagger_{\mathbf{p},\lambda}$ are operators. Since for scalar fields we have understood that imposing on $a_{\mathbf{p},\lambda}, a^\dagger_{\mathbf{p},\lambda}$ the commutation relation of the harmonic oscillator allows us to interpret the states of the Fock space as particles, we impose the commutation relations

$$\boxed{[a_{\mathbf{p},\lambda}, a^\dagger_{\mathbf{q},\lambda'}] = (2\pi)^3\delta^{(3)}(\mathbf{p}-\mathbf{q})\delta_{\lambda\lambda'}} \tag{4.70}$$

and

$$[a_{\mathbf{p},\lambda}, a_{\mathbf{q},\lambda'}] = [a^\dagger_{\mathbf{p},\lambda}, a^\dagger_{\mathbf{q},\lambda'}] = 0\,. \tag{4.71}$$

Equations (4.70) and (4.71) are our defining rules for the quantization of the electromagnetic field in the radiation gauge. We now want to understand what this definition means in terms of the commutation relations of the fields A^i with their conjugate momenta. The momentum conjugate to A_0 is

$$\Pi_0 = \frac{\delta}{\delta(\partial_0 A_0)}\left(-\frac{1}{4}F_{\mu\nu}F^{\mu\nu}\right) = 0\,. \tag{4.72}$$

In this quantization scheme A_0 and Π_0 are equal to zero, and are not dynamical variables. The momentum conjugate to A_i is instead

$$\Pi^i = \frac{\delta}{\delta(\partial_0 A_i)}\left(-\frac{1}{4}F_{\mu\nu}F^{\mu\nu}\right) = \frac{\delta}{\delta(\partial_0 A_i)}\left(-\frac{1}{2}F_{0i}F^{0i}\right) = -F^{0i} = E^i\,. \tag{4.73}$$

One can verify that from the commutation relations (4.70) it follows that

$$[A^i(t,\mathbf{x}), E^j(t,\mathbf{y})] = -i\int \frac{d^3k}{(2\pi)^3}\, e^{i\mathbf{k}\cdot(\mathbf{x}-\mathbf{y})}\left(\delta^{ij} - \frac{k^i k^j}{\mathbf{k}^2}\right). \tag{4.74}$$

In the derivation one uses the relation

$$\frac{1}{2}\sum_{\lambda=1,2}\left(\epsilon^i(\mathbf{k},\lambda)\epsilon^{j\,*}(\mathbf{k},\lambda) + \epsilon^{i\,*}(-\mathbf{k},\lambda)\epsilon^j(-\mathbf{k},\lambda)\right) = \delta^{ij} - \frac{k^i k^j}{\mathbf{k}^2}\,. \tag{4.75}$$

This identity can be verified choosing a frame where $\mathbf{k} = (0,0,k)$. In this frame we can choose as orthogonal vectors the linear polarization vectors, i.e. $\epsilon(\mathbf{k},1) = (1,0,0)$ and $\epsilon(\mathbf{k},2) = (0,1,0)$, and eq. (4.75) is then trivially checked. The validity in any frame follows from the fact that both sides transform as tensors under rotations.[11] The integral on the right-hand side of eq. (4.74) is called a "transverse" Dirac delta and is denoted by $\delta^{ij}_{\rm tr}(\mathbf{x}-\mathbf{y})$, so we can write

$$[A^i(t,\mathbf{x}), E^j(t,\mathbf{y})] = -i\delta^{ij}_{\rm tr}(\mathbf{x}-\mathbf{y})\,. \tag{4.76}$$

Note that, were it not for the term $k^i k^j/\mathbf{k}^2$, the integral over d^3k in eq. (4.74) would give an ordinary Dirac delta, and we would have the standard equal time commutation relations.[12] The necessity of the term $k^i k^j/\mathbf{k}^2$ can be understood taking the divergence with respect to $\mathbf{x}$ of both sides of eq. (4.74); on the left-hand side we find $[\nabla\cdot\mathbf{A}(t,\mathbf{x}), \mathbf{E}(t,\mathbf{y})]$. However, since we have imposed the gauge condition $\nabla\cdot\mathbf{A} = 0$, this must vanish. Indeed, the divergence of the right-hand side vanishes thanks to the additional term $-k^i k^j/\mathbf{k}^2$ since, taking the divergence, inside the integral we get

$$k^i\left(\delta^{ij} - \frac{k^i k^j}{\mathbf{k}^2}\right) = k^j - k^j = 0\,. \tag{4.77}$$

For the same reason, taking now the divergence of eq. (4.76) with respect to $\mathbf{y}$, we get $[A^i(t,\mathbf{x}), \nabla\cdot\mathbf{E}(t,\mathbf{y})] = 0$. This means that $\nabla\cdot\mathbf{E}$ commutes

[11] The linear polarizations are real and therefore in this basis the above identity simplifies to $\sum_{\lambda=1,2}\epsilon^i(\mathbf{k},\lambda)\epsilon^j(\mathbf{k},\lambda) = \delta^{ij} - \frac{k^i k^j}{\mathbf{k}^2}$. However the form (4.75) holds also choosing as a basis the circular polarizations, i.e. $\epsilon(\mathbf{k},1) = (1/\sqrt{2})(1,i,0)$ and $\epsilon(\mathbf{k},2) = (1/\sqrt{2})(1,-i,0)$.

[12] Observe that E^i is the momentum conjugate to $A_i = -A^i$. Therefore the sign in eq. (4.76) is in agreement with eq. (4.1).

with all operators and therefore, even in the quantum theory of the free electromagnetic field, it is a c-number, so it is consistent to impose $\nabla\cdot\mathbf{E}=0$ as an operator equation. Classically, $\nabla\cdot\mathbf{E}=0$ is just a Maxwell equation in the absence of sources.[13]

We can now proceed with the standard construction of the Fock space. We define the vacuum of the Fock space from

$$a_{\mathbf{p},\lambda}|0\rangle = 0 \tag{4.78}$$

for all $\mathbf{p}$ and $\lambda=1,2$; the Fock space is then generated acting with the creation operators $a^\dagger_{\mathbf{p},\lambda}$. The quantum Hamiltonian is obtained normal ordering the classical expression (3.158),

$$H=\frac{1}{2}\int d^3x\, :\mathbf{E}^2+\mathbf{B}^2:=\int\frac{d^3k}{(2\pi)^3}\sum_{\lambda=1,2}\omega_{\mathbf{k}}\,a^\dagger_{\mathbf{k},\lambda}a_{\mathbf{k},\lambda}\,, \tag{4.79}$$

where $\omega_k=|\mathbf{k}|$. The momentum is obtained from the normal ordering of (3.159),

$$\mathbf{P}=\int d^3x\, :\mathbf{E}\times\mathbf{B}:=\int\frac{d^3k}{(2\pi)^3}\sum_{\lambda=1,2}\mathbf{k}\,a^\dagger_{\mathbf{k},\lambda}a_{\mathbf{k},\lambda}\,. \tag{4.80}$$

This shows that the state $a^\dagger_{\mathbf{k},\lambda}|0\rangle$ describes a particle with energy $\omega_{\mathbf{k}}$, momentum $\mathbf{k}$ and two polarization states $\lambda=1,2$. Since the dispersion relation is $\omega_{\mathbf{k}}=|\mathbf{k}|$, it has zero mass. To compute its spin we must first compute the angular momentum operator of the electromagnetic field using the Noether theorem, and then we can study its action on the one-particle states.

[13]In particular, it is the equation of motion obtained performing the variation with respect to A_0; note that to choose the gauge $A_0=0$ means that we set $A_0=0$ in the solutions of the equations of motion, not directly in the action, otherwise we lose this equation of motion. The classical solutions are defined by the fact that the action must be stationary with respect to all fields, including A_0. Observe also that the equation of motion $\nabla\cdot\mathbf{E}=0$ contains no time derivative. Therefore it is not an equation that determines the evolution of an initial field configuration, but rather a constraint on the possible initial field configurations.

The reader who wishes to skip the explicit calculation can go directly to eq. (4.88).

It is instructive to perform the computation explicitly. We consider a rotation in the (jk) plane and we call J^{jk} the associated conserved charge. The angular momentum along the i axis J^i is then given by $J^i=(1/2)\epsilon^{ijk}J^{jk}$. From the Noether's theorem J^{jk} is given by the integral of the $\mu=0$ component of a current $j^{\mu(jk)}$, given by eq. (3.31)

$$J^{jk}=\int d^3x\, j^{0(jk)}=\int d^3x\left[\frac{\partial\mathcal{L}}{\partial(\partial_0A_i)}(a^{\nu(jk)}\partial_\nu A_i-F_i^{(jk)})-a^{0(jk)}\mathcal{L}\right]. \tag{4.81}$$

For a space-time rotation the coefficients $a^{\mu(jk)}$ (here we denote them by a lower case letter in order not to create confusion with the gauge field) have been found in eq. (3.48), $a^\mu_{(\rho\sigma)}=\delta^\mu_\rho x_\sigma-\delta^\mu_\sigma x_\rho$ so in particular $a^{0(jk)}=0$ and $a^{l(jk)}=\delta^{lj}x^k-\delta^{lk}x^j$. The coefficients $a^{l(jk)}$ measure the variation of the vector x^l under rotation, $\delta x^l=\omega^{jk}a^l_{(jk)}$. The coefficients $F^{l(jk)}$ similarly measure the variation of the gauge field A^i under a rotation, and since A^i is a spatial vector its transformation law is the same as x^i, so that $F^{i(jk)}=\delta^{ij}A^k-\delta^{ik}A^j$. Therefore

$$\begin{aligned}J^{jk}&=\int d^3x\,\partial_0A^i\left[-(\delta^{lj}x^k-\delta^{lk}x^j)\partial^lA^i-(\delta^{ij}A^k-\delta^{ik}A^j)\right]\\&=\int d^3x\left[\partial_0A^i(x^j\partial^k-x^k\partial^j)A^i-(A^k\partial_0A^j-A^j\partial_0A^k)\right].\end{aligned} \tag{4.82}$$

The first term in the bracket is clearly the contribution from the orbital angular momentum. More precisely, it is the matrix element of the orbital angular momentum operator $L^{jk} = i(x^j\partial^k - x^k\partial^j)$ with the same scalar product used in the Klein–Gordon case; compare with eq. (3.50) and the discussion below it. We are now interested in the second term which, according to the discussion after eq. (2.84), is the spin part S^{ij}. Inserting the expansion (4.69) and performing the normal ordering we get

$$\begin{aligned} S^{ij} &= \int d^3x \, : A^i\partial_0 A^j - A^j\partial_0 A^i : \\ &= \int d^3x \frac{d^3k}{(2\pi)^3\sqrt{2\omega_{\mathbf{k}}}} \frac{d^3q}{(2\pi)^3\sqrt{2\omega_{\mathbf{q}}}} \sum_{\lambda'\lambda''} (i\omega_{\mathbf{q}}) \\ &\times \Big\{ : \left[\epsilon^i(\mathbf{k},\lambda')a_{\mathbf{k},\lambda'}e^{-ikx} + \epsilon^{i\,*}(\mathbf{k},\lambda')a^\dagger_{\mathbf{k},\lambda'}e^{ikx} \right] \\ &\times \left[-\epsilon^j(\mathbf{q},\lambda'')a_{\mathbf{q},\lambda''}e^{-iqx} + \epsilon^{j\,*}(\mathbf{q},\lambda'')a^\dagger_{\mathbf{q},\lambda''}e^{iqx} \right] : \Big\} - (i\leftrightarrow j) \,. \end{aligned} \tag{4.83}$$

The integration over $\mathbf{x}$ of the various terms gives $(2\pi)^3\delta^{(3)}(\mathbf{k}\pm\mathbf{q})$; then one finds that the terms $\sim aa$ and $a^\dagger a^\dagger$ cancel, while the terms obtained exchanging $(i\leftrightarrow j)$ gives a factor of two, so

$$S^{ij} = i\int \frac{d^3q}{(2\pi)^3} \sum_{\lambda'\lambda''} \left[\epsilon^i(\mathbf{q},\lambda'')\epsilon^{j\,*}(\mathbf{q},\lambda') - \epsilon^{i\,*}(\mathbf{q},\lambda')\epsilon^j(\mathbf{q},\lambda'') \right] a^\dagger_{\mathbf{q},\lambda'}a_{\mathbf{q},\lambda''} \,. \tag{4.84}$$

Now we apply this operator to the one-particle state $a^\dagger_{\mathbf{k},\lambda}|0\rangle$. Using

$$a_{\mathbf{q},\lambda''}a^\dagger_{\mathbf{k},\lambda}|0\rangle = [a_{\mathbf{q},\lambda''}\,,\, a^\dagger_{\mathbf{k},\lambda}]\,|0\rangle = (2\pi)^3\delta^{(3)}(\mathbf{k}-\mathbf{q})\delta_{\lambda''\lambda}|0\rangle \tag{4.85}$$

we find

$$S^{ij}a^\dagger_{\mathbf{k},\lambda}|0\rangle = i\sum_{\lambda'=1,2} \left[\epsilon^i(\mathbf{k},\lambda)\epsilon^{j\,*}(\mathbf{k},\lambda') - \epsilon^{i\,*}(\mathbf{k},\lambda')\epsilon^j(\mathbf{k},\lambda) \right] a^\dagger_{\mathbf{k},\lambda'}|0\rangle \,. \tag{4.86}$$

We choose $\mathbf{k} = (0,0,k)$ and we compute the spin along the z axis, i.e. $S^3 = S^{12}$. As a basis for the polarization vectors we choose the linear polarizations, $\boldsymbol{\epsilon}(\mathbf{k},1) = (1,0,0), \boldsymbol{\epsilon}(\mathbf{k},2) = (0,1,0)$; in components $\epsilon^i(\mathbf{k},\lambda) = \delta^i_\lambda$ and

$$S^3 a^\dagger_{\mathbf{k},\lambda}|0\rangle = i\sum_{\lambda'=1,2} (\delta^1_\lambda\delta^2_{\lambda'} - \delta^2_\lambda\delta^1_{\lambda'})a^\dagger_{\mathbf{k},\lambda'}|0\rangle \,. \tag{4.87}$$

The final result of this calculation is therefore

$$\begin{aligned} S^3 a^\dagger_{\mathbf{k},1}|0\rangle &= i a^\dagger_{\mathbf{k},2}|0\rangle \\ S^3 a^\dagger_{\mathbf{k},2}|0\rangle &= -i a^\dagger_{\mathbf{k},1}|0\rangle \,, \end{aligned} \tag{4.88}$$

with $\mathbf{k} = (0,0,k)$. We see that the linear polarizations are not eigenstates of the helicity. The eigenstates are given by the circular polarizations,

$$\begin{aligned} S^3 a^\dagger_{\mathbf{k},+}|0\rangle &= +a^\dagger_{\mathbf{k},+}|0\rangle \\ S^3 a^\dagger_{\mathbf{k},-}|0\rangle &= -a^\dagger_{\mathbf{k},-}|0\rangle \end{aligned} \tag{4.89}$$

where

$$a^\dagger_{\mathbf{k},\pm} = \frac{1}{\sqrt{2}}(a^\dagger_{\mathbf{k},1} \pm i a^\dagger_{\mathbf{k},2}) \,. \tag{4.90}$$

The conclusion is that *the states $a^\dagger_{\mathbf{k},\pm}|0\rangle$ describe particles with momentum* $\mathbf{k}$, energy $\omega_k = |\mathbf{k}|$, *mass zero, spin 1, and helicity* ± 1. These quanta are the photons. The fact that massless particles are helicity eigenstates and that there is no state with $J_z = 0$ is in agreement with our general discussion of the representation of the Poincaré group in Section 2.7.2.

Our quantization procedure did not maintain Lorentz covariance, since we broke it from the beginning with our gauge choice. We must therefore now ask whether at the end Lorentz invariance is recovered. The fact that we found a particle which fits within the representations of the Poincaré group already indicates that the final result is compatible with Poincaré (and therefore Lorentz) invariance. To make sure that indeed the theory has Lorentz invariance what we actually have to do is to construct all generators of the Poincaré group in terms of the creation and annihilation operators. We already wrote explicitly the energy, momentum and the spin part of the angular momentum in eqs. (4.79), (4.80) and (4.84) and the reader can complete it computing the orbital part of the angular momentum, and the boost generator. Using the commutation relations of the creation and annihilation operators one can then check that these generators indeed close the Lorentz algebra, and that the one-particle states, under the transformations generated by these generators, transform as expected for a spin-1 massless particle. This proves the covariance of the quantization in the radiation gauge.[14]

Finally, we can define on the photon states the operations of parity and charge conjugation. Concerning parity, we have understood that the physical photon states are described by a vector field $\mathbf{A}(t,\mathbf{x})$, subject to the condition $\boldsymbol{\nabla}\cdot\mathbf{A} = 0$. The gauge field $\mathbf{A}$ is a true vector, as follows for instance from the fact that the electric field is a true vector and the magnetic field is a pseudovector, so under parity it transforms as

$$\mathbf{A}(t,\mathbf{x}) \to -\mathbf{A}(t,-\mathbf{x}) . \tag{4.91}$$

Expanding each of the three components $\mathbf{A}(t,\mathbf{x})$ in spherical harmonics, under parity the terms with orbital angular momentum L get the usual factor $(-1)^L$ from the transformation of the spherical harmonic Y_{LM} under $\mathbf{x} \to -\mathbf{x}$, plus an overall minus sign from the fact that $\mathbf{A}$ is a vector. In terms of photon states, this means that

$$P|\gamma;\mathbf{k},\mathbf{s}\rangle = -|\gamma;-\mathbf{k},\mathbf{s}\rangle , \tag{4.92}$$

where $\mathbf{k},\mathbf{s}$ are the momentum and spin of the photon γ. Therefore the intrinsic parity of a physical photon state is -1.

We saw in eq. (4.63) that the fermionic current changes sign under charge conjugation. Therefore, if we define C on the gauge field as

$$CA^\mu C = -A^\mu , \tag{4.93}$$

then charge conjugation is a symmetry of the QED Lagrangian. On the creation and annihilation operators of the photon, the above equation means that

$$Ca_{\mathbf{p},\lambda}C = -a_{\mathbf{p},\lambda} . \tag{4.94}$$

[14] A note for the advanced reader. In general, it is quite common in field theory that, in the quantization procedure, a symmetry of the classical Lagrangian is not explicitly preserved in the intermediate steps, and at the end of the quantization procedure one must check whether the symmetry is still present in the quantum theory. It turns out that it is not at all automatic that such a symmetry is recovered. If this does not happen, the symmetry is called *anomalous*, and the theory is said to have an *anomaly*. In QFT this can only happen as a consequence of the divergences of the interacting theory, that will be the subject of the next chapter, and in a free field theory, such as the free electromagnetic field that we are considering here, no anomaly can appear. For this reason, the recovery of Lorentz invariance in the quantization of the free electromagnetic field is guaranteed. It is however interesting to observe that in string theory there is no distinction between a free Lagrangian and an interaction term, i.e. the free propagation of the string fixes also the interaction. It is possible to quantize the theory in a way very similar to the quantization in radiation gauge of the electromagnetic field, breaking the explicit Lorentz covariance, and one finds that at the end Lorentz invariance is recovered only if the theory lives in 26 space-time dimensions (for the bosonic string) or in 10 space-time dimensions (for superstrings). See Polchinski (1998), Section 1.3, for a clear discussion.

By definition, we take $C|0\rangle = +|0\rangle$. Then, since $C^2 = 1$, we have

$$Ca_{\mathbf{p},\lambda}|0\rangle = (Ca_{\mathbf{p},\lambda}C)\,C|0\rangle = -a_{\mathbf{p},\lambda}|0\rangle\,. \tag{4.95}$$

Therefore the photon has charge conjugation -1.

4.3.2 Covariant quantization

In this section we take a different route for the quantization of the electromagnetic field. We do not want to spoil the covariance, so we do not impose the radiation gauge and we accept working with the redundant field A_μ. However, if we try to perform straightforwardly a covariant quantization of the Maxwell Lagrangian,

$$\mathcal{L} = -\frac{1}{4}F_{\mu\nu}F^{\mu\nu}\,, \tag{4.96}$$

we fail immediately. In fact, in a naive covariant quantization, we would first of all define the conjugate momenta as

$$\Pi^\mu(x) = \frac{\partial\mathcal{L}}{\partial(\partial_0 A_\mu)}\,, \tag{4.97}$$

Consider first the spatial components. The momentum conjugate to A^i is $\Pi_i = -\Pi^i$, with $\Pi^i = E^i$. Equation (4.1) would therefore suggest imposing the equal time commutation relations[15]

$$[A^i(t,\mathbf{x}),\Pi^j(t,\mathbf{y})] = -i\delta^{ij}\delta^{(3)}(\mathbf{x}-\mathbf{y}) \tag{4.98}$$

(the minus sign is due to the fact that the momentum conjugate to A^i is Π_i, and $\Pi_i = -\Pi^i$), while

$$[A^i(t,\mathbf{x}),A^j(t,\mathbf{y})] = 0\,. \tag{4.99}$$

These commutation relations have the covariant generalization

$$[A^\mu(t,\mathbf{x}),A^\nu(t,\mathbf{y})] = 0\,, \tag{4.100}$$

$$[A^\mu(t,\mathbf{x}),\Pi^\nu(t,\mathbf{y})] = i\eta^{\mu\nu}\delta^{(3)}(\mathbf{x}-\mathbf{y})\,. \tag{4.101}$$

The metric $\eta_{\mu\nu}$ is forced upon us from the condition of Lorentz covariance, since the left-hand side is a tensor. In a covariant quantization, one would therefore use eqs. (4.97), (4.100) and (4.101) as the starting point. However, eqs. (4.97) and (4.101) are incompatible, because in the Maxwell Lagrangian there is no dependence on $\partial_0 A_0$ and therefore Π^0 vanishes identically, and cannot have a non-trivial commutator with A^0.

To tackle this problem we proceed as follows. We start from a modified Lagrangian,

$$\mathcal{L}' = -\frac{1}{4}F_{\mu\nu}F^{\mu\nu} - \frac{1}{2}(\partial_\mu A^\mu)^2\,. \tag{4.102}$$

This Lagrangian at first sight seems to describe a very different theory compared to the Maxwell Lagrangian (4.96). Indeed, the Lagrangian

[15] Observe that in the quantization in radiation gauge of the previous section the commutator $[A^i(t,\mathbf{x}),\Pi^j(t,\mathbf{y})]$ was rather given in terms of a transverse Dirac delta, see eq. (4.76). This was a consequence of our gauge fixing, which eliminated from the beginning the longitudinal polarization vector, i.e. the vector $\boldsymbol{\epsilon}^i(\mathbf{k},3)$ which, in the frame where $\mathbf{k}=(0,0,k)$, has the form $\boldsymbol{\epsilon}^i(\mathbf{k},3) = (0,0,1)$. Because of this, the sum over the polarization gave the transverse tensor $\delta^{ij} - k^ik^j/\mathbf{k}^2$, see eq. (4.75). The difference with the covariant quantization is that now we are not fixing the gauge, and we keep for the moment all polarization vectors. Therefore, the sum over the spatial polarization vectors now gives δ^{ij} rather than $\delta^{ij} - k^ik^j/\mathbf{k}^2$.

(4.102) is not even gauge invariant. For the moment we postpone the question of what the Lagrangian (4.102) has to do with (4.96), and we proceed to its quantization. The conjugate momenta can now be defined straightforwardly

$$\Pi^\mu(x) = \frac{\partial \mathcal{L}'}{\partial(\partial_0 A_\mu)} \tag{4.103}$$

so that $\Pi^i = -F^{0i} = E^i$ as in the usual Maxwell Lagrangian (4.96), while $\Pi^0 = -\partial_\mu A^\mu$ is non-vanishing. It therefore makes perfectly sense to impose the canonical commutation relations (4.101).

The equation of motion derived from eq. (4.102) is simply $\Box A_\mu = 0$. The operators A_μ can therefore be expanded as

$$A_\mu(x) = \int \frac{d^3p}{(2\pi)^3\sqrt{2\omega_{\mathbf{p}}}} \sum_{\lambda=0}^{3} \left[\epsilon_\mu(\mathbf{p},\lambda)a_{\mathbf{p},\lambda}e^{-ipx} + \epsilon_\mu^*(\mathbf{p},\lambda)a^\dagger_{\mathbf{p},\lambda}e^{ipx}\right], \tag{4.104}$$

and the equation of motion $\Box A_\mu = 0$ translates into $p^2 = 0$. The important difference compared to the canonical quantization discussed in the previous section is that now there is no constraint on ϵ^μ; in the canonical quantization the conditions $\epsilon^0 = 0$ and $p_\mu \epsilon^\mu = 0$ came from the gauge choice, while here we start from a Lagrangian which is not even gauge invariant, and no constraint has been imposed on A_μ. Therefore we have four independent solutions for $\epsilon^\mu(\mathbf{p},\lambda)$, labeled by $\lambda = 0,1,2,3$. In the frame where $p^\mu = (p,0,0,p)$ we will choose as a basis

$$\begin{aligned} \epsilon^\mu(\mathbf{p},0) &= (1,0,0,0)\,, \quad \epsilon^\mu(\mathbf{p},1) = (0,1,0,0)\,, \\ \epsilon^\mu(\mathbf{p},2) &= (0,0,1,0)\,, \quad \epsilon^\mu(\mathbf{p},3) = (0,0,0,1)\,, \end{aligned} \tag{4.105}$$

or, more compactly, $\epsilon^\mu(\mathbf{p},\lambda) = \delta^\mu_\lambda$. In a generic frame the form of $\epsilon^\mu(\mathbf{p},\lambda)$ is found performing the appropriate Lorentz transformation. The two vectors $\epsilon^\mu(\mathbf{p},1)$ and $\epsilon^\mu(\mathbf{p},2)$ satisfy $\epsilon_\mu p^\mu = 0$, i.e. they are transverse. Instead $\epsilon^\mu(\mathbf{p},0)$ and $\epsilon^\mu(\mathbf{p},3)$ have $\epsilon_\mu p^\mu \neq 0$.

From the expansion (4.104) and the canonical commutation relations (4.101) it follows that

$$\boxed{[a_{\mathbf{p},\lambda}, a^\dagger_{\mathbf{q},\lambda'}] = -(2\pi)^3\delta^{(3)}(\mathbf{p}-\mathbf{q})\eta_{\lambda\lambda'}\,,} \tag{4.106}$$

with $\lambda, \lambda' = 0,1,2,3$, and $[a_{\mathbf{p},\lambda}, a_{\mathbf{q},\lambda'}] = [a^\dagger_{\mathbf{p},\lambda}, a^\dagger_{\mathbf{q},\lambda'}] = 0$. The crucial new point here is that the commutator (4.106) with $\lambda = \lambda' = 0$ has the "wrong" sign, because $\eta_{00} = +1$. The consequence of this sign apparently is a disaster. Consider the states

$$|\mathbf{p},\lambda\rangle \equiv (2\omega_{\mathbf{p}})^{1/2}\, a^\dagger_{\mathbf{p},\lambda}|0\rangle \tag{4.107}$$

and try to interpret them as one-particle states. The norm of these states is

$$\langle\mathbf{p},\lambda|\mathbf{p},\lambda\rangle = (2\omega_{\mathbf{p}})\langle 0|a_{\mathbf{p},\lambda}a^\dagger_{\mathbf{p},\lambda}|0\rangle = (2\omega_{\mathbf{p}})\langle 0|[a_{\mathbf{p},\lambda}, a^\dagger_{\mathbf{p},\lambda}]|0\rangle = -\eta_{\lambda\lambda}2\omega_{\mathbf{p}}V \tag{4.108}$$

(where, in a finite volume, $(2\pi)^3\delta^{(3)}(0) = V$, see eq. (4.7)). Therefore, the state created by the oscillator with $\lambda = 0$ has a negative norm! Since the scalar products in quantum mechanics are interpreted as probabilities, a Fock space with a scalar product which is not positive definite has no probabilistic interpretation.

On the other hand, we must observe that the states that create the problem have no counterpart in the canonical quantization of the electromagnetic field discussed in the previous section; we have seen in fact that the physical states are only those associated with transverse polarization vectors. The states created by $a^\dagger_{\mathbf{p},0}$ and by $a^\dagger_{\mathbf{p},3}$ (in the frame where $\mathbf{p}$ is along the third axis) are unphysical.

We must now recall that the theory that we have quantized so far is not electrodynamics, because of the extra term $(\partial_\mu A^\mu)^2$ in the Lagrangian. The basic idea of the covariant quantization of the electromagnetic field, or Gupta–Bleuler quantization, is to start from the apparently different theory (4.102) and to recover a quantum theory of the electromagnetic field imposing a restriction on the Fock space: we define the subspace of *physical states* requiring that for any two physical states $|\text{phys}\rangle$, $|\text{phys}'\rangle$

$$\langle \text{phys}'|\partial_\mu A^\mu|\text{phys}\rangle = 0\,. \tag{4.109}$$

In other words, $\partial_\mu A^\mu = 0$, rather than being imposed at the level of the Lagrangian (4.102), is recovered as an operator equation on physical states, and we expect that the quantization of the theory (4.102), supplemented with the condition (4.109), is equivalent to the canonical quantization studied in the previous section. We must therefore study whether imposing the condition (4.109) is sufficient to eliminate the states with negative norm from the physical space and whether the final result is a Fock space of transverse photons, as we expect from the previous section.

We first observe that the operator $\partial_\mu A^\mu$ can be separated into its positive and negative frequency parts,

$$\partial_\mu A^\mu = (\partial_\mu A^\mu)^+ + (\partial_\mu A^\mu)^- \tag{4.110}$$

where $(\partial_\mu A^\mu)^+$ contains only the positive frequency part, i.e. the annihilation operators, and $(\partial_\mu A^\mu)^-$ contains the creation operators,

$$(\partial_\mu A^\mu)^+ = -i\int \frac{d^3p}{(2\pi)^3\sqrt{2\omega_{\mathbf{p}}}}\sum_{\lambda=0}^{3} p_\mu \epsilon^\mu(\mathbf{p},\lambda) a_{\mathbf{p},\lambda} e^{-ipx} \tag{4.111}$$

$$(\partial_\mu A^\mu)^- = i\int \frac{d^3p}{(2\pi)^3\sqrt{2\omega_{\mathbf{p}}}}\sum_{\lambda=0}^{3} p_\mu \epsilon^{\mu *}(\mathbf{p},\lambda) a^\dagger_{\mathbf{p},\lambda} e^{ipx}\,. \tag{4.112}$$

Since $(\partial_\mu A^\mu)^- = (\partial_\mu A^\mu)^{+\dagger}$, eq. (4.109) is satisfied if we define the physical states from the condition

$$\boxed{(\partial_\mu A^\mu)^+|\text{phys}\rangle = 0\,,} \tag{4.113}$$

since then automatically $\langle \text{phys}|(\partial_\mu A^\mu)^- = 0$ for any physical state. Equation (4.113) will be taken as the definition of the physical subspace. Since it has the form of a linear operator applied to a state, it preserves the linear structure of the physical Hilbert space: if $|\text{phys}_1\rangle$ and $|\text{phys}_2\rangle$ are physical states, then $\alpha|\text{phys}_1\rangle + \beta|\text{phys}_2\rangle$ is a physical state.

Let us examine what this condition means for one-particle states. We consider a state $|\psi\rangle = \sum_\lambda c_\lambda a^\dagger_{\mathbf{k},\lambda}|0\rangle$, i.e. the most general superposition of polarization states with a given momentum $\mathbf{k}$, and we choose $\mathbf{k}$ along the third axis, $k^\mu = (k,0,0,k)$. On this state the physical state condition (4.113) becomes $c_0 + c_3 = 0$. We see that the two transverse photons $a^\dagger_{\mathbf{k},1}|0\rangle$ and $a^\dagger_{\mathbf{k},2}|0\rangle$, and any linear combination of them, are physical states. This is good news, since we know from the previous section that these are the true degrees of freedom of the photon. Consider now the subspace generated by $a^\dagger_{\mathbf{k},0}|0\rangle$ and $a^\dagger_{\mathbf{k},3}|0\rangle$. We see that neither $a^\dagger_{\mathbf{k},0}|0\rangle$ nor $a^\dagger_{\mathbf{k},3}|0\rangle$ are physical states. This is also good news, since the state $a^\dagger_{\mathbf{k},0}|0\rangle$ is just the negative norm state, and $a^\dagger_{\mathbf{k},3}|0\rangle$, even if it has a positive norm, does not correspond to a physical polarization state. However the combination

$$|\phi\rangle = (a^\dagger_{\mathbf{k},0} - a^\dagger_{\mathbf{k},3})|0\rangle \tag{4.114}$$

has $c_0 = +1, c_3 = -1$ and satisfies the physical state condition $c_0 + c_3 = 0$. Therefore the most general one-particle state of the physical subspace, with momentum $\mathbf{k}$, is of the form

$$|\psi_T\rangle + c|\phi\rangle \tag{4.115}$$

where $|\psi_T\rangle$ is an arbitrary linear combination of the transverse states $a^\dagger_{\mathbf{k},1}|0\rangle$ and $a^\dagger_{\mathbf{k},2}|0\rangle$, $|\phi\rangle$ is given by eq. (4.114) and c is an arbitrary constant. The question now is: what shall we do with $|\phi\rangle$, which has no counterpart in the canonical quantization?

First of all, observe that $|\phi\rangle$ has zero norm,

$$\begin{aligned}\langle\phi|\phi\rangle &= \langle 0|(a_{\mathbf{k},0} - a_{\mathbf{k},3})(a^\dagger_{\mathbf{k},0} - a^\dagger_{\mathbf{k},3})|0\rangle = \langle 0|a_{\mathbf{k},0}a^\dagger_{\mathbf{k},0} + a_{\mathbf{k},3}a^\dagger_{\mathbf{k},3}|0\rangle \\ &= \langle 0|[a_{\mathbf{k},0}, a^\dagger_{\mathbf{k},0}] + [a_{\mathbf{k},3}, a^\dagger_{\mathbf{k},3}]|0\rangle = 0\,,\end{aligned} \tag{4.116}$$

because the commutator $[a_{\mathbf{k},0}, a^\dagger_{\mathbf{k},0}]$ has the opposite sign of $[a_{\mathbf{k},3}, a^\dagger_{\mathbf{k},3}]$. This means that $|\phi\rangle$ *is orthogonal to all physical states*, since it is trivially orthogonal to all states of the form $|\psi_T\rangle$, and is also orthogonal to itself. Therefore all scalar products of $|\psi_T\rangle + c|\phi\rangle$ with any other physical state are the same as the scalar product of $|\psi_T\rangle$.

Let us next look at the contribution of $|\phi\rangle$ to the energy and momentum. The energy and momentum in the covariant quantization are found as usual from the Noether theorem and are

$$H = \int \frac{d^3k}{(2\pi)^3}\, \omega_{\mathbf{k}} \left[-a^\dagger_{\mathbf{k},0}a_{\mathbf{k},0} + \sum_{\lambda=1,2,3} a^\dagger_{\mathbf{k},\lambda}a_{\mathbf{k},\lambda}\right], \tag{4.117}$$

$$P = \int \frac{d^3k}{(2\pi)^3} \, \mathbf{k} \left[-a^\dagger_{\mathbf{k},0} a_{\mathbf{k},0} + \sum_{\lambda=1,2,3} a^\dagger_{\mathbf{k},\lambda} a_{\mathbf{k},\lambda} \right] , \tag{4.118}$$

and the minus sign in front of $a^\dagger_{\mathbf{k},0} a_{\mathbf{k},0}$ is simply a consequence of Lorentz covariance; the terms in brackets can in fact be written as $-\eta^{\lambda\lambda'} a^\dagger_{\mathbf{k},\lambda} a_{\mathbf{k},\lambda'}$. Computing the matrix element of the energy and momentum operators between physical states, the contribution from the term $-a^\dagger_{\mathbf{k},0} a_{\mathbf{k},0}$ cancels that from the term $a^\dagger_{\mathbf{k},3} a_{\mathbf{k},3}$. In fact, the condition $c_0 + c_3 = 0$ can be rewritten as

$$(a_{\mathbf{k},0} - a_{\mathbf{k},3}) |\psi\rangle = 0 . \tag{4.119}$$

Then

$$\langle \text{phys}' | - a^\dagger_{\mathbf{k},0} a_{\mathbf{k},0} + a^\dagger_{\mathbf{k},3} a_{\mathbf{k},3} | \text{phys} \rangle = \langle \text{phys}' | (-a^\dagger_{\mathbf{k},0} + a^\dagger_{\mathbf{k},3}) a_{\mathbf{k},3} | \text{phys} \rangle = 0 , \tag{4.120}$$

where the first equality follows from eq. (4.119) and the second from its hermitian conjugate. This means that the contribution to the energy and to the momentum comes only from the transverse oscillators, and therefore, for one-particle states, it is determined completely by the transverse part $|\psi_T\rangle$ in eq. (4.115), and is independent of $c|\phi\rangle$.

In conclusion, the states $|\psi_T\rangle + c|\phi\rangle$ and $|\psi_T\rangle$ have the same energy, momentum (and angular momentum, as can be checked similarly), and they have the same scalar product with all physical states. Therefore they are physically indistinguishable. We therefore introduce an equivalence relation, saying that $|\psi_T\rangle$ is equivalent to $|\psi'_T\rangle$,

$$|\psi_T\rangle \sim |\psi'_T\rangle \tag{4.121}$$

if for some constant c

$$|\psi'_T\rangle = |\psi_T\rangle + c|\phi\rangle . \tag{4.122}$$

We then identify the photons as the equivalence classes with respect to this relation. As a representative of the equivalence class we can conveniently take the purely transverse state $|\psi_T\rangle$. In any case, no physical result depends on this choice. The photon is described by two transverse degrees of freedom, and the energy, momentum (and angular momentum) coincide with those found performing the quantization in the radiation gauge. The generic multiparticle state is obtained tensoring this physical one-particle state. As long as we quantize the free theory, this gives a consistent quantization scheme. In an interacting theory, one must however be careful to check that the interaction between physical states does not produce unphysical states.

Summary of chapter

- The basic principle of the canonical quantization of a free scalar field is to promote the field ϕ and its conjugate momentum to operators, and to impose the equal time commutation relation (4.1).

Expanding the field in plane waves, the coefficients $a_{\mathbf{p}}$ of the expansion become operators, and their complex conjugates $a^*_{\mathbf{p}}$ become the hermitian conjugate operators $a^\dagger_{\mathbf{p}}$. The commutation relation between $a_{\mathbf{p}}$ and $a^\dagger_{\mathbf{p}}$ is given by eq. (4.3) and shows that a free scalar field theory is equivalent to a collection of harmonic oscillators, one for each degree of freedom, labeled by the momentum $\mathbf{p}$.

- The Fock space is constructed in eqs. (4.8) and (4.9). It describes a multiparticle space. The operator $a_{\mathbf{p}}$, acting on a state of the Fock space, destroys a particle with momentum $\mathbf{p}$, while $a^\dagger_{\mathbf{p}}$ creates it. This is a crucial aspect of quantum field theory. The transition amplitudes between different states of the Fock space (that we will learn to compute in the following chapters) describe processes in which the number and the type of particle changes, something which is impossible to describe using only first-quantized wave equations.
- The Hamiltonian and momentum operators are obtained from the classical expressions, performing the normal ordering. Equations (4.16) and (4.20) show that the state $a^\dagger_{\mathbf{p}}|0\rangle$ is a one-particle state with momentum $\mathbf{p}$ and energy $E_{\mathbf{p}} = \sqrt{\mathbf{p}^2 + m^2}$.
- The quantization of complex fields gives rise to two different kinds of quanta, i.e. each particle has its antiparticle, which has the same mass but opposite $U(1)$ charge.
- Spinor fields are quantized imposing anticommutation relations, eq. (4.31), and obey Fermi–Dirac statistics.
- The quantization of the electromagnetic field has a complication due to gauge invariance. One can choose between: (i) a description in which only the physical degrees of freedom appear, at the price of dealing in the intermediate steps with equations which are not explicitly Lorentz covariant (Section 4.3.1), and (ii) a description where Lorentz covariance is explicit, but in the intermediate steps we must deal with spurious degrees of freedom (Section 4.3.2). In any case, one ends up with a fully Lorentz and Poincaré invariant theory, describing a massless particle with spin 1 and two helicity states, the photon.

Exercises

(4.1) Positronium is a hydrogen-like bound state of an electron and positron.

(i) Show that the parity of a positronium state with orbital angular momentum L is $P = (-1)^{L+1}$.

(ii) The charge conjugation operator $\hat{C}$ exchanges the electron with the positron, and therefore positronium is an eigenstate of the charge conjugation operator, $\hat{C}|\text{pos}\rangle = C|\text{pos}\rangle$. Show that, on a positronium state with angular momentum L and total spin S, the eigenvalue is $C = (-1)^{L+S}$.

(iii) In QED parity and charge conjugation are conserved, and therefore the positronium states have well-defined values of C and P. From the results obtained above it follows that positronium states also have L, S defined. The state with $S = 0$ is called para-positronium while $S = 1$ is called ortho-positronium.

Show that the ground state of para-positronium can decay into two photons while the ground state of ortho-positronium cannot decay into two photons but can decay into three photons. We will compute explicitly the decay rate for the annihilation in two photons in Exercise 7.2.

(4.2) Show that the quantity $E_{\mathbf{p}}\,\delta^{(3)}(\mathbf{p} - \mathbf{k})$ is Lorentz invariant and therefore the one-particle states have a Lorentz-invariant normalization.

(4.3) Show that under charge conjugation $\bar{\Psi}\gamma^{\mu}\Psi$ changes sign, where Ψ is the quantized field operator.

(4.4) Consider the Proca Lagrangian

$$L = -\frac{1}{4}F_{\mu\nu}F^{\mu\nu} + \frac{1}{2}m^2 A_{\mu}A^{\mu}\,, \tag{4.123}$$

with $m \neq 0$.

(i) Verify that this theory is not gauge invariant and show that the equations of motion derived from this Lagrangian are

$$(\Box + m^2)A^{\mu} = 0\,, \qquad \partial_{\mu}A^{\mu} = 0\,. \tag{4.124}$$

(ii) Perform the canonical quantization and verify that the theory describes a massive spin-1 particle.

(4.5) (i) Let H be the second quantized Hamiltonian of a free real scalar field, see eq. (4.16), and β a number. Prove the identity

$$e^{-\beta H}a^{\dagger}_{\mathbf{p}} = a^{\dagger}_{\mathbf{p}}e^{-\beta(H+E_{\mathbf{p}})}\,. \tag{4.125}$$

(ii) According to the rules of quantum mechanics, in a mixed state described by a density matrix ρ the expectation value of any operator $\mathcal{O}$ is given by Tr $(\rho\mathcal{O})$, when ρ is normalized by Tr $\rho = 1$. On a thermal state with temperature $T = 1/\beta$ the density matrix is $\rho = e^{-\beta H}/\mathrm{Tr}\, e^{-\beta H}$ and therefore the thermal expectation values are defined as

$$\langle\mathcal{O}\rangle_{\beta} = \frac{\mathrm{Tr}\,\mathcal{O}e^{-\beta H}}{\mathrm{Tr}\,e^{-\beta H}}\,, \tag{4.126}$$

where the trace is over the Fock space. Using the result obtained in (i), show that

$$\langle a^{\dagger}_{\mathbf{p}}a_{\mathbf{q}}\rangle_{\beta} = \frac{(2\pi)^3\delta^{(3)}(\mathbf{p} - \mathbf{q})}{e^{\beta E_{\mathbf{p}}} - 1} \tag{4.127}$$

and therefore, in a finite volume V,

$$\langle a^{\dagger}_{\mathbf{p}}a_{\mathbf{p}}\rangle_{\beta} = \frac{V}{e^{\beta E_{\mathbf{p}}} - 1}\,. \tag{4.128}$$

This shows that $a^{\dagger}_{\mathbf{p}}/\sqrt{V}$ create quanta that obey the Bose–Einstein distribution.

(iii) Repeat the exercise for anticommuting operators and verify that one obtains Fermi–Dirac statistics.

(4.6) (i) Consider a gas of N electrons in a box of volume V. Show that, in the ground state, the electrons fill all states with a momentum $\mathbf{p}$ such that $|\mathbf{p}\,| \leqslant p_F$, with p_F given by

$$\frac{N}{V} = \frac{p_F^3}{3\pi^2}\,. \tag{4.129}$$

The state with all levels filled up to p_F is called the Fermi vacuum and p_F is called the Fermi momentum. We denote the Fermi vacuum by $|0\rangle_F$ to distinguish it from the Fock vacuum $|0\rangle$, where all levels are empty. The Fermi vacuum is the vacuum state of the system subject to the constraint of a fixed number of particle, N, while the Fock vacuum is the vacuum state of the system with no constraint on the particle number.

(ii) Let $a_{\mathbf{p},s}, a^{\dagger}_{\mathbf{p},s}$ be the usual annihilation and creation operators of the electron, introduced in Section 4.2.1. By definition $a_{\mathbf{p},s}|0\rangle = 0$ for all $\mathbf{p}$. Verify that it is not true that $a_{\mathbf{p},s}|0\rangle_F = 0$ for all $\mathbf{p}$. As a consequence, $a^{\dagger}_{\mathbf{p},s}$ is not the appropriate operator to describe the excitation above $|0\rangle_F$. Define

$$\begin{aligned} A_{\mathbf{p}\,,s} &= \theta(|\mathbf{p}\,| - p_F)a_{\mathbf{p},s} + \theta(p_F - |\mathbf{p}\,|)a^{\dagger}_{-\mathbf{p}\,,-s} \\ A^{\dagger}_{\mathbf{p}\,,s} &= \theta(|\mathbf{p}\,| - p_F)a^{\dagger}_{\mathbf{p},s} + \theta(p_F - |\mathbf{p}\,|)a_{-\mathbf{p}\,,-s}\,, \end{aligned} \tag{4.130}$$

where θ is the step function. Verify that $A_{\mathbf{p}\,,s}|0\rangle_F = 0$ for all $\mathbf{p}$ and that $A_{\mathbf{p}\,,s}$ and $A^{\dagger}_{\mathbf{q}\,,r}$ still satisfy canonical anticommutation relations. Give a physical interpretation of the action of $A^{\dagger}_{\mathbf{q}\,,r}$ on $|0\rangle_F$.

(iii) Equation (4.130) is a special case of a Bogoliubov transformation, which can be defined both on bosonic and on fermionic operators, as

$$\begin{aligned} A_{\mathbf{p}\,,s} &= \alpha_{\mathbf{p}}\,a_{\mathbf{p},s} - \beta_{\mathbf{p}}\,a^{\dagger}_{-\mathbf{p}\,,-s} \\ A^{\dagger}_{\mathbf{p}\,,s} &= \alpha^{*}_{\mathbf{p}}\,a^{\dagger}_{\mathbf{p},s} - \beta^{*}_{\mathbf{p}}\,a_{-\mathbf{p}\,,-s}\,, \end{aligned} \tag{4.131}$$

with $\alpha_{\mathbf{p}}\,, \beta_{\mathbf{p}}$ complex coefficients (for the spin zero case just omit the spin index s). Show that, in the bosonic case, the condition that $A_{\mathbf{p},s}$ and $A^\dagger_{\mathbf{p},s}$ satisfy the canonical commutation relations requires that

$$|\alpha_{\mathbf{p}}|^2 - |\beta_{\mathbf{p}}|^2 = 1\,,$$
$$\alpha_{\mathbf{p}}\beta_{-\mathbf{p}} - \alpha_{-\mathbf{p}}\beta_{\mathbf{p}} = 0\,, \qquad (4.132)$$

while for fermions

$$|\alpha_{\mathbf{p}}|^2 + |\beta_{\mathbf{p}}|^2 = 1\,,$$
$$\alpha_{\mathbf{p}}\beta_{-\mathbf{p}} + \alpha_{-\mathbf{p}}\beta_{\mathbf{p}} = 0\,. \qquad (4.133)$$

(iv) Considering for notational simplicity the spin zero case, let $|0,a\rangle$ be the vacuum state annihilated by the operators $a_{\mathbf{p}}$ and $|0,A\rangle$ be the vacuum state annihilated by the operators $A_{\mathbf{p}}$. Correspondingly we have two different type of particles, the "a"-particles obtained acting with $a^\dagger_{\mathbf{p}}$ on $|0,a\rangle$ and the "A"-particles obtained acting with $A^\dagger_{\mathbf{p}}$ on $|0,A\rangle$. The respective particle number operators (setting for simplicity the spatial volume $V=1$) are $A^\dagger_{\mathbf{p}}A_{\mathbf{p}}$ and $a^\dagger_{\mathbf{p}}a_{\mathbf{p}}$. Let $|n_{\mathbf{p}}\rangle$ be an eigenstate of the operator $a^\dagger_{\mathbf{p}}a_{\mathbf{p}}$ with eigenvalue $n_{\mathbf{p}}$ and define

$$N_{\mathbf{p}} \equiv \langle n_{\mathbf{p}}|A^\dagger_{\mathbf{p}}A_{\mathbf{p}}|n_{\mathbf{p}}\rangle\,. \qquad (4.134)$$

Show that

$$N_{\mathbf{p}} = n_{\mathbf{p}} + 2|\beta_{\mathbf{p}}|^2\left(n_{\mathbf{p}} + \frac{1}{2}\right)\,. \qquad (4.135)$$

In particular, on the vacuum of the "a"-particles, $N_{\mathbf{p}} = |\beta_{\mathbf{p}}|^2$. This means that, in terms of "A"-particles, the vacuum of the "a"-particles is a multiparticle state.

Bogoliubov transformations of this type are used in condensed matter physics, in the context of supefluidity or superconductivity, and also in cosmology, to compute particle production by gravitational fields.

Perturbation theory and Feynman diagrams

5

In Chapter 4 we studied the quantization of free fields. We now introduce the interaction. In the canonical quantization, perturbation theory is developed more easily using the Hamiltonian formalism (the Lagrangian formalism is instead more useful in the path integral quantization that will be discussed in Chapter 9). We therefore consider a general field theory with a Hamiltonian

$$H = H_0 + H_{\text{int}} \tag{5.1}$$

where H_0 is the free Hamiltonian and H_{int} is the interaction term. The interaction term will be considered small. For instance in QED

$$H_{\text{int}} = \int d^3x\, \mathcal{H}_{\text{int}} = -\int d^3x\, \mathcal{L}_{\text{int}} \tag{5.2}$$

with

$$\mathcal{L}_{\text{int}} = -eA_\mu \bar{\Psi}\gamma^\mu \Psi \tag{5.3}$$

as discussed in Section 3.5.4. (Note that the identity $\mathcal{H}_{\text{int}} = -\mathcal{L}_{\text{int}}$ holds only when the interaction Lagrangian does not contain derivatives of the fields.) The smallness of the interaction follows from the fact that the parameter which turns out to be relevant for the perturbative expansion is $\alpha = (e^2/4\pi) \simeq 1/137$.

A useful toy model for learning the basic techniques is a quartic self-interaction of a scalar field. In this case H_0 corresponds to the free Klein–Gordon theory, and

$$\mathcal{H}_{\text{int}} = \frac{\lambda}{4!}\phi^4\,, \tag{5.4}$$

with λ a dimensionless coupling constant. Perturbation theory will be meaningful in the weak coupling regime, $\lambda \ll 1$.

5.1 The S-matrix

In the Schrödinger picture we consider a state $|a\rangle(t)$ which, at an initial time T_i, is an eigenstate of a set of commuting operators, with eigenvalues labeled collectively by a. Typically, a will be the set of momenta and spins of the incoming particles. Let us denote $|a\rangle(T_i)$ simply by

$|a\rangle$. Similarly, we consider a state $|b\rangle(t)$ that, at a final time T_f, is an eigenstate with eigenvalues b, and we denote $|b\rangle(T_f)$ simply as $|b\rangle$.

The state $|a\rangle(t)$ evolves as $|a\rangle(t) = e^{-iH(t-T_i)}|a\rangle$ and therefore at the final time T_f it has evolved into $e^{-iH(T_f-T_i)}|a\rangle$. The amplitude for the process in which the initial state $|a\rangle$ evolves into the final state $|b\rangle$ is therefore given by

$$\langle b|e^{-iH(T_f-T_i)}|a\rangle\,. \tag{5.5}$$

In the limit $T_f - T_i \to \infty$ the evolution operator $e^{-iH(T_f-T_i)}$, with H the second quantized Hamiltonian of field theory, is called the S-matrix. Therefore S is an operator that maps an initial state to a final state,

$$|a\rangle \to S|a\rangle\,, \tag{5.6}$$

and the scattering amplitudes are given by its matrix elements, $\langle b|S|a\rangle$. Observe that S is a unitary operator, $SS^\dagger = S^\dagger S = 1$. In fact, if $|a\rangle$ is an initial state, normalized as $\langle a|a\rangle = 1$, and $|n\rangle$ is a complete set of states, the probability that $|a\rangle$ evolves into $|n\rangle$, summed over all $|n\rangle$, must be 1,

$$\sum_n |\langle n|S|a\rangle|^2 = 1\,. \tag{5.7}$$

On the other hand we can write

$$\sum_n |\langle n|S|a\rangle|^2 = \sum_n \langle a|S^\dagger|n\rangle\langle n|S|a\rangle = \langle a|S^\dagger S|a\rangle\,, \tag{5.8}$$

since $|n\rangle$ is a complete set and therefore $\sum_n |n\rangle\langle n| = 1$. This means that $\langle a|S^\dagger S|a\rangle = 1$ for $|a\rangle$ arbitrary, and we conclude that $S^\dagger S = SS^\dagger = 1$. We see that the unitarity of the S-matrix expresses the conservation of probability. It is also convenient to define the T matrix, separating the identity operator,

$$S = 1 + iT\,. \tag{5.9}$$

In terms of T the condition $SS^\dagger = 1$ becomes

$$-i(T - T^\dagger) = TT^\dagger\,. \tag{5.10}$$

Denoting the matrix element $\langle b|T|a\rangle$ by T_{ba}, and inserting a complete set of states, the above equation reads

$$-i(T_{ba} - T^*_{ab}) = \sum_n T_{bn}T^*_{an}\,, \tag{5.11}$$

and in particular, if $a = b$,

$$2\,\mathrm{Im}\,T_{aa} = \sum_n |T_{an}|^2\,. \tag{5.12}$$

Therefore unitarity relates the imaginary part of the diagonal matrix element T_{aa} to the squared modulus of $|T_{an}|^2$, summed over all possible intermediate states.

In quantum field theory the Heisenberg representation is often more useful than the Schrödinger representation. The reason is that in QFT the operators are just the fields, so in the Heisenberg representation the quantum fields depend both on $\mathbf{x}$ and t while in the Schrödinger representation they depend only on $\mathbf{x}$. The Heisenberg representation is therefore more natural from the point of view of Lorentz covariance.

Given a state $|a\rangle(t)$ in the Schrödinger representation, in the Heisenberg picture we define the state $|a\rangle_{\rm H}$ as $|a\rangle_{\rm H} = e^{iHt}|a\rangle(t)$. If A is an operator in the Schrödinger representation, the corresponding Heisenberg operator A_H is defined as $A_{\rm H}(t) = e^{iHt}Ae^{-iHt}$. Since $|a\rangle(t)$ evolves with e^{-iHt}, and A is time-independent, by definition in the Heisenberg picture the states $|a\rangle_{\rm H}$ are independent of t while the operators A_H evolve with time. Writing $|a\rangle_{\rm H} = e^{iHt}|a\rangle(t)$ at time $t = T_i$ and recalling that we denoted $|a\rangle(T_i)$ simply as $|a\rangle$, we can write

$$|a, T_i\rangle_{\rm H} = e^{iHT_i}|a\rangle\,. \tag{5.13}$$

Note that, even if it is time-independent, the Heisenberg state $|a\rangle_{\rm H}$ carries a label T_i which was implicit in the definition of $|a\rangle$, and therefore we have denoted it as $|a, T_i\rangle_{\rm H}$. This label tells us of what Heisenberg operator the state $|a, T_i\rangle_{\rm H}$ is an eigenvector. For instance, suppose that in the Schrödinger representation the state $|x_0\rangle$, at $t = t_0$, is an eigenvector of the position operator $\hat{x}$, and let $\hat{x}_{\rm H}(t) = e^{iHt}\hat{x}e^{-iHt}$. Then the state $|x_0, t_0\rangle_{\rm H} = e^{iHt}|x_0\rangle(t)$ is an eigenvector of the Heisenberg position operator $\hat{x}_{\rm H}(t_0)$ but it is not an eigenvector of the operator $\hat{x}_{\rm H}(t_1)$ with $t_1 \neq t_0$.

Similarly to eq. (5.13) (and omitting hereafter the subscript "H" on states in the Heisenberg representation), we have

$$|b, T_f\rangle = e^{iHT_f}|b\rangle\,, \tag{5.14}$$

and in terms of the states in the Heisenberg picture the matrix element (5.5) is written as

$$\langle b|S|a\rangle = \langle b, T_f|a, T_i\rangle\,. \tag{5.15}$$

5.2 The LSZ reduction formula

Consider a generic S-matrix element written in the Heisenberg picture,

$$\langle \mathbf{p}_1, \mathbf{p}_2, \ldots, \mathbf{p}_n; T_f|\mathbf{k}_1, \mathbf{k}_2, \ldots, \mathbf{k}_m; T_i\rangle\,. \tag{5.16}$$

It is understood that at the end of the computation $T_f \to +\infty$ and $T_i \to -\infty$. For notational simplicity we consider a single species of neutral scalar particle, so the states are labeled just by their momenta, but all our considerations can be generalized to particles with spin. Our first step will be to relate this matrix element to the expectation value of some operator on the vacuum state.

We begin by observing that the expansion of a *free* real scalar field in terms of creation and annihilation operators, eq. (4.2), can be inverted to give

$$(2E_{\mathbf{k}})^{1/2} a_{\mathbf{k}} = i \int d^3x \, e^{ikx} \overleftrightarrow{\partial_0} \phi_{\text{free}} \,, \tag{5.17}$$

$$(2E_{\mathbf{k}})^{1/2} a_{\mathbf{k}}^{\dagger} = -i \int d^3x \, e^{-ikx} \overleftrightarrow{\partial_0} \phi_{\text{free}} \,, \tag{5.18}$$

as one easily verifies substituting eq. (4.2) in the above equations and performing the integration over d^3x. Note that in eqs. (5.17) and (5.18) the integrands are time-dependent but the integrals are independent of t. We have denoted the field by ϕ_{free} to stress that eqs. (5.17) and (5.18) hold only if the field is free. When the field is not free, it cannot be expanded in terms of creation and annihilation operators as in eq. (4.2), and eqs. (5.17) and (5.18) do not hold.

However, as $t \to -\infty$ we intuitively expect that the theory reduces to a free theory, since all incoming particles are infinitely far apart and, if the interaction decreases sufficiently fast with the distance, there will be no difference between a free and an interacting theory.[1] These intuitive considerations are formalized by the hypothesis that, as $t \to -\infty$,

$$\phi(x) \to Z^{1/2} \phi_{\text{in}}(x) \,, \tag{5.19}$$

where $\phi_{\text{in}}(x)$ is a free field and Z is a c-number, known as wave function renormalization. We will discuss later the physical meaning of Z, and how to compute it. Similarly we assume that, as $t \to +\infty$,

$$\phi(x) \to Z^{1/2} \phi_{\text{out}}(x) \,, \tag{5.20}$$

with ϕ_{out} again a free field, and the same constant Z. The limits in eqs. (5.19) and (5.20) must be understood in the weak sense, i.e. they are assumed to hold not as operator equations, but only when we take matrix elements.[2]

We now consider eq. (5.18) with ϕ_{in} playing the role of the free field ϕ_{free}. As we observed above, the integrand in eq. (5.18) is time-dependent, but the result of the integration is independent of t. We can therefore perform it at $t \to -\infty$, and use eq. (5.19) to write

$$\begin{aligned}(2E_{\mathbf{k}})^{1/2} a_{\mathbf{k}}^{\dagger,(\text{in})} &= -i \int_{t \to -\infty} d^3x \, e^{-ikx} \overleftrightarrow{\partial_0} \phi_{\text{in}} \\ &= -i Z^{-1/2} \lim_{t \to -\infty} \int d^3x \, e^{-ikx} \overleftrightarrow{\partial_0} \phi \,, \end{aligned} \tag{5.21}$$

where the superscript "in" means that the operator $a_{\mathbf{k}}^{\dagger}$ acts on the space of initial states at $T_i = -\infty$. Similarly, we define creation operators acting on the final states as

$$\begin{aligned}(2E_{\mathbf{k}})^{1/2} a_{\mathbf{k}}^{\dagger,(\text{out})} &= -i \int_{t \to +\infty} d^3x \, e^{-ikx} \overleftrightarrow{\partial_0} \phi_{\text{out}} \\ &= -i Z^{-1/2} \lim_{t \to +\infty} \int d^3x \, e^{-ikx} \overleftrightarrow{\partial_0} \phi \,. \end{aligned} \tag{5.22}$$

[1] The most important example where the interaction does not decrease at large distances is the interaction of quarks in QCD. As a consequence, quarks are not seen as free particles (they are "confined" inside hadrons), and the free particles seen at $t \to \pm\infty$ are rather the hadrons. We will discuss in Problem 8.2 how to proceed in these cases.

[2] Using a technique known as Källen–Lehmann representation (see Weinberg (1995), Section 10.7) one can show that eq. (5.19) cannot hold as an operator equation, since otherwise one would find that $Z = 1$ and that ϕ is a free field; see, e.g., Itzykson and Zuber (1980), Section 5.1.2.

Observe that in eqs. (5.21) and (5.22) the final integral depends on time, since it is performed with ϕ rather than with a free field; $a_{\mathbf{k}}^{\dagger,(\mathrm{in})}$ is defined taking the limit $t \to -\infty$ of this integral while $a_{\mathbf{k}}^{\dagger,(\mathrm{out})}$ is defined taking the limit $t \to +\infty$, and the relation between in and out creation operators is non-trivial. Recalling our normalization (4.10) for one-particle states, we see that we can eliminate the particle with momentum $\mathbf{k}_1$ from the initial state writing

$$\begin{aligned}
&\langle \mathbf{p}_1, \mathbf{p}_2, \dots, \mathbf{p}_n; T_f | \mathbf{k}_1, \mathbf{k}_2, \dots \mathbf{k}_m; T_i \rangle \\
&= (2E_{\mathbf{k}_1})^{1/2} \langle \mathbf{p}_1, \mathbf{p}_2, \dots, \mathbf{p}_n; T_f | a_{\mathbf{k}_1}^{\dagger,(\mathrm{in})} | \mathbf{k}_2, \dots, \mathbf{k}_m; T_i \rangle \qquad (5.23) \\
&= -iZ^{-1/2} \lim_{t \to -\infty} \int d^3x \, e^{-ik_1 x} \langle \mathbf{p}_1, \mathbf{p}_2, \dots, \mathbf{p}_n; T_f | \overleftrightarrow{\partial_0} \phi | \mathbf{k}_2, \dots, \mathbf{k}_m; T_i \rangle \,.
\end{aligned}$$

The idea is to iterate the process removing all particles from the initial and final states. We perform the computation in detail.

The reader uninterested in the derivation can just take note of the definition of time-ordered product in eq. (5.32) and then can jump directly to eq. (5.40).

First of all, eq. (5.23) can be written in an explicitly covariant form. We use the fact that, for any integrable function $f(t, \mathbf{x})$, we have the identity

$$\left(\lim_{t \to +\infty} - \lim_{t \to -\infty} \right) \int d^3x \, f(t, \mathbf{x}) = \int_{-\infty}^{\infty} dt \, \frac{\partial}{\partial t} \int d^3x \, f(t, \mathbf{x}) \,. \qquad (5.24)$$

Applying this identity to the function $f(t, \mathbf{x}) = -iZ^{-1/2} e^{-ikx} \overleftrightarrow{\partial_0} \phi$ and using eqs. (5.21) and (5.22) we find

$$(2E_{\mathbf{k}})^{1/2} (a_{\mathbf{k}}^{\dagger,(\mathrm{in})} - a_{\mathbf{k}}^{\dagger,(\mathrm{out})}) = iZ^{-1/2} \int d^4x \, \partial_0 (e^{-ikx} \overleftrightarrow{\partial_0} \phi) \,. \qquad (5.25)$$

The integral in this equation can be written in a covariant form observing that

$$\begin{aligned}
\int d^4x \, \partial_0 (e^{-ikx} \overleftrightarrow{\partial_0} \phi) &= \int d^4x \, \partial_0 (e^{-ikx} \partial_0 \phi - \phi \, \partial_0 e^{-ikx}) \\
&= \int d^4x \, (e^{-ikx} \partial_0^2 \phi - \phi \partial_0^2 e^{-ikx}) \\
&= \int d^4x \left[e^{-ikx} \partial_0^2 \phi - \phi (\nabla^2 - m^2) \, e^{-ikx} \right] , \qquad (5.26)
\end{aligned}$$

where in the last line we used the fact that $k^2 = m^2$, since k^μ is the four-momentum of an initial or final particle with mass m, and therefore $\partial_0^2 e^{-ikx} = (\nabla^2 - m^2) e^{-ikx}$. It is understood that our initial and final particle states, which we have written simply as states with definite momentum, i.e. plane waves, will be convoluted to form wave packets, so at each given time they are localized in space. This means that we can integrate ∇^2 twice by parts (while ∂_0 cannot be integrated by parts, since ϕ is not localized in time), and we find

$$(2E_{\mathbf{k}})^{1/2} (a_{\mathbf{k}}^{\dagger,(\mathrm{in})} - a_{\mathbf{k}}^{\dagger,(\mathrm{out})}) = iZ^{-1/2} \int d^4x \, e^{-ikx} (\Box + m^2) \phi(x) \,. \qquad (5.27)$$

Therefore

$$\begin{aligned}
&(2E_{\mathbf{k}_1})^{1/2} \langle \mathbf{p}_1, \mathbf{p}_2, \dots, \mathbf{p}_n; T_f | a_{\mathbf{k}_1}^{\dagger,(\mathrm{in})} - a_{\mathbf{k}_1}^{\dagger,(\mathrm{out})} | \mathbf{k}_2, \dots, \mathbf{k}_m; T_i \rangle \qquad (5.28) \\
&= iZ^{-1/2} \int d^4x \, e^{-ik_1 x} (\Box + m^2) \langle \mathbf{p}_1, \mathbf{p}_2, \dots, \mathbf{p}_n; T_f | \phi(x) | \mathbf{k}_2, \dots, \mathbf{k}_m; T_i \rangle \,.
\end{aligned}$$

The operator $a^{\dagger,(\text{out})}_{\mathbf{k}_1}$ acts on the state to its left, destroying an out particle with momentum $\mathbf{k}_1$. We assume that none of the initial momenta $\mathbf{p}_j$ coincides with a final momentum $\mathbf{k}_i$. This eliminates processes in which one of the particles behaves as a "spectator" and does not interact with the other particles.[3] Then $a^{\dagger,(\text{out})}_{\mathbf{k}_1}$ acting on the state on its left gives zero, because the particle that it would annihilate is absent, and the left-hand side of eq. (5.28) coincides with the expression that appears in eq. (5.23).

[3] In the language of Feynman diagrams that we will explain below, this means that we can restrict to connected diagrams.

The conclusion is that we can remove the particle with momentum $\mathbf{k}_1$ from the initial state, at the price of inserting the operator

$$iZ^{-1/2}\int d^4x\, e^{-ik_1x}(\Box+m^2)\phi(x) \tag{5.29}$$

in the matrix element, i.e.

$$\begin{aligned}&\langle\mathbf{p}_1,\mathbf{p}_2,\ldots,\mathbf{p}_n;T_f|\mathbf{k}_1,\mathbf{k}_2,\ldots,\mathbf{k}_m;T_i\rangle \qquad (5.30)\\ &= iZ^{-1/2}\int d^4x\, e^{-ik_1x}(\Box+m^2)\langle\mathbf{p}_1,\mathbf{p}_2,\ldots,\mathbf{p}_n;T_f|\phi(x)|\mathbf{k}_2,\ldots,\mathbf{k}_m;T_i\rangle\,.\end{aligned}$$

Now we would like to iterate the procedure, eliminating all initial and final particles and remaining with the vacuum expectation value of some combination of fields. For instance, we next eliminate the final particle with momentum $\mathbf{p}_1$. Following the same strategy adopted before, we write

$$\begin{aligned}&\langle\mathbf{p}_1,\mathbf{p}_2,\ldots,\mathbf{p}_n;T_f|\phi(x)|\mathbf{k}_2,\ldots,\mathbf{k}_m;T_i\rangle\\ &= (2E_{\mathbf{p}_1})^{1/2}\langle\mathbf{p}_2,\ldots,\mathbf{p}_n;T_f|a^{(\text{out})}_{\mathbf{p}_1}\phi(x)|\mathbf{k}_2,\ldots,\mathbf{k}_m;T_i\rangle\,.\end{aligned} \tag{5.31}$$

We now define the *time-ordered product*, or simply the T-product, of two fields as follows,

$$T\{\phi(y)\phi(x)\} = \begin{cases}\phi(y)\phi(x) & y^0>x^0\\ \phi(x)\phi(y) & y^0<x^0\end{cases} \tag{5.32}$$

or

$$T\{\phi(y)\phi(x)\} = \theta(y^0-x^0)\phi(y)\phi(x)+\theta(x^0-y^0)\phi(x)\phi(y)\,, \tag{5.33}$$

where $\theta(x^0)$ is the step function: $\theta(x^0)=1$ if $x^0>0$ and $\theta(x^0)=0$ if $x^0<0$. Taking the hermitian conjugate of eq. (5.21) we see that $a^{(\text{in})}_{\mathbf{p}_1}$ is constructed in terms of $\phi(y)$ with $y^0\to-\infty$, and therefore

$$T\{a^{(\text{in})}_{\mathbf{p}_1}\phi(x)\} = \phi(x)a^{(\text{in})}_{\mathbf{p}_1}\,. \tag{5.34}$$

Similarly, $a^{(\text{out})}_{\mathbf{p}_1}$ is constructed in terms of $\phi(y)$ with $y^0\to+\infty$ and

$$T\{a^{(\text{out})}_{\mathbf{p}_1}\phi(x)\} = a^{(\text{out})}_{\mathbf{p}_1}\phi(x)\,. \tag{5.35}$$

We can use this to write the right-hand side of eq. (5.31) as

$$(2E_{\mathbf{p}_1})^{1/2}\langle\mathbf{p}_2,\ldots,\mathbf{p}_n;T_f|T\{(a^{(\text{out})}_{\mathbf{p}_1}-a^{(\text{in})}_{\mathbf{p}_1})\phi(x)\}|\mathbf{k}_2,\ldots,\mathbf{k}_m;T_i\rangle\,. \tag{5.36}$$

In fact, the first term in the T-product is the same as the original expression in eq. (5.31), while the second gives zero since we have seen that $T\{a^{(\text{in})}_{\mathbf{p}_1}\phi(x)\} = \phi(x)a^{(\text{in})}_{\mathbf{p}_1}$ and then $a^{(\text{in})}_{\mathbf{p}_1}$ annihilates the state on its right (recall that we are assuming that the final momenta $\mathbf{p}_j$ are different from any of the initial momenta $\mathbf{k}_i$).

The advantage of the form (5.36) is that the combination $a^{(\text{out})}_{\mathbf{p}_1} - a^{(\text{in})}_{\mathbf{p}_1}$ is given in terms of a covariant expression involving the ϕ field, which is just the hermitian conjugate of eq. (5.27),

$$(2E_{\mathbf{p}_1})^{1/2}(a^{(\text{out})}_{\mathbf{p}_1} - a^{(\text{in})}_{\mathbf{p}_1}) = iZ^{-1/2}\int d^4y\, e^{ip_1y}(\Box_y + m^2)\phi(y)\,. \qquad (5.37)$$

Therefore

$$\begin{aligned}&\langle \mathbf{p}_1, \mathbf{p}_2, \ldots, \mathbf{p}_n; T_f|\phi(x)|\mathbf{k}_2, \ldots, \mathbf{k}_m; T_i\rangle \qquad (5.38)\\ &= iZ^{-1/2}\int d^4y\, e^{ip_1y}(\Box_y + m^2)\langle \mathbf{p}_2, \ldots, \mathbf{p}_n; T_f|T\{\phi(y)\phi(x)\}|\mathbf{k}_2, \ldots, \mathbf{k}_m; T_i\rangle\,,\end{aligned}$$

where $\Box_y = \frac{\partial}{\partial y^\mu}\frac{\partial}{\partial y_\mu}$.[4] Putting together eqs. (5.30) and (5.38) we find the result of eliminating the particles with momenta $\mathbf{k}_1$ and $\mathbf{p}_1$,

$$\begin{aligned}&\langle \mathbf{p}_1, \mathbf{p}_2, \ldots, \mathbf{p}_n; T_f|\mathbf{k}_1, \mathbf{k}_2, \ldots, \mathbf{k}_m; T_i\rangle \qquad (5.39)\\ &= (iZ^{-1/2})^2\int d^4x\, e^{-ik_1x}(\Box_x + m^2)\int d^4y\, e^{+ip_1y}(\Box_y + m^2)\\ &\quad\times\langle \mathbf{p}_2, \ldots, \mathbf{p}_n; T_f|T\{\phi(y)\phi(x)\}|\mathbf{k}_2, \ldots, \mathbf{k}_m; T_i\rangle\,.\end{aligned}$$

The procedure can now be iterated in a straightforward way, and the result is

$$\begin{aligned}&\langle \mathbf{p}_1, \ldots, \mathbf{p}_n; T_f|\mathbf{k}_1, \ldots, \mathbf{k}_m; T_i\rangle\\ &= (iZ^{-1/2})^{n+m}\int\prod_{i=1}^{m} d^4x_i\prod_{j=1}^{n} d^4y_j\, \exp(i\sum_{j=1}^{n} p_jy_j - i\sum_{i=1}^{m} k_ix_i)\\ &\quad\times(\Box_{x_1} + m^2)\ldots(\Box_{y_n} + m^2)\langle 0|T\{\phi(x_1)\ldots\phi(y_n)\}|0\rangle\,, \qquad (5.40)\end{aligned}$$

where the T-product $T\{\phi(x_1)\ldots\phi(y_n)\}$ by definition orders the $n + m$ fields $\phi(x_1), \ldots, \phi(y_m)$ according to decreasing times, so that larger times are leftmost. The vacuum at $t = \pm\infty$ is the perturbative vacuum, i.e. the vacuum used in the construction of the Fock space of the free theory.[5]

As we explained in Section 5.1, $\langle \mathbf{p}_1 \ldots \mathbf{p}_n; T_f|\mathbf{k}_1 \ldots \mathbf{k}_m; T_i\rangle$ is the matrix element in the Heisenberg representation. In the Schrödinger representation we write instead

$$\langle \mathbf{p}_1 \ldots \mathbf{p}_n|S|\mathbf{k}_1 \ldots \mathbf{k}_m\rangle\,. \qquad (5.41)$$

We have also defined the operator T from $S = 1 + iT$. Since in eq. (5.40) we restricted to the situation in which no initial and final momenta coincide, the matrix element of the identity operator between these states vanishes, and we have actually computed the matrix element of iT, i.e. of the non-trivial part of the evolution operator,

$$\begin{aligned}&\langle \mathbf{p}_1 \ldots \mathbf{p}_n|iT|\mathbf{k}_1 \ldots \mathbf{k}_m\rangle\\ &= (iZ^{-1/2})^{n+m}\int\prod_{i=1}^{m} d^4x_i\prod_{j=1}^{n} d^4y_j\, \exp(i\sum_{j=1}^{n} p_jy_j - i\sum_{i=1}^{m} k_ix_i)\\ &\quad\times(\Box_{x_1} + m^2)\ldots(\Box_{y_n} + m^2)\langle 0|T\{\phi(x_1)\ldots\phi(y_n)\}|0\rangle\,. \qquad (5.42)\end{aligned}$$

[4] A very technical remark: writing eq. (5.38) we have extracted $\Box_y$ from the T-product; strictly speaking this is not correct, because $\partial/\partial y^0$ does not commute with the theta function that enters in the definition of the T-product, since $\partial_x\theta(x) = \delta(x)$. However, a simple calculation shows that the additional term is proportional to $\delta(x^0 - y^0)[\partial_0\phi(y), \phi(x)] \sim \delta^{(4)}(x - y)$, and the inclusion of this Dirac delta (and of its derivatives, coming from acting on it with the $\Box_x$ operators present in the LSZ formula) modifies the final result for the LSZ formula, eq. (5.46), by the addition of terms which are polynomial in the four-momenta. Since however both the left-hand side and the right-hand side of eq. (5.46) are pole-like in the four-momenta, i.e. proportional to factors $1/(p^2 - m^2)$, the addition of a regular term is irrelevant when we go on mass shell, i.e. when we set $p^2 = m^2$; see the discussion below eq. (5.46).

[5] Observe that initial one-particle states are defined from $|\mathbf{k}\rangle = (2E_{\mathbf{k}})^{1/2}a_{\mathbf{k}}^{\dagger,(\text{in})}|0\rangle$ and final states from $|\mathbf{k}\rangle = (2E_{\mathbf{k}})^{1/2}a_{\mathbf{k}}^{\dagger,(\text{out})}|0\rangle$, with the same state $|0\rangle$ in both cases, including its phase.

We now define the N-point Green's function

$$G(x_1,\ldots,x_N) = \langle 0|T\{\phi(x_1)\ldots\phi(x_N)\}|0\rangle\,. \tag{5.43}$$

In terms of its Fourier transform $\tilde{G}$, we have

$$G(x_1,\ldots,x_N) = \int\prod_{i=1}^{N}\frac{d^4k_i}{(2\pi)^4}\,e^{-i\sum_{i=1}^{N}x_ik_i}\,\tilde{G}(k_1,\ldots,k_N)\,. \tag{5.44}$$

Using

$$\begin{aligned}(\Box_{x_j}&+m^2)G(x_1,\ldots,x_N)\\ &= -\int\prod_{i=1}^{N}\frac{d^4k_i}{(2\pi)^4}\,(k_j^2-m^2)e^{-i\sum_{i=1}^{N}x_ik_i}\tilde{G}(k_1,\ldots,k_N)\,,\end{aligned} \tag{5.45}$$

eq. (5.42) can be rewritten as

$$\begin{aligned}&\prod_{i=1}^{m}\int d^4\,x_ie^{-ik_ix_i}\prod_{j=1}^{n}\int d^4\,y_je^{+ip_jy_j}\\ &\quad\times\langle 0|T\{\phi(x_1)\ldots\phi(x_m)\phi(y_1)\ldots\phi(y_n)\}|0\rangle\\ &=\left(\prod_{i=1}^{m}\frac{i\sqrt{Z}}{k_i^2-m^2}\right)\left(\prod_{j=1}^{n}\frac{i\sqrt{Z}}{p_j^2-m^2}\right)\langle\mathbf{p}_1\ldots\mathbf{p}_n|iT|\mathbf{k}_1\ldots\mathbf{k}_m\rangle\,.\end{aligned} \tag{5.46}$$

This is the Lehmann–Symanzik–Zimmermann (LSZ) reduction formula. It is important to understand the meaning of the factors $k_i^2-m^2$ and $p_j^2-m^2$ in the denominator. Of course for a physical particle with four-momentum p^μ we have $p^2-m^2=0$ (which is often expressed saying that the particle is "on mass shell"). The meaning of these factors is that we must first compute the left-hand side of eq. (5.46) working off mass shell, i.e. without using any relation between p_0^2 and $\mathbf{p}^2$. In the limit in which we send the particles on mass shell, the left-hand side develops poles of the form $1/(k_i^2-m^2)$ for each incoming particle and $1/(p_j^2-m^2)$ for each outgoing particle. These factors cancel the same pole factors which appear explicitly on the right-hand side, and we remain with an equation between quantities that are finite when the particles are on mass shell.

We have therefore succeeded in relating the scattering amplitude to the vacuum expectation value of a time-ordered product of fields. In the next section we will see how the latter can be computed order by order in perturbation theory.

5.3 Setting up the perturbative expansion

At the classical level, the field $\phi(x)$ satisfies a complicated non-linear equation of motion, determined by the full Lagrangian $L_0+L_{\rm int}$ which corresponds to the full Hamiltonian $H_0 + H_{\rm int}$. The exact form of the solution will in general be very difficult to obtain, but certainly it will not be given just by simple plane waves, and so $\phi(x)$ does not have a simple expansion in plane waves with coefficients that in the quantum theory can be interpreted as creation and annihilation operators. In order to set up the perturbative expansion, we want to relate ϕ to a field ϕ_I whose time evolution is instead determined just by the free Hamiltonian H_0. We therefore define a quantum field $\phi_I(t, \mathbf{x})$ stating that, if at some reference time $t = t_0$ it is equal to $\phi_I(t_0, \mathbf{x})$, then at generic t it is given by

$$\phi_I(t, \mathbf{x}) = e^{iH_0(t-t_0)} \phi_I(t_0, \mathbf{x}) e^{-iH_0(t-t_0)} . \tag{5.47}$$

The field ϕ_I that evolves with the free Hamiltonian is called the *interaction picture* field. By definition this is a free field and we can expand it as

$$\phi_I(t, \mathbf{x}) = \int \frac{d^3p}{(2\pi)^3 \sqrt{2E_{\mathbf{p}}}} \left(a_{\mathbf{p}} e^{-ipx} + a_{\mathbf{p}}^\dagger e^{ipx} \right) , \tag{5.48}$$

with $a_{\mathbf{p}}, a_{\mathbf{p}}^\dagger$ the usual destruction and annihilation operators. Now we want to express the full Heisenberg field ϕ in terms of ϕ_I. At time t_0 the field $\phi(x)$ is a given function of the spatial coordinates, $\phi(t_0, \mathbf{x}) = f(\mathbf{x})$. Let $\phi_I(t, \mathbf{x})$ be the interaction picture field that at $t = t_0$ is equal to the same function $f(\mathbf{x})$. Then (setting $t - t_0 \equiv \tau$)

$$\begin{aligned} \phi(t, \mathbf{x}) &= e^{iH\tau} \phi(t_0, \mathbf{x}) e^{-iH\tau} \\ &= e^{iH\tau} e^{-iH_0\tau} \left[e^{+iH_0\tau} \phi(t_0, \mathbf{x}) e^{-iH_0\tau} \right] e^{iH_0\tau} e^{-iH\tau} \\ &= e^{iH\tau} e^{-iH_0\tau} \phi_I(t, \mathbf{x}) e^{iH_0\tau} e^{-iH\tau} . \end{aligned} \tag{5.49}$$

It is therefore useful to define the *time evolution operator*

$$U(t, t_0) \equiv e^{iH_0(t-t_0)} e^{-iH(t-t_0)} , \tag{5.50}$$

which evolves from time t_0 to time t, and is unitary. Then

$$\boxed{\phi(t, \mathbf{x}) = U^\dagger(t, t_0) \phi_I(t, \mathbf{x}) U(t, t_0) .} \tag{5.51}$$

Note that $U(t, t_0) \neq \exp\{i(H_0 - H)(t - t_0)\} = \exp\{-iH_{\rm int}(t - t_0)\}$ because H_0 and $H_{\rm int}$ do not commute, and therefore we cannot combine the exponentials in this way. However, one can observe that

$$i\frac{\partial U}{\partial t} = e^{iH_0(t-t_0)} (H - H_0) e^{-iH(t-t_0)} = e^{iH_0(t-t_0)} H_{\rm int} e^{-iH_0(t-t_0)} U(t, t_0) . \tag{5.52}$$

We define the interaction picture Hamiltonian H_I as

$$\boxed{H_I(t) = e^{iH_0(t-t_0)} H_{\rm int} e^{-iH_0(t-t_0)} .} \tag{5.53}$$

The solution of eq. (5.52) with the boundary condition $U(t_0, t_0) = 1$ is

$$\boxed{U(t, t_0) = T\left\{ \exp\left[-i \int_{t_0}^t dt' H_I(t') \right] \right\} .} \tag{5.54}$$

We recall that the exponential of an operator is defined by its Taylor expansion. Then the time-ordering T of the exponential means that all terms in the Taylor expansion are time ordered. The fact that this is a solution of eq. (5.52) can be checked expanding the exponential and comparing order by order in H_I. Equations (5.51), (5.53) and (5.54) express the field $\phi(t, \mathbf{x})$ in terms of the interaction picture field.

Our task is to compute the n-point Green's function, i.e.

$$\langle 0|\phi(x_1)\phi(x_2)\dots\phi(x_n)|0\rangle \tag{5.55}$$

when the x_i are T-ordered, i.e. $t_1 > t_2 > \dots > t_n$. Then, using eq. (5.51), we can rewrite it as

$$\begin{aligned}&\langle 0|\left(U^\dagger(t_1,t_0)\phi_I(x_1)U(t_1,t_0)\right)\left(U^\dagger(t_2,t_0)\phi_I(x_2)U(t_2,t_0)\right)\dots\\&\dots\left(U^\dagger(t_n,t_0)\phi_I(x_n)U(t_n,t_0)\right)|0\rangle\,.\end{aligned} \tag{5.56}$$

Observe now that $U^\dagger(t_2, t_0) = U(t_0, t_2)$, since the hermitian conjugation changes $i \to -i$ in the exponent in eq. (5.54), and this is reabsorbed inverting the integration limits. Furthermore, $U(t_1, t_0)U(t_0, t_2) = U(t_1, t_2)$. A simple derivation of this identity (valid independently of the ordering between t_0, t_1, t_2) is obtained observing that

$$i\frac{\partial}{\partial t}[U(t,t_0)U(t_0,t_2)] = i\left[\frac{\partial}{\partial t}U(t,t_0)\right]U(t_0,t_2) = H_I(t)[U(t,t_0)U(t_0,t_2)]. \tag{5.57}$$

Therefore the equation satisfied by $[U(t, t_0)U(t_0, t_2)]$ is the same as that satisfied by $U(t, t_0)$, eq. (5.52), but the boundary condition is

$$[U(t,t_0)U(t_0,t_2)]|_{t=t_0} = U(t_0,t_2)\,, \tag{5.58}$$

since $U(t_0, t_0) = 1$. The solution with this boundary condition is

$$U(t,t_0)U(t_0,t_2) = T\left\{\exp\left[-i\int_{t_2}^{t}dt' H_I(t')\right]\right\}. \tag{5.59}$$

However, this is nothing but $U(t, t_2)$, as we see comparing with eq. (5.54).

Using these identities we can combine the various factors U, $U^\dagger$ in eq. (5.56) and we get

$$\begin{aligned}&\langle 0|U^\dagger(t_1,t_0)\phi_I(x_1)U(t_1,t_2)\phi_I(x_2)U(t_2,t_3)\dots\\&\dots U(t_{n-1},t_n)\phi_I(x_n)U(t_n,t_0)|0\rangle\,.\end{aligned} \tag{5.60}$$

We now introduce a new variable t with a very large value, so that $t \gg t_1 > t_2 > \dots > t_n \gg -t$. Using $U(t_n, t_0) = U(t_n, -t)U(-t, t_0)$ and $U^\dagger(t_1, t_0) = U(t_0, t_1) = U(t_0, t)U(t, t_1) = U^\dagger(t, t_0)U(t, t_1)$, we rewrite (5.60) as

$$\begin{aligned}&\langle 0|U^\dagger(t,t_0)\;[U(t,t_1)\phi_I(x_1)U(t_1,t_2)\phi_I(x_2)U(t_2,t_3)\dots\\&\dots U(t_{n-1},t_n)\phi_I(x_n)U(t_n,-t)]\;U(-t,t_0)|0\rangle\,.\end{aligned} \tag{5.61}$$

Observe that the term in brackets is automatically time-ordered. In fact, e.g. $U(t_1, t_2)$ contains powers of the integral of $H_I(t)$ between t_1 and t_2, time ordered, so terms like $\phi_I(x_1)U(t_1,t_2)\phi_I(x_2)$ are sums of terms of the form $\phi_I(x_1)H_I(t'_1)\dots H_I(t'_k)\phi_I(x_2)$ with $t_1 > t'_1 > \dots > t'_k > t_2$, so everything is automatically time ordered. Therefore the term in square brackets can be rewritten as

$$[\dots] = T\{\phi_I(x_1)\dots\phi_I(x_n)U(t,t_1)U(t_1,t_2)\dots U(t_n,-t)\}\,. \tag{5.62}$$

We have rewritten the various factors in a convenient order, since anyway the order in which they appeared in eq. (5.61) is implemented by the T-product symbol. Now however we can combine the U factors into a single factor $U(t,-t)$. We therefore arrive at

$$\langle 0|U^\dagger(t,t_0)T\left\{\phi_I(x_1)\dots\phi_I(x_n)\exp\left[-i\int_{-t}^{t}dt'H_I(t')\right]\right\}U(-t,t_0)|0\rangle\,. \tag{5.63}$$

Note that the T-product symbol in eq. (5.54) need not be repeated inside (5.63) because the outmost T-product symbol already instructs to time order all its arguments.

These manipulations hold for t_0 arbitrary. We now chose $t_0 = -t$ and we send $t \to \infty$. Then $U(-t,t_0) = 1$ while $U^\dagger(t,t_0) \to U^\dagger(\infty,-\infty)$. The term $\langle 0|U^\dagger(\infty,-\infty)$ in eq. (5.63) is the hermitian conjugate of $U(\infty,-\infty)|0\rangle$, which is the state obtained evolving the vacuum state from time $-\infty$ to $+\infty$. Physically it is clear that, if the vacuum state is stable, applying to it the evolution operator $U(\infty,-\infty)$ we still find the vacuum. Recall however that in quantum mechanics state vectors that differ by a phase still represent the same physical state. Therefore we will have in general

$$U(\infty,-\infty)|0\rangle = e^{i\alpha}|0\rangle\,, \tag{5.64}$$

with α a phase. The explicit form of this phase can be obtained taking the scalar product of the above equation with $\langle 0|$ and using the explicit form (5.54) of the evolution operator,

$$e^{i\alpha} = \langle 0|T\left\{\exp\left[-i\int_{-\infty}^{+\infty}dt'H_I(t')\right]\right\}|0\rangle\,. \tag{5.65}$$

In eq. (5.63) the hermitian conjugate, $\langle 0|U^\dagger(\infty,-\infty) = e^{-i\alpha}\langle 0|$ appears, and from eq. (5.65) we have

$$e^{-i\alpha} = \left(\langle 0|T\left\{\exp\left[-i\int_{-\infty}^{+\infty}dt'H_I(t')\right]\right\}|0\rangle\right)^{-1}\,. \tag{5.66}$$

So we finally get our basic formula

$$\begin{aligned}&\langle 0|T\{\phi(x_1)\dots\phi(x_n)\}|0\rangle\\ &= \frac{\langle 0|T\left\{\phi_I(x_1)\dots\phi_I(x_n)\exp\left[-i\int d^4x\,\mathcal{H}_I\right]\right\}|0\rangle}{\langle 0|T\left\{\exp\left[-i\int d^4x\,\mathcal{H}_I\right]\right\}|0\rangle}\,,\end{aligned} \tag{5.67}$$

where we have written $\int_{-\infty}^{\infty} dt\, H_I = \int d^4x\, \mathcal{H}_I$. The left-hand side of eq. (5.67) is the Green's function which enters in the LSZ formula, and the right-hand side shows how we can compute it in terms of the *free* field ϕ_I. Observe furthermore that $\mathcal{H}_I$ is expressed very simply in terms of ϕ_I, since the functional dependence of $\mathcal{H}_I$ on ϕ_I is exactly the same as the functional dependence of $\mathcal{H}_{\rm int}$ on ϕ. To understand this point, consider for instance a scalar field theory with the quartic self-interaction, $\mathcal{H}_{\rm int} = (\lambda/4!)\phi^4$. Then, using eqs. (5.53) and (5.47),

$$\begin{aligned} \mathcal{H}_I(t) &= e^{iH_0(t-t_0)} \left(\frac{\lambda}{4!}\phi^4\right) e^{-iH_0(t-t_0)} \\ &= \frac{\lambda}{4!} \left[e^{iH_0(t-t_0)}\phi\, e^{-iH_0(t-t_0)}\right] \left[e^{iH_0(t-t_0)}\phi\, e^{-iH_0(t-t_0)}\right] \\ &\quad \times \left[e^{iH_0(t-t_0)}\phi\, e^{-iH_0(t-t_0)}\right] \left[e^{iH_0(t-t_0)}\phi\, e^{-iH_0(t-t_0)}\right] \\ &= \frac{\lambda}{4!}\phi_I^4 \,. \end{aligned} \tag{5.68}$$

At this point the perturbative strategy is clear. We expand the exponential in eq. (5.67) in powers of $\mathcal{H}_I$, and we are left with the task of computing time-ordered products of free fields. In principle, it is clear that these can be written in terms of creation and annihilation operators, and therefore we know how to compute them. Such a brute force computation, however, would quickly become too cumbersome, and in the following we will study a technique, based on Wick's theorem and Feynman graphs, which makes these computations feasible. In the next section we will start from the simplest case, the T-product of two fields.

5.4 The Feynman propagator

We want to compute the Feynman propagator, defined as

$$\langle 0|T\{\phi_I(x)\phi_I(y)\}|0\rangle \,. \tag{5.69}$$

When we study perturbation theory we always use the interaction picture field ϕ_I. The original field ϕ which evolves with the full Hamiltonian will never appear again. Therefore, to make the notation simpler, *from now on we will denote the interaction picture field simply by* ϕ, omitting the subscript "I".

We first separate ϕ into its creation and annihilation parts,

$$\phi(x) = \phi^+(x) + \phi^-(x) \,, \tag{5.70}$$

where

$$\phi^+(x) = \int \frac{d^3p}{(2\pi)^3\sqrt{2E_{\mathbf{p}}}}\, a_{\mathbf{p}} e^{-ipx} \,, \tag{5.71}$$

$$\phi^-(x) = \int \frac{d^3p}{(2\pi)^3\sqrt{2E_{\mathbf{p}}}}\, a_{\mathbf{p}}^\dagger e^{+ipx} \,. \tag{5.72}$$

Of course $\phi^+(x)|0\rangle = 0$ and $\langle 0|\phi^-(x) = 0$. Consider first the case $x^0 > y^0$. Then

$$\begin{aligned} T\{\phi(x)\phi(y)\} &= \phi^+(x)\phi^+(y) + \phi^+(x)\phi^-(y) + \phi^-(x)\phi^+(y) \\ &\quad +\phi^-(x)\phi^-(y) \\ &= \phi^+(x)\phi^+(y) + \phi^-(y)\phi^+(x) + \phi^-(x)\phi^+(y) \\ &\quad +\phi^-(x)\phi^-(y) + [\phi^+(x), \phi^-(y)] \\ &= \,:\phi(x)\phi(y): + [\phi^+(x), \phi^-(y)]\,, \end{aligned} \tag{5.73}$$

where as usual the colons denote normal ordering. Similarly for $y^0 > x^0$ we get $T\{\phi(x)\phi(y)\} =: \phi(x)\phi(y) : +[\phi^+(y), \phi^-(x)]$. Therefore

$$T\{\phi(x)\phi(y)\} =: \phi(x)\phi(y) : + D(x-y)\,, \tag{5.74}$$

where

$$D(x-y) = \theta(x^0 - y^0)[\phi^+(x), \phi^-(y)] + \theta(y^0 - x^0)[\phi^+(y), \phi^-(x)]\,. \tag{5.75}$$

Now observe that the expectation value of the normal ordered term $: \phi(x)\phi(y) :$ is zero, because there is always either one annihilation operator acting on $|0\rangle$ or a creation operator acting on $\langle 0|$. Instead the commutator of $a_{\mathbf{p}}$ and $a^\dagger_{\mathbf{p}}$ is a c-number, so also $D(x-y)$ is a c-number, and $\langle 0|D(x-y)|0\rangle = D(x-y)\langle 0|0\rangle = D(x-y)$. Therefore $D(x-y)$ is just the Feynman propagator,

$$\langle 0|T\{\phi(x)\phi(y)\}|0\rangle = D(x-y)\,. \tag{5.76}$$

Computing the commutators we find

$$D(x-y) = \int \frac{d^3p}{(2\pi)^3}\frac{1}{2E_{\mathbf{p}}}\left(\theta(x^0-y^0)e^{-ip(x-y)} + \theta(y^0-x^0)e^{ip(x-y)}\right)\,. \tag{5.77}$$

The integral over d^3p can be computed explicitly, but it is more useful to rewrite (5.77) as a four-dimensional integral,

$$\boxed{D(x-y) = \int \frac{d^4p}{(2\pi)^4}\frac{i}{p^2 - m^2 + i\epsilon}e^{-ip(x-y)}\,,} \tag{5.78}$$

where $\epsilon \to 0^+$. To prove the equivalence of eqs. (5.77) and (5.78), observe that the integral in eq. (5.78) can be written as

$$\int \frac{d^3p}{(2\pi)^3}e^{i\mathbf{p}\cdot(\mathbf{x}-\mathbf{y})}\int_{-\infty}^{+\infty}\frac{dp^0}{2\pi}\frac{i}{(p^0)^2 - E_{\mathbf{p}}^2 + i\epsilon}e^{-ip^0(x^0-y^0)}\,, \tag{5.79}$$

where $E_{\mathbf{p}} = +(\mathbf{p}^2 + m^2)^{1/2}$. The integral over p^0 can be computed going in the complex p^0-plane. The $i\epsilon$ factor displaces slightly the poles from the real axis. The poles are at $\pm p^0 \simeq E_{\mathbf{p}}(1 - i\epsilon/(2E_{\mathbf{p}}^2))$. Thus the pole at $p^0 = E_{\mathbf{p}}$ is slightly displaced below the real axis and the pole at $p^0 = -E_{\mathbf{p}}$ is slightly displaced above the real axis, as shown in Fig. 5.1.

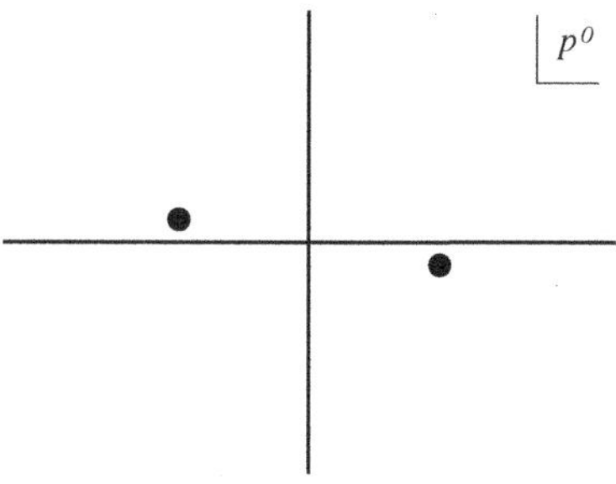

Fig. 5.1 The position of the poles in the complex p^0-plane.

For $x^0 - y^0 > 0$ we can close the contour in the lower half-plane and we only get the contribution of the pole at $p^0 = E_{\mathbf{p}}$. The pole is encircled clockwise, so it gives

$$\frac{i}{2\pi}(-2\pi i)\,\frac{e^{-iE_{\mathbf{p}}(x^0-y^0)}}{2E_{\mathbf{p}}} = +\frac{1}{2E_{\mathbf{p}}}e^{-iE_{\mathbf{p}}(x^0-y^0)}\,. \tag{5.80}$$

If instead $x^0 - y^0 < 0$ we can close the contour in the upper half-plane and we only get the contribution of the pole at $p^0 = -E_{\mathbf{p}}$, which now is encircled counterclockwise, so it gives

$$\frac{i}{2\pi}(2\pi i)\,\frac{e^{iE_{\mathbf{p}}(x^0-y^0)}}{-2E_{\mathbf{p}}} = +\frac{1}{2E_{\mathbf{p}}}e^{iE_{\mathbf{p}}(x^0-y^0)}\,. \tag{5.81}$$

In both cases, we have reproduced eq. (5.77) (in the second term we must also rename the integration variable $\mathbf{p} \to -\mathbf{p}$).

From eq. (5.78) we read off the Feynman propagator in momentum space,

$$\boxed{\tilde{D}(p) = \frac{i}{p^2 - m^2 + i\epsilon}\,.} \tag{5.82}$$

It is also easy to see that the Feynman propagator is just a Green's function of the operator $\Box + m^2$. In fact, from eq. (5.78),

$$\begin{aligned}(\Box_x + m^2)D(x-y) &= \int \frac{d^4p}{(2\pi)^4}\,\frac{i}{p^2 - m^2 + i\epsilon}\,(-p^2 + m^2)e^{-ip(x-y)}\\ &= -i\delta^{(4)}(x-y)\,.\end{aligned} \tag{5.83}$$

Observe that this result holds independently of the prescription for going around the poles. There are in principle four different prescriptions (each of the two poles can be slightly displaced above or below the real axis) and different Green's functions are obtained with different prescriptions, and obey different boundary conditions (see Exercise 5.1).

5.5 Wick's theorem and Feynman diagrams

Wick's theorem is a very useful tool for reducing the expectation value of a generic T-product of fields

$$\langle 0|T\{\phi(x_1)\ldots\phi(x_n)\}|0\rangle \tag{5.84}$$

to a combination of Feynman propagators. It generalizes the identity (5.74) to an n-point function, and states that $T\{\phi(x_1)\ldots\phi(x_n)\}$ is equal to the normal ordered product $:\phi(x_1)\ldots\phi(x_n):$ plus all possible combinations of normal ordering and *contractions* of fields, where a contraction of two fields $\phi(x_1), \phi(x_2)$ is defined to be equal to the Feynman propagator $D(x_1 - x_2)$. For instance (using the notation $\phi(x_i) = \phi_i$ and

$D(x_i - x_j) = D_{ij}$),

$$T\{\phi_1\phi_2\phi_3\phi_4\} = :\phi_1\phi_2\phi_3\phi_4: +D_{12}:\phi_3\phi_4: +D_{13}:\phi_2\phi_4: \\ +D_{14}:\phi_2\phi_3: +D_{23}:\phi_1\phi_4: +D_{24}:\phi_1\phi_3: \\ +D_{34}:\phi_1\phi_2: +D_{12}D_{34} + D_{13}D_{24} + D_{14}D_{23}\,. \tag{5.85}$$

The proof of the theorem can be given by induction on the number of fields, see, e.g. Peskin and Schroeder (1995), pages 88–90. When we take the vacuum expectation value, all terms with a normal ordering factor give zero, and only the terms where all fields have been contracted survive. Thus in our example

$$\langle 0|T\{\phi_1\phi_2\phi_3\phi_4\}|0\rangle = D_{12}D_{34} + D_{13}D_{24} + D_{14}D_{23}\,. \tag{5.86}$$

The above equation has a vivid physical interpretation. If we interpret $D(x_1 - x_2)$ as the amplitude for the propagation of a particle from the space-time point x_1 to x_2, then $D(x_1 - x_2)D(x_3 - x_4)$ is the amplitude for the process in which one particle goes from x_1 to x_2 and another from x_3 to x_4, without interacting with each other. We can now associate a Feynman graph in position space to each non-vanishing contribution. We simply draw a line connecting points x_i and x_j for each propagator $D(x_i - x_j)$. For instance, the term $D(x_1 - x_2)D(x_3 - x_4)$ can be associated to the (rather trivial) Feynman diagram in position space given in Fig. 5.2.

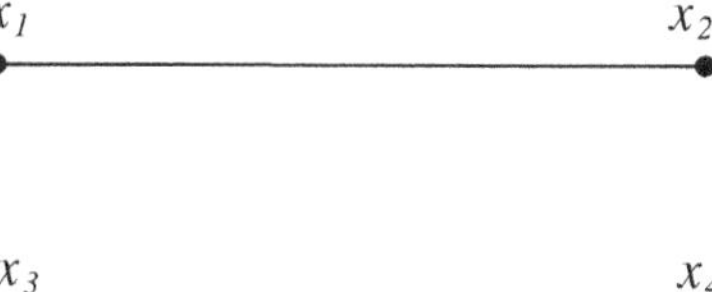

Fig. 5.2 The diagrammatic representation of $D(x_1 - x_2)D(x_3 - x_4)$.

When we expand the exponential in eq. (5.67) in powers of $\mathcal{H}_I$, each term $\mathcal{H}_I$ contains fields at the same space-time point. As we will see explicitly below, this gives rise to less trivial Feynman graphs.

The best way to understand all this machinery is to put it to work, and to start computing. Therefore, in the next two subsections, we will perform a few computations in all details. The important point that will emerge, however, is that it is not necessary every time to go through the rather involved steps that we will present, since the results can be summarized very compactly by a set of rules, the Feynman rules, that allow us to associate to each amplitude a set of Feynman diagrams, and to write down almost immediately the contribution of each Feynman diagram. Still, once in a lifetime, it might be useful to go through all the detailed steps, before starting to use the Feynman rules as an automatic machinery. The reader who does not wish to follow the computations in all details can go quickly through the next two subsections and find a summary of Feynman rules in Section 5.5.3.

5.5.1 A few very explicit computations

We begin with the scattering amplitude for a process with two initial particles with momenta $\mathbf{k}_1, \mathbf{k}_2$ into two final particles with momenta $\mathbf{p}_1, \mathbf{p}_2$, in the theory with $\mathcal{H}_I = (\lambda/4!)\phi^4$. The general formulas (5.46, 5.67) give

$$\left(\prod_{i=1}^{2} \frac{i\sqrt{Z}}{p_i^2 - m^2}\right)\left(\prod_{j=1}^{2} \frac{i\sqrt{Z}}{k_j^2 - m^2}\right)\langle \mathbf{p}_1\mathbf{p}_2|iT|\mathbf{k}_1\mathbf{k}_2\rangle$$

$$= \int d^4x_1 d^4x_2 d^4x_3 d^4x_4 e^{i(p_1x_1+p_2x_2-k_1x_3-k_2x_4)}$$

$$\times \frac{\langle 0|T\{\phi(x_1)\phi(x_2)\phi(x_3)\phi(x_4)\exp\left[-i\frac{\lambda}{4!}\int d^4x\phi^4\right]\}|0\rangle}{\langle 0|T\{\exp\left[-i\frac{\lambda}{4!}\int d^4x\phi^4\right]\}|0\rangle}. \quad (5.87)$$

We work at the lowest non-trivial order in perturbation theory in λ. As we will see later, in $\lambda\phi^4$ theory $Z = 1 + O(\lambda^2)$ and therefore, since we will work up to $O(\lambda)$, we set $Z = 1$.

Zero-order term: First of all there is a term of order λ^0, which is given simply setting $\lambda = 0$ in eq. (5.87). Of course, if there is no coupling, there is no scattering, and we must find a trivial amplitude at this order. Let us nevertheless check explicitly how this comes out. At $\lambda = 0$, using eq. (5.86),

$$\int \prod_{i=1}^{4} d^4x_i\, e^{i(p_1x_1+p_2x_2-k_1x_3-k_2x_4)} \langle 0|T\{\phi(x_1)\phi(x_2)\phi(x_3)\phi(x_4)\}|0\rangle$$

$$= \int \prod_{i=1}^{4} d^4x_i\, e^{i(p_1x_1+p_2x_2-k_1x_3-k_2x_4)} \left(D(x_1-x_2)D(x_3-x_4)+\ldots\right), \quad (5.88)$$

where the dots represent the other two terms in eq. (5.86). In terms of Feynman graphs in position space, the contribution written explicitly is shown in Fig. 5.2 and represents a particle traveling from x_1 to x_2 and a second particle traveling from x_3 to x_4, without interacting with the first (and similarly for the other two terms denoted by the dots in eq. (5.88)). Changing integration variables to $x = x_1 - x_2$ and $X = (x_1 + x_2)/2$ and similarly for x_3, x_4 we have

$$\int d^4x_1 d^4x_2 d^4x_3 d^4x_4 e^{i(p_1x_1+p_2x_2-k_1x_3-k_2x_4)} D(x_1-x_2)D(x_3-x_4)$$

$$= \left[\int d^4x d^4X e^{i(p_1+p_2)X+i(p_1-p_2)x/2} D(x)\right]$$

$$\times \left[\int d^4x d^4X e^{-i(k_1+k_2)X-i(k_1-k_2)x/2} D(x)\right]$$

$$= (2\pi)^4\delta^{(4)}(p_1+p_2)(2\pi)^4\delta^{(4)}(k_1+k_2)\,\frac{i}{p_1^2-m^2}\,\frac{i}{k_1^2-m^2}. \quad (5.89)$$

Comparing with eq. (5.87) we see that on the right-hand side we have obtained only two pole factors,[6] which originated from the two propagators in momentum space, while on the left-hand side of eq. (5.87) there are four pole factors. Therefore only two of them cancel, and we get a contribution to the T-matrix element

$$\langle \mathbf{p}_1\mathbf{p}_2|iT|\mathbf{k}_1\mathbf{k}_2\rangle = -(2\pi)^8(p_2^2-m^2)(k_2^2-m^2)\delta^{(4)}(p_1+p_2)\delta^{(4)}(k_1+k_2) \quad (5.90)$$

and when we go on mass shell this gives zero. The same happens for the other contributions indicated by the dots in eq. (5.88). Therefore at zero order in λ there is no contribution to the scattering amplitude, as expected.

This is a general situation which repeats for the $n \to m$ scattering amplitudes, and disconnected graphs do not contribute, simply because the Feynman graphs do not provide enough pole factors to cancel those that appear in the LSZ formula.[7]

Term $O(\lambda)$: The first non-trivial contribution comes expanding the exponential in eq. (5.87) to first order in λ. Let us for the moment neglect the

[6] Observe that the factor $i\epsilon$ in $p^2-m^2+i\epsilon$ should be kept when p is an integration variable, because in this case it gives the prescription for going around the poles, but if p is the momentum on an external leg, and is therefore fixed, we can set directly $\epsilon = 0$.

[7] There is a subtlety here connected with graphs known as tadpoles, that we will introduce below. In principle inserting tadpole graphs in the lines of a disconnected diagram we can obtain the appropriate number of pole factors. However, the tadpoles are simply reabsorbed in the mass renormalization, as we will see in Section 5.6.

denominator in eq. (5.87). Then at this order we have

$$\left(\prod_{i=1}^{2}\frac{i}{p_i^2-m^2}\right)\left(\prod_{j=1}^{2}\frac{i}{k_j^2-m^2}\right)\langle\mathbf{p}_1\mathbf{p}_2|iT|\mathbf{k}_1\mathbf{k}_2\rangle$$
$$=\int d^4x_1d^4x_2d^4x_3d^4x_4\exp\{i(p_1x_1+p_2x_2-k_1x_3-k_2x_4)\}$$
$$\times\left(-i\frac{\lambda}{4!}\right)\int d^4x\,\langle 0|T\{\phi(x_1)\phi(x_2)\phi(x_3)\phi(x_4)\phi^4(x)\}|0\rangle\,. \qquad (5.91)$$

As before, the disconnected graphs are unable to cancel all pole factors and therefore, when we go on mass shell, they give zero. The only connected Feynman graphs are obtained contracting each of the four $\phi(x_i)$ with one of the four $\phi(x)$; there are 4! possible contractions of this type, and therefore the right-hand side of eq. (5.91) is equal to

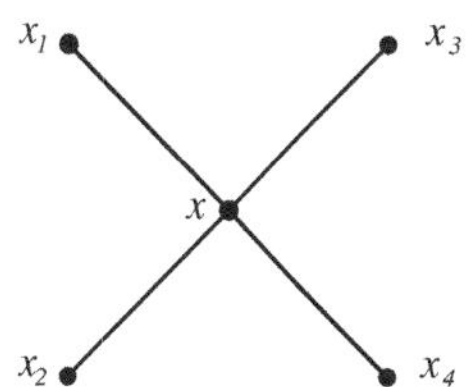

Fig. 5.3 The Feynman diagram in position space representing the $2\to 2$ scattering amplitude at $O(\lambda)$.

$$\int d^4x\,d^4x_1d^4x_2d^4x_3d^4x_4\exp\{i(p_1x_1+p_2x_2-k_1x_3-k_2x_4)\}$$
$$\times(-i\lambda)D(x_1-x)D(x_2-x)D(x_3-x)D(x_4-x)\,. \qquad (5.92)$$

We can represent it with the Feynman graph shown in Fig. 5.3. Setting $y_i = x_i - x$ and performing first the integration over the y_i we get

$$(-i\lambda)\tilde{D}(p_1)\tilde{D}(p_2)\tilde{D}(k_1)\tilde{D}(k_2)\int d^4x\,e^{i(p_1+p_2-k_1-k_2)x} \qquad (5.93)$$
$$=(-i\lambda)(2\pi)^4\delta^{(4)}(p_1+p_2-k_1-k_2)\left(\prod_{i=1}^{2}\frac{i}{p_i^2-m^2}\right)\left(\prod_{j=1}^{2}\frac{i}{k_j^2-m^2}\right).$$

Therefore

$$\langle\mathbf{p}_1\mathbf{p}_2|iT|\mathbf{k}_1\mathbf{k}_2\rangle=(-i\lambda)(2\pi)^4\delta^{(4)}(p_1+p_2-k_1-k_2)\,. \qquad (5.94)$$

We note several aspects which can be generalized to all Feynman graphs:

- Only connected graphs contribute.
- For connected graphs, the propagators associated to the external legs cancel exactly the pole factors in the LSZ formula.
- Each interaction vertex gives a factor $-i\lambda$.
- There is an integration over a variable x which gives the overall energy–momentum conservation.

Finally, we have to consider the effect of the denominator in eq. (5.87). The denominator gives only vacuum-to-vacuum graphs, i.e. Feynman diagrams with no external lines. However each contribution from the numerator can be "dressed" with all possible vacuum-to-vacuum graphs, considering all possible disconnected graph made with the original graph plus all possible vacuum-to-vacuum graphs. This is shown in the first line of Fig. 5.4, where we give examples of disconnected graphs with four external legs. As shown graphically in the second line of the figure, the connected graph with four external legs factorizes, and the term in the parentheses is the sum of all vacuum-to-vacuum graphs. This is nothing but the perturbative expansion of the denominator, so the term in parentheses exactly cancels the denominator. This means that we can simply set the vacuum expectation value at the denominator to one, if at the same time we neglect in the numerator all disconnected diagrams in which a disconnected component is a vacuum diagram.

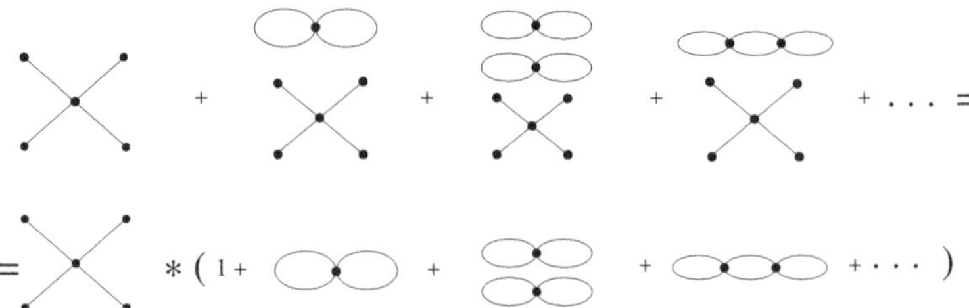

Fig. 5.4 A diagrammatic representation of the effect of vacuum-to-vacuum graphs.

The Feynman graphs that we have considered until now have no internal lines. To illustrate what happens with internal lines it is simpler to consider a theory with $H_I = (\lambda/3!)\phi^3$. In this case three lines instead of four meet at each vertex. To compute the $2 \to 2$ scattering amplitude we consider

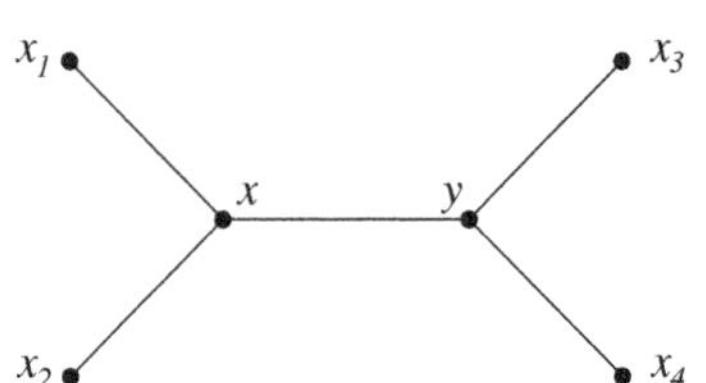

Fig. 5.5 The Feynman graph in position space corresponding to the contraction described in the text.

$$\int d^4x_1 d^4x_2 d^4x_3 d^4x_4 \exp\{i(p_1x_1 + p_2x_2 - k_1x_3 - k_2x_4)\}$$
$$\times \langle 0|T\{\phi(x_1)\phi(x_2)\phi(x_3)\phi(x_4) \exp\left[-i\frac{\lambda}{3!}\int d^4x\, \phi^3(x)\right]\}|0\rangle_c\,. \quad (5.95)$$

We have seen that the denominator in eq. (5.67) is canceled by disconnected graphs in which a disconnected component is a vacuum-to-vacuum graph, while disconnected graphs with external legs in more than one disconnected component do not contribute because they do not provide the right pole factors. We therefore omit the vacuum-to-vacuum amplitude in the denominator hereafter, and we add a subscript $\langle 0|\ldots|0\rangle_c$ to remind us that we must consider only connected graphs.

In this theory at $O(\lambda)$ there is no contribution to the four-point amplitude, because expanding the exponential to first order in λ we get the Green's function $\langle 0|T\{\phi(x_1)\phi(x_2)\phi(x_3)\phi(x_4)\phi^3(x)\}|0\rangle$, which has an odd number of fields, so we cannot contract all of them and the vacuum expectation value is zero. Therefore the leading term is $O(\lambda^2)$, and is

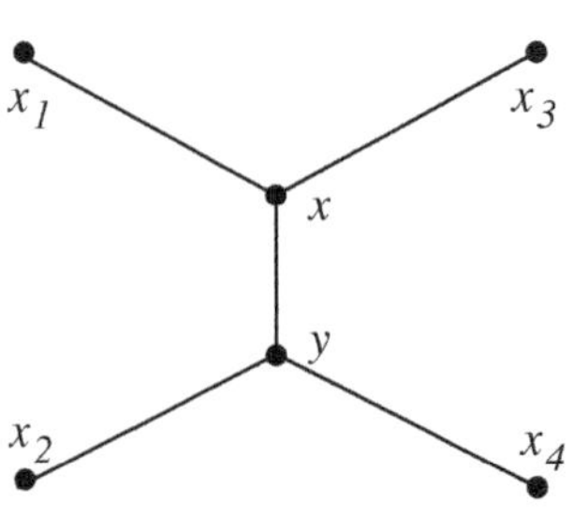

Fig. 5.6 A Feynman graph corresponding to a different set of contractions.

$$\int d^4x_1 d^4x_2 d^4x_3 d^4x_4 \exp\{i(p_1x_1 + p_2x_2 - k_1x_3 - k_2x_4)\} \quad (5.96)$$
$$\times \frac{1}{2!}\left(-i\frac{\lambda}{3!}\right)^2 \int d^4x \int d^4y \langle 0|T\{\phi(x_1)\phi(x_2)\phi(x_3)\phi(x_4)\phi^3(x)\phi^3(y)\}|0\rangle_c\,.$$

The new aspect here is that, in order to contract all fields and to obtain a connected diagram, we must contract one field $\phi(x)$ with one $\phi(y)$, where x, y are the positions of the two vertices. We therefore have graphs with two vertices connected by an internal line, while in the previous example all lines were external, i.e. related to the initial or final particles. To understand how to treat these graphs we consider the case in which we contract $\phi(x_1)$ and $\phi(x_2)$ each with one of the $\phi(x)$, $\phi(x_3)$ and $\phi(x_4)$ each with one of the $\phi(y)$, and the remaining $\phi(x)$ with the remaining $\phi(y)$, as in Fig. 5.5. (With different contractions, we can also obtain the graph in Fig. 5.6, and an analogous graph with x_3 and x_4 interchanged.) Taking into account the combinatorial factor, the factor $1/2!$ from the expansion of the exponential, and the fact that there is an equal contribution with x and y interchanged, the graph in Fig. 5.5 gives

$$\int d^4x_1 d^4x_2 d^4x_3 d^4x_4 \exp\{i(p_1x_1 + p_2x_2 - k_1x_3 - k_2x_4)\} \quad (5.97)$$

$$\times \int d^4x \int d^4y (-i\lambda)^2 D(x_1 - x) D(x_2 - x) D(x - y) D(y - x_3) D(y - x_4)$$
$$= (-i\lambda)^2 \tilde{D}(p_1)\tilde{D}(p_2)\tilde{D}(k_1)\tilde{D}(k_2) \int d^4x \int d^4y\, e^{i(p_1+p_2)x - i(k_1+k_2)y} D(x-y)$$
$$= (-i\lambda)^2 (2\pi)^4 \delta^{(4)}(p_1 + p_2 - k_1 - k_2) \tilde{D}(p_1)\tilde{D}(p_2)\tilde{D}(p_1+p_2)\tilde{D}(k_1)\tilde{D}(k_2)\,.$$

Again we find the momentum space propagators associated to the external legs, which are canceled by the pole terms in the LSZ formula, and the energy–momentum conservation. The new factor is $\tilde{D}(p_1 + p_2)$, associated to the internal line.

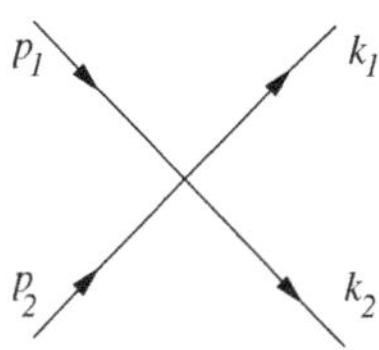

Fig. 5.7 The graph in momentum space corresponding to Fig. 5.3.

It now becomes clear that it is more convenient to work in momentum space, rather than in position space. Then to each line of a graph is associated the momentum space propagator given in eq. (5.82). The external lines give factors which cancel the pole terms in the LSZ formula. The Feynman diagrams in momentum space corresponding to Figs. 5.3 and 5.5 are shown in Figs. 5.7 and 5.8.

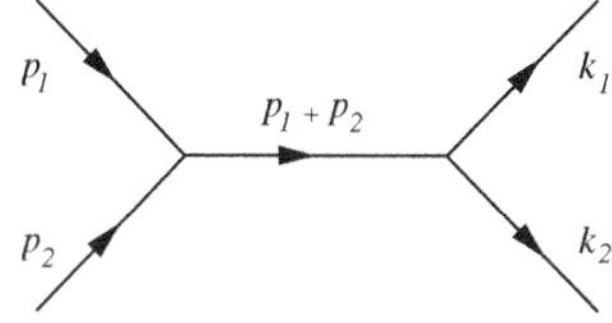

Fig. 5.8 The graph in momentum space corresponding to Fig. 5.5.

At this point it is not difficult to see how to compute the most general amplitude associated with *tree graphs* (a tree graph is a graph with internal and external lines, but with no closed internal loop). The technique can be summarized by the Feynman rules in momentum space, for a scalar field theory (some generalization will be needed for fermions, gauge fields, etc., and will be discussed later):

- Draw all connected graphs corresponding to the given initial and final states. The number of lines that meet at each vertex is determined by the interaction term; e.g. three lines in ϕ^3 theory and four lines in ϕ^4 theory. Disconnected graphs do not contribute.
- To each external leg is associated a factor which compensates the pole factor in the LSZ reduction formula, eq. (5.46). Therefore we can simply omit all these factors from the graph, and we will obtain directly the matrix element of iT. This is often expressed saying that we consider the graphs "with external legs amputated".
- There is an overall Dirac delta imposing energy–momentum conservation. In order not to write explicitly the Dirac delta each time we compute a Feynman graph, it is convenient to define a matrix element $\mathcal{M}_{fi}$ from

$$\langle \mathbf{p}_1 \dots \mathbf{p}_n | iT | \mathbf{k}_1 \dots \mathbf{k}_m \rangle = (2\pi)^4 \delta^{(4)} \left(\sum_i p_i - \sum_j k_j \right) i\mathcal{M}_{fi}\,. \quad (5.98)$$

  The labels i, f refer to the initial and final states or more explicitly, for a scalar theory, $\mathcal{M}_{fi} = \mathcal{M}(\mathbf{p}_1, \dots, \mathbf{p}_n; \mathbf{k}_1, \dots, \mathbf{k}_m)$ (more generally, the initial and final states are labeled also by the spin states of the initial and final particles).
- Energy–momentum conservation must be imposed separately at each vertex. Note for instance that in the internal line in Fig. 5.8 we had two external momenta p_1 and p_2 flowing into a vertex, and the momentum associated to the internal line flowing out of this vertex is $p_1 + p_2$. The "virtual particle" associated with this internal line decays in the final states with momenta k_1, k_2, and the overall Dirac delta ensures $p_1 + p_2 = k_1 + k_2$, so momentum is conserved also at the other vertex.
- To each vertex associate a factor $-i$ times the coupling constant.

- To each internal line associate a propagator, with the value of the four-momentum given by energy–momentum conservation.
- There is a combinatorial factor which combines the number of equivalent contractions, the factors $1/n!$ from the expansion of the exponential at order n, and numerical factors associated to the definition of the coupling constant, such as the factors $1/4!$ in $(\lambda/4!)\phi^4$.

In the next subsection we will understand what happens when there are internal loops in a graph.

5.5.2 Loops and divergences

We have by now understood how to write a Feynman graph when there are internal lines whose momentum is fixed by energy–momentum conservation at a vertex. Consider however the $O(\lambda^2)$ corrections to the $2 \to 2$ scattering amplitude in $\lambda\phi^4$. The possible connected graphs that can be constructed with four external legs (since we study a $2 \to 2$ amplitude), four lines meeting at each vertex (since we are in ϕ^4 theory) and two vertices (since we want to study the terms $O(\lambda^2)$) are given in Fig. 5.9. The three graphs correspond to different types of contractions, which we will discuss in more detail below. The important new point is that the momenta of the internal lines are not completely fixed imposing energy–momentum conservation at each vertex. We can assign an arbitrary momentum k to one internal line and the other has a momentum fixed by energy–momentum conservation at each vertex. The fact that not all momenta are fixed takes place each time we have loops in a Feynman graph. We must therefore understand how to treat these graphs. Again, we discuss some examples in full detail.

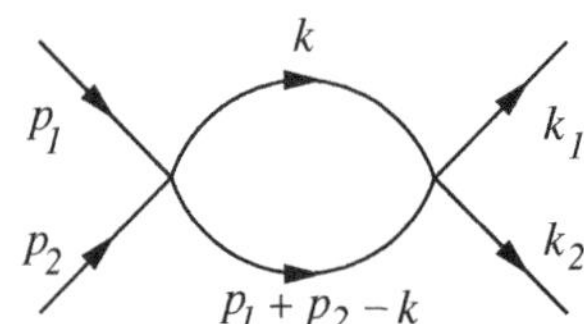

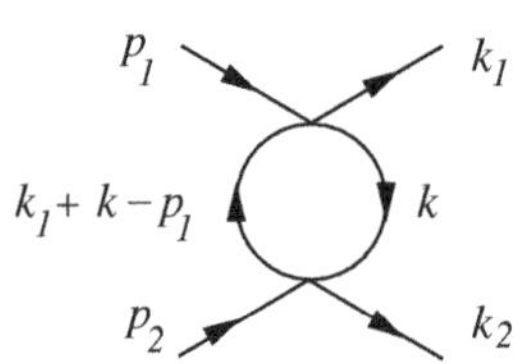

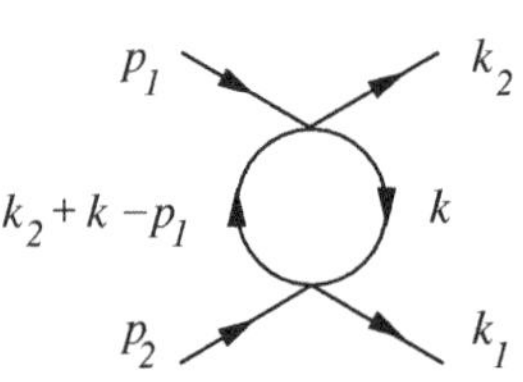

Fig. 5.9 The three contributions to the one-loop $2 \to 2$ amplitude in momentum space.

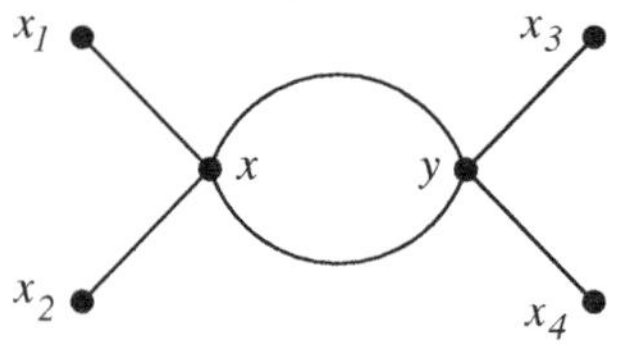

Fig. 5.10 A contribution to the one-loop amplitude in position space.

Let us work out the $2 \to 2$ amplitude in $\lambda\phi^4$ at one loop order. We start as usual from

$$\int d^4x_1 d^4x_2 d^4x_3 d^4x_4 \exp\{i(p_1x_1 + p_2x_2 - k_1x_3 - k_2x_4)\}$$
$$\times \langle 0|T\{\phi(x_1)\phi(x_2)\phi(x_3)\phi(x_4) \exp\left[-i\frac{\lambda}{4!}\int d^4x\phi^4\right]\}|0\rangle_c . \quad (5.99)$$

The term of order λ^2 is

$$\frac{1}{2!}\int d^4x_1 d^4x_2 d^4x_3 d^4x_4 \exp\{i(p_1x_1 + p_2x_2 - k_1x_3 - k_2x_4)\} \quad (5.100)$$
$$\times \left(-i\frac{\lambda}{4!}\right)^2 \int d^4x \int d^4y \langle 0|T\{\phi(x_1)\phi(x_2)\phi(x_3)\phi(x_4)\phi^4(x)\phi^4(y)\}|0\rangle_c .$$

The Feynman graph in position space shown in Fig. 5.10 is obtained contracting $\phi(x_1)$ with one of the $\phi(x)$, $\phi(x_2)$ with another $\phi(x)$ (there are $4\cdot 3$ possible ways to do it), $\phi(x_3)$ with one $\phi(y)$, $\phi(x_4)$ with another of the $\phi(y)$ (again $4\cdot 3$ combinations), and the remaining two of the $\phi(x)$ with the remaining two $\phi(y)$ (there are 2 possible ways to do it). An equal contribution is obtained exchanging $x \leftrightarrow y$, i.e. contracting $\phi(x_1)$

and $\phi(x_2)$ with two of the $\phi(y)$ and $\phi(x_3), \phi(x_4)$ with two of the $\phi(x)$; this gives an additional factor of 2, so we finally get

$$\begin{aligned}
&\frac{1}{2}(-i\lambda)^2 \int d^4x_1 d^4x_2 d^4x_3 d^4x_4\, d^4x\, d^4y\; e^{i(p_1x_1+p_2x_2-k_1x_3-k_2x_4)} \\
&\times D(x_1-x)D(x_2-x)D(x-y)D(x-y)D(y-x_3)D(y-x_4) \\
&= \frac{1}{2}(-i\lambda)^2 \tilde{D}(p_1)\tilde{D}(p_2)\tilde{D}(k_1)\tilde{D}(k_2) \\
&\times \int d^4x\, d^4y\; e^{i(p_1+p_2)x-i(k_1+k_2)y} D^2(x-y) \\
&= \frac{1}{2}(-i\lambda)^2 \tilde{D}(p_1)\tilde{D}(p_2)\tilde{D}(k_1)\tilde{D}(k_2)(2\pi)^4\delta^{(4)}(p_1+p_2-k_1-k_2) \\
&\times \int \frac{d^4k}{(2\pi)^4}\, \tilde{D}(k)\tilde{D}(p_1+p_2-k)\,. \qquad (5.101)
\end{aligned}$$

Similar expressions are obtained if x_1 and x_3 are joined to x while x_2 and x_4 to y, and if x_1 and x_4 are joined to x while x_2 and x_3 to y. These three possibilities in position space give rise to the three graphs of Fig. 5.9 in momentum space. In eq. (5.101) we recognize the usual factors associated to the external legs and the overall energy–momentum conservation. The new aspect here is the integration over the momentum k of the internal line which was not fixed by the energy–momentum conservation at the vertices.

We can therefore add to our list of Feynman rules in momentum space:

- associate a propagator to each internal line in a loop, use momentum conservation at the vertices to reduce the number of independent momenta, and integrate over the remaining unfixed momenta, with the measure $d^4k/(2\pi)^4$.

We define

$$\mathcal{A}(p) \equiv \frac{(-i\lambda)^2}{2} \int \frac{d^4k}{(2\pi)^4} \frac{i}{k^2-m^2+i\epsilon} \frac{i}{(p-k)^2-m^2+i\epsilon}\,. \qquad (5.102)$$

The $2 \to 2$ scattering amplitude at one loop level is then

$$i\mathcal{M}_{2\to 2} = -i\lambda + \mathcal{A}(p_1+p_2) + \mathcal{A}(p_1-k_1) + \mathcal{A}(p_1-k_2)\,, \qquad (5.103)$$

where the three contributions correspond to the three ways in which the process can take place, shown in Fig. 5.9.

Now however we discover that Feynman diagrams containing loops can be divergent! Indeed, the integral in eq. (5.102) diverges at large k, and is an example of an *ultraviolet (UV) divergence.* To study this integral we proceed as follows. First of all, we will limit for simplicity to the calculation of $\mathcal{A}(p)$ when $p = 0$.[8] Recall that the $i\epsilon$ prescription means that, in the complex k^0-plane, the pole at $k^0 > 0$ is below the real axis and the pole at $k^0 < 0$ is above the real axis, see Fig. 5.1. Therefore we can change the integration path in the complex k^0 plane, rotating counterclockwise from the real axis to the imaginary axis.

[8] For this graph, this is sufficient for extracting the divergent part, because the divergence comes from the region $k \to \infty$, where $(p-k)^2 \to k^2$, and there are no subleading divergencies.

This is called the *Wick rotation.* The integration variable is the complex variable k^0; on the real axis we write $dk^0 = dk^0_M$ while on the imaginary axis we write $dk^0 = idk^0_E$, where k^0_M and k^0_E are real variables, and the subscript denotes Minkowskian and Euclidean space, respectively. Then we obtain

$$\mathcal{A}(0) = i\frac{\lambda^2}{2}\int \frac{d^4k}{(2\pi)^4}\frac{1}{(k^2+m^2)^2}\,. \tag{5.104}$$

Here k is a Euclidean momentum, $k^2 = (k^0_E)^2 + \mathbf{k}^2$. The overall factor of i comes from the rotation.[9]

[9] To check the sign consider the complex z-plane, with $z = x + iy$; let C_1 denote the real axis running from $x = -\infty$ toward $x = +\infty$ and C_2 the imaginary axis again oriented from $y = -\infty$ to $y = +\infty$; then, closing the contour at infinity in the first and third quadrants, we have $\int_{C_1-C_2} dz = 0$ so that $\int_{C_1} dz = \int_{C_2} dz$. But $\int_{C_1} dz = \int dx$ and $\int_{C_2} dz = i\int dy$.

For our purposes the exact computation of the integral is not necessary. The important point is that the integral diverges in the UV. We introduce a cutoff Λ stating that we integrate only over Euclidean momenta with $k^2 < \Lambda^2$, and we extract the divergent part:

$$\begin{aligned}\mathcal{A}(0) &= i\frac{\lambda^2}{2}\int \frac{d^4k}{(2\pi)^4}\frac{1}{k^4} + \text{finite parts}\\ &= i\frac{\lambda^2}{2}\frac{1}{(2\pi)^4}\left(2\pi^2\right)\int^{\Lambda}\frac{dk}{k} + \text{finite parts}\\ &= i\lambda^2\frac{1}{16\pi^2}\log\Lambda + \text{finite parts}\,.\end{aligned} \tag{5.105}$$

The factor $2\pi^2$ is the solid angle in four dimensions. At p finite the calculation is more complicated, and depending on the value of p one can have poles also in the first and third quadrant. In general, when one has to perform more complicated computations, it is not convenient to use as regularization a cutoff over Euclidean momenta; there are techniques, in particular dimensional regularization, which are much more convenient (see Chapter 7). In any case, in this graph the divergent part is independent of p and the p dependence only enters in the finite parts. Since the divergence is independent of p, the three graphs in Fig. 5.9 give the same contribution to the divergence. As a result, at one-loop, the $2 \to 2$ scattering amplitude in $\lambda\phi^4$ is

$$i\mathcal{M}_{2\to 2} = -i\lambda + i\lambda^2(\beta_0 \log\Lambda + \text{finite parts})\,, \tag{5.106}$$

with

$$\beta_0 = \frac{3}{16\pi^2}\,. \tag{5.107}$$

The sign of β_0 plays a very important role and will be discussed in detail in Section 5.9.

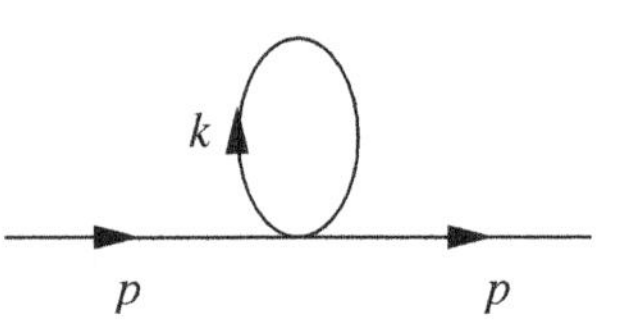

Fig. 5.11 A tadpole graph.

These divergences are typical of loop graphs. In the next section we will understand how they can be cured. First however we discuss another example of divergence, considering the two-point function in $\lambda\phi^4$ theory. In momentum space, at zero order in λ, this is just the Feynman propagator $\tilde{D}(p)$. At order λ we have the graph shown in Fig. 5.11. This graph is known as a tadpole graph.

Using the Feynman rules in momentum space and performing the Wick rotation $k^0 \to ik^0$, this is given by

$$-iB \equiv \frac{-i\lambda}{2} \int \frac{d^4k}{(2\pi)^4} \frac{1}{k^2+m^2} = -i\frac{\lambda}{32\pi^2} \left(\Lambda^2 - m^2 \log \frac{\Lambda^2+m^2}{m^2} \right), \tag{5.108}$$

so this term has a quadratic and a logarithmic divergence, both coming from the large k integration region, so they are again UV divergences. Observe also that the tadpole graph (with external legs amputated) is independent of the external momentum p. Actually, one can even resum the whole class of graphs shown in Fig. 5.12. Including also the propagators from the external legs, the result for the two-point function in momentum space is

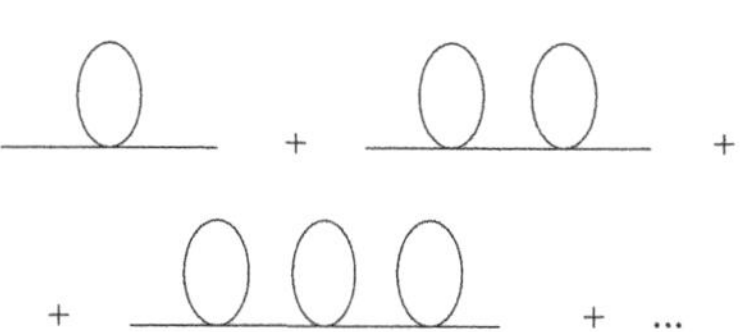

Fig. 5.12 The resummation of tadpole graphs.

$$\begin{aligned} &\tilde{D}(p) + \tilde{D}(p)(-iB)\tilde{D}(p) + \tilde{D}(p)(-iB)\tilde{D}(p)(-iB)\tilde{D}(p) + \dots \\ &= \tilde{D}(p) \left(1 + (-iB\tilde{D}(p)) + (-iB\tilde{D}(p))^2 + \dots \right) \\ &= \tilde{D}(p) \frac{1}{1+iB\tilde{D}(p)} = \frac{i}{p^2-m^2} \left(\frac{1}{1-\frac{B}{p^2-m^2}} \right) = \frac{i}{p^2-m^2-B} . \end{aligned} \tag{5.109}$$

We see that the net effect is to shift the mass from m^2 to m^2+B. We will make use of this fact when we study the renormalization of the theory. At $O(\lambda^2)$ there are further contributions to the two-point function. One possible graph is shown in Fig. 5.13. Iterating graphs of this type we obtain again a geometric series, see Fig. 5.14, so again the result goes in the denominator.

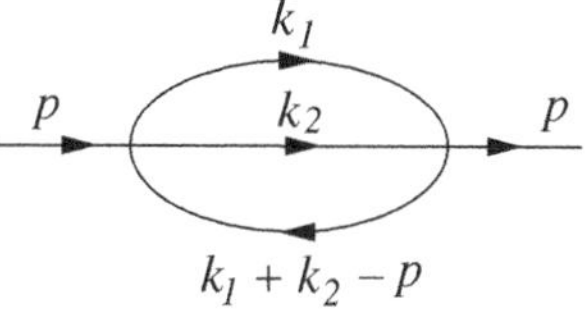

Fig. 5.13 A two-loop correction to the propagator.

The Feynman graph in Fig. 5.13 gives

$$i\frac{\lambda^2}{6} \int \frac{d^4k_1}{(2\pi)^4} \frac{d^4k_2}{(2\pi)^4} \frac{1}{[(p-k_1-k_2)^2-m^2](k_1^2-m^2)(k_2^2-m^2)} . \tag{5.110}$$

The integral is somewhat more difficult to compute, compared to the previous examples, and the result turns out to be proportional to

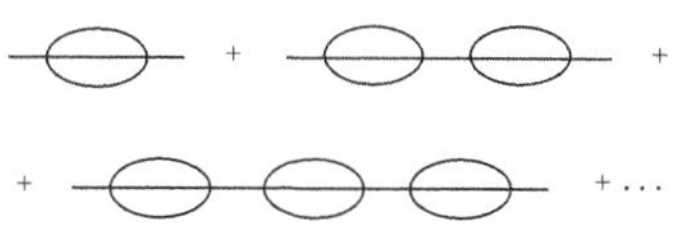

Fig. 5.14 The resummation of the graph shown in Fig. 5.13.

$$\lambda^2 p^2 \left(\log \frac{\Lambda^2}{p^2} + C \right), \tag{5.111}$$

so in this case the divergence depends on p^2. After resumming the geometrical series, the result for the two-point function in momentum space becomes of the form

$$\frac{i}{A(\Lambda,p^2)p^2 - m^2 - B(\Lambda)} \tag{5.112}$$

with $A(\Lambda,p^2) = 1 + \lambda^2(c_1 \log \frac{\Lambda^2}{p^2} + c_2)$, and c_1, c_2 some constants.

5.5.3 Summary of Feynman rules for a scalar field

It is now clear that it is not necessary to go each time explicitly through the process of developing the exponential of the interaction Hamiltonian to the desired order and performing the contractions using the Wick theorem, since the result can always be summarized by a simple set of rules. First of all, we have seen that it is usually more convenient to work in momentum space, and we summarize here the Feynman rules in momentum space. We consider for definiteness a real scalar field with an interaction term

$$\mathcal{H}_{\text{int}} = \frac{\lambda}{n!}\phi^n . \tag{5.113}$$

The first step is to draw all connected graphs corresponding to the given initial and final states, with n lines meeting at each vertex. For each graph proceed as follows:

- Neglect the external legs.
- Energy–momentum conservation must be imposed separately at each vertex.
- To each vertex associate a factor $-i\lambda$.
- To each internal line with momentum p associate a propagator

$$\tilde{D}(p) = \frac{i}{p^2 - m^2 + i\epsilon} . \tag{5.114}$$

- Integrate over the four-momenta k_i which are not fixed by energy–momentum conservation at each vertex, with a measure $d^4k_i/(2\pi)^4$.
- Include the appropriate combinatorial factor, which combines a factor $1/N!$ from the expansion of the exponential at order N, the number of equivalent contractions, and the factor $1/n!$ from the normalization of the coupling in the theory with $\mathcal{H}_{\text{int}} = (\lambda/n!)\phi^n$.

The sum of the contributions of all Feynman diagrams gives $i\mathcal{M}_{fi}$. This is related to the matrix element of the T operator by

$$\langle \mathbf{p}_1 \ldots \mathbf{p}_n | iT | \mathbf{k}_1 \ldots \mathbf{k}_m \rangle = (2\pi)^4 \delta^{(4)}\left(\sum_i p_i - \sum_j k_j\right) i\mathcal{M}_{fi} , \tag{5.115}$$

and T is related to the S-matrix by $S = 1 + iT$.

5.5.4 Feynman rules for fermions and gauge bosons

We now want to understand the Feynman rules for more interesting theories, like QED, which contains fermions and gauge bosons. The derivation is conceptually similar to what we have seen for the scalar fields. We will therefore just collect the relevant results, referring to Peskin and Schroeder (1995), Sections 4.7 and 4.8 for the derivations.

The fermion propagator. Wick's theorem can be generalized to fermionic fields, if we define the T-product of two Dirac fields as

$$T\{\Psi(x)\bar{\Psi}(y)\} = \begin{cases} \Psi(x)\bar{\Psi}(y) & x^0 > y^0 \\ -\bar{\Psi}(y)\Psi(x) & x^0 < y^0 . \end{cases} \tag{5.116}$$

The Feynman propagator for the Dirac field is

$$S(x-y) = \langle 0 | T\{\Psi(x)\bar{\Psi}(y)\} | 0 \rangle . \tag{5.117}$$

Observe that $S(x-y)$ is a 4×4 matrix in the Dirac indices. It can be computed explicitly expanding Ψ in creation and destruction operators and using the anticommutation relations (4.34). The result is

$$S(x-y) = \int \frac{d^4p}{(2\pi)^4} \tilde{S}(p) e^{-ip(x-y)}, \tag{5.118}$$

where the momentum space propagator is

$$\tilde{S}(p) = \frac{i(\not{p}+m)}{p^2 - m^2 + i\epsilon}. \tag{5.119}$$

Observe that

$$(\not{p}-m)(\not{p}+m) = \gamma^\mu\gamma^\nu p_\mu p_\nu - m^2 = p^2 - m^2, \tag{5.120}$$

since $p_\mu p_\nu$ is symmetric under $\mu \leftrightarrow \nu$ and therefore we can replace $\gamma^\mu\gamma^\nu \to (1/2)\{\gamma^\mu, \gamma^\nu\} = \eta^{\mu\nu}$. Then we can multiply by $(\not{p}-m)$ both the numerator and the denominator in eq. (5.119) (where dividing by $(\not{p}-m)$ means to multiply by the 4×4 Dirac matrix $(\not{p}-m)^{-1}$) and rewrite $\tilde{S}(p)$ in the form

$$\boxed{\tilde{S}(p) = \frac{i}{\not{p}-m},} \tag{5.121}$$

where the prescription for going around the poles is understood. An alternative way to compute the fermion propagator is to observe that, from the definition, $S(x-y)$ is a Green's function of the Dirac operator,

$$(i\not{\partial} - m)S(x-y) = i\delta^{(4)}(x-y). \tag{5.122}$$

It is then straightforward to check that eqs. (5.118) and (5.121) provide a solution, and the prescription for going around the poles is the same as in the scalar case, and corresponds to the Feynman propagator.

The photon propagator. By definition the photon propagator is

$$D_{\mu\nu}(x-y) = \langle 0|T\{A_\mu(x)A_\nu(y)\}|0\rangle. \tag{5.123}$$

Using the covariant quantization of the gauge field discussed in Section 4.3.2, the calculation is a simple generalization of the calculation for the KG field, with the only difference that now we have four different creation operators $a^\dagger_{\mathbf{p},\lambda}$ labeled by a Lorentz index $\lambda = 0, \dots 3$, and the commutator is given by eq. (4.106). Then the propagator in momentum space is simply

$$\boxed{\tilde{D}_{\mu\nu}(k) = \frac{-i}{k^2 + i\epsilon}\eta_{\mu\nu}.} \tag{5.124}$$

In other words, the spatial components A^i have the same propagator as a massless scalar field, while A^0 has the "wrong" sign. A more general form of the photon propagator will be given in Chapter 7.

The interaction vertex. While the propagators are fixed by the kinetic terms, i.e. by the free theory, the interaction vertices depend of course on the specific theory that we are considering. In QED the interaction term in the Hamiltonian is $eA_\mu \bar{\Psi}\gamma^\mu\Psi$. Let us recall from Section 4.2 that the expansion of the field $\Psi, \bar{\Psi}$ in terms of creation and annihilation operators is (see eqs. (4.32) and (4.33))

$$\Psi(x) = \int \frac{d^3p}{(2\pi)^3\sqrt{2E_{\mathbf{p}}}} \sum_{s=1,2} \left(a_{\mathbf{p},s}u^s(p)e^{-ipx} + b^\dagger_{\mathbf{p},s}v^s(p)e^{+ipx}\right) , \tag{5.125}$$

$$\bar{\Psi}(x) = \int \frac{d^3p}{(2\pi)^3\sqrt{2E_{\mathbf{p}}}} \sum_{s=1,2} \left(b_{\mathbf{p},s}\bar{v}^s(p)e^{-ipx} + a^\dagger_{\mathbf{p},s}\bar{u}^s(p)e^{+ipx}\right) , \tag{5.126}$$

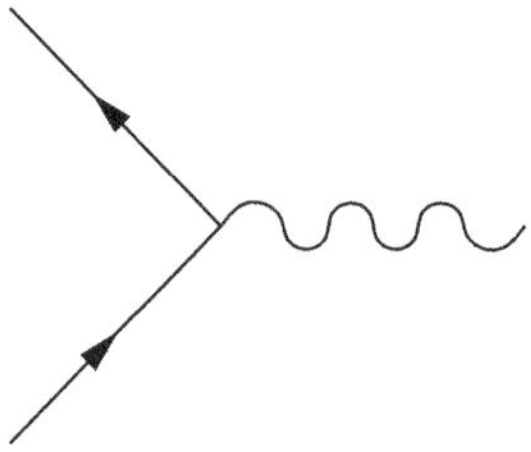

Fig. 5.15 The QED vertex: the solid lines represent the fermions and the wavy line the photon.

where $a_{\mathbf{p},s}$ destroys an electron (in a spin state labeled by s), $a^\dagger_{\mathbf{p},s}$ creates an electron, $b_{\mathbf{p},s}$ destroys a positron and $b^\dagger_{\mathbf{p},s}$ creates a positron. Therefore Ψ can destroy an electron or create a positron while $\bar{\Psi}$ can destroy a positron or create an electron. Similarly the gauge field, in the covariant quantization, has the expansion (4.104),

$$A_\mu(x) = \int \frac{d^3p}{(2\pi)^3\sqrt{2\omega_{\mathbf{p}}}} \sum_{\lambda=0}^{3} \left[\epsilon_\mu(\mathbf{p},\lambda)a_{\mathbf{p},\lambda}e^{-ipx} + \epsilon^*_\mu(\mathbf{p},\lambda)a^\dagger_{\mathbf{p},\lambda}e^{+ipx}\right] , \tag{5.127}$$

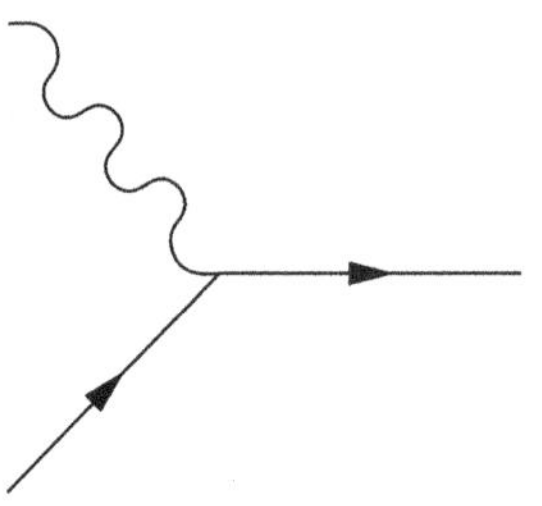

Fig. 5.16 The same interaction vertex, describing $e^-\gamma \to e^-$.

and can destroy or create a photon. Therefore in $eA_\mu\bar{\Psi}\gamma^\mu\Psi$ there are all possible terms with two fermion lines and one photon line, which conserve the electric charge: for instance, we can destroy an electron with Ψ and create it back with $\bar{\Psi}$ while at the same time emitting a photon, corresponding to a vertex $e^- \to e^-\gamma$; or we can absorb the photon, corresponding to a vertex $e^-\gamma \to e^-$; or we can destroy an electron with Ψ, destroy a positron with $\bar{\Psi}$ and create a photon, $e^+e^- \to \gamma$, etc.

All these possibilities are summarized associating a factor

$$\boxed{-ie\gamma^\mu} \tag{5.128}$$

to the interaction vertex of Fig. 5.15. As in the scalar field theory, the factor $-i$ in eq. (5.128) comes from the fact that in the T-product appears the exponential of $-iH_I$. In Fig. 5.15 the solid line can represent either an electron propagating in the direction of the arrow or a positron propagating in the opposite direction. If we imagine that time runs from left to right, then Fig. 5.15 actually describes the process $e^+e^- \to \gamma$, while $e^-\gamma \to e^-$ will be drawn as in Fig. 5.16, etc.[10]

The interaction vertex is proportional to γ^μ and therefore is a matrix in the Dirac indices and carries a Lorentz index.

[10] Observe that for the physical process $e^+e^- \to \gamma$ the matrix element $\mathcal{M}_{fi}$ is non-vanishing, $i\mathcal{M}_{fi} = ie\gamma^\mu$ but the matrix element of iT is zero because the Dirac delta in eq. (5.98) cannot be satisfied, so the process is forbidden by energy–momentum conservation. However, the vertex of Fig. 5.15 enters as a building block in all other Feynman diagrams of QED.

The external lines. In the case of the scalar field, acting with the field operator ϕ on the vacuum to create a particle brings a factor e^{ipx} while destroying a particle brings a factor e^{-ipx}, see eqs. (4.21) and (4.22). This is the origin of the factors $e^{ip_ix_i}$ for each final particle and $e^{-ik_jy_j}$

for each initial particle in the LSZ formula (5.46). From eqs. (5.125), (5.126) and (5.127) we see that for fermions and gauge bosons, together with the exponential factors (which, as we have seen, combine to give an overall energy–momentum conservation and transform the position space propagators into momentum space propagators) there are further factors associated to the external legs. Namely

- A factor $\epsilon_\mu^*(k)$ for each final photon with momentum k and polarization given by $\epsilon_\mu(k)$.
- A factor $\epsilon_\mu(k)$ for each initial photon with momentum k and polarization given by $\epsilon_\mu(k)$.
- A factor $u^s(p)$ for each initial electron with momentum p and spin state s.
- A factor $v^s(p)$ for each final positron with momentum p and spin state s.
- A factor $\bar{u}^s(p)$ for each final electron with momentum p and spin state s.
- A factor $\bar{v}^s(p)$ for each initial positron with momentum p and spin state s.

In other words, to each initial particle is associated its wave function, and to each final particle is associated the complex conjugate of the wave function (or the Dirac adjoint, for Dirac spinors). For an elementary scalar field the wave function is just the plane wave e^{-ipx} while for particles with spin there is also the spin wave function, e.g. $\epsilon_\mu(k)$ for a photon or $u^s(p)$ for an electron.

Closed fermionic loops. Finally, from the anticommuting nature of fermionic fields, it follows that for each closed fermionic loop there is an additional minus sign.

5.6 Renormalization

The basic idea of renormalization is the following. We have seen that some diagrams give divergent contributions. The first step is therefore to *regularize* the theory. For instance, we can put a cutoff Λ over the modulus of the Euclidean momenta, as we have done above (for technical reasons, especially in gauge theories, there are more convenient choices of the regularization scheme; however, for understanding the general ideas, we will use this cutoff). Eventually we want to send $\Lambda \to \infty$ but, as long as we take Λ finite, our theory has a dependence on the cutoff, and therefore we begin by defining the theory at finite Λ admitting that even the couplings, the masses and the fields depend on Λ, in a way which for the moment we leave unspecified. In $\lambda\phi^4$ theory we will therefore write the Lagrangian at finite Λ as

$$\mathcal{L} = \frac{1}{2}(\partial\phi_0)^2 - \frac{1}{2}m_0^2\phi_0^2 - \frac{\lambda_0}{4!}\phi_0^4\,, \tag{5.129}$$

where the subscript 0 indicates that these quantities depend on the cutoff Λ: $\phi_0 = \phi_0(x;\Lambda), m_0 = m_0(\Lambda), \lambda_0 = \lambda_0(\Lambda)$. We call $\phi_0(x;\Lambda)$ the *bare field*, $m_0(\Lambda)$ the *bare mass* and $\lambda_0(\Lambda)$ the *bare coupling*.

Consider first the two-point amplitude. In this new notation the two-point function of the bare field, eq. (5.112), is written as

$$\langle 0|T\{\phi_0(x,\Lambda)\phi_0(y,\Lambda)\}|0\rangle_c$$
$$= \int \frac{d^4p}{(2\pi)^4} \frac{i}{A(\Lambda,p^2)p^2 - m_0^2(\Lambda) - B(\Lambda)} e^{-ip(x-y)} . \tag{5.130}$$

The $i\epsilon$ prescription in the denominator is understood. We saw in eq. (5.108) that $B(\Lambda)$ is divergent as Λ^2 at the one-loop level,

$$B(\Lambda) = \frac{\lambda_0(\Lambda)}{32\pi^2} \left(\Lambda^2 + O(\log\Lambda) + \text{finite parts}\right) , \tag{5.131}$$

and A diverges as $\log\Lambda$ at the two-loop level,

$$A(\Lambda,p^2) = 1 + \lambda_0^2(\Lambda) \left(c_1 \log\frac{\Lambda^2}{p^2} + c_2\right) . \tag{5.132}$$

Observe that A and B also have an implicit, and as yet unspecified, dependence on Λ through $\lambda_0(\Lambda)$. For simplicity, let us at first examine eq. (5.130) at the one-loop level. Then the two-point function in momentum space is

$$\frac{i}{p^2 - m_0^2(\Lambda) - B(\Lambda)} , \tag{5.133}$$

i.e. $A = 1$, and we recall that B is independent of p^2. The basic idea is that neither $m_0(\Lambda)$ nor $B(\Lambda)$ are physically observable. Rather, the physical or *renormalized mass* m_R is defined by

$$m_R^2 = m_0^2(\Lambda) + B(\Lambda) . \tag{5.134}$$

In other words, we fix the physical mass requiring that the propagator has a pole at $p^2 = m_R^2$. Since $m_0(\Lambda)$ is a parameter completely in our hands, we *choose* it such that it cancels the divergence in $B(\Lambda)$, and it leaves us with a value of m_R finite and equal to the measured physical value.

At the two-loop level the situation is slightly more complicated because there is also the divergence coming from A. However, we still define m_R as the position of the pole of the propagator, i.e. by the condition

$$[A(\Lambda,p^2)p^2 - m_0^2(\Lambda) - B(\Lambda)]_{|p^2=m_R^2} = 0 . \tag{5.135}$$

This is one condition, and is not yet sufficient to eliminate the two divergencies coming from A and B. However, expanding the function $A(p^2)p^2 - m_0^2 - B$ near $p^2 = m_R^2$, we find that close to the pole

$$\int d^4x\, e^{ipx} \langle 0|T\{\phi_0(x,\Lambda)\phi_0(0,\Lambda)\}|0\rangle_c = \frac{iZ}{p^2 - m_R^2} + \dots , \tag{5.136}$$

where

$$Z = Z\left(\lambda_0(\Lambda), \frac{\Lambda}{m_R}\right) \equiv \left[\left(\frac{d}{d(p^2)} A(\Lambda, p^2)p^2\right)_{|p^2=m_R^2}\right]^{-1}, \qquad (5.137)$$

and the dots represent terms that are finite for $p^2 = m_R^2$. For later reference, we have also written explicitly that Z depends also on the bare coupling λ_0. Furthermore, being dimensionless, Z can depend on Λ and m_R only through the combination Λ/m_R. Now we define the *renormalized field* ϕ_R from

$$\phi_0(x, \Lambda) = Z^{1/2}\left(\lambda_0(\Lambda), \frac{\Lambda}{m_R}\right) \phi_R(x)\,. \qquad (5.138)$$

By definition ϕ_R is independent of the cutoff, and this fixes the dependence of ϕ_0 on Λ. The factor $Z^{1/2}$ is called the wave function renormalization, or field renormalization. We see from eq. (5.136) that in terms of ϕ_R the two-point function is the same as that of a free field with mass m_R, and therefore Z is the same factor that appeared in the LSZ reduction formula (5.46). In other words, Z disappears from the LSZ formula if, instead of using the bare field ϕ_0, as we did in eq. (5.46), we use the "physical", renormalized field. Thus, after mass and wave function renormalization, the on-shell two-point function is finite.[11]

[11] The fact that to see the wave function renormalization we had to go to two-loops is a peculiarity of $\lambda\phi^4$ theory. In general theories $Z \neq 1$ already at one loop.

Now that we have made the two-point function finite, we turn our attention to the four-point function. At one-loop there are two types of divergences in the four-point function. The first is associated to the graphs in Fig. 5.9, and we have seen that it is a logarithmic divergence. The second is associated with graphs like Fig. 5.17, i.e. tadpoles on external legs. The crucial point is that the divergence due to tadpoles is automatically cured by the renormalization of the two-point function, i.e. by the mass and wave function renormalization, because it is a divergence that concerns only a subgraph of Fig. 5.17, corresponding to a two-point function, and we have already made the two-point function finite. The graphs in Fig. 5.9 give instead a genuinely new divergence, and we computed it in eq. (5.106). Actually, to renormalize the divergence, we have first to be more careful in specifying the kinematical configuration. A simple choice is to consider the scattering amplitude in the limit of zero spatial momentum, i.e. $p_1 = p_2 = k_1 = k_2 = (m_R, 0)$. With this choice it is clear that, in eq. (5.106), the only scale that can be combined with Λ to give a dimensionless argument of the logarithm is m_R, so we rewrite eq. (5.106) (with our new notation λ_0 for the bare coupling)

Fig. 5.17 A tadpole on an external line.

$$\begin{aligned} i\mathcal{M}_{2\to 2}(\mathbf{p}_i = \mathbf{k}_i = 0) &\equiv -i\lambda_R \qquad (5.139)\\ &= -i\lambda_0(\Lambda)\left[1 - \lambda_0(\Lambda)\left(\beta_0 \log\frac{\Lambda}{m_R} + \text{finite parts}\right)\right] + O(\lambda_0^3)\,. \end{aligned}$$

Now, λ_R is the quantity that is measured performing a scattering experiment and which therefore must be finite. We call it the *physical*, or *renormalized* coupling. We therefore choose the parameter $\lambda_0(\Lambda)$, which is completely in our hands, requiring that λ_R is finite, and equal to the desired value.

If we consider the scattering amplitude in a different kinematical regime, for instance when $(p_1 + p_2)^2 \equiv q^2 \gg m_R^2$, we find instead an amplitude

$$i\mathcal{M}_{2\to 2}(q^2) = -i\lambda_0(\Lambda)\left[1 - \lambda_0(\Lambda)\left(\frac{\beta_0}{2}\log\frac{\Lambda^2}{q^2} + \text{finite parts}\right)\right] + O(\lambda_0^3). \tag{5.140}$$

This result follows from the explicit computation, but it is easily understood observing that in the limit $q^2 \gg m_R^2$ the relevant dimensional scale is provided by q^2 rather than m_R^2, so it is this scale that combines with Λ to provide a dimensionless argument of the logarithm. Writing $\log(\Lambda^2/q^2) = \log(\Lambda^2/m_R^2) + \log(m_R^2/q^2)$ and using the definition of λ_R from eq. (5.139), we get

$$i\mathcal{M}_{2\to 2}(q^2) = -i\lambda_R\left[1 + \lambda_R\frac{\beta_0}{2}\log\frac{q^2}{m_R^2}\right] + O(\lambda_R^3), \tag{5.141}$$

where we could replace λ_0^2 with λ_R^2 since terms $O(\lambda_0^3)$ are neglected anyway.

The important point that we understand from eq. (5.141) is that, once we have made finite the four-point amplitude at a given value of the external momenta, it is finite for all momenta.

In this way we have cured the divergences of the two-point and four-point amplitudes. We can continue and examine the six-point amplitude and so on. In principle, what can happen is that in the Feynman graphs that determine the six-point amplitude there are divergent subgraphs that are automatically cured by the renormalization of the two- and four-point functions plus possibly some genuinely new divergence. Figure 5.18 shows an example of a graph contributing to the six-point amplitude with a divergent subgraph that is automatically cured by the renormalization of the four-point function. Observe that it is possible to separate this graph into two disconnected parts by cutting a single line, along the dashed line in Fig. 5.18. Such graphs are called one-particle reducible, and cannot carry genuinely new divergences.

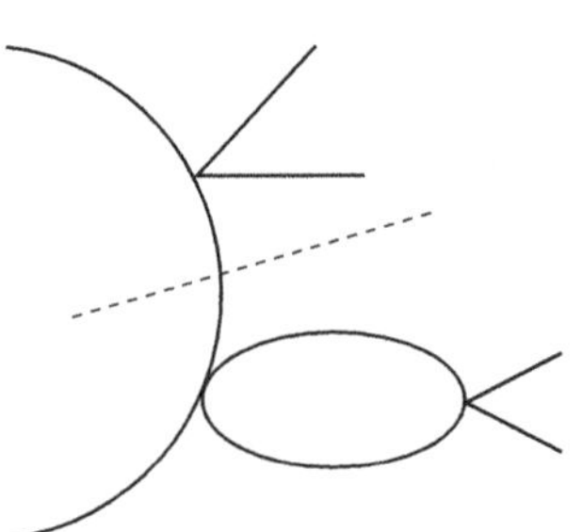

Fig. 5.18 A one-particle reducible graph. Cutting along the dashed line, the graph is separated in two disconnected pieces.

If instead it is not possible to make the graph disconnected by cutting just one line, the graph is called one-particle irreducible (often abbreviated 1PI), and can in principle carry genuinely new divergences.

In the case of $\lambda\phi^4$ theory we saw that in the four-point function there were graphs cured by the renormalization of the two-point function and a genuinely new divergence which required the renormalization of λ_R. If this were the case also for the six-point function, after the renormalization of the field, of the mass and of λ we would still be left with a divergent result for the six-point function. To cure it, we could introduce a new term proportional to ϕ^6 in the Lagrangian, with a new bare coupling $\lambda_{(6),0}(\Lambda)$. This would give a further contribution to the six-point amplitude, and we could choose $\lambda_{(6),0}(\Lambda)$ so that it cancels the divergence that was left. This means that we should again fix the renormalized coupling $\lambda_{(6),R}$ by comparison with experiment.

If this process never terminates, and each time that we consider a new amplitude with a larger number of external legs we must introduce a new coupling and fix the amplitude at a certain energy by comparison with experiment, then the theory that we have constructed by this renormalization procedure is finite, because all divergences have been reabsorbed, but (apparently) has little predictive power, because we have introduced an infinite number of parameters, to be fixed by experiment. Then the theory is called non-renormalizable. Actually, even non-renormalizable theories can be very useful, but we postpone their discussion until Section 5.8.

If instead at some point the process terminates, we just need to eliminate the divergences from a few amplitudes, fixing a few parameters by comparison with experiment, and then all the other amplitudes are automatically finite. In this case the theory is called renormalizable.

The criterion for understanding when a theory is renormalizable turns out to be quite simple and holds not only for the scalar theory that we are considering, but more generally (although the actual proof of renormalizability for gauge theories, and especially for non-abelian gauge theory, is far from trivial!). Consider for example a theory with interaction Hamiltonian $\lambda\phi^n$, with $n \geqslant 3$, integer, in four space-time dimensions. In a general Feynman diagram there will be an integration over d^4k for each loop in the graph and a factor of the type $1/((k-p)^2 - m^2)$ for each propagator on an internal line, where k is one of the integration variables (or a linear combination of integration variables) and p is a combination of external momenta. So each loop integration carries four powers of momenta at the numerator and each internal line two powers at the denominator. In this theory the *superficial degree of divergence* D is then defined as

$$D = 4L - 2N_i \tag{5.142}$$

where L is the number of loops and N_i the number of internal lines. The number of loops can be expressed as

$$L = N_i - V + 1 \tag{5.143}$$

where V is the number of vertices of the graph. We can check it observing that the simplest tree level graph has one vertex and no internal line, so $V = 1, N_i = 0$ and eq. (5.143) correctly gives $L = 0$. Adding a second vertex and connecting it to the first with just one propagator still gives a tree level graph, and eq. (5.143) still correctly gives $L = 0$, since we have increased both N_i and V by one. Similarly we construct the most general tree level graph adding each time one vertex and one internal line. Instead, each time a new vertex is joined by two lines, we have added a loop and correctly eq. (5.143) shows that L increases by one.

Finally, if the theory is $\lambda\phi^n$, there are n lines at each vertex, so

$$2N_i + N_e = nV \tag{5.144}$$

where N_e is the number of external lines and the factor 2 reflects the fact that one internal line connects two vertices. Combining these expressions we find

$$D = (n-4)V + 4 - N_e \,. \tag{5.145}$$

If $D \geqslant 0$ we expect that the diagram is divergent, unless some numerical cancellation in the leading term appears ($D = 0$ corresponds to $\int d^4k/k^4$, i.e. to a logarithmic divergence). If $D < 0$ the diagram is not necessarily convergent. In fact, the various integrations and propagators could be distributed in such a way that there is a divergent subgraph. However, these divergences are cured by the renormalization of the Green's function with a smaller number of points, and do not bother us. Genuinely new divergences instead do not appear if $D < 0$. The condition for renormalizability therefore is that only a finite number of Green's functions have $D \geqslant 0$. Consider first the case $n = 4$, i.e. $\lambda\phi^4$ theory. Then eq. (5.145) gives $D = 4 - N_e$. Therefore the only genuinely divergent graphs are those with no external legs, i.e. the vacuum-to-vacuum amplitude which diverges as Λ^4 (and that will be examined more closely in Section 5.7 in connection with the cosmological constant problem), the two-point function that, as we have seen, diverges as Λ^2, and the four-point function that diverges as $\log\Lambda$. After renormalizing these divergences, there is no other divergence to be cured, so the theory is renormalizable. Similarly, a theory $\lambda\phi^3$ is renormalizable since $(n-4)V = -V$ gives a negative contribution to D, and therefore helps the convergence.

If instead $n > 4$, for each N_e given there are graphs with a sufficiently large number of vertices which have $D \geqslant 0$. Therefore all Green's functions, at a sufficiently large order in perturbation theory, have genuinely new divergences, and the theory is not renormalizable.

The criterion $n \leqslant 4$ for $\lambda\phi^n$ theory can be understood in a way that can be generalized to other theories. The field ϕ has dimensions of mass, since the action is dimensionless, the kinetic term is $\sim \int d^4x(\partial\phi)^2$, and $\partial \sim 1/\text{length} = \text{mass}$. Requiring that $\int d^4x\, \lambda_n\phi^n$ is dimensionless, we see that the coupling λ_n has dimensions of $(\text{mass})^{4-n}$. Then the criterion $n \leqslant 4$ means that:

Terms in the Lagrangian whose coefficients have either a positive mass dimension or are dimensionless are renormalizable. Terms with negative mass dimension are not renormalizable.

In this form the criterion for renormalizability turns out to hold quite generally, and is not restricted to ϕ^n theories; the proofs however can be very complicated and depend on the details of the theory. The intuitive reason for this is however easily understood. If the coupling constant has for instance dimensions $1/M^2$ (as would be the case for a term ϕ^6 in four space-time dimensions) each new vertex brings in a new factor $1/M^2$. For dimensional reasons, this must be compensated by some parameter with dimensions of mass squared. Barring cancellations, we therefore expect to find divergences with higher and higher powers of Λ^2/M^2.

5.7 Vacuum energy and the cosmological constant problem

In Section 4.1.1 we found that the vacuum energy is divergent. It was our first example of a divergence in field theory, and we simply disposed of it eliminating the infinity by hand, with the physical argument that only energy differences can be observed. This physical argument is however incorrect when we include gravity, since in general relativity any form of energy contributes to the gravitational interaction. To understand this point, we recall a few basic facts of general relativity and of cosmology (see e.g. Kolb and Turner (1990)).

The energy density of the vacuum, $\rho_{\rm vac}$, is conveniently written as

$$\rho_{\rm vac} = \frac{\Lambda}{8\pi G_N}\,, \tag{5.146}$$

where Λ is called the cosmological constant (in this section we reserve the symbol Λ for the cosmological constant and we denote the cutoff by $\Lambda_{\rm cut}$). This normalization is chosen so that the Einstein equations of general relativity, in the presence of a cosmological constant, read[12]

$$G_{\mu\nu} = 8\pi G T_{\mu\nu} + \Lambda g_{\mu\nu} \tag{5.147}$$

[12] The signs depend on a number of conventions on the metric signature, Riemann tensors, etc. Here we are following the conventions used for instance in Kolb and Turner (1990) and Landau and Lifshitz, vol. II (1979).

where $g_{\mu\nu}$ is the metric, $G_{\mu\nu}$ is the Einstein tensor, which contains up to two derivatives of the metric, and $T_{\mu\nu}$ is the energy–momentum tensor of matter. The Universe on very large scales is in a first approximation homogeneous, and can be described by the Friedmann–Robertson–Walker (FRW) metric,

$$ds^2 = dt^2 - R^2(t)(dx^2 + dy^2 + dz^2) \tag{5.148}$$

i.e. $g_{\mu\nu} = (1, -R^2, -R^2, -R^2)$ in these coordinates; $R(t)$ is known as the scale factor (actually we have restricted for simplicity to a spatially flat Universe, otherwise the spatial part of the metric is more complicated). The expansion of the Universe is encoded in the fact that $R(t)$ is growing. The energy–momentum tensor of a fluid of matter is of the form $T^{\mu}{}_{\nu} = (\rho, -p, -p, -p)$, where ρ is the energy density and p is the pressure. Then, using the above form of the metric, one can show that eq. (5.147) implies an equation for the acceleration $\ddot{R}$,

$$\frac{\ddot{R}}{R} = -\frac{4\pi G_N}{3}(\rho + 3p)\,. \tag{5.149}$$

For ordinary matter ρ and p are positive, therefore $\rho+3p > 0$ and $\ddot{R} < 0$: *the effect of ordinary matter is to produce a deceleration of the expansion of the Universe.* Instead, we see from eq. (5.147) that the effect of Λ on the equation for $\ddot{R}$ is formally equivalent to that of a fluid with energy–momentum tensor $T^{\mu}{}_{\nu} = (\Lambda/8\pi G)\delta^{\mu}_{\nu} = \rho_{\rm vac}\delta^{\mu}_{\nu}$, and therefore to a fluid with $\rho = \rho_{\rm vac}$ and $p = -\rho_{\rm vac}$. This means that $\rho + 3p = -2\rho_{\rm vac} < 0$: *a positive vacuum energy density contributes to accelerate, rather than decelerate the Universe!*

A detection of the vacuum energy is therefore in principle possible from cosmological measurements. Without entering into the details that are beyond the scope of this course, we just mention that the most important observations have been obtained from type Ia supernovae and from the fluctuations of the cosmic microwave background (CMB). The result is usually expressed introducing the quantity

$$\Omega_\Lambda = \frac{\rho_{\rm vac}}{\rho_c}, \tag{5.150}$$

where ρ_c is the critical density for closing the Universe. The combination of supernovae and CMB measurement indicates a non-zero value $\Omega_\Lambda \simeq 0.7 \pm 0.1$. Using the known value of ρ_c, this means that

$$\rho_{\rm vac}^{1/4} \simeq 2 \times 10^{-3}\,{\rm eV}. \tag{5.151}$$

This result stimulates us to look in more detail into the problem of the vacuum energy, using the language of renormalization that we have just developed. As we saw in Section 4.1.1, when we compute the vacuum energy density we find a result that diverges quartically with the cutoff,

$$\rho_{\rm vac} = c\Lambda_{\rm cut}^4, \tag{5.152}$$

with c some numerical constant. According to the general rules of renormalization of QFT, we then say that the correct starting point is a Lagrangian which also has a bare vacuum energy density. So, for instance, for a $\lambda\phi^4$ theory we would generalize eq. (5.129) to

$$\mathcal{L} = \frac{1}{2}(\partial\phi_0)^2 - \frac{1}{2}m_0^2\phi_0^2 - \frac{\lambda_0}{4!}\phi_0^4 - \rho_0, \tag{5.153}$$

where the bare vacuum energy ρ_0 depends on the cutoff $\Lambda_{\rm cut}$, similarly to m_0, λ_0, ϕ_0. Computing the vacuum energy density with this Lagrangian we now find

$$\rho_{\rm vac} = \rho_0(\Lambda_{\rm cut}) + c\,\Lambda_{\rm cut}^4. \tag{5.154}$$

The bare quantity ρ_0 is a parameter completely in our hands; we can choose it positive or negative, as we wish, and we fix it requiring that the physical energy density of the vacuum $\rho_{\rm vac}$ has the value determined by experiment. It is important to make clear two points:

- In field theory quantities like the cosmological constant, the elementary particle masses, the couplings (such as the fine structure coupling), etc., cannot be predicted. They are just fixed, by the renormalization procedure, to the experimentally observed values. Questions such as why the cosmological constant has a certain value, or why the electron mass is about 0.5 MeV, or why the fine structure constant is about 1/137 strictly speaking are ill defined in the framework of quantum field theory.[13]
- The bare quantities, such as $\rho_0(\Lambda_{\rm cut})$, are objects which are useful in the intermediate steps of the calculations, but they have no physical meaning. They are just chosen so that they cancel the divergences and leave us with the desired renormalized quantity.

[13] In string theory, at least as far as the masses and couplings are concerned, the question of why they have certain values becomes equivalent to why the extra dimensions have a certain geometry (compare with Exercise 3.6) or why certain fields acquire a given vacuum expectation value, so in this sense they become at least meaningful dynamical questions, which however presently we do not know how to answer. Instead, for the cosmological constant, even string theory presently does not shed any new light.

For these reasons, a possible attitude could be simply to take notice of the value (5.151) and observe that in QFT we can fix the renormalized vacuum energy density to this value. However, this value of the vacuum energy density is probably trying to tell us something very important that we do not yet understand. There are two main reasons for this.

(i) A fine tuning argument: even if strictly speaking in the framework of QFT we are not really allowed to ask why a renormalized quantity has a given value, still we can make the following observation. We know experimentally that, at least up to an energy scale of the order of one TeV, Nature is well described by a quantum field theory, the Standard Model. This means that in the Standard Model the cutoff $\Lambda_{\rm cut}$ can be taken to be at least 1 TeV ($= 10^{12}$ eV), and eq. (5.154) numerically reads something like

$$(2\times 10^{-3}{\rm eV})^4 = \rho_0(\Lambda_{\rm cut}) + c\,(10^{12}{\rm eV})^4\,. \tag{5.155}$$

Therefore $\rho_0(\Lambda_{\rm cut})$ must be chosen so that it cancels something of order $10^{48}\,{\rm eV}^4$, leaving something of order $10^{-11}\,{\rm eV}^4$. Even if $\rho_0(\Lambda_{\rm cut})$ is a parameter that we can choose at will, and of no physical meaning, still this requires an incredible fine tuning, at the level of 60 decimal figures, and in this sense it seems very unnatural. Furthermore, this fine tuning does not show up for most other observables in QFT, and is really specific to the vacuum energy (and to the Higgs mass, see below). The point is that most quantities have logarithmic, rather than power-like divergences. For instance, in QED at one-loop the renormalized electron mass m is related to the bare electron mass $m_0(\Lambda_{\rm cut})$ by

$$m = m_0\left(1+\frac{3\alpha}{4\pi}\log\frac{\Lambda_{\rm cut}^2}{m_0^2}\right)\,. \tag{5.156}$$

The cutoff $\Lambda_{\rm cut}$ appears only inside the log, and in front of the log we have α which is small. Therefore, even if we are so bold as to push the cutoff to the Planck scale, $\Lambda_{\rm cut}\sim 10^{19}$ GeV, with $m_0\sim$ MeV we have $(3\alpha/4\pi)\log\Lambda^2/m_0^2\sim 0.1$, so this is really a small correction and to reproduce a physical electron mass $m\simeq 0.5$ MeV we must indeed take a value m_0 of the same order of magnitude. Therefore here there is no fine tuning problem.[14]

(ii) The really crucial point, however, is that the milli-eV scale indicated by the experimental results does not remind us of anything meaningful in particle physics (except possibly neutrino masses, but there is no reason why the mass of any specific particle should be related to $\rho_{\rm vac}$), so it is difficult to see how such a value could be derived from fundamental physics. And especially, why, of all possible energy scales, it turns out that this value is just comparable to the energy density needed for closing the Universe?

[14] The only other important situation where one is confronted with a fine tuning problem similar to that of the vacuum energy is when we consider the renormalization of the mass of a scalar field. As we have seen in eqs. (5.131) and (5.134), the divergence in the mass of a scalar field is quadratic, rather than logarithmic. If we say that the cutoff is given by the Planck scale $M_{\rm Pl} = 10^{19}$ GeV, this poses a fine tuning problem for the Higgs field, which is a scalar field predicted by the Standard Model, and is expected to have a mass around a few hundred GeV. In fact in this case the bare mass m_0 should satisfy something like $(10^2{\rm GeV})^2 = m_0^2 + (10^{19}{\rm GeV})^2$. However, the problem here is different from the cosmological constant problem, since it really depends on the form of the vacuum fluctuations from the TeV scale up to the Planck scale, about which we know nothing experimentally. In particular, in supersymmetric extensions of the Standard Model this fine tuning problem disappears because above a few TeV the contribution of the vacuum energy due to bosons is canceled by the contribution of fermions.

5.8 The modern point of view on renormalizability

The fact that quantum field theory is plagued by divergences was already realized around 1929–1930 by Heisenberg and Pauli and, quite understandably, it was considered a major problem. It was then realized (by Dyson in 1949, building on work of Feynman, Schwinger, Tomonaga, and others) that in some theories these divergences can be reabsorbed into the redefinition of a finite number of parameters. These theories were then called renormalizable, and considered "honest" theories, while non-renormalizable theories were considered intrinsically sick.

The modern point of view (largely stimulated by the work of K. Wilson) is quite different. First of all, it is important to realize that the problem with non-renormalizable theories is not mathematical consistency, but rather predictivity. As we have seen in Section 5.6, we can reabsorb the infinities of a non-renormalizable theory, but the price is that any N-point amplitude $A_N(p_1, \ldots, p_N)$, at a sufficiently large order n in perturbation theory, will develop divergences which are not automatically cured by the renormalization of amplitude with less than N external legs. To cure it we must therefore introduce a new term, and a new coupling, in the Lagrangian, and then fix this new coupling comparing the contribution to A_N at perturbative order n with the experimental value.

The fact that in principle we have to fix an infinite number of quantities by comparison with experiment is however only an apparent disaster. The point is that, as we have seen, in a typical non-renormalizable theory the coupling is not dimensionless, but rather has the dimensions of inverse powers of mass. We suppose for definiteness that the coupling λ is dimensionally the inverse of a mass squared, so that it can be written as $1/M^2$, for some mass-scale M; we also assume for simplicity that we have just one typical momentum (or energy) scale, and we denote it by E. Then the renormalized perturbative expansion of an N-point amplitude A_N up to order $(\lambda^2)^n$ reads

$$A_N(E) = A_N^0(E)\left(1 + c_1\frac{E^2}{M^2} + c_2\frac{E^4}{M^4} + \ldots + c_n\frac{E^{2n}}{M^{2n}}\right). \qquad (5.157)$$

The quantities $c_1, \ldots c_{n-1}$ are finite and calculable, once we have renormalized the amplitudes with less than N points. Because of the genuinely new divergence at some order n, the coefficient c_n must instead be fixed by comparison with experiment, and this is the source of the loss of predictivity. However, we can now realize that at low energy, $E \ll M$, the lack of predictivity on c_n becomes completely irrelevant, because anyhow c_n is multiplied by the very small quantity $(E/M)^{2n}$. Therefore, if we want to make computations with a given accuracy, we can just renormalize a corresponding *finite* number of amplitudes and, as long as $E \ll M$, our ignorance about higher-order divergences is beyond the desired accuracy.

This means that:

Non-renormalizable theories are perfectly acceptable low-energy theories.

At $E \sim M$ the expansion (5.157) blows up, signaling that the theory is no longer meaningful, and a more complete theory must take its place. We therefore see that non-renormalizable theories have a built-in scale M that provides their limit of validity.

Renormalizable theories, in contrast, have no built-in mass-scale M which tells us from the beginning that we cannot trust them at $E > M$, and in principle they are mathematically consistent and predictive at any energy scale. The distinction between renormalizable and non-renormalizable theories, however, can be more mathematical than physical. For instance QED is a renormalizable theory, and so we could naively hope that it correctly describes Nature up to arbitrary energy scales. But in fact, experimentally we know that at a mass-scale $M \sim 100$ GeV QED merges with weak interactions in a larger theory, the Standard Model, and so pure QED is in any case a low-energy approximation to a more complete theory.

In other words, renormalizability is related to the behavior of the theory at infinitely large energies or, equivalently, at infinitesimally small distances. As such, it reflects rather formal mathematical properties of the field theory, since we can never test a theory up to infinitely large energies, not only practically, but even in principle, since quantum gravity must come into play at the latest at the Planck scale, $M_{\rm Pl} \sim 10^{19}$ GeV, i.e. at distances $l \sim 10^{-33}$ cm.

Rather than focus on the concept of renormalizability, the modern approach focuses on the concept of *effective field theory*. The really crucial point is that if we want to compute, with a given precision, processes that take place at a given length-scale l (or, equivalently, at a corresponding energy scale E), we do not need to know the full theory at infinitely small distances or infinitely high energies. To study what happens in an atom, at $l \sim 10^{-8}$ cm, we do not need to know what happens at the scale $l \sim 10^{-17}$ cm, typical of weak interactions, except if we are looking for extremely fine effects, and certainly we do not need to know what happens at $l \sim 10^{-33}$ cm where quantum gravity becomes important. Once we have fixed the scale l which we are interested in and the level of precision that we want to get from our computations, all we need to know is the effective theory down to a lenghtscale l_* a few orders of magnitude smaller than l. How many orders of magnitude depends, of course, on the level of precision at which we aim.

In Chapter 8 we will discuss the Fermi theory of weak interactions, which is the low-energy limit of the electroweak theory. We will see that it is an extremely useful low-energy theory, despite the fact that it is not renormalizable.

A very important example of a non-renormalizable theory is the theory obtained quantizing general relativity. The coupling constant in this case is the Newton constant which dimensionally, in our units $\hbar = c = 1$, is the inverse of mass squared, $G_N = 1/M_{\rm Pl}^2$. Therefore quantum gravity is not renormalizable. However, the loss of predictivity comes into play only in processes that probe space-time down to distances of order of the Planck length $l_{\rm Pl} = 1/M_{\rm Pl} \simeq 10^{-33}$ cm. Such small scales will never be reached in accelerator physics, and probably our best hopes to get information on such small length-scales, or correspondingly high energies, come from some relics of the Big Bang. In any case, in "normal" conditions, classical general relativity gives a completely adequate description, and the lack or renormalizability of the quantum theory is irrelevant.

In Section 9.5 we will come back to the effective Lagrangian approach, and we will see that it is rooted in the concept of universality for critical phenomena.

5.9 The running of coupling constants

A surprising effect of the renormalization procedure is that, after renormalization, the coupling "constants" are not constant at all, but they depend on the energy. We have already seen this in eq. (5.141), where we computed the $2 \to 2$ scattering amplitude when the center of mass energy squared is q^2. This is nothing but the effective coupling constant at $E^2 = q^2$, and we see that it is different from the value λ_R at $E = 0$. We found in eq. (5.141) that at energies $E \gg m_R$ the amplitude can be written as

$$i\mathcal{M}_{2\to 2}(q^2) = -i\lambda_{\rm eff}(E)\,, \tag{5.158}$$

where the effective one-loop coupling constant $\lambda_{\rm eff}(E)$ is given by

$$\lambda_{\rm eff}(E) = \lambda_R + \lambda_R^2 \beta_0 \log\frac{E}{m_R} + O(\lambda_R^3)\,. \tag{5.159}$$

The same formula holds in the limit $E \ll m_R$, as one can see from explicit calculation. We see that the sign of the coefficient β_0 plays an especially important role. If $\beta_0 > 0$ the coupling increases in the UV (until it becomes large and we cannot trust the perturbative expression). If instead $\beta_0 < 0$ the coupling becomes smaller when we go to higher energies.

Theories where the coupling becomes small in the UV are called *asymptotically free*. Asymptotic freedom means that at large energies (and therefore at short distances) the fields which appear in the Lagrangian can be treated perturbatively. The other side of the coin, however, is that in an asymptotically free theory at low energies (i.e. large distances) the coupling becomes strong, and the perturbative treatment is inadequate. In this regime, the degrees of freedom that one actually observes are not described by the fields that enter in the Lagrangian, but are rather composite objects built from these more fundamental degrees of freedom, and bound by the strong interaction.

The most important example of this situation is QCD. As we will see in Chapter 10, in QCD the basic fields which enter the Lagrangian are quarks and gluons, and at distances $l \ll 1$ fm their interaction can be treated perturbatively. At $l \sim 1$ fm, however, the interaction becomes strong and quarks and gluons do not even appear as free particles, but rather they are confined into hadrons.

In this section we study the dependence of the coupling constants on the energy (the "running of the coupling constants" as it is usually called) in more detail. The rest of this section is more advanced and can be skipped at a first reading.

First of all, let us be more general: rather than focusing on the four-point function, we consider a generic n-point function. A general *renormalized* n-point function Γ_R depends on the external momenta p_i (or simply on just one invariant q^2 if we choose a simpler kinematical situation, rather than the most general), on the renormalized coupling g_R (for a general theory, not necessarily of a scalar field, we use the notation g rather than λ for the coupling; we also assume for simplicity that there is just one coupling, but the generalization is straightforward), and on the scale μ used to define the renormalization procedure,

$$\Gamma_R = \Gamma_R(p_i; g_R, \mu) \,. \tag{5.160}$$

In the previous section we renormalized the theory choosing $\mu = m_R$: we first defined m_R as the position of the pole in the propagator; we then defined the renormalized fields requiring that their two-point function has residue $+i$ at the pole $p^2 = m_R^2$. We finally fixed the finite value of the four-point function at zero momentum, i.e. when the square of the center-of-mass energy s was equal to $4m_R^2$. So m_R was always our mass-scale used to fix the finite values of the renormalized quantities.

However we can be more general, and use a generic mass-scale μ to define the renormalization procedure. For instance, we can decide to fix the value of the renormalized constant looking at the four-point amplitude at a value $q^2 = \mu^2 \gg m_R^2$, or even at a space-like value $q^2 = -\mu^2$. It is quite useful to keep μ generic and see what are the consequences of a change in μ.

The relation between Γ_R and the bare n-point function Γ_0 is[15]

$$\boxed{\Gamma_R(p_i; g_R, \mu) = Z^{-n/2}\left(g_0(\Lambda), \frac{\Lambda}{\mu}\right) \Gamma_0(p_i; g_0(\Lambda), \Lambda) \,.} \tag{5.161}$$

[15] We assume for simplicity that in the theory there is only one type of field. Otherwise, the wave function renormalization factors depend on the field, and in eq. (5.161) there will be a factor $Z^{-1/2}$ for each field.

The important point is that Γ_R, by definition, does not depend on the cutoff Λ. We have adjusted the bare coupling, bare mass and field renormalization just in such a way that all renormalized Green's functions are finite. In the following we will be interested in the situation where the typical energy scales are much bigger than the masses, and we will neglect all mass dependences.

Γ_0 instead has been computed in the regularized theory, using the bare coupling and the bare masses, and therefore depends on the cutoff Λ both explicitly and, implicitly, through the bare coupling $g_0(\Lambda)$ and the bare masses, while of course it is independent of μ, because the scale μ enters only afterwards, when we fix the value of the renormalized Green's function with a renormalization prescription at momenta given by μ, e.g. at $q^2 = \mu^2$ for the four-point function.

The factor $Z^{-n/2}$ is the contribution from the wave function renormalization of the n fields. As we have seen, it is obtained computing first the two-point amplitude in the bare theory. The result of this first step is therefore a function of Λ, of $g_0(\Lambda)$ and of p^2; then Z is defined fixing the value of the numerator of the two-point function at a given value of $p^2 = \mu^2$. For instance, in Section 5.6 we fixed it choosing $\mu = m_R$ and requiring that the residue of the pole of the propagator at $p^2 = m_R^2$ is $+i$. So, in general, Z is a function of $\Lambda, g_0(\Lambda)$ and μ (and not of p^2), see e.g. eq. (5.137). Since Z is a dimensionless quantity, it can only depend on Λ and μ through their ratio Λ/μ, in the high-energy limit in which all masses can be neglected.

Since Γ_R is independent of Λ, we can write

$$\Lambda\frac{d\Gamma_R}{d\Lambda} = 0 \tag{5.162}$$

and using eq. (5.161) we obtain

$$\boxed{\left[\Lambda\frac{\partial}{\partial\Lambda} + \beta(g_0)\frac{\partial}{\partial g_0} - n\eta(g_0)\right]\Gamma_0(p_i; g_0, \Lambda) = 0\,,} \tag{5.163}$$

where

$$\boxed{\beta(g_0) \equiv \Lambda\frac{dg_0}{d\Lambda}\,,} \tag{5.164}$$

$$\boxed{\eta(g_0) \equiv \frac{1}{2}\Lambda\frac{d}{d\Lambda}\log Z\,.} \tag{5.165}$$

Equation (5.163) is called a renormalization group equation, and eqs. (5.164) and (5.165) define the beta function and the eta function of the theory.[16]

[16] In principle, η depends also on Λ/μ. However Z depends on Λ/μ through terms $\sim \log\Lambda/\mu$, which after taking the derivative with respect to $\log\Lambda$ are independent of μ. There are also subleading terms $\log\log\Lambda/\mu$ in Z; after taking the derivative these become $1/(\log\Lambda/\mu)$ and disappear in the limit $\Lambda/\mu \to \infty$, leaving a finite function $\eta(g_0)$.

Equations (5.163)–(5.165) can be solved by the method of characteristics. We introduce a dilatation parameter u and the solution is given by

$$\Gamma_0(p_i; g_0, \frac{\Lambda}{u}) = Z_{\text{eff}}^{-n/2}(u)\Gamma_0(p_i; g_{\text{eff}}(u), \Lambda) \tag{5.166}$$

where $g_{\text{eff}}(u)$ is defined as the solution of the equation

$$u\frac{dg_{\text{eff}}}{du} = \beta(g_{\text{eff}}(u)) \tag{5.167}$$

with the initial condition $g_{\text{eff}}(1) = g_0$, and $Z_{\text{eff}}(u)$ is defined as the solution of

$$\frac{1}{2}u\frac{d}{du}\log Z_{\text{eff}} = \eta(g_{\text{eff}}(u)) \tag{5.168}$$

with the initial condition $Z_{\text{eff}}(1) = 1$. We see that g_{eff} plays the role of an effective bare coupling constant, and a change in the cutoff is equivalent to a change in g_{eff} and in Z_{eff}. To study what happens as we remove the cutoff we must take the limit $u \to 0$. Equation (5.167) can be written in the integrated form

$$\int_{g_0}^{g_{\text{eff}}(u)} \frac{dg'}{\beta(g')} = \log u\,. \tag{5.169}$$

We see that in the limit $u \to 0$ the integral on the left-hand side must diverge, and this is possible only if, as $u \to 0$, $g_{\text{eff}}(u)$ approaches a zero of the beta function.

In general the renormalization of the coupling has the form

$$g_R = g_0 - \beta_0 g_0^2 \log\Lambda + O(g_0^3)\,. \tag{5.170}$$

The dependence of g_0 on the cutoff is obtained inverting the above relation,

$$g_0 = g_R + \beta_0 g_R^2 \log\Lambda + O(g_R^3) \tag{5.171}$$

and therefore

$$\beta(g_0) \equiv \frac{dg_0}{d\log\Lambda} = \beta_0 g_0^2 + O(g_0^3)\,. \tag{5.172}$$

This shows that there is always a zero of the beta function at $g_0 = 0$, and it is possible to remove the cutoff while at the same time sending $g_0(\Lambda) \to 0$.

In other words, given a regularized theory, with a cutoff in momentum space or, for example, on a space-time lattice (which is another possible UV regulator) we find the limit $\Lambda \to \infty$ (or the continuum limit in the case of a lattice) tuning the bare couplings toward a zero of the beta function. This is a way to see things that has very important applications to statistical mechanics and critical phenomena, as well as in lattice gauge theory, and we will come back to it in Section 9.5.

There is another way to extract information from eq. (5.161), which is more useful from the point of view of particle physics. We rather write the equation as $\Gamma_0 = Z^{n/2}\Gamma_R$ and we use the fact that Γ_0 is independent of the renormalization point μ. Instead Γ_R depends on μ explicitly, and also through the renormalized mass and coupling. Let us again neglect all mass terms. Then we write

$$0 = \mu\frac{d\Gamma_0}{d\mu} = \left[\mu\frac{\partial}{\partial\mu} + \beta(g_R)\frac{\partial}{\partial g_R} + n\gamma(g_R)\right]\Gamma_R(p_i; g_R, \mu)\,. \tag{5.173}$$

where now

$$\beta(g_R) = \mu\frac{dg_R}{d\mu} \tag{5.174}$$

and

$$\gamma(g_R) = \frac{1}{2}\mu\frac{d}{d\mu}\log Z\,. \tag{5.175}$$

Equation (5.173) is the *Callan–Symanzik* equation. This equation is formally very similar to the one previously studied, but now we have the renormalized coupling g_R rather than the bare coupling g_0. It tells us how Γ_R changes if we change the renormalization point μ. The reason why this equation is very useful is that, in the high-energy limit where all masses can be neglected, using dimensional arguments the dependence on μ can be translated into a dependence on the energy. In fact, if d_Γ is the mass dimension of Γ_R, then $\Gamma_R(p_i; g_R, \mu)$ for dimensional reasons must have the form

$$\Gamma_R(p_i; g_R, \mu) = \mu^{d_\Gamma} F\left(g_R, \frac{p_i}{\mu}\right) \tag{5.176}$$

with F a dimensionless function. Using again the method of characteristics, eq. (5.173) implies that

$$\Gamma_R\left(p_i; g_R, \frac{\mu}{u}\right) = Z_{\rm eff}^{-n/2}(u)\Gamma_R(p_i; g_{\rm eff}(u), \mu) \tag{5.177}$$

where now $g_{\rm eff}(u)$ is defined as the solution of the equation

$$u\frac{dg_{\rm eff}}{du} = \beta(g_{\rm eff}(u)) \tag{5.178}$$

with the initial condition $g_{\rm eff}(1) = g_R$, where g_R is the value of the renormalized coupling constant at the reference scale μ. Similarly, $Z_{\rm eff}(u)$ is defined as the solution of

$$\frac{1}{2}u\frac{d}{du}\log Z_{\rm eff} = -\gamma(g_{\rm eff}(u)) \tag{5.179}$$

with the initial condition $Z_{\text{eff}}(1) = 1$. Using eq. (5.176) we see that

$$\Gamma_R\left(p_i; g_R, \frac{\mu}{u}\right) = \frac{\mu^{d_\Gamma}}{u^{d_\Gamma}} F\left(g_R, \frac{up_i}{\mu}\right) = \frac{1}{u^{d_\Gamma}} \Gamma_R(up_i; g_R, \mu) . \tag{5.180}$$

Combining eqs. (5.177) and (5.180) we find

$$\Gamma_R(up_i; g_R, \mu) = u^{d_\Gamma} Z_{\text{eff}}^{-n/2}(u) \Gamma_R(p_i; g_{\text{eff}}(u), \mu) \tag{5.181}$$

Writing eq. (5.179) in the integrated form, we can rewrite the above expression as

$$\Gamma_R(up_i; g_R, \mu) = u^{d_\Gamma} \exp\left\{ n \int_0^{\log u} \gamma(g_{\text{eff}})(u') d\log u' \right\} \Gamma_R(p_i; g_{\text{eff}}(u), \mu) . \tag{5.182}$$

We see that the rescaling of energies (in the limit when masses can be neglected) is summarized by two effects: first of all, naive dimensional analysis does not work anymore. Instead of a simple overall factor u^{d_Γ} we get also a modification determined by the γ function; γ is then called the *anomalous dimension*. Its origin is in the divergencies of field theory, which force us to introduce a new mass-scale (the cutoff, which is sent to infinity and is replaced by the renormalization point μ), and spoiled naive dimensional analysis. Second, the coupling g_R at the scale μ is replaced by $g_{\text{eff}}(u)$, or $g_{\text{eff}}(E)$ with $E = u\mu$, which is called the *running coupling constant*. Therefore $g_{\text{eff}}(E)$ plays the role of an effective renormalized coupling constant. We see that the beta function $\beta(g)$ contains important information. In particular, the sign of the beta function near $g = 0$ is crucial, as we will see below and in Section 9.5.

In the case of $\lambda\phi^4$ theory the explicit one-loop computation is very simple, since there is no wave function renormalization at one loop, and the result comes just from the graph of Fig. 5.9. We computed it in Section 5.5.2 where we found

$$\lambda_R = \lambda_0 + \lambda_0^2 \frac{3}{16\pi^2} \log \Lambda , \tag{5.183}$$

Therefore the one-loop beta function is

$$\beta(\lambda_0) = \beta_0 \lambda_0^2 + O(\lambda_0^3) , \qquad \beta_0 = \frac{3}{16\pi^2} . \tag{5.184}$$

Limiting ourselves to one-loop, the explicit integration of

$$E \frac{d}{dE} \lambda_{\text{eff}} = \beta_0 \lambda_{\text{eff}}^2 , \tag{5.185}$$

with the initial condition $\lambda_{\text{eff}}(E = \mu) = \lambda_*$, gives

$$\lambda_{\text{eff}}(E) = \frac{\lambda_*}{1 - \beta_0 \lambda_* \log(E/\mu)} . \tag{5.186}$$

Comparing with eq. (5.159) we see that the renormalization group (RG) analysis has provided a resummation of a whole class of logarithmic terms. Expanding eq. (5.186) in powers of λ_* we get

$$\lambda_{\text{eff}}(E) = \lambda_* \left[1 + \sum_{n=1}^{\infty} c_n(E) \lambda_*^n \right] \tag{5.187}$$

with

$$c_n(E) = \left(\beta_0 \log \frac{E}{\mu} \right)^n . \tag{5.188}$$

We might ask what we have really gained, since we used only the one-loop beta function, so one might think that eq. (5.187) is not justified beyond the term $n = 1$ that we already knew. However, we will see in Exercise 5.4 that the effect of higher-loop corrections to the beta function is to produce additional terms proportional to $\log\log E$ in the denominator of eq. (5.186), see eq. (5.198). Including these corrections, the coefficients c_n are modified from the value given in eq. (5.188) to a value of the form

$$c_n(E) = \left[\beta_0 \log\frac{E}{\mu} + O(\log\log E)\right]^n . \tag{5.189}$$

Then the term in c_n proportional to $(\log E)^n$ is not affected by higher-order corrections to the beta function, while, at each order n, in c_n there are also terms of order $(\log E)^{n-1}$ and smaller, that are missed using only the one-loop beta function. Therefore the resummation (5.186) is useful when $\log(E/\mu) \gg 1$, since in this case at each order n we have picked the term which dominates at high energies. In turn, this means that eq. (5.186) is really useful only when $\beta_0 < 0$, since in this case when $\log(E/\mu) \gg 1$ we have $\lambda_{\rm eff}(E) \ll 1$ and perturbation theory is consistent. Equation (5.186) is called the *leading logarithms* approximation.

However, in the case of $\lambda\phi^4$ we have $\beta_0 > 0$ and therefore the running coupling increases in the UV. Formally, eq. (5.186) would even predict that $\lambda_{\rm eff}(E)$ diverges at

$$E = \mu \exp\left\{\frac{1}{\beta_0\lambda_*}\right\} . \tag{5.190}$$

Of course this result should not be taken literally, because as soon as $\lambda_{\rm eff}(E)$ becomes of order one the whole perturbative expansion, even if improved by RG, blows up and cannot be trusted anymore. The correct conclusion, instead, is that, even if we started with $\lambda_* \ll 1$, there is a critical energy at which the theory enters in a strong coupling regime.

Besides our toy model $\lambda\phi^4$, it turns out, more importantly, that $\beta_0 > 0$ in QED. This means that in QED the fine structure constant increases with energy, and formally there even exists an energy scale where it becomes strong. The energy similar to eq. (5.190), where formally the one-loop running coupling diverges, is called the Landau pole. However, the running is very slow, since it is logarithmic, and long before the theory enters in the strong coupling regime, we arrive at the electroweak scale, where QED in isolation is no longer the correct theory. From this point on, the evolution of α must be studied in the context of the Standard Model and possibly of its high-energy extensions.

From its low-energy value $\alpha = 1/137.035\,999\,11(46)$, the fine structure constant at the Z^0 mass $M_Z = 91.1876 \pm 0.0021$ GeV grows only to the value[17]

$$\alpha(M_Z) = \frac{1}{127.918 \pm 0.018} . \tag{5.191}$$

If instead $\beta_0 < 0$, we see from eq. (5.186) that the running coupling becomes smaller and smaller at high energies (and therefore the perturbative result is more and more accurate). As we mentioned at the beginning of this section, this property is known as *asymptotic freedom* and is one of the most important features of quantum chromodynamics, the theory of strong interactions. The running of the strong coupling, which is denoted by α_s, is well-verified experimentally, see Fig. 5.19, taken from S. Eidelman et al., Phys. Lett. B592, 1 (2004).

[17] To be precise, this is the value of the fine structure constant, renormalized in the so-called $\overline{\rm MS}$-scheme. We do not enter into the details of how this is defined. See e.g. Peskin and Schroeder (1995), page 377.

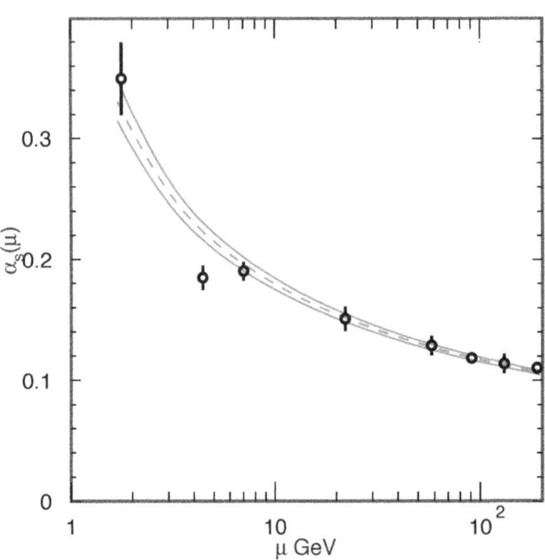

Fig. 5.19 Summary of the values of $\alpha_s(\mu)$ at the values of μ where they are measured. The lines show the central values and the $\pm 1\sigma$ limits of the average. The figure clearly shows the decreases in $\alpha_s(\mu)$ with increasing μ. The data are, in increasing order of μ: τ width, Υ decays, deep inelastic scattering, e^+e^- event rate at 22 GeV from the JADE data, shapes at TRISTAN at 58 GeV, Z width, and e^+e^- event shapes at 135 GeV and 189 GeV. From S. Eidelman et al., Phys. Lett. B592, 1 (2004).

Summary of chapter

- The aim of this chapter was to set up the formalism for computing the transition amplitudes between initial and final states, with an arbitrary number of incoming and outgoing particles, i.e. to compute the S-matrix elements (5.16).
- The first step is the LSZ formula (5.46), which expresses the S-matrix elements in terms of the vacuum expectation value of a T-product of fields. In eq. (5.46), ϕ is the Heisenberg operator that evolves with the full Hamiltonian, including the interaction term, so it is not a free field. The T-product is defined in eq. (5.32).
- The next step is given by eq. (5.67), where we write the vacuum expectation value of the T-product of the fields ϕ (which evolve with the full Hamiltonian) in terms of a *free* field ϕ_I (the interaction picture field) and of the interaction picture Hamiltonian, $\mathcal{H}_I$, which is a function of the free fields ϕ_I. The perturbative expansion is the expansion of the exponential in eq. (5.67) in powers of the interaction Hamiltonian. Observe that from now on the interaction picture field ϕ_I will be denoted simply by ϕ, while the fully interacting field will not appear again.
- Expanding the exponential of the Hamiltonian we are left with the task of computing vacuum expectation values of free fields. The actual computation is enormously simplified by the use of Wick's theorem and of Feynman rules (Section 5.5). A basic role is played by the Feynman propagator (Section 5.4), since Wick's theorem reduces the vacuum expectation value of the product of N fields, with N arbitrary, to a sum of products of Feynman propagators.

All contributions can be represented graphically by Feynman diagrams.

- We can now perform explicit calculations, and we discover that Feynman graphs containing closed loops are in general divergent. To cure these divergences one first regularizes the theory, introducing a cutoff. The couplings, fields and masses which appear in the Lagrangian are then given a dependence on the cutoff (and are now called bare couplings, bare fields and bare masses) chosen so that the divergences coming from the loops are canceled, and physical observables (like the renormalized masses and couplings) are finite.
- Theories where the divergences can be reabsorbed in a finite number of parameters are called renormalizable, and are mathematically consistent and predictive, in principle at any energy scale. Non-renormalizable theories, however, can still be perfectly acceptable low-energy theories, and have an intrinsic mass-scale above which we know that they must be replaced by a more fundamental theory.
- As a consequence of the renormalization procedure, the physical (i.e. renormalized) coupling constants are not at all constant. Rather, their value depends on the energy scale at which they are measured. If the coupling goes to zero in the high-energy limit the theory is called asymptotically free. The theory of strong interactions, QCD, turns out to be an asymptotically free theory.

Further reading

- The subject of perturbative expansion, Feynman diagrams, renormalization, etc. is treated in detail in all QFT textbooks. Excellent and very detailed discussions are given in Itzykson and Zuber (1980), Peskin and Schroeder (1995) and Weinberg (1995). At a simpler level, a clear book is Mandl and Shaw (1984).
- The measurement of the cosmological constant discussed in Section 5.7 are very important and are very likely to become even more important in the future. A book discussing the modern developments in cosmology is S. Dodelson, *Modern Cosmology*, Elsevier, San Diego 2003. See, e.g. Section 2.4.5 for a discussion of vacuum energy and type Ia supernovae. Beautiful recent experimental results on type Ia supernova are reported in A. Riess *et al.*, astro-ph/0402512.

  Fluctuations in the cosmic microwave background (CMB) have been measured with great accuracy by the WMAP experiment, see e.g. C. L. Bennett *et al.*, First Year Wilkinson Microwave Anisotropy Probe (WMAP) Observations: Preliminary Maps and Basic Results, Astrophys. J. Suppl. **148** (2003) 1 [arXiv:astro-ph/0302207].

Exercises

(5.1) Using the definition (5.69) and the expression of the T-product in terms of the theta function, show that the Feynman propagator is a Green's function of the KG operator, i.e. it satisfies

$$(\Box_x + m^2)D(x-y) = -i\delta^{(4)}(x-y)\,. \qquad (5.192)$$

Show that this is true in general for the T-product of N-fields (5.43), justifying the fact that we have called them the N-point Green's function.

Solve eq. (5.192) going in momentum space and study to what boundary conditions corresponds each prescription for going around the poles.

(5.2) Consider a scalar theory in d space-time dimensions whose action has the standard kinetic term $\int d^dx(\partial\phi)^2$ and an interaction term $\int d^dx\,\lambda\phi^n$. According to the counting argument presented in the text, for what values of n and d is the theory renormalizable?

(5.3) In the text we have written the interaction as $\lambda\phi^4$. We could instead study the theory with interaction term $:\lambda\phi^4:$. Compare the perturbative expansion in the two cases. In particular, show that the mass renormalization is different, and with the interaction term $:\lambda\phi^4:$ it vanishes at $O(\lambda)$.

(5.4) In QCD the perturbative expansion is an expansion in powers of g^2 and, at the two-loop level,

$$\frac{dg}{d\log E} = -\beta_0 g^3 - \beta_1 g^5 + O(g^7)\,, \qquad (5.193)$$

with $\beta_0 > 0$ and $\beta_1 > 0$. In terms of $\alpha_s = g^2/(4\pi)$,

$$\frac{d\alpha_s}{d\log E} = -b_0\alpha_s^2 - b_1\alpha_s^3 + O(\alpha_s^4)\,, \qquad (5.194)$$

with $b_0 > 0$ and $b_1 > 0$.

(i) Neglect the term $O(\alpha_s^3)$ and verify that the solution is

$$\alpha(E) = \frac{\alpha(\mu)}{1 + b_0\alpha(\mu)\log(E/\mu)}\,. \qquad (5.195)$$

where μ is a mass-scale of reference, used to fix the initial condition. Define a new mass-scale $\Lambda_{\rm QCD}$ as

$$\Lambda_{\rm QCD} = \mu \exp\left\{-\frac{1}{b_0\alpha(\mu)}\right\}\,. \qquad (5.196)$$

Verify that eq. (5.195) can be rewritten as

$$\alpha(E) = \frac{1}{b_0 \log(E/\Lambda_{\rm QCD})}\,. \qquad (5.197)$$

Therefore the coupling is small (and the approximation of neglecting the term $O(\alpha_s^3)$ in eq. (5.194) is justified) when $E \gg \Lambda_{\rm QCD}$. Experimentally, $\Lambda_{\rm QCD} \simeq 200$ MeV, and typically the perturbative calculations are valid at $E > 1$ GeV.

(ii) Using eq. (5.197) as a lowest-order solution, show that the solution of eq. (5.194) at two loops, i.e. including the term $O(\alpha_s^3)$, can be written as

$$\alpha(E) = \frac{1}{b_0\log(E/\Lambda_{\rm QCD}) + \frac{b_1}{b_0}\log\log(E/\Lambda_{\rm QCD})}\,, \qquad (5.198)$$

after a suitable redefinition of $\Lambda_{\rm QCD}$.

Cross-sections and decay rates

6

In the previous chapter we have understood how to compute matrix elements between initial and final states and how to make them finite. We will see in this chapter how to use these matrix elements to compute scattering cross-sections and decay rates.

6.1 Relativistic and non-relativistic normalizations

It is useful first of all to clarify a difference between the relativistic and non-relativistic normalization of one-particle states. To make the argument cleaner, we first consider a system in a cubic box with spatial volume $V = L^3$. At the end of the computation V will be sent to infinity. Momenta are therefore discrete; for instance, if we use periodic boundary conditions, $\mathbf{p} = 2\pi\mathbf{n}/L$ with $\mathbf{n} = (n_x, n_y, n_z)$ a vector with integer components. In non-relativistic quantum mechanics a one-particle state with momentum $\mathbf{p}$ in the coordinate representation is given by a plane wave

$$\psi_{\mathbf{p}}(\mathbf{x}) = C\, e^{i\mathbf{p}\cdot\mathbf{x}} \tag{6.1}$$

and the normalization constant is fixed by the condition that there is one particle in the volume V,

$$\int_V d^3x\, |\psi_{\mathbf{p}}(\mathbf{x})|^2 = 1\,. \tag{6.2}$$

This fixes $C = 1/\sqrt{V}$. Wave functions with different momenta are orthogonal, and therefore

$$\int_V d^3x\, \psi^*_{\mathbf{p}_1}(\mathbf{x})\psi_{\mathbf{p}_2}(\mathbf{x}) = \delta_{\mathbf{p}_1,\mathbf{p}_2}\,. \tag{6.3}$$

Writing $\psi_{\mathbf{p}}(\mathbf{x}) = \langle\mathbf{x}|\mathbf{p}\rangle$ and using the completeness relation $\int d^3x\, |\mathbf{x}\rangle\langle\mathbf{x}| = 1$, we can rewrite this as

$$\langle\mathbf{p}_1|\mathbf{p}_2\rangle^{(NR)} = \delta_{\mathbf{p}_1,\mathbf{p}_2}\,. \tag{6.4}$$

The superscript (NR) reminds us that the states have been normalized according to the conventions of non-relativistic quantum mechanics.

In relativistic QFT this normalization is not the most convenient, because the spatial volume V is not relativistically invariant, and therefore the condition "one-particle per volume V" is not invariant. We have already introduced in eq. (4.11) a more convenient Lorentz-invariant normalization; in a finite box, using eq. (4.6), it reads

$$\langle \mathbf{p}_1 | \mathbf{p}_2 \rangle^{(R)} = 2E_{\mathbf{p}_1} V \, \delta_{\mathbf{p}_1,\mathbf{p}_2} \,. \tag{6.5}$$

Therefore the difference between the relativistic and non-relativistic normalization of the one-particle states is

$$|\mathbf{p}\rangle^{(R)} = (2E_{\mathbf{p}}V)^{1/2} |\mathbf{p}\rangle^{(NR)} \,, \tag{6.6}$$

and of course for a multiparticle state

$$|\mathbf{p}_1, \ldots, \mathbf{p}_n\rangle^{(R)} = \left(\prod_{i=1}^{n} (2E_{\mathbf{p}_i} V)^{1/2} \right) |\mathbf{p}_1, \ldots, \mathbf{p}_n\rangle^{(NR)} \,. \tag{6.7}$$

We denote by M_{fi} the scattering amplitude between the initial state with momenta $\mathbf{q}_1, \ldots, \mathbf{q}_m$ and the final state with momenta $\mathbf{p}_1, \ldots, \mathbf{p}_n$, with non-relativistic normalization, and by $\mathcal{M}_{fi}$ the same matrix element with relativistic normalization of the states. Then

$$M_{fi} = \prod_{i=1}^{n} (2E_{\mathbf{p}_i} V)^{-1/2} \prod_{j=1}^{m} (2E_{\mathbf{q}_j} V)^{-1/2} \mathcal{M}_{fi} \,. \tag{6.8}$$

We saw in Chapter 5 that the S-matrix can be written as $S = 1 + iT$ and that it is convenient to extract a factor $(2\pi)^4 \delta^{(4)}(P_i - P_f)$ from T, where P_i and P_f are the total initial and final four-momenta. Then

$$S = 1 + (2\pi)^4 \delta^{(4)}(P_i - P_f) iM \,. \tag{6.9}$$

If we take the matrix element of S between the initial state $|i\rangle$ and the final state $\langle f|$, taken with the non-relativistic normalization, then the matrix element of the operator M is just M_{fi} while the matrix element of the identity operator is just a Kronecker delta, because of eq. (6.4),

$$S_{fi} = \delta_{fi} + (2\pi)^4 \delta^{(4)}(P_i - P_f) i M_{fi} \,. \tag{6.10}$$

6.2 Decay rates

Consider first the case in which the initial state is a single particle with four-momentum p and mass M, and the final state is given by n particles with four-momenta p_i and masses m_i, $i = 1, \ldots n$. We are therefore considering a decay process. Assume for the moment that all particles are distinguishable.

The rules of quantum mechanics tell us that the probability for this process is obtained by taking the squared modulus of the amplitude and summing over all possible final states. In eq. (6.10) the term δ_{fi} gives of course zero because the initial and final states are different. When we take the square of the other term we are confronted with the square of the delta function. To compute it, we recall that we are working in a finite spatial volume and, from eq. (4.6),

$$(2\pi)^3\delta^{(3)}(0) = V\,. \tag{6.11}$$

Similarly, we regularize also the time interval, saying that time runs from $-T/2$ to $T/2$ (at the end of the computation $T \to \infty$) so that

$$(2\pi)^4\delta^{(4)}(0) = VT \tag{6.12}$$

and

$$|(2\pi)^4\delta^{(4)}(P_i - P_f)iM_{fi}|^2 = (2\pi)^4\delta^{(4)}(P_i - P_f)\,VT|M_{fi}|^2\,. \tag{6.13}$$

We must now sum this expression over all final states. Since we are working in a finite volume V, this is the sum over the possible discrete values of the momenta of the final particles. In the large-volume limit for each particle we can write, using eq. (4.5)

$$\sum_{\mathbf{p}_i} \to \frac{V}{(2\pi)^3}\int d^3p_i\,. \tag{6.14}$$

It is interesting to understand this result physically, observing that in statistical mechanics the integration measure over the phase space is

$$\frac{d^3x_i d^3p_i}{(2\pi)^3} \tag{6.15}$$

for each particle. The factor $(2\pi)^3$ is simply the volume of the cells of the phase space, $h^3 = (2\pi\hbar)^3$, in units $\hbar = 1$. In our case the particles are momentum eigenstates and are completely delocalized in space, so the scattering amplitude depends on the momentum but not on the positions of the particles, and we can integrate over d^3x_i, obtaining the volume factor.

In conclusion, the *probability* dw for a decay in which in the final state the i-th particle has momentum between p_i and $p_i + dp_i$ is

$$dw = (2\pi)^4\delta^{(4)}(P_i - P_f)\,VT|M_{fi}|^2\prod_{i=1}^{n}\frac{Vd^3p_i}{(2\pi)^3}\,. \tag{6.16}$$

This is the probability that the decay takes place at any time between $-T/2$ and $T/2$. We are more interested in the *decay rate* $d\Gamma$, which is the decay probability per unit time, and therefore is obtained dividing by T,

$$d\Gamma = (2\pi)^4\delta^{(4)}(P_i - P_f)\,V|M_{fi}|^2\prod_{i=1}^{n}\frac{Vd^3p_i}{(2\pi)^3}\,, \tag{6.17}$$

and by construction has a finite limit for $T \to \infty$. To get rid of the divergent V factors we now express this in terms of the matrix element with relativistic normalization. Using eq. (6.8) we see that the volume factors cancel, and we are left with

$$d\Gamma = (2\pi)^4\delta^{(4)}(p - \sum_i p_i)\,\frac{1}{2E_p}|\mathcal{M}_{fi}|^2\prod_{i=1}^{n}\frac{d^3p_i}{(2\pi)^3(2E_i)}\,, \tag{6.18}$$

where E_p is the energy of the initial particle and E_i of the final particles. Various observations are in order:

- Energy–momentum conservation is guaranteed by the Dirac delta. In particular, if $M < \sum_i m_i$ the process is forbidden.
- The factors d^3p_i/E_i are relativistically invariant.
- $\mathcal{M}_{fi}$ is computed with the relativistic normalization of the states and therefore is just the matrix element that we learned to compute in Chapter 5, and it is relativistically invariant.
- The factor $1/(2E_p)$ reduces to $1/(2M)$ in the rest frame of the decaying particle. In a generic frame in which the particle has speed v, we have $E_p = \gamma M$ with $\gamma = (1 - v^2)^{-1/2}$ and therefore the rate Γ is smaller by a factor γ. The lifetime of a particle is the inverse of its total decay rate (i.e. of the rate $d\Gamma$ integrated over momenta and summed over all possible decay modes). Therefore the factor γ is nothing but the relativistic dilatation of time.

It is useful to define the (differential) *n-body phase space* $d\Phi^{(n)}$,

$$d\Phi^{(n)} \equiv (2\pi)^4\delta^{(4)}(P_i - P_f)\prod_{i=1}^{n}\frac{d^3p_i}{(2\pi)^3 2E_i}\,. \tag{6.19}$$

Equation (6.18) can therefore be written as

$$d\Gamma = \frac{1}{2E_p}\,|\mathcal{M}_{fi}|^2\,d\Phi^{(n)}\,. \tag{6.20}$$

Finally, observe that if n of the final particles are identical, configurations that differ by a permutation are not distinct and therefore the phase space is reduced by a factor $n!$.

6.3 Cross-sections

Consider a beam of particles with mass m_1, number density (that is, number of particles/unit volume) n_1^0 and velocity v_1 impinging on a target made of particles with mass m_2 and number density n_2^0, at rest. The superscript 0 on n_i^0 is meant to stress that these are the number densities in a specific frame, that with particle 2 at rest. Assume for simplicity that both types of particles have a uniform distribution (it is not difficult to generalize to a non-uniform distribution, as one can have in a typical beam). The number of scattering events, N, that take place per unit volume and per unit time will be proportional to the incoming flux $n_1^0 v_1$ and to the density of targets n_2^0. The proportionality constant is, by definition, the cross-section:

$$dN = \sigma v_1 n_1^0 n_2^0 \, dV dt \,. \tag{6.21}$$

Dimensional analysis shows immediately that σ has the dimensions of an area. Equation (6.21) holds in the rest frame of the particles of type 2. We want to write a similar expression in a generic frame. First of all, in a generic frame, we define the cross-section as the *Lorentz invariant quantity* that, in the rest frame of particle 2, is given by eq. (6.21). The number dN is Lorentz invariant (its integral is the number of clicks of the detector, so it is clearly independent of the reference frame), and $dV dt = d^4x$ is also invariant. Therefore we must find a Lorentz-invariant expression that, in the rest frame of particle 2, reduces to $v_1 n_1^0 n_2^0$. This is given by (see, e.g. Landau and Lifshitz, vol. II (1979), Section 12)

$$n_1 n_2 \sqrt{(\mathbf{v_1} - \mathbf{v_2})^2 - (\mathbf{v_1} \times \mathbf{v_2})^2} \,, \tag{6.22}$$

where n_1, n_2 are the number densities of the two types of particles in the frame where their respective velocities are $\mathbf{v}_1$ and $\mathbf{v}_2$ (note that the number density is not invariant, but transforms as the inverse of a spatial volume). If the particles are collinear, we simply have

$$dN = \sigma |v_1 - v_2| n_1 n_2 \, dV dt \,. \tag{6.23}$$

It is convenient to define the quantity

$$I \equiv \sqrt{(p_1 p_2)^2 - m_1^2 m_2^2} = E_1 E_2 \sqrt{(\mathbf{v_1} - \mathbf{v_2})^2 - (\mathbf{v_1} \times \mathbf{v_2})^2} \,, \tag{6.24}$$

so that

$$dN = \sigma \frac{I}{V E_1 E_2} (n_1 V)(n_2 dV) dt \,. \tag{6.25}$$

Integrating over dV, $n_2 dV$ gives the total number N_2 of particles of type 2, while $n_1 V = N_1$. Then the total number of events per unit particle of type 1, per unit particle of type 2 in a total time T is given by $\sigma IT/(V E_1 E_2)$. However, this is nothing but the probability of the event, i.e. the square of the matrix element, summed over all final states. Therefore

$$\sigma = \frac{V E_1 E_2}{IT} (2\pi)^4 \delta^{(4)}(P_i - P_f) \, VT \int |M_{fi}|^2 \prod_{i=1}^{n} \frac{V d^3 p_i}{(2\pi)^3} \,, \tag{6.26}$$

or, in differential form,

$$d\sigma = \frac{V^2 E_1 E_2}{I} (2\pi)^4 \delta^{(4)}(P_i - P_f) \, |M_{fi}|^2 \prod_{i=1}^{n} \frac{V d^3 p_i}{(2\pi)^3} \,. \tag{6.27}$$

We now pass from $|M_{fi}|^2$ to $|\mathcal{M}_{fi}|^2$. The two initial particles bring a factor $1/[(2E_1 V)(2E_2 V)]$ so the overall factor $V^2 E_1 E_2$ cancels, and the

n final particles bring each one a factor $1/(2E_iV)$ so that also the volume factors in Vd^3p_i cancel. The final result is then

$$d\sigma = (2\pi)^4\delta^{(4)}(P_i - P_f)\,\frac{1}{4I}\,|\mathcal{M}_{fi}|^2\prod_{i=1}^{n}\frac{d^3p_i}{(2\pi)^3 2E_i}\,. \tag{6.28}$$

Observe that in the above expression the factors I, $\mathcal{M}_{fi}$ and d^3p_i/E_i are separately Lorentz invariant, so the Lorentz invariance of the cross-section is evident. The term $4I$ is called the flux factor. In terms of the phase space (6.19), eq. (6.28) reads

$$d\sigma = \frac{1}{4I}\,|\mathcal{M}_{fi}|^2\,d\Phi^{(n)}\,. \tag{6.29}$$

6.4 Two-body final states

Consider first the decay of a particle of mass M into two particles of masses m_1, m_2. Since the phase space is Lorentz invariant, we can compute it in the frame that we prefer, and of course the simplest choice is the rest frame of the initial particle. Then

$$d\Phi^{(2)} = (2\pi)^4\delta(M - E_1 - E_2)\delta^{(3)}(\mathbf{p}_1 + \mathbf{p}_2)\frac{d^3p_1}{(2\pi)^3 2E_1}\frac{d^3p_2}{(2\pi)^3 2E_2}\,. \tag{6.30}$$

We have six integration variables and four Dirac deltas, so we can reduce this to only two integrations. We can perform explicitly the integration over d^3p_2 using the Dirac delta $\delta^{(3)}(\mathbf{p}_1 + \mathbf{p}_2)$, and we are left with a phase space which is still differential with respect to d^3p_1,

$$d\Phi^{(2)} = \frac{1}{(2\pi)^2}\frac{1}{4E_1E_2}\delta(M - E_1 - E_2)d^3p_1\,. \tag{6.31}$$

Of course here $E_2^2 = \mathbf{p}_2^2 + m_2^2$ where now $\mathbf{p}_2$ has become a notation for $-\mathbf{p}_1$ instead of being an independent integration variable. We now write $d^3p_1 = p_1^2dp_1d\Omega$, where $d\Omega$ is the infinitesimal solid angle and $p_1 = |\mathbf{p}_1|$, and we integrate over p_1 using the conservation of energy, i.e.

$$d\Phi^{(2)} = \frac{1}{(2\pi)^2}\,d\Omega\int_0^\infty \frac{1}{4E_1E_2}p_1^2dp_1\delta\left(M - \sqrt{p_1^2 + m_1^2} - \sqrt{p_1^2 + m_2^2}\right)\,. \tag{6.32}$$

The integral is easily performed using the identity

$$\delta(f(x)) = \frac{1}{|f'(x_0)|}\delta(x - x_0) \tag{6.33}$$

where x_0 is the zero of $f(x)$ (if there is more than one zero we must sum over all of them, but in this case there is only one zero in the integration domain $p_1 \geqslant 0$) and we find

$$d\Phi^{(2)} = \frac{1}{32\pi^2M^2}\left[M^4 + (m_1^2 - m_2^2)^2 - 2M^2(m_1^2 + m_2^2)\right]^{1/2}d\Omega\,. \tag{6.34}$$

In the limit $m_1 = m_2 \equiv m$ this simplifies further to

$$d\Phi^{(2)} = \frac{1}{32\pi^2}\sqrt{1-\frac{4m^2}{M^2}}\,d\Omega\,, \tag{6.35}$$

where we have assumed that the two particles are distinguishable. If instead they are identical, the phase space is reduced by a factor 1/2!. Another common situation is $m_1 = m, m_2 = 0$, in which case

$$d\Phi^{(2)} = \frac{1}{32\pi^2}\left(1-\frac{m^2}{M^2}\right)d\Omega\,. \tag{6.36}$$

Observe that the phase space goes to zero when the decay products have the maximum mass compatible with the conservation of energy, i.e. at $m = M/2$ in eq. (6.35) and at $m = M$ in eq. (6.36). Using eq. (6.20) we can write the differential decay rate for a two-body decay, $d\Gamma/d\Omega$, where $d\Omega = d\cos\theta\, d\phi$. In the rest frame of the decaying particle, it is

$$d\Gamma = \frac{1}{64\pi^2 M^3}\left[M^4 + (m_1^2 - m_2^2)^2 - 2M^2(m_1^2 + m_2^2)\right]^{1/2} |\mathcal{M}_{fi}|^2 d\Omega\,. \tag{6.37}$$

In principle, $\mathcal{M}_{fi}$ depends on the angles θ, ϕ. If the decaying particle has spin, it is convenient to choose the direction of the spin as the polar axis. In the absence of external fields we have cylindrical symmetry around this axis (the symmetry could be broken, for instance, by an external magnetic field pointing in a direction different from the spin of the particle), and in this case $\mathcal{M}_{fi}$ does not depend on ϕ and the integration over $d\phi$ simply gives a factor 2π. If furthermore the particle has spin zero, there is no preferred direction and the decay is isotropic, i.e. $\mathcal{M}_{fi}$ is independent also of θ, and the integration over $d\Omega$ gives simply a factor 4π.

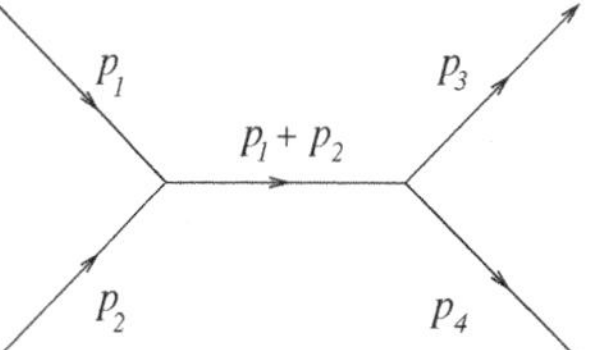

Fig. 6.1 An s-channel amplitude.

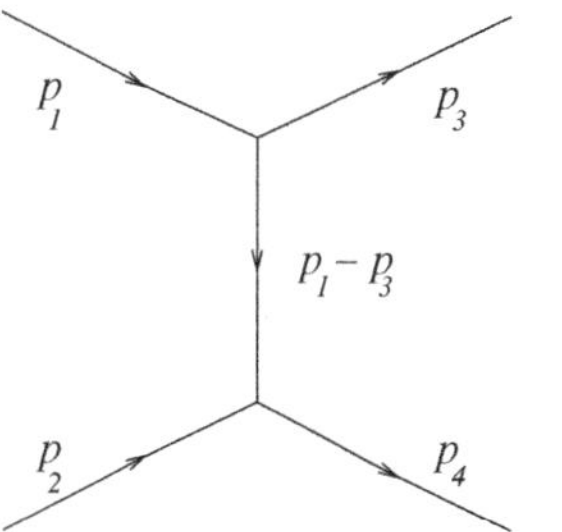

Fig. 6.2 A t-channel amplitude.

Consider now a scattering process $2 \to 2$. We consider an initial state with two particles with masses m_1, m_2 and four-momenta p_1, p_2, and a final state with two particles with masses m_3, m_4 and four-momenta p_3, p_4. It is useful to introduce the Mandelstam variables s, t and u,

$$\begin{aligned} s &= (p_1 + p_2)^2\,, \\ t &= (p_1 - p_3)^2\,, \\ u &= (p_1 - p_4)^2\,. \end{aligned} \tag{6.38}$$

These variables are clearly Lorentz invariant, and satisfy the relation

$$s + t + u = m_1^2 + m_2^2 + m_3^2 + m_4^2\,, \tag{6.39}$$

as one verifies immediately from the definitions, using energy–momentum conservation, $p_1 + p_2 = p_3 + p_4$.

It is convenient to work in the center of mass (CM), where the incoming particles have four-momenta $p_1 = (E_1, \mathbf{p})$ and $p_2 = (E_2, -\mathbf{p})$, with $E_{1,2}^2 = \mathbf{p}^2 + m_{1,2}^2$. Computing s in the CM we find $s = (E_1 + E_2)^2$, so the center-of-mass energy is $\sqrt{s}$. Observe that in a Feynman graph like Fig. 6.1 the momentum of the intermediate particle is p_1+p_2, so its propagator is a function of s; for instance, if the intermediate particle is a scalar with mass m, the propagator in Fig. 6.1 can be written as $i/(s-m^2)$. Instead in Fig. 6.2 the propagator of the intermediate particle is $i/(t-m^2)$. For this reason, the amplitudes in Figs. 6.1 and 6.2 are referred to as the s-channel and t-channel amplitude, respectively. The u-channel amplitude is obtained exchanging p_3 and p_4 in Fig. 6.2. In the CM, we can also write $p_3 = (E_3, \mathbf{p}'), p_4 = (E_4, -\mathbf{p}')$. Energy conservation gives immediately

$$|\mathbf{p}'| = \frac{1}{2\sqrt{s}}\left[s^2 + (m_3^2 - m_4^2)^2 - 2s(m_3^2 + m_4^2)\right]^{1/2}, \tag{6.40}$$

and the same calculation performed in the case of two-body decay gives

$$d\Phi^{(2)} = \frac{1}{(2\pi)^2}\frac{|\mathbf{p}'|}{4\sqrt{s}}d\Omega\,. \tag{6.41}$$

With a simple computation (see the solutions to Exercise 7.1) one can show that the flux factor I, evaluated in the CM, becomes

$$I = |\mathbf{p}|\sqrt{s} \tag{6.42}$$

and therefore the $2 \to 2$ differential scattering cross-section is

$$\boxed{d\sigma = \frac{1}{64\pi^2 s}\,|\mathcal{M}_{fi}|^2\,\frac{|\mathbf{p}'|}{|\mathbf{p}|}d\Omega\,.} \tag{6.43}$$

For elastic scattering ($m_1 = m_3, m_2 = m_4$) we have $|\mathbf{p}'| = |\mathbf{p}|$ and

$$\boxed{d\sigma_{\text{elas}} = \frac{1}{64\pi^2 s}\,|\mathcal{M}_{fi}|^2 d\Omega\,.} \tag{6.44}$$

The above formulas are valid also for particles with spin, if the initial and final spin states are known; in this case the initial state has the form $|i\rangle = |\mathbf{p}_1, \mathbf{s}_1; \ldots; \mathbf{p}_n, \mathbf{s}_n\rangle$, and similarly for the final state, so the only modification in the above equations is that the labels i, f in $\mathcal{M}_{fi}$ include also the spin degrees of freedom.

However, experimentally it is more common that we do not know the initial spin configuration and we accept in the detector all final spin configurations; in this case, to compare with experiment, we must *sum* the right-hand side of eq. (6.29) over the final spin configurations and *average* it over the initial spin configurations. Defining

$$\overline{|\mathcal{M}_{fi}|^2} \equiv \sum_{\text{initial spins}}\;\sum_{\text{final spins}} |\mathcal{M}_{fi}|^2 \tag{6.45}$$

all formulas for the cross-sections are modified with the replacement

$$|\mathcal{M}_{fi}|^2 \to \frac{1}{(2s_a+1)(2s_b+1)}\overline{|\mathcal{M}_{fi}|^2}\,, \tag{6.46}$$

where s_a, s_b are the spins of the two initial particles. For a decay rate Γ the average over the spin s of the initial particle brings instead a factor $1/(2s+1)$.

6.5 Resonances and the Breit–Wigner distribution

Consider the scattering $2 \to 2$ in a theory with a cubic interaction vertex, as for instance in a scalar theory with $\mathcal{L}_{\text{int}} = \lambda\phi^3$. At tree level the amplitude is given by the sum of three Feynman diagrams, the s-channel amplitude of Fig. 6.1, the t-channel amplitude of Fig. 6.2, and the u-channel amplitude obtained from Fig. 6.2 exchanging p_3 with p_4. We focus on the s-channel amplitude. We denote by m the (renormalized) mass of the ϕ field and by $p = p_1 + p_2$ the total initial four momentum. Then $p^2 = (E_1 + E_2)^2 - (\mathbf{p_1} + \mathbf{p_2})^2 \equiv s$ is the square of the CM energy, and therefore the physical region is defined by

$$p^2 > (2m)^2 \tag{6.47}$$

and in the physical region the propagator $i/(p^2 - m^2)$ of the internal line in Fig. 6.1 is always finite. In other words, the internal line is off-shell, which is also expressed saying that it represents a "virtual" particle, rather than a real particle.

However, in other situations it is possible to have one or more *real* particles in the intermediate state. Consider for instance the theory described by two light real scalar fields[1] ϕ_1 and ϕ_2 and one heavy scalar field Φ, with Lagrangian

[1]We take two different light fields ϕ_1, ϕ_2 to avoid the small complication of identical particles; we might as well consider a single light real scalar field ϕ with coupling $\phi^2\Phi$, but we should be careful to insert a factor $1/2!$ in the phase space, because of identical particles, and the appropriate combinatorial factors in the amplitudes.

$$\mathcal{L} = \frac{1}{2}\left[(\partial\phi_1)^2 - m^2\phi_1^2 + (\partial\phi_2)^2 - m^2\phi_2^2 + (\partial\Phi)^2 - M^2\Phi^2\right] + g\phi_1\phi_2\Phi\,. \tag{6.48}$$

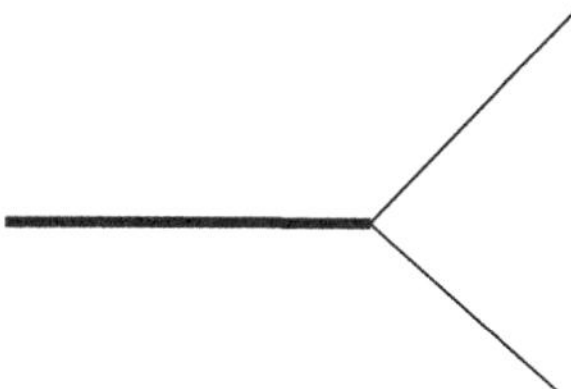

Fig. 6.3 The vertex for a $\Phi\phi_1\phi_2$ interaction. The heavy line is the Φ field and the thin lines represent one field ϕ_1 and one field ϕ_2.

The Feynman rules for this theory are very simple. The propagator of the fields ϕ_i is $i/(p^2-m^2)$, the propagator of Φ is $i/(p^2-M^2)$, and there is an interaction vertex shown in Fig. 6.3, equal to $-ig$. After dressing the propagators with their loop corrections, the masses m and M appearing in the propagators become the physical, renormalized masses, as we saw in Chapter 5 (apart from a crucial subtlety to be explained soon). The $\phi_1\phi_2 \to \phi_1\phi_2$ scattering amplitude in the s-channel is described at tree level by the diagram in Fig. 6.4, and the Feynman rules give

$$i\mathcal{M}_{2\to 2} = (-ig)^2 \frac{i}{p^2 - M^2}\,, \tag{6.49}$$

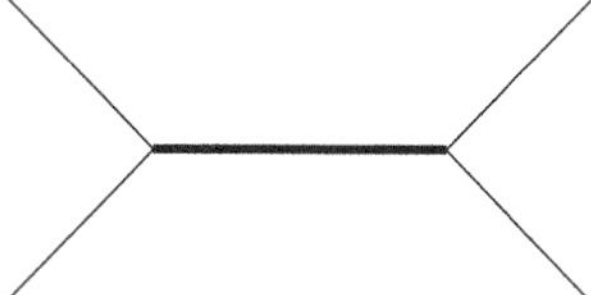

Fig. 6.4 The diagram for $2 \to 2$ scattering in the s-channel.

with $p^2 = s = (p_1+p_2)^2$. The physical region corresponds to $p^2 \geqslant (2m)^2$ since each of the two incoming particles have at least an energy equal to its mass. Therefore, if $M^2 < (2m)^2$, in the physical region we always have $p^2 > M^2$, so $p^2 - M^2$ is always non-zero and the amplitude is finite. However, if $M^2 > 4m^2$, we apparently have a divergent amplitude (and therefore a divergent cross-section) at a physical value of the energy. The divergence appears when the momenta of the incoming particles are such that $p^2 = M^2$, i.e. when the internal line represents an on-shell particle. This divergence means that in the case $M > 2m$ the amplitude (6.49) cannot be correct, and we must have missed something.

The origin of this unphysical divergence is that we have neglected that, if $M > 2m$, the particle described by the field Φ is unstable, because it can decay into two particles of mass m through the graph in Fig. 6.3. This graph gives an amplitude $i\mathcal{M}_{\Phi\to\phi_1\phi_2} = -ig$, independently of the masses of the particles. However, if $M < 2m$, the Dirac delta in the phase space is never satisfied and the decay rate is zero. If $M > 2m$ instead the phase space opens up and we have a non-zero decay rate.

To understand the physics, let us first consider what happens when we have an unstable particle in non-relativistic quantum mechanics (see Landau and Lifshitz, vol. III (1977), Section 134). When we study the Schrödinger equation in three spatial dimensions, we obtain real eigenvalues for the energy operator under the assumption that the wave function vanishes at infinity. For a decay process, instead, we have an outgoing spherical wave at infinity, and since this boundary condition is complex, the eigenvalues of the Hamiltonian are also complex. If we write them in the form

$$E = E_0 - i\frac{\Gamma}{2} \tag{6.50}$$

the time-dependence of the wave function is

$$\psi \sim e^{-iEt} = e^{-iE_0 t - \frac{\Gamma}{2}t} . \tag{6.51}$$

Therefore the probability $|\psi|^2$ decays as $\exp(-\Gamma t)$ and we see that $1/\Gamma$ is the lifetime, and therefore Γ is just the decay rate discussed in Section 6.2.

Fig. 6.5 The one-loop correction to the mass of Φ.

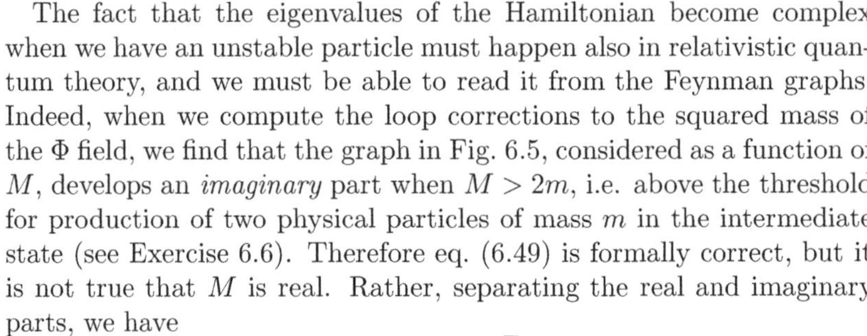

The fact that the eigenvalues of the Hamiltonian become complex when we have an unstable particle must happen also in relativistic quantum theory, and we must be able to read it from the Feynman graphs. Indeed, when we compute the loop corrections to the squared mass of the Φ field, we find that the graph in Fig. 6.5, considered as a function of M, develops an *imaginary* part when $M > 2m$, i.e. above the threshold for production of two physical particles of mass m in the intermediate state (see Exercise 6.6). Therefore eq. (6.49) is formally correct, but it is not true that M is real. Rather, separating the real and imaginary parts, we have

$$M = M_R - i\frac{\Gamma}{2} , \tag{6.52}$$

where M_R is the renormalized mass. For the simple Lagrangian that we have considered, it is straightforward to verify that the imaginary part of M is indeed equal to $-i\Gamma/2$, with Γ the decay rate of the process $\Phi \to \phi_1\phi_2$. We can just compute explicitly the imaginary part of the graph in Fig. 6.5 and compare it with the decay rate computed from the graph in Fig. 6.3, using the general formulas for the decay rate given in Sections 6.2 and 6.4. The *optical theorem* states that this is a general result, and the imaginary part of M is always equal to $-\Gamma/2$, where Γ is the total decay rate (if there are many possible decay channels they all contribute to the imaginary part).

The case $\Gamma \ll M_R$ is particularly interesting; this means that the intermediate particle has a lifetime $1/\Gamma$ much bigger than the time that the light takes to travel a distance equal to its Compton wavelength $1/M_R$. In this case it makes sense to consider it has a real intermediate state, which is produced in the collision, lives for some time and then decays. In this case we call this intermediate particle a *resonance.* In this limit we can approximate $M^2 = (M_R - i\Gamma/2)^2 \simeq M_R^2 - iM_R\Gamma$ and the amplitude (6.49) becomes

$$i\mathcal{M}_{2\to 2} \simeq \frac{-ig^2}{E^2 - M_R^2 + iM_R\Gamma}\,, \tag{6.53}$$

with E the total CM energy. We see that, thanks to the imaginary contribution, at $E = M_R$ the amplitude is no longer divergent. However, it is much larger than far from the resonance. In fact, when we are far from the resonance, i.e. when $E = cM_R$ with c a numerical constant not too close to one, the modulus of the amplitude is of order g^2/M_R^2. Instead, at the resonance, it becomes of order $g^2/M_R\Gamma$. Since $\Gamma \ll M_R$, this is a much bigger value than far from the resonance.

At $E \simeq M_R$ we can further approximate $E^2 - M_R^2 = (E - M_R)(E + M_R) \simeq 2M_R(E - M_R)$ and the amplitude becomes

$$i\mathcal{M}_{2\to 2} \simeq \left(\frac{-ig^2}{2M_R}\right)\frac{1}{E - M_R + i(\Gamma/2)}\,. \tag{6.54}$$

We can now compute the elastic cross-section near the resonance, using eq. (6.44), and observing that the t- and u-channel amplitudes can be neglected since they are not resonant. Then

$$\frac{d\sigma}{d\Omega} \simeq \left(\frac{g^4}{(16\pi)^2 M_R^4}\right)\frac{1}{(E - M_R)^2 + (\Gamma^2/4)}\,. \tag{6.55}$$

The width Γ can also be easily computed explicitly using eq. (6.20) with $i\mathcal{M}_{\Phi\to\phi_1\phi_2} = -ig$ and the phase space (6.35). This gives

$$\Gamma = \frac{g^2}{2M_R}\frac{1}{32\pi^2}\sqrt{1 - \frac{4m^2}{M^2}}\,4\pi = \frac{g^2}{16\pi M_R}\sqrt{1 - \frac{4m^2}{M^2}}\,. \tag{6.56}$$

It is convenient to use this relation to eliminate g in favor of Γ from eq. (6.55), since Γ is the quantity directly observed, while g can be an effective coupling of no fundamental significance if the resonance is a bound state of more fundamental components. Integrating the cross-section over $d\Omega$ we find, at $E \simeq M_R$,

$$\sigma(E) \simeq \frac{4\pi}{M_R^2 - 4m^2}\frac{\Gamma^2}{(E - M_R)^2 + (\Gamma^2/4)}\,. \tag{6.57}$$

At $E = M_R$, conservation of energy gives $M_R = 2\sqrt{m^2 + \mathbf{k}^2}$, where $\mathbf{k}$ is the momentum of the final particles in the CM. Therefore $M_R^2 - 4m^2 = 4\mathbf{k}^2$, and we can rewrite eq. (6.57) as

$$\boxed{\sigma(E) \simeq \frac{\pi}{|\mathbf{k}|^2}\frac{\Gamma^2}{(E - M_R)^2 + (\Gamma^2/4)}\,.} \tag{6.58}$$

This is the *Breit–Wigner distribution*, when the initial and final states of the process are the same. We can generalize it observing that, if the initial and final states are different, the factor Γ^2 in the numerator is replaced by $\Gamma_{R\to i}\Gamma_{R\to f}$ where $\Gamma_{R\to i}$ and $\Gamma_{R\to f}$ are the decay rates of the resonance R into the initial and final states, respectively, simply because the factor of g^2 in the numerator becomes $g_{Ri}g_{Rf}$ where g_{Ri} is the effective coupling of the initial state to the resonance R, and g_{Rf} is the effective coupling to the final state. Instead the factor Γ^2 in the denominator remains the total decay rate.

For the moment, we have limited ourselves to the case where the initial and final particles, described by ϕ_1, ϕ_2, are scalars, and we have also assumed that the resonance Φ is a scalar. If instead the resonance has spin J we must sum over the $2J+1$ possible spin states of the resonance, and we therefore have an overall factor $2J+1$. Furthermore, if the two initial particles have spin s_a and s_b, and we know their spin state, we simply use the partial width $\Gamma_{R\to i}$ for these spin states. If, as it is more common, we do not know the initial polarizations, we average over the initial spins inserting a factor

$$\frac{1}{(2s_a+1)(2s_b+1)}\sum_{s_a,s_b} . \tag{6.59}$$

We reabsorb $\sum_{s_a,s_b}$ in $\Gamma_{R\to i}$, which therefore becomes the width for the process $R \to i$ summed over all possible spin configurations of the state i. Similarly, we sum over the final polarizations, reabsorbing the sum in the redefinition of $\Gamma_{R\to f}$. In conclusion, the general form of the Breit–Wigner distribution is

$$\boxed{\sigma(E) \simeq \frac{2J+1}{(2s_a+1)(2s_b+1)}\frac{\pi}{|\mathbf{k}|^2}\frac{\Gamma_{R\to i}\Gamma_{R\to f}}{(E-M_R)^2+(\Gamma^2/4)} .} \tag{6.60}$$

As an example, consider the scattering

$$e^+e^- \to Z^0 \to f\bar{f} \tag{6.61}$$

where $e^{\pm}$ are the electron and positron, Z^0 is one of the vector bosons of the Standard Model, and $f, \bar{f}$ a fermion–antifermion pair. In this case $s_a = s_b = 1/2$ while $J = 1$. Since $M_Z \gg m_e$, $M_Z^2 \simeq 4|\mathbf{k}|^2$ and eq. (6.60) gives, for the cross-section at the Z^0 peak,

$$\begin{aligned}\sigma_{\text{peak}} &= \frac{3}{2\cdot 2}\left(\frac{4\pi}{M_Z^2}\right)\frac{\Gamma(Z^0\to e^+e^-)\Gamma(Z^0\to f\bar{f})}{\Gamma_Z^2/4}\\ &= \frac{12\pi}{M_Z^2}\frac{\Gamma(Z^0\to e^+e^-)\Gamma(Z^0\to f\bar{f})}{\Gamma_Z^2} .\end{aligned} \tag{6.62}$$

6.6 Born approximation and non-relativistic scattering

In the non-relativistic limit, the computations performed with the Feynman diagrams must reproduce the results of non-relativistic quantum mechanics, where the interaction between particles is described by a potential $V(\mathbf{x})$. The question that we want to answer in this section is the following: given the field theory Lagrangian, what is the potential $V(\mathbf{x})$ experienced by the particles in the non-relativistic limit?

We begin by recalling the basic formulas of scattering theory in non-relativistic quantum mechanics (see e.g. Landau and Lifshitz, vol. III (1977), Section 126): the *elastic* scattering cross-section for a particle of mass m in the potential $V(\mathbf{x})$ has the general form

$$\boxed{\frac{d\sigma}{d\Omega} = |f(\theta)|^2\,,} \tag{6.63}$$

where θ is the scattering angle. The scattering amplitude $f(\theta)$ can be computed considering V as a small perturbation of the free Hamiltonian. The result to first order in V is called the Born approximation, and is given by

$$\boxed{f(\theta) = -\frac{m}{2\pi}\int d^3x\, e^{-i\mathbf{q}\cdot\mathbf{x}}\, V(\mathbf{x})\,.} \tag{6.64}$$

We denote the initial momentum by $\mathbf{k}$, the final momentum by $\mathbf{k}'$ (with $|\mathbf{k}| = |\mathbf{k}'|$ since we consider an elastic process) and the transferred momentum by $\mathbf{q} = \mathbf{k}' - \mathbf{k}$. The scattering angle θ is related to $q = |\mathbf{q}|$ and to $k = |\mathbf{k}|$ by $q = 2k\sin(\theta/2)$. In a central potential $V(r)$ the angular integral in eq. (6.64) is easily performed explicitly, and

$$f(\theta) = -\frac{2m}{q}\int_0^\infty dr\, rV(r)\sin qr\,. \tag{6.65}$$

Let us compare these results with the relativistic formalism that we have developed. For definiteness, we consider the scattering of a non-relativistic particle of momentum $\mathbf{k}$ and mass m, with $|\mathbf{k}| \ll m$, off a heavy target A, with mass $M_A \gg m$. We can think for instance of an electron scattering off an atom. Since $\mathbf{k} \ll m \ll M_A$, we can neglect the recoil of the atom. We limit ourselves to elastic scattering.[2] We assume that the incoming particle and the particle A interact through the exchange of a massless or massive boson; it would be a photon in the electron–atom case, but we can treat similarly the exchange of a massive vector particle, or of a scalar particle. At tree level, the interaction is described by the Feynman diagram in Fig. 6.6a. The fact that we neglect the recoil of the scattering center is represented writing the Feynman diagram as in Fig. 6.6b.[3]

[2] The case where the atom is left in an excited state, and therefore the collision is inelastic, is discussed in Problem 6.2.

[3] We are considering a theory in which there is a (light particle)–(light particle)–boson vertex and a (heavy particle)–(heavy particle)–boson vertex, but no (light particle)–(heavy particle)–boson vertex, so there are no s-channel and u-channel amplitudes, but only the t-channel amplitude of Fig. 6.6a.

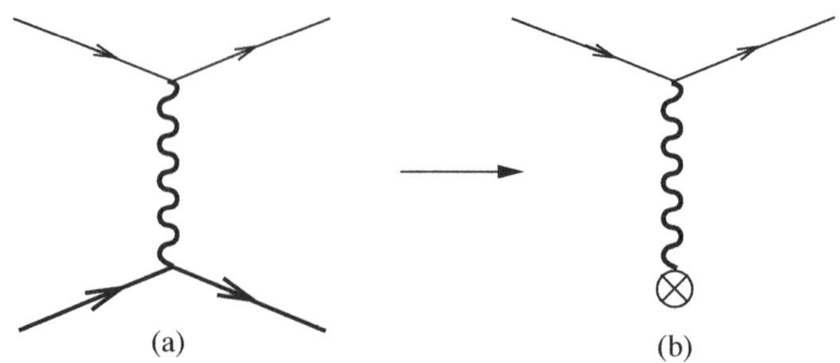

Fig. 6.6 (a): the scattering of a light particle off a heavy target. In the limit in which we neglect the recoil of the target, the graph is drawn as in Fig. (b), which represents more generally the scattering of a particle in an external potential.

We start from our basic formula for elastic scattering, eq. (6.44). By assumption M_A is much larger than the electron energy, so $s \simeq M_A^2$ and eq. (6.44) becomes

$$d\sigma_{\rm elas} \simeq \frac{1}{64\pi^2 M_A^2} |\mathcal{M}_{fi}|^2 d\Omega \,. \tag{6.66}$$

Since we want to compare with the non-relativistic equations, it is convenient to use the non-relativistic normalization of the matrix element. We denote by $|\mathbf{k}\,, A\rangle$ a state with an electron with momentum $\mathbf{k}$ and the atom in a state labeled by A. According to eq. (6.7) we have

$$|\mathbf{k}\,, A\rangle^{(R)} = (2E_{\mathbf{k}})^{1/2}(2E_A)^{1/2}|\mathbf{k}\,, A\rangle^{(NR)} \simeq (2m)^{1/2}(2M_A)^{1/2}|\mathbf{k}\,, A\rangle^{(NR)} \,. \tag{6.67}$$

We work in the rest frame of the atom, so $E_A = M_A$, and we have used the fact that the incoming particle is non-relativistic, so $E_{\mathbf{k}} \simeq m$. For notational convenience we have also set the spatial volume V equal to one, since we have already checked explicitly in Section 6.1 that the volume factors cancel in the final expression for the cross-section. As in Section 6.1, we denote by M_{fi} the matrix element with non-relativistic normalization. Then, from eq. (6.67),

$$\mathcal{M}_{fi} = (\langle \mathbf{k}', A|T_{fi}|\mathbf{k}\,, A\rangle)^{(R)} \simeq (2m)(2M_A)\, M_{fi} \,, \tag{6.68}$$

and eq. (6.66) becomes

$$\frac{d\sigma_{\rm elas}}{d\Omega} \simeq \left(\frac{m}{2\pi}\right)^2 |M_{fi}|^2 \,. \tag{6.69}$$

Comparing with eqs. (6.63) and (6.64) we see that we can identify the non-relativistic scattering amplitude $f(\theta)$ with $(m/2\pi)M_{fi}$.[4]

[4] Comparing the cross-sections, the relative phase between $f(\theta)$ and M_{fi} remains undetermined. The correct phase is a plus sign, as we have written, and can be found either comparing directly the amplitudes, or checking that in this way one obtains the Coulomb potential with the correct sign, as we will do below.

We therefore arrive at the following conclusion. The interaction potential is an intrinsically non-relativistic concept, since it describes an instantaneous (rather than a retarded) interaction. In QFT, in the fully relativistic regime, it makes no sense. However, in the non-relativistic limit, it is recovered with the following procedure: 1) Compute the scattering amplitude $\mathcal{M}_{fi}$ using the Feynman diagrams. 2) Transform it into the amplitude with non-relativistic normalization of the states, M_{fi}. This is related to the non-relativistic scattering amplitude $f(\theta)$ by $M_{fi} = (2\pi/m)f(\theta)$. Using eq. (6.64),

$$M_{fi}(\mathbf{q}) = -\int d^3x\, e^{-i\mathbf{q}\cdot\mathbf{x}}\, V(\mathbf{x}) \,, \tag{6.70}$$

or, inverting the Fourier transform,

$$V(\mathbf{x}) = -\int \frac{d^3q}{(2\pi)^3} M_{fi}(\mathbf{q}) e^{i\mathbf{q}\cdot\mathbf{x}} \,. \tag{6.71}$$

We apply this formula to the electromagnetic scattering of an electron with charge e off a positively charged ion with charge $-Ze$ and spin 1/2, and we consider the case where the initial and final spin states are equal, so we will not write the spin labels explicitly. The vertex factor associated to the electron is $-ie\gamma^\nu$ while the vertex associated to the ion is $+iZe\gamma^\mu$. Therefore the Feynman diagram of Fig. 6.6a gives

$$i\mathcal{M}_{fi} = [\bar{u}_A(+iZe\gamma^\mu)u_A] \, \frac{-i\eta_{\mu\nu}}{q^2} \, [\bar{u}_e(-ie\gamma^\nu)u_e] \,, \tag{6.72}$$

where u_A is the wave function of the ion and u_e of the electron. The momentum transferred by the photon is $q^\mu = (q^0, \mathbf{q})$ and, since we are considering elastic scattering, we have $q^0 = 0$ and $q^2 = -\mathbf{q}^2$. Then

$$\mathcal{M}_{fi} = \frac{Ze^2}{\mathbf{q}^2} \, (\bar{u}_A\gamma^\mu u_A)\, (\bar{u}_e\gamma_\mu u_e) \,. \tag{6.73}$$

Since the particle A is at rest, it is convenient to use the standard representation for the γ matrices, as discussed in Sections 3.4.2 and 3.6. For a particle with mass M_A at rest we found $u_L = u_R = \sqrt{M_A}\,\xi$, with $\xi^\dagger\xi = 1$, see eq. (3.103). As we found in eq. (3.95), the spinor in the standard representation is given in terms of u_L, u_R by

$$u = \frac{1}{\sqrt{2}} \begin{pmatrix} u_R + u_L \\ u_R - u_L \end{pmatrix} , \tag{6.74}$$

so, for the particle A at rest,

$$u_A = \sqrt{2M_A} \begin{pmatrix} \xi \\ 0 \end{pmatrix} . \tag{6.75}$$

Then $\bar{u}_A\gamma^0 u_A = u_A^\dagger u_A = 2M_A\xi^\dagger\xi = 2M_A$ and, from the form of the γ matrices in the standard representation, eq. (3.96), $\bar{u}_A\gamma^i u_A = 0$. Therefore

$$(\bar{u}_A\gamma^\mu u_A)\,(\bar{u}_e\gamma_\mu u_e) = (\bar{u}_A\gamma^0 u_A)\,(\bar{u}_e\gamma_0 u_e) \simeq (2M_A)(2m_e) \,, \tag{6.76}$$

where we have set $\bar{u}_e\gamma_0 u_e \simeq 2m_e$ since the electron is non-relativistic. We see that the contribution of the wave functions on the external lines is just what is needed to convert $\mathcal{M}_{fi}$ into M_{fi}, and we find

$$M_{fi}(\mathbf{q}) = \frac{Ze^2}{\mathbf{q}^2} \,. \tag{6.77}$$

Using eq. (6.71) and performing the angular integration similarly to eq. (6.65), the potential is therefore given by

$$V(\mathbf{x}) = -\int \frac{d^3q}{(2\pi)^3}\, M_{fi}(\mathbf{q}) e^{i\mathbf{q}\cdot\mathbf{x}} = -\frac{4\pi}{(2\pi)^3 r} \int_0^\infty dq\, q M_{fi}(q) \sin qr \,, \tag{6.78}$$

where here we have used the notation $q = |\mathbf{q}|$. The integral is performed using the identity $\int_0^\infty dx\,(\sin x)/x = \pi/2$, and we finally find

$$V(r) = -\frac{Z\alpha}{r}\,, \tag{6.79}$$

which is the standard Coulomb potential. The same calculation can be performed if we consider the exchange of a massive boson of mass μ. The result is now proportional to the Fourier transform of the massive propagator.[5] For elastic scattering we have $q^0 = 0$ and therefore $q^2 = (q^0)^2 - \mathbf{q}^2 = -\mathbf{q}^2$, so

[5] Actually, in the case of a massive gauge boson the propagator is not proportional to $\eta_{\mu\nu}$ but to $\eta_{\mu\nu} - (q_\mu q_\nu/m^2)$, see eq. (8.26). The reader can verify that the additional term $q_\mu q_\nu/m^2$ in our case gives zero since $q = p' - p$ and then $q_\mu \bar{u}(p')\gamma^\mu u(p) = 0$, using the fact that the spinors $u(p), \bar{u}(p')$ are solutions of the Dirac equation.

$$\frac{1}{q^2 - \mu^2} = -\frac{1}{\mathbf{q}^2 + \mu^2}\,. \tag{6.80}$$

Therefore the r-dependence of the potential is now given by

$$\int \frac{d^3q}{(2\pi)^3}\frac{1}{\mathbf{q}^2 + \mu^2}\,e^{i\mathbf{q}\mathbf{x}} \sim \frac{e^{-\mu r}}{r}\,. \tag{6.81}$$

A potential of this form is called a *Yukawa potential.* This is a short-range potential, with a characteristic interaction range equal to the Compton wavelength $1/\mu$ of the particle exchanged. We conclude that:

exchange of a massless boson $\leftrightarrow$ *long-range Coulomb potential*

exchange of a massive boson $\leftrightarrow$ *short-range Yukawa potential.*

As for the sign of the potential, we have seen that in the case of the exchange of a vector boson only the component $\mu = \nu = 0$ of the propagator contributes, because in the non-relativistic limit only the $\mu = 0$ component survives in $\bar{u}\gamma^\mu u$. Therefore the factor coming from the propagator is

$$D_{00} = \frac{-i\eta_{00}}{q^2 - \mu^2} = +\frac{i}{|\mathbf{q}|^2 + \mu^2}\,. \tag{6.82}$$

For the exchange of a scalar particle instead the factor coming from the propagator is

$$D = \frac{i}{q^2 - \mu^2} = -\frac{i}{|\mathbf{q}|^2 + \mu^2}\,. \tag{6.83}$$

The interaction mediated by a vector particle is repulsive for particles with the same charge and attractive for particles with opposite charge, as we have checked in the computation above. We see comparing eqs. (6.82) and (6.83) that the interaction mediated by a scalar particle is instead attractive for particles with the same charge. In fact, the strong interaction between nucleons, at distances larger than the fermi, can be thought of as mediated by the pion, which is scalar and massive, and therefore the strong interaction between nucleons is attractive, and short-ranged.

More complicated potentials can be obtained with the simultaneous exchange of more than one boson. For instance, in the language of Feynman diagrams, van der Waals forces at large distances arise from the exchange of two photons between atoms which are electrically neutral but have an electric dipole moment, see Landau and Lifshitz, vol. IV (1982), Section 85.

6.7 Solved problems

Problem 6.1. Three-body kinematics and phase space

In this problem we investigate various aspects of the kinematics of three-body final states. Consider the decay of a particle with four-momentum p in its rest frame, with $p = (M, 0)$, into three particles with momenta p_i and masses m_i, $i = 1, 2, 3$. Let E_1, E_2, E_3 be the energies of the decay products in the rest frame of the decaying particle. The differential three-body phase space is

$$d\Phi^{(3)} = \frac{d^3p_1}{(2\pi)^3 2E_1} \frac{d^3p_2}{(2\pi)^3 2E_2} \frac{d^3p_3}{(2\pi)^3 2E_3} (2\pi)^4 \delta^{(4)}(p_1 + p_2 + p_3 - p)\,, \qquad (6.84)$$

where $E_i^2 = p_i^2 + m_i^2$. The Dirac delta can be used to integrate over the spatial momentum $\mathbf{p}_3$, so that

$$d\Phi^{(3)} = \frac{1}{8(2\pi)^5} \frac{d^3p_1 d^3p_2}{E_1 E_2 E_3} \delta(E_1 + E_2 + E_3 - M)\,, \qquad (6.85)$$

where $E_3 = (p_3^2 + m_3^2)^{1/2}$ and now $\mathbf{p}_3$ is a notation for $-(\mathbf{p}_1 + \mathbf{p}_2)$. The matrix elements $|\mathcal{M}_{fi}|^2$ must be integrated with this measure. To proceed further, we must know the dependence of $|\mathcal{M}_{fi}|^2$ on $\mathbf{p}_1, \mathbf{p}_2$. The simplest case is the decay of a spin-0 particle into spin-0 particles, in the absence of external fields. In this case there is no preferred direction in space, and the matrix element can only depend on the angle θ between $\mathbf{p}_1$ and $\mathbf{p}_2$. Then eq. (6.85) becomes

$$\begin{aligned} d\Phi^{(3)} &= \frac{1}{8(2\pi)^5} \frac{4\pi p_1^2 dp_1}{E_1 E_2 E_3} 2\pi p_2^2 dp_2 \, d\cos\theta \, \delta(E_1 + E_2 + E_3 - M) \qquad (6.86) \\ &= \frac{1}{32\pi^3} \frac{p_1 dp_1}{E_1 E_2 E_3} (p_2 dp_2) (p_1 p_2 \, d\cos\theta) \, \delta(E_1 + E_2 + E_3 - M)\,. \end{aligned}$$

Now we use the identity $E_1 dE_1 = p_1 dp_1$, which follows from $E_1^2 = p_1^2 + m_1^2$, and similarly $E_2 dE_2 = p_2 dp_2$. Furthermore,

$$E_3^2 = (\mathbf{p}_1 + \mathbf{p}_2)^2 + m_3^2 = p_1^2 + p_2^2 + 2p_1 p_2 \cos\theta + m_3^2\,. \qquad (6.87)$$

Therefore, at p_1, p_2 fixed, we have $E_3 dE_3 = p_1 p_2 d\cos\theta$. In eq. (6.86) it is then convenient to perform the integration in $d\cos\theta$ as the innermost, so we can rewrite eq. (6.86) as

$$d\Phi^{(3)} = \frac{1}{32\pi^3} dE_1 dE_2 dE_3 \, \delta(E_1 + E_2 + E_3 - M)\,, \qquad (6.88)$$

and use the Dirac delta to eliminate E_3. In conclusion, for spin-0 particles, and in the absence of external fields,

$$\boxed{d\Phi^{(3)} = \frac{1}{32\pi^3} dE_1 dE_2\,.} \qquad (6.89)$$

Of course this expression is valid only in the region of the (E_1, E_2) plane where energy–momentum conservation is satisfied, otherwise the Dirac delta gives zero. To determine this region, we first introduce the *Mandelstam variables* s, t, u for the decay of a particle with four-momentum p into three particles with four-momenta p_1, p_2, p_3,

$$s = (p - p_1)^2 = (p_2 + p_3)^2\,, \qquad (6.90)$$

$$t = (p - p_2)^2 = (p_1 + p_3)^2\,, \qquad (6.91)$$

$$u = (p - p_3)^2 = (p_1 + p_2)^2\,. \qquad (6.92)$$

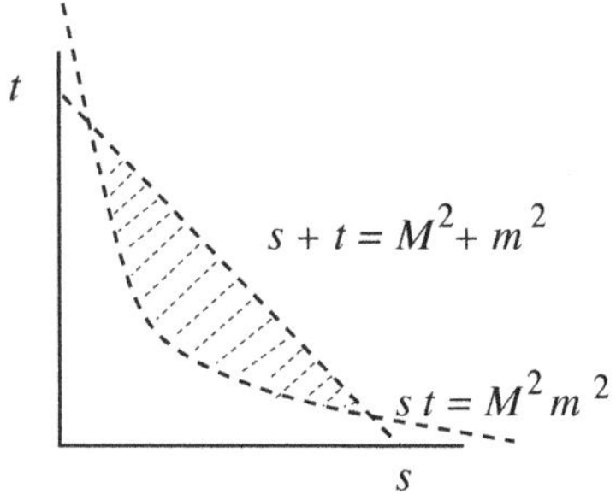

Fig. 6.7 The allowed region of phase space (shaded area) when two final particles are massless.

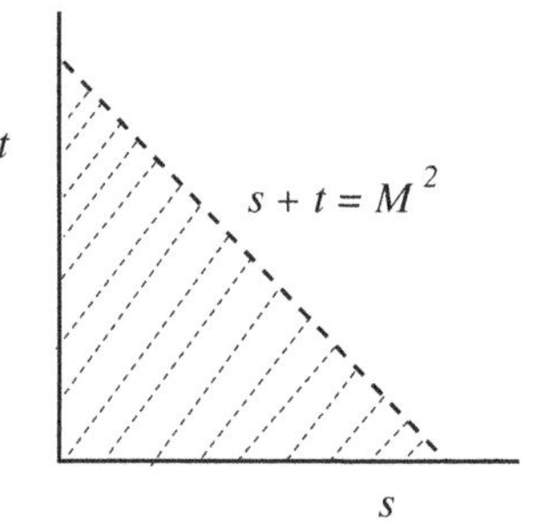

Fig. 6.8 The Dalitz plot when all three final particles are massless becomes a triangle.

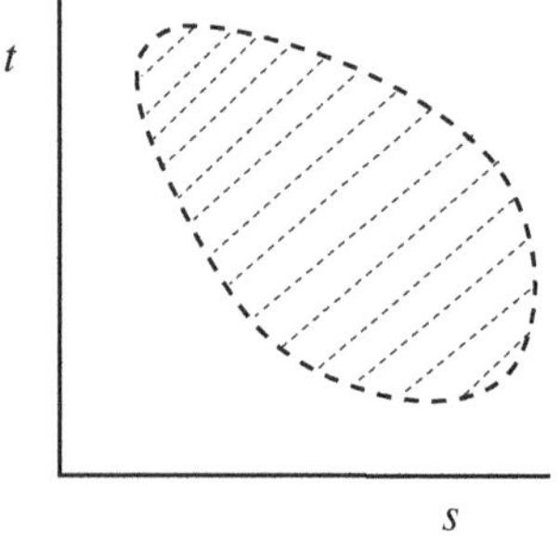

Fig. 6.9 The generic form of the Dalitz plot when all three final particles are massive.

These variables are Lorentz invariant by definition, and can be written in terms of the center-of-mass energies E_1, E_2, E_3 as

$$s = M^2 + m_1^2 - 2ME_1\,, \tag{6.93}$$

$$t = M^2 + m_2^2 - 2ME_2\,, \tag{6.94}$$

$$u = M^2 + m_3^2 - 2ME_3\,. \tag{6.95}$$

s is also called the *invariant mass* of the $(2,3)$ pair, and is denoted also as m_{23}^2, and similarly $t = m_{13}^2$ and $u = m_{12}^2$. From $E_1 + E_2 + E_3 = M$ it follows that

$$s + t + u = M^2 + m_1^2 + m_2^2 + m_3^2\,. \tag{6.96}$$

Therefore the three Mandelstam variables are not independent. We choose s and t as the independent variables. Since $ds = -2MdE_1$ and $dt = -2MdE_2$, the phase space can be rewritten as

$$d\Phi^{(3)} = \frac{dsdt}{16M^2(2\pi)^3}\,. \tag{6.97}$$

We now find the kinematical limits on s, t. First of all s attains its maximum value when E_1 has its minimum value, see eq. (6.93), i.e. when $E_1 = m_1$, so $s_{\max} = (M - m_1)^2$. This corresponds to the configuration in which, in the rest frame of the decaying particle, the initial particle decays into particle 1 at rest, while particles 2 and 3 have opposite momenta, with the modulus of momentum fixed by energy conservation.

The minimum value of s is found instead writing $s = (p_2 + p_3)^2 = m_2^2 + m_3^2 + 2(E_2E_3 - \mathbf{p}_2 \cdot \mathbf{p}_3)$. Since s is invariant, we can compute it in the frame that we prefer. In the CM of the pair $(2,3)$ we have $\mathbf{p}_3 = -\mathbf{p}_2$, and in this frame $s = m_2^2 + m_3^2 + 2(E_2E_3 + |\mathbf{p}_2| \cdot |\mathbf{p}_3|)$, which shows that the minimum value is obtained, in this frame, for $\mathbf{p}_2 = \mathbf{p}_3 = 0$, so that $E_2 = m_2, E_3 = m_3$, and $s = (m_2 + m_3)^2$. Since s is Lorentz invariant, this is the minimum value in any frame. In conclusion, the limits on s are

$$(m_2 + m_3)^2 \leqslant s \leqslant (M - m_1)^2\,. \tag{6.98}$$

Now, fixing s within these limits, we look for the limits on t. We therefore look for a relation which expresses the conservation of energy and momentum and which is written only in terms of s and t. We start from $E_3^2 = p_3^2 + m_3^2$ and we use the conservation of energy, $E_3 = M - E_1 - E_2$ and of momentum, $\mathbf{p}_3 = -\mathbf{p}_1 - \mathbf{p}_2$, to write

$$(M - E_1 - E_2)^2 = m_3^2 + \mathbf{p}_1^2 + \mathbf{p}_2^2 + 2\mathbf{p}_1 \cdot \mathbf{p}_2\,. \tag{6.99}$$

The limiting cases correspond to

$$\mathbf{p}_1 \cdot \mathbf{p}_2 = \pm|\mathbf{p}_1| \cdot |\mathbf{p}_2| = \pm\sqrt{(E_1^2 - m_1^2)(E_2^2 - m_2^2)}\,. \tag{6.100}$$

Inserting this into eq. (6.99), we find that the limiting curve in the (s,t) plane is given by

$$M^2 + 2E_1E_2 + m_1^2 + m_2^2 - m_3^2 - 2M(E_1 + E_2) = \pm 2\sqrt{(E_1^2 - m_1^2)(E_2^2 - m_2^2)}\,. \tag{6.101}$$

Using eqs. (6.93) and (6.94) we can eliminate E_1, E_2 in favor of s, t,

$$E_1 = \frac{M^2 + m_1^2 - s}{2M}\,, \qquad E_2 = \frac{M^2 + m_2^2 - t}{2M}\,. \tag{6.102}$$

We examine first this curve in the limiting case $m_1 = m_2 = 0$ (we then denote m_3 simply by m). The curve with the plus sign becomes simply $t+s = M^2+m^2$ while that with the minus sign becomes $st = m^2M^2$. If $m \neq 0$, the resulting region of the (s,t) plane is shown in Fig. 6.7, while if even $m = 0$ the area degenerates to a triangle, see Fig. 6.8. The plot of the phase space region allowed by energy–momentum conservation is known as the *Dalitz plot*. If all three masses are different from zero the Dalitz plot has the generic form shown in Fig. 6.9. Observe that the number of cusps in the limiting curve is equal to the number of massless final particles.

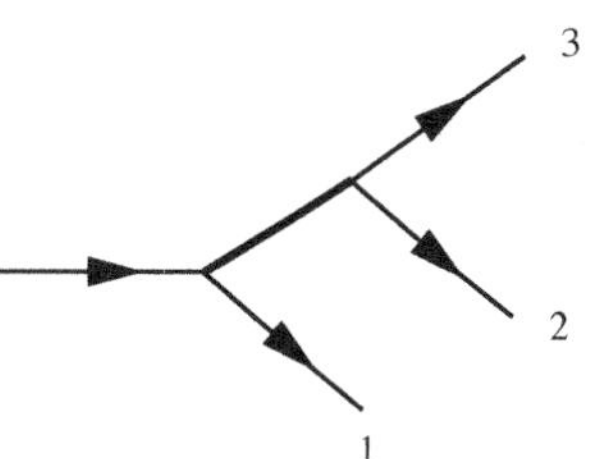

Fig. 6.10 A three-particle decay going through a resonant intermediate state.

The usefulness of this representation is that the phase space is uniform in the Dalitz plot, see eq. (6.97), and therefore any non-uniformity in the distribution of events is due to the matrix element. This allows us to identify immediately possible resonances. Suppose for instance that the decay of the initial particle proceeds through an intermediate resonance that subsequently decays into particles 2 and 3 as in Fig. 6.10. As we saw in Section 6.5, the process will be greatly enhanced when the kinematic invariant $\sqrt{s}$ (i.e. the invariant mass m_{23} of the $(2,3)$ pair) is equal to the mass m_R of the resonance. Therefore the distribution of the experimental events will be mostly localized in a band corresponding to this value of m_{23}, rather than being distributed more or less uniformly over the whole Dalitz plot, and it might look as in Fig. 6.11. This is the way in which many resonances are discovered. For example, the D^0 meson is a particle with mass $m_D = 1864.6 \pm 0.5$ MeV, spin zero and a lifetime $\tau = (410.3\pm1.5)\times10^{-15}$ s. Among its decay modes, one finds a three-body decay $D^0 \to K^-\pi^+\pi^0$. Displaying the various events collected by the detector on a Dalitz plot, one finds a band of the type shown in Fig. 6.11 when on the horizontal axis we plot the invariant mass of the $K^-\pi^+$ system, and the band is localized at $m^2_{K^-\pi^+} \simeq (892\,\text{MeV})^2$. This shows that the process goes through a resonance, known as $K^*(892)^0$, i.e. $D^0 \to K^*(892)^0\pi^0$ and subsequently $K^*(892)^0 \to K^-\pi^+$.

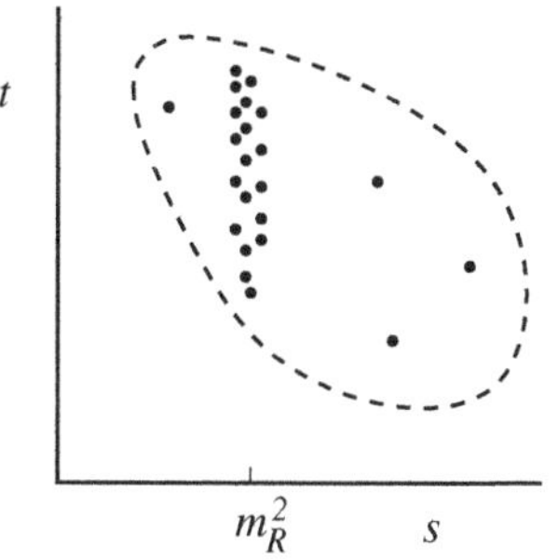

Fig. 6.11 If the three-body decays proceed through a resonance of mass m_R, the experimental events are concentrated on a band around $s = m_R^2$, of width equal to the resonance width, rather than being distributed more or less uniformly in the Dalitz plot.

The same considerations can be applied to a scattering process of two particles into three particles. In this case the initial state, in the CM, has four-momentum $p = (E_{\text{CM}}, 0)$, where E_{CM} is the total energy in the CM, and all considerations above go through with the replacement $M \to E_{\text{CM}}$. So, for instance, in the scattering process of kaons on protons, $K^-p \to \Lambda^0\pi^+\pi^-$, with the kaon momentum of the order of the GeV and the protons at rest, one finds that the events are concentrated in two bands, as in Fig. 6.12, corresponding to the two processes $K^-p \to \Sigma^+\pi^-$, followed by $\Sigma^+ \to \Lambda^0\pi^+$, and $K^-p \to \Sigma^-\pi^+$, followed by $\Sigma^- \to \Lambda^0\pi^-$.

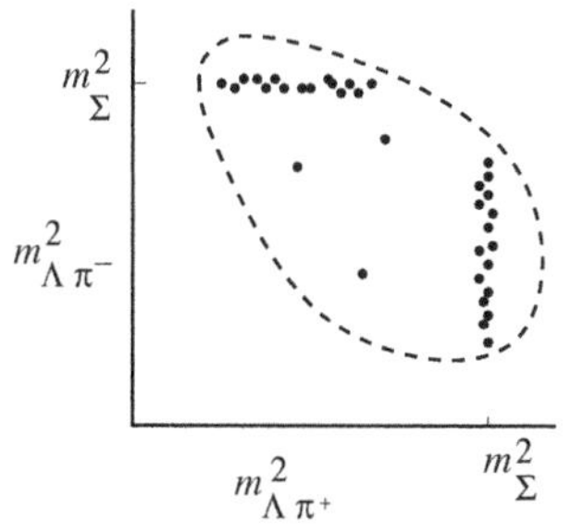

Fig. 6.12 The distribution of events in $K^-p \to \Lambda^0\pi^+\pi^-$, showing that the process goes through the resonances $\Sigma^\pm$.

Problem 6.2. Inelastic scattering of non-relativistic electrons on atoms

In this problem we study the process in which a non-relativistic electron scatters on an atom A, leaving the atom in an excited state A^*, i.e. $e^-A \to e^-A^*$. We start from eq. (6.43) for the inelastic cross-section. We work in the rest frame of the atom A, and we denote its mass by M_A. As in Section 6.6, we use $s \simeq M_A^2$, since the mass of the atom A is much bigger than the electron energy. We use the matrix element with non-relativistic normalization, see eq. (6.8), and therefore eq. (6.43) becomes

$$\frac{d\sigma}{d\Omega} = \left(\frac{m}{2\pi}\right)^2 \frac{p'}{p} |M_{fi}|^2 . \tag{6.103}$$

We use the notation $p = |\mathbf{p}|, p' = |\mathbf{p}'|$. We saw in Section 6.6 that, for *elastic* scattering,

$$M_{fi}(\mathbf{q}) = -\int d^3x\, e^{-i\mathbf{q}\cdot\mathbf{x}}\, V(\mathbf{x}) = -\langle \mathbf{p}'|\hat{V}|\mathbf{p}\rangle\,, \tag{6.104}$$

where the hat denotes the quantum operator, and $|\mathbf{p}\rangle$, $|\mathbf{p}'\rangle$ are the incoming and outgoing electron states with non-relativistic normalization, so $\langle \mathbf{x}|\mathbf{p}\rangle = e^{i\mathbf{p}\cdot\mathbf{x}}$, and $\mathbf{q} = \mathbf{p}' - \mathbf{p}$. We set all volume factors equal to one for simplicity, since they cancel anyhow at the end. The last equality in eq. (6.104) is easily proved inserting two complete sets of states,

$$\langle \mathbf{p}'|\hat{V}|\mathbf{p}\rangle = \int d^3x d^3x' \langle \mathbf{p}'|\mathbf{x}'\rangle\langle \mathbf{x}'|\hat{V}|\mathbf{x}\rangle\langle \mathbf{x}|\mathbf{p}\rangle\,, \tag{6.105}$$

and using $\langle \mathbf{x}'|\hat{V}|\mathbf{x}\rangle = V(\mathbf{x})\langle \mathbf{x}'|\mathbf{x}\rangle = V(\mathbf{x})\delta^{(3)}(\mathbf{x} - \mathbf{x}')$. In eq. (6.104) the atom is treated as an external scattering center, without internal dynamics. If we take into account the fact that the internal state of the atom changes, we must insert in the initial and final state also the atomic state, so for inelastic scattering eq. (6.104) must be replaced by

$$M_{fi} = -\langle \mathbf{p}' A^*|\hat{V}|\mathbf{p}\, A\rangle\,. \tag{6.106}$$

For simplicity, we limit ourselves to the Coulomb interaction of the incoming electrons with the atomic electrons and with the nucleus, neglecting all spin interactions, and the form factor of the nucleus. Then

$$\hat{V}(\mathbf{x},\mathbf{x}_a) = -\alpha\left(\frac{Z}{|\mathbf{x}|} - \sum_{a=1}^{Z}\frac{1}{|\mathbf{x} - \mathbf{x}_a|}\right)\,, \tag{6.107}$$

where $\mathbf{x}$ is the position operator of the incoming electron, $\mathbf{x}_a$ of the atomic electrons, and we have considered a neutral atom with Z electrons. The wave function of the incoming electron is $e^{i\mathbf{p}\cdot\mathbf{x}}$ and of the outgoing electron is $e^{i\mathbf{p}'\cdot\mathbf{x}}$. Therefore, with $\mathbf{q} = \mathbf{p}' - \mathbf{p}$, the matrix element (6.106) becomes

$$\begin{aligned} M_{fi} &= -\int d^3x\, e^{-i\mathbf{q}\cdot\mathbf{x}}\langle A^*|\hat{V}(\mathbf{x},\mathbf{x}_a)|A\rangle \\ &= +\alpha\int d^3x\, e^{-i\mathbf{q}\cdot\mathbf{x}}\left[\frac{Z}{|\mathbf{x}|}\langle A^*|A\rangle - \sum_{a=1}^{Z}\langle A^*|\frac{1}{|\mathbf{x} - \mathbf{x}_a|}|A\rangle\right]\,. \end{aligned} \tag{6.108}$$

We now observe that

$$\int d^3x\, e^{-i\mathbf{q}\cdot\mathbf{x}}\frac{1}{|\mathbf{x} - \mathbf{x}_a|} = e^{-i\mathbf{q}\cdot\mathbf{x}_a}\int d^3x\, e^{-i\mathbf{q}\cdot\mathbf{x}}\frac{1}{|\mathbf{x}|} \tag{6.109}$$

and we use

$$\int d^3x\, e^{-i\mathbf{q}\cdot\mathbf{x}}\frac{1}{|\mathbf{x}|} = \frac{4\pi}{|\mathbf{q}|^2} \tag{6.110}$$

(this equality can be proved more easily adding a factor $e^{-\epsilon|\mathbf{x}|}$ in the integrand, to assure the convergence; in polar coordinates the integral is then elementary, and at the end we take the limit $\epsilon \to 0^+$). Therefore

$$M_{fi} = \frac{4\pi\alpha}{q^2}\left[Z\langle A^*|A\rangle - \sum_{a=1}^{Z}\langle A^*|e^{-i\mathbf{q}\cdot\mathbf{x}_a}|A\rangle\right]\,, \tag{6.111}$$

where $q = |\mathbf{q}|$. We now perform a multipole expansion, expanding $e^{-i\mathbf{q}\cdot\mathbf{x}_a}$, and we retain terms up to quadratic order. The expansion is valid when $qa \ll 1$, where a is the atomic size. Then

$$M_{fi} = \frac{4\pi\alpha}{q^2}\left[iq^i\langle A^*|\sum_{a=1}^{Z}x_a^i|A\rangle + \frac{1}{2}q^iq^j\langle A^*|\sum_{a=1}^{Z}x_a^ix_a^j|A\rangle + O(q^3)\right]\,, \tag{6.112}$$

where in x_a^i the index a labels the atomic electrons, while $i = 1, 2, 3$ is the spatial index. We introduce the dipole and the quadrupole operators

$$D^i = \sum_{a=1}^{Z} x_a^i \,, \tag{6.113}$$

$$Q^{ij} = \sum_{a=1}^{Z} \left(x_a^i x_a^j - \frac{1}{3}\delta^{ij} r_a^2 \right) , \tag{6.114}$$

with $r_a = |\mathbf{x}_a|$. Then

$$M_{fi} = \frac{4\pi\alpha}{q^2}\left[\frac{1}{6}q^2\langle A^*|\sum_a r_a^2|A\rangle + iq^i\langle A^*|D^i|A\rangle + \frac{1}{2}q^iq^j\langle A^*|Q^{ij}|A\rangle + O(q^3)\right] . \tag{6.115}$$

When we take the modulus squared of this amplitude, there is no interference between the dipole and the other two terms (the scalar term $\sim r^2$ and the quadrupole), since the dipole contributes only to transitions which change the parity, while both the scalar and the quadrupole are non-vanishing only between states with the same parity.[6] Let us denote more explicitly $|A\rangle = |nLM\rangle$, where L is the orbital angular momentum, $M = L_z$, and n denotes collectively all the other quantum numbers, e.g. n is the principal quantum number in the hydrogen atom. Observe that, since we are neglecting the spin–orbit coupling, L is separately conserved. Similarly, we write $|A^*\rangle = |n'L'M'\rangle$. Putting together eqs. (6.103) and (6.115), and taking into account that the interference term involving the dipole vanishes, the cross-section is the sum of an even-parity term and the dipole term,

[6] The angular momentum selection rules instead are as follows: the dipole is a vector, and as such it could mediate transitions with $\Delta L = 0, \pm 1$. However parity eliminates $\Delta L = 0$, since x^i is a true vector, and we are left with $\Delta L = \pm 1$. Similarly the quadrupole is a spin-2 operator and can mediate transitions with $\Delta L = 0, \pm 2$, whereas $\Delta L = \pm 1$ are eliminated by parity. The scalar r^2 of course mediates only transitions with $\Delta L = 0, \Delta M = 0$. Therefore in a transition with $\Delta L = 0, \Delta M = 0$ the scalar and quadrupole can interfere.

$$\left(\frac{d\sigma}{d\Omega}\right)_{\rm even} = \frac{p'}{p} m^2\alpha^2 \left| \frac{1}{3}\langle n'L'M'|\sum_a r_a^2|nLM\rangle + \frac{q^iq^j}{q^2}\langle n'L'M'|Q^{ij}|nLM\rangle \right|^2 , \tag{6.116}$$

$$\left(\frac{d\sigma}{d\Omega}\right)_{\rm dipole} = \frac{p'}{p}\frac{4m^2\alpha^2}{q^4} q^iq^j \langle n'L'M'|D^i|nLM\rangle\langle nLM|D^j|n'L'M'\rangle \,. \tag{6.117}$$

These expressions can be simplified observing that typically we do not know the value of M before the transition and we are not interested in a specific value of M' after. Therefore in the cross-section we average over the initial value of M and we sum over the final values. Summing over M, the interference term between the scalar and quadrupole in eq. (6.116) disappears. In fact, the scalar operator $\sum_a r_a^2$ has non-vanishing matrix elements only if $M = M', L = L'$, and its matrix element is independent of M. Therefore the M-dependence in the interference term is completely contained in $\langle n'LM|Q^{ij}|nLM\rangle$. Summing over M, we find

$$\sum_{M=-L}^{L} \langle n'LM|Q^{ij}|nLM\rangle = 0 \,. \tag{6.118}$$

This could be shown by explicit computation, but it is much easier to observe that $\langle n'LM|Q^{ij}|nLM\rangle$ is a spatial tensor with two indices; before performing the sum over M, we have at our disposal the tensor δ^{ij}, and the direction n^i of the quantization axis, so the result will be a combination of δ^{ij} and n^in^j. After we sum over M, any dependence on the direction of the quantization axis disappears, and the result can depend only on δ^{ij}. However, Q^{ij} is a traceless tensor, and therefore the left-hand side of eq. (6.118) cannot be proportional to δ^{ij}. This means that it must vanish.

Therefore the cross-section splits into a sum of scalar, dipole and quadrupole terms,

$$\left(\frac{d\sigma}{d\Omega}\right)_{\text{scalar}} = \frac{p'}{p}\frac{m^2\alpha^2}{9}\,|\langle n'LM|\sum_a r_a^2|nLM\rangle|^2\,\delta_{LL'}\delta_{MM'}\,, \tag{6.119}$$

$$\left(\frac{d\sigma}{d\Omega}\right)_{\text{dipole}} = \frac{p'}{p}\frac{4m^2\alpha^2}{q^4}\frac{q^iq^j}{2L+1}\sum_{M,M'}\langle n'L'M'|D^i|nLM\rangle\langle nLM|D^j|n'L'M'\rangle\,, \tag{6.120}$$

$$\left(\frac{d\sigma}{d\Omega}\right)_{\text{quad}} = \frac{p'}{p}\frac{m^2\alpha^2}{q^4}\frac{q^iq^jq^kq^l}{2L+1}\sum_{M,M'}\langle n'L'M'|Q^{ij}|nLM\rangle\langle nLM|Q^{kl}|n'L'M'\rangle\,. \tag{6.121}$$

Performing the sum over M, M' allows us to simplify further the dipole and quadrupole cross-sections. Again we use the fact that the choice of the quantization axis becomes irrelevant and there is no preferred direction. Therefore, in the dipole cross-section, the quantity

$$\sum_{M,M'}\langle n'L'M'|D^i|nLM\rangle\langle nLM|D^j|n'L'M'\rangle$$

must be proportional to δ^{ij}, since it is a tensor and there is no other quantity that can appear in the final result. We denote the proportionality constant by $D^2/3$,

$$\sum_{M,M'}\langle n'L'M'|D^i|nLM\rangle\langle nLM|D^j|n'L'M'\rangle = \frac{1}{3}\delta^{ij}D^2\,, \tag{6.122}$$

so that by definition $D^2 = \sum_{M,M'}|\langle n'L'M'|D^i|nLM\rangle|^2$. In order to simplify the quadrupole term we introduce the notation

$$T^{ijkl} = \sum_{M,M'}\langle n'L'M'|Q^{ij}|nLM\rangle\langle nLM|Q^{kl}|n'L'M'\rangle\,. \tag{6.123}$$

The advantage of summing over M, M' is that even the apparently complicated tensor structure of T^{ijkl} is fully determined by symmetry considerations. Again, we use the fact that T^{ijkl} is a tensor and δ^{ij} is the only tensor at our disposal, since we have no preferred direction (it is easy to see that ϵ^{ijk} cannot enter, both because of parity and because it is impossible to use it to construct a tensor with the symmetry properties of T^{ijkl}). Therefore we must have $T^{ijkl} = c_1\delta^{ij}\delta^{kl} + c_2\delta^{ik}\delta^{jl} + c_3\delta^{il}\delta^{jk}$. Since $Q^{ij} = Q^{ji}$, T^{ijkl} must satisfy $T^{ijkl} = T^{jikl}$, and similarly for the second pair of indices. This implies that, apart from an overall constant, $T^{ijkl} \sim \delta^{ik}\delta^{jl} + \delta^{il}\delta^{jk} - c\delta^{ij}\delta^{kl}$. The constant c is fixed observing that $\sum_i Q^{ii} = 0$ and therefore $\sum_i T^{iikl} = 0$. This gives $c = 2/3$. Defining

$$Q^2 = \sum_{i,j} T^{ijij}\,, \tag{6.124}$$

we therefore have

$$T^{ijkl} = \frac{Q^2}{10}\left(\delta^{ik}\delta^{jl} + \delta^{il}\delta^{jk} - \frac{2}{3}\delta^{ij}\delta^{kl}\right)\,. \tag{6.125}$$

(The factor $1/10$ comes contracting i with k and j with l in the above equation.) In conclusion, all the information about the atomic structure has been condensed in just D^2 for dipole transitions and Q^2 for quadrupole transitions.

Inserting eqs. (6.122) and (6.125) into eqs. (6.120) and (6.121), respectively, we find

$$\left(\frac{d\sigma}{d\Omega}\right)_{\text{dipole}} = \frac{p'}{p}\frac{4}{3}\frac{1}{2L+1}\frac{m^2\alpha^2 D^2}{q^2}, \tag{6.126}$$

$$\left(\frac{d\sigma}{d\Omega}\right)_{\text{quad}} = \frac{p'}{p}\frac{2}{15}\frac{1}{2L+1} m^2\alpha^2 Q^2. \tag{6.127}$$

The total cross-section is obtained integrating over $d\Omega$. The angular dependence is hidden in the transferred momentum q while p' is fixed by the conservation of energy. Therefore the quadrupole cross-section, which is independent of q, simply gets a factor of 4π after integration over $d\Omega$. For the dipole cross-section we observe that, since $\mathbf{q} = \mathbf{p}' - \mathbf{p}$, we have

$$q^2 = p'^2 + p^2 - 2pp'\cos\theta, \tag{6.128}$$

and therefore

$$qdq = -pp'd\cos\theta. \tag{6.129}$$

and the limits on q are $q_{\text{min}} = p - p'$ (we are considering excitations of the atom due to the collision, so $p > p'$) and $q_{\text{max}} = p + p'$. Then

$$\begin{aligned}
\sigma_{\text{dipole}} &= \int d\Omega \left(\frac{d\sigma}{d\Omega}\right)_{\text{dipole}} \\
&= 2\pi \int_{p-p'}^{p+p'} q\,dq \left(\frac{1}{pp'}\frac{d\sigma}{d\Omega}\right)_{\text{dipole}} \\
&= \frac{8\pi}{3}\frac{1}{2L+1}\frac{m^2\alpha^2 D^2}{p^2}\log\frac{p+p'}{p-p'}.
\end{aligned} \tag{6.130}$$

Summary of chapter

- The calculation of scattering cross-sections and of decay rates is made of two parts: (1) The dynamical part, which is the computation of the matrix element $\mathcal{M}_{fi}$. When a perturbative approach is applicable, $\mathcal{M}_{fi}$ can be computed using the Feynman diagram technique discussed in Chapter 5; (2) The kinematical part, i.e. the summation over the final states, with the appropriate factors for the initial state.
- The kinematics of the final state is contained in the phase space, eq. (6.19). The basic equation for computing the decay width of a particle is eq. (6.20), while scattering cross-sections are computed using eqs. (6.29) and (6.24). Explicit formulas for two-body and three-body final states are given in Section 6.4 and in Problem 6.1.
- When the values of the initial momenta are such that an internal line becomes on-shell, the cross-section is enhanced. In this case the intermediate state can be seen as a real particle which is formed, lives for a certain time, and then decays. Such a particle is called a resonance. The resonant cross-section is described by the Breit–Wigner distribution, eq. (6.60).

Further reading

- Many useful results on the topics of this chapter can be found in the old but still beautiful series of books by Landau and Lifshitz. For the definition of cross-sections, decay rates, phase space, etc. see vol. II (Classical Field Theory), Section 12 and vol. IV (Relativistic Field Theory), Section 65. For resonances in non-relativistic quantum mechanics see vol. III (Quantum Mechanics), Section 134. For emission of radiations by atoms or molecules, diffusion of light, interactions between electrons and between atoms with Feynman diagram techniques, Chapters 5, 6, 9 and 10 of Landau and Lifshitz, vol. IV (1982) give an unmatched source of explicit calculations.
- A rich source of solved problems in particle physics and field theory, including many examples of calculations of scattering processes, decays, etc., is Di Giacomo, Paffuti and Rossi (1994).

Exercises

(6.1) Consider the differential cross-section for the $2 \to 2$ scattering, given in eq. (6.43). Show that it can be rewritten, in terms of the Mandelstam variable t and of the flux factor I, as

$$d\sigma = \frac{1}{64\pi I^2}\,|\mathcal{M}_{fi}|^2 dt\,. \tag{6.131}$$

Observe that all factors are explicitly Lorentz invariant. (Assume cylindrical symmetry to perform the integration over $d\phi$.)

(6.2) Consider a $2 \to 2$ elastic scattering process for two particles of masses m_1 and m_2. In the CM, let $\mathbf{p}\,'$ be the final momentum of the particle 1, E' its energy, v_2 the modulus of the initial velocity of the particle 2, and θ the scattering angle. Perform the Lorentz transformation to the laboratory frame, where the particle 2 is initially at rest and check that the final energy of the particle 1, in the lab frame, is

$$E_{\rm lab} = \gamma_2(E' + v_2|\mathbf{p}\,'|\cos\theta) \tag{6.132}$$

with $\gamma_2 = (1 - v_2^2)^{-1/2}$, and therefore

$$dE_{\rm lab} = \gamma_2 v_2 |\mathbf{p}\,'|\, d\cos\theta\,. \tag{6.133}$$

Use this to show that the two-body phase space can be rewritten as

$$d\Phi^{(2)} = \frac{1}{8\pi\gamma_2 v_2\sqrt{s}}\, dE_{\rm lab} \tag{6.134}$$

where $\sqrt{s}$ is the center of mass energy, and we assumed cylindrical symmetry around the beam axis. The two-body phase space is therefore uniform with respect to the lab energy $E_{\rm lab}$, between the kinematical limits $E_{\rm min} \leqslant E_{\rm lab} \leqslant E_{\rm max}$, with

$$E_{\rm min} = \gamma_2(E' - v_2|\mathbf{p}\,'|)$$
$$E_{\rm max} = \gamma_2(E' + v_2|\mathbf{p}\,'|)\,. \tag{6.135}$$

(6.3) (i) Consider the decay of an excited atomic state A^* into a lower state A, with emission of a photon, $A^* \to A\gamma$. Verify that the phase space can be written as

$$d\Phi^{(2)} = \frac{\omega}{16\pi^2 M_A}\, d\Omega\,, \tag{6.136}$$

where ω is the energy of the photon and M_A the mass of the atom.

(ii) Verify that the decay width can be written as

$$d\Gamma = |M_{fi}|^2 \frac{\omega}{8\pi^2}\, d\Omega\,, \tag{6.137}$$

where M_{fi} is the matrix element with the normalization of one particle per unit volume for the atomic states.

(iii) Consider a scattering process $A\gamma \to A^*\gamma'$. Show that

$$d\sigma = \frac{1}{16\pi^2}\frac{\omega'}{\omega}|M_{fi}|^2 d\Omega\,, \tag{6.138}$$

where ω, ω' are the energy of the initial and final photon, respectively.

(6.4) Consider a two-photon decay of an atomic state, $A^* \to A\gamma_1\gamma_2$. Show that the decay width can be written as

$$\frac{d\Gamma}{d\omega_1} = \frac{1}{8(2\pi)^5}\,\omega_1(\omega-\omega_1)\int d\Omega_1 d\Omega_2 |M_{fi}|^2\,, \tag{6.139}$$

where ω_1, ω_2 are the energies of the two photons, $\omega = E_{A^*} - E_A = \omega_1 + \omega_2$, and $d\Omega_1, d\Omega_2$ are the solid angles of the two photons.

(6.5) Denote by $d\Phi^{(n)}(P; p_1, \dots, p_n)$ the n-body phase space, with $p_1 + \dots p_n = P$. Show that

$$d\Phi^{(n)}(P; p_1, \dots, p_n) = \int_0^\infty \frac{d\mu^2}{2\pi} \tag{6.140}$$
$$\times d\Phi^{(j)}(q; p_1, \dots, p_j) d\Phi^{(n-j+1)}(P; p_{j+1}, \dots, p_n, q),$$

where $\mu^2 = q_0^2 - \mathbf{q}^2$. Discuss the physical meaning of this recursive representation of the phase space.

(6.6) (i) In the theory with Lagrangian given in eq. (6.48), show that the Feynman diagram in Fig. 6.5 develops an imaginary part when $M > 2m$, and compute it.

(ii) Denoting the result of the Feynman diagram of Fig. 6.5 by $i\mathcal{M}$, show that eq. (6.52) predicts that the decay rate Γ for the process $\Phi \to \phi_1\phi_2$ is related to $\mathcal{M}$ by

$$\Gamma = \frac{1}{M_R}\,\mathrm{Im}\mathcal{M}\,. \tag{6.141}$$

(iii) Verify the correctness of the above relation (which is a form of the optical theorem) computing Γ explicitly.

7 Quantum electrodynamics

7.1 The QED Lagrangian

Quantum electrodynamics (QED) describes the interaction between electrons (or any other charged spin 1/2 particle, like muons) and photons. It is convenient to quantize the photons using the covariant quantization of Section 4.3.2. Actually, it is also useful to generalize slightly the Lagrangian used in Section 4.3.2: instead of eq. (4.102), we describe the free electromagnetic field by

$$\mathcal{L}_{\rm em} = -\frac{1}{4}F_{\mu\nu}F^{\mu\nu} - \frac{1}{2\xi}\left(\partial_\mu A^\mu\right)^2 , \tag{7.1}$$

with ξ a generic parameter. In Section 4.3.2 we set $\xi = 1$, but it can be shown that for any ξ, after requiring that $\partial_\mu A^\mu$ vanishes between physical states, the spectrum of the theory is given by the two transverse polarization states of the photon. Basically this comes out because the only role of the term $(1/2\xi)(\partial A)^2$ is to break gauge invariance and to allow us to define the momentum conjugate to A_0. Then, between physical states, the operator $\partial_\mu A^\mu$ vanishes and the matrix elements between physical states obtained with eq. (7.1) are independent of ξ. Of course intermediate steps, like the equal time commutation relations between A_μ and the conjugate momenta, or the propagator, do depend on ξ. In the interacting theory, it will turn out that the dependence on ξ vanishes if A_μ is coupled to matter respecting gauge invariance, so in particular A_μ must be coupled to a conserved current.

It is sometimes useful to work with ξ generic, and to check the correctness of the computation verifying that in the end ξ cancels in the matrix elements between physical states. Also, in different problems, different choices of ξ can simplify the calculation. The term $(1/2\xi)(\partial A)^2$ is called the *gauge fixing* term and ξ is the gauge fixing parameter; the choice $\xi = 1$ is called the *Feynman gauge*, and is typically the simplest choice. Sometimes also the choice $\xi = 0$ (Landau gauge) is useful; the Lagrangian is singular in this limit, but we will see below that the photon propagator is well defined at $\xi = 0$.

The interaction between the photon and the electron is written in terms of the covariant derivative, as explained in Section 3.5.4. QED is then described by the Lagrangian

$$\mathcal{L}_{\rm QED} = \bar{\Psi}(i\partial\!\!\!/ - m)\Psi - \frac{1}{4}F_{\mu\nu}F^{\mu\nu} - \frac{1}{2\xi}\left(\partial_\mu A^\mu\right)^2 - eA_\mu\bar{\Psi}\gamma^\mu\Psi . \tag{7.2}$$

The Feynman rules of QED have already been given in Section 5.5.4. We just add that, if we use a generic $\xi \neq 1$, the photon propagator becomes

$$\tilde{D}_{\mu\nu}(k) = \frac{-i}{k^2 + i\epsilon} \left(\eta_{\mu\nu} - (1-\xi)\frac{k_\mu k_\nu}{k^2} \right) . \tag{7.3}$$

We now discuss the symmetries of the QED action. From the point of view of space-time symmetries, QED has of course Poincaré invariance. We have also seen that the coupling is constructed in such a way that the theory is invariant under gauge transformation, i.e. *local* $U(1)$ transformations

$$\Psi(x) \to e^{ie\theta(x)}\Psi(x) , \tag{7.4}$$
$$A_\mu(x) \to A_\mu(x) - \partial_\mu\theta . \tag{7.5}$$

The presence of this local symmetry implies also the existence of the corresponding *global* $U(1)$ symmetry with θ a constant parameter,

$$\Psi(x) \to e^{ie\theta}\Psi(x) , \tag{7.6}$$
$$A_\mu(x) \to A_\mu(x) . \tag{7.7}$$

There is therefore an associated conserved Noether current, which is $\bar{\Psi}\gamma^\mu\Psi$, and a $U(1)$ charge which is conserved by the electromagnetic interaction. To understand the meaning of this charge, observe that $Q = \int d^3x\, j^0$, with $j^0 = \bar{\Psi}\gamma^0\Psi$. In the Lagrangian density j^0 is coupled to A^0, and in the Hamiltonian density j^0 enters as $+eA_0 j^0$. Since in classical electrodynamics A_0 is the electrostatic potential, we see that j^0 is the electric charge density, measured in units of e, and therefore Q is the electric charge, again in units of e. As we saw in Section 4.2, for electrons $Q = 1$ and for positrons $Q = -1$, see eq. (4.43).

Gauge invariance implies that the photon is massless: a mass term for the photon would correspond to a term $m_\gamma^2 A_\mu A^\mu$ in the Lagrangian, but this is forbidden since it is not invariant under gauge transformations. If gauge invariance were broken we should expect a photon mass of the order of the symmetry-breaking scale. However, the experimental bound on the photon mass is extraordinarily tight, $m_\gamma < 2 \times 10^{-16}$ eV.

QED also has important discrete symmetries. Consider first the parity operation P. We have defined the action of P on a quantized spinor field in eq. (4.58): $\Psi(x) \to \eta_a\gamma^0\Psi(x')$. From this it follows that the current $\bar{\Psi}\gamma^\mu\Psi$ is a true four-vector, i.e. under parity the spatial components change sign and the temporal component is invariant. Since also ∂_μ is a true four-vector, the kinetic term of the fermion is invariant under parity. Similarly, the gauge field A_μ is a true four-vector, so both the kinetic term of the gauge field and the interaction term are invariant under P. Therefore QED *is invariant under parity.*

Another important symmetry is charge conjugation, C. In Section 4.2 we defined the operation of charge conjugation on the quantized Dirac spinors. We saw in Exercise 4.3 that, using the fact that the quantized Dirac fields anticommute, the operator $\bar{\Psi}\gamma^\mu\Psi$ changes sign under charge conjugation,

$$C\bar{\Psi}\gamma^\mu\Psi C = -\bar{\Psi}\gamma^\mu\Psi . \tag{7.8}$$

Observe that, even if it involves complex conjugation, on the quantized fields C is defined as a linear (rather than antilinear) operator. Its action on the quantized Dirac field is determined by its action on the creation and annihilation operators $a_{\mathbf{p},s}, b_{\mathbf{p},s}$ given in eq. (4.59), regardless of the fact that the coefficients of $a_{\mathbf{p},s}, b_{\mathbf{p},s}$ in the expansion of Ψ are the complex functions $u^s(p)e^{-ipx}$ and $v^s(p)e^{ipx}$. Therefore $Ci\Psi C = iC\Psi C$, so also $i\bar{\Psi}\gamma^\mu\Psi$ changes sign under charge conjugation. Then the kinetic term transforms as

$$Ci\bar{\Psi}\gamma^\mu\partial_\mu\Psi C = -i(\partial_\mu\bar{\Psi})\gamma^\mu\Psi\,, \tag{7.9}$$

(since the term $\partial_\mu\Psi$, after the action of C, becomes proportional to $\partial_\mu\Psi^*$ and is then anticommuted to the left where it combines with a γ^0 to give $\partial_\mu\bar{\Psi}$) and, after integrating ∂_μ by parts, the kinetic term in the action is invariant. Since the interaction term is proportional to $A_\mu\bar{\Psi}\gamma^\mu\Psi$ and $\bar{\Psi}\gamma^\mu\Psi$ changes sign, if we define the charge conjugation on A_μ as $CA_\mu(x)C = -A_\mu(x)$, QED *is invariant under charge conjugation*, as we already saw in Section 4.3.2. The photon is then an eigenstate of charge conjugation, with eigenvalue -1,

$$C|\gamma\rangle = -|\gamma\rangle\,. \tag{7.10}$$

As an example of the use of these invariance principles, we examine the electromagnetic decay of the neutral pion. In general, a particle can be an eigenstate of charge conjugation only if it is electrically neutral. Consider the three pions $\pi^\pm, \pi^0$. Apart from an arbitrary phase, we have $C|\pi^+\rangle = |\pi^-\rangle$ and $C|\pi^-\rangle = |\pi^+\rangle$. Instead, the π^0 is neutral, and therefore $C|\pi^0\rangle = \eta|\pi^0\rangle$; C has been defined on spinors so that $C^2 = 1$ (see Section 4.2.3) and of course $C^2 = 1$ also on the gauge field, so C^2 is the identity operator. So, even if π^0, at the fundamental level, is a possibly complicated bound state of fermions (in terms of quarks $\pi^0 = u\bar{u} + d\bar{d}$), we know that C^2 is the identity operator also when we apply it to the π^0, and therefore $\eta^2 = 1$ and η can only take the values ± 1. To see which one is the actual value, we observe that π^0 decays electromagnetically as

$$\pi^0 \to 2\gamma\,. \tag{7.11}$$

The two-photon state is an eigenstate of C with eigenvalue $(-1)^2 = +1$ and since the electromagnetic interaction conserves C, this must also be the value of C for the π^0, i.e. $\eta = +1$. In turn, this means that the electromagnetic decay $\pi^0 \to 3\gamma$ is forbidden because it violates C. Experimentally the decay into three photons is not observed, and the limit is

$$\frac{\Gamma(\pi^0 \to 3\gamma)}{\Gamma(\pi^0 \to 2\gamma)} < 3.1 \times 10^{-8}\,. \tag{7.12}$$

Finally, the QED action is invariant under time reversal T, and therefore also under CPT, in agreement with the CPT theorem, see Section 4.2.3.

7.2 One-loop divergences

In Section 5.6 we defined the superficial degree of divergence D for a scalar field theory, and we saw that the condition for renormalizability is that only a limited number of Green's functions have $D \geqslant 0$. In QED, or in general in the presence of fermions, the definition of D must be modified, since the fermionic propagator decreases as $1/p$ rather than $1/p^2$. We denote by $N_f^{\text{ext}}, N_\gamma^{\text{ext}}$ the number of external fermionic and photonic lines respectively, by $N_f^{\text{int}}, N_\gamma^{\text{int}}$ the number of internal fermionic and photonic lines, by V the number of vertices in the graph and by L the number of loops. Then, repeating the arguments of Section 5.6, the superficial degrees of divergence is defined in QED as

$$D = 4L - 2N_\gamma^{\text{int}} - N_f^{\text{int}}\,. \tag{7.13}$$

The number of loops is related to the total number of internal lines $N_f^{\text{int}} + N_\gamma^{\text{int}}$ as in eq. (5.143),

$$L = N_f^{\text{int}} + N_\gamma^{\text{int}} - V + 1 \tag{7.14}$$

and the fact that to each vertex are associated two fermionic lines and one photonic line means that

$$2V = 2N_f^{\text{int}} + N_f^{\text{ext}}\,, \qquad V = 2N_\gamma^{\text{int}} + N_\gamma^{\text{ext}}\,. \tag{7.15}$$

Combining these expressions, we find

$$D = 4 - N_\gamma^{\text{ext}} - \frac{3}{2}N_f^{\text{ext}}\,. \tag{7.16}$$

This means that only the Green's functions with $N_\gamma^{\text{ext}} + \frac{3}{2}N_f^{\text{ext}} \leqslant 4$ are potentially dangerous. Furthermore, some of the potentially dangerous Green's functions are actually finite or even zero. Consider in fact the Green's functions with no external electron line and an arbitrary number $N_\gamma^{\text{ext}} = n$ of external photon lines. They correspond to

$$\langle 0|A_{\mu_1}(x_1)\dots A_{\mu_n}(x_n)\exp\left\{-i\int d^4x\,\mathcal{H}_{QED}\right\}|0\rangle_c\,. \tag{7.17}$$

We have seen that the QED Hamiltonian is invariant under charge conjugation, $C\mathcal{H}_{QED}C = \mathcal{H}_{QED}$. Inserting multiple factors $C^2 = 1$ in the above expression and using $C|0\rangle = |0\rangle$, we find

$$\begin{aligned}
&\langle 0|A_{\mu_1}(x_1)\dots A_{\mu_n}(x_n)\exp\left\{-i\int d^4x\,\mathcal{H}_{QED}\right\}|0\rangle \\
&= \langle 0|(CA_{\mu_1}(x_1)C)\dots(CA_{\mu_n}(x_n)C)(C\exp\left\{-i\int d^4x\,\mathcal{H}_{QED}\right\}C)|0\rangle \\
&= (-1)^n\langle 0|A_{\mu_1}(x_1)\dots A_{\mu_n}(x_n)\exp\left\{-i\int d^4x\,\mathcal{H}_{QED}\right\}|0\rangle\,.
\end{aligned} \tag{7.18}$$

Therefore the Green's functions with no external fermion lines and with an odd number of external photon lines are identically zero, to all orders in perturbation theory (Furry's theorem).

Thus, we are finally left with the following potentially dangerous[1] Green's functions: if $N_f^{\text{ext}} = 0$ we can have

(i) $N_\gamma^{\text{ext}} = 0$. This is a vacuum diagram, and can be cured simply by normal ordering the Hamiltonian.

(ii) $N_\gamma^{\text{ext}} = 2$. This is a divergence in the photon propagator.

(iii) $N_\gamma^{\text{ext}} = 4$. This is a light–light scattering amplitude.

If instead $N_f^{\text{ext}} = 2$ (recall that N_f^{ext} must be even, as we see from the first equation in (7.15), and as dictated by charge conservation) we can have only

(i) $N_\gamma^{\text{ext}} = 0$ (the fermion propagator), and

(ii) $N_\gamma^{\text{ext}} = 1$ (the interaction vertex).

If $N_f^{\text{ext}} \geqslant 4$ all Green's functions have instead $D < 0$. Let us discuss these UV divergences at the one-loop level. The corresponding one-loop diagrams are shown in Figs. 7.1–7.4. Using for simplicity the Feynman gauge $\xi = 1$, the graph in Fig. 7.1 is given by

$$i\Sigma(p) \equiv \int \frac{d^4k}{(2\pi)^4}(-ie\gamma_\mu)\left(-\frac{i\eta^{\mu\nu}}{k^2}\right)\frac{i(\not{p}-\not{k}+m)}{(p-k)^2-m^2+i\epsilon}(-ie\gamma_\nu)\,. \tag{7.19}$$

The correction to the photon propagator is given by the graph in Fig. 7.2,

$$i\Pi_{\mu\nu}(q) \equiv (-1)\int\frac{d^4k}{(2\pi)^4} \tag{7.20}$$
$$\times \text{Tr}\left[(-ie\gamma_\mu)\frac{i(\not{k}+m)}{k^2-m^2+i\epsilon}(-ie\gamma_\nu)\frac{i(\not{k}-\not{q}+m)}{(k-q)^2-m^2+i\epsilon}\right]\,.$$

The minus sign comes from the fermionic loop, and writing explicitly the Dirac indices one can check that they combine to give a trace. The correction to the interaction vertex in Fig. 7.3 is

$$-i\Gamma_\mu(p,q) \equiv \int\frac{d^4k}{(2\pi)^4}(-ie\gamma_\nu)\left(-\frac{i\eta^{\nu\rho}}{k^2}\right)(-ie\gamma_\rho)\frac{i(\not{p}'+\not{k}+m)}{(p'+k)^2-m^2+i\epsilon}$$
$$\times(-ie\gamma_\mu)\frac{i(\not{p}+\not{k}+m)}{(p+k)^2-m^2+i\epsilon}\,, \tag{7.21}$$

with $q+p=p'$. The one-loop light–light scattering amplitude of Fig. 7.4 is instead given by (the $+i\epsilon$ are understood)

$$A_{\mu\nu\rho\sigma}(k_1,k_2,k_3,k_4) = (-1)(-ie)^4 i^4\int\frac{d^4p}{(2\pi)^4}\,\text{Tr}\left[\gamma^\mu\frac{(\not{p}+m)}{p^2-m^2}\right.$$
$$\times\gamma^\nu\frac{(\not{p}-\not{k}_3+m)}{(p-k_3)^2-m^2}\gamma^\rho\frac{(\not{p}-\not{k}_1-\not{k}_2+m)}{(p-k_1-k_2)^2-m^2}$$
$$\left.\times\,\gamma^\sigma\frac{(\not{p}-\not{k}_1+m)}{(p-k_1)^2-m^2}\right]\,, \tag{7.22}$$

with $k_1+k_2=k_3+k_4$. The integrals in eqs. (7.19), (7.20) and (7.21) are UV divergent, so to make sense of them we must specify a regularization procedure. Putting a cutoff Λ in Euclidean momentum space, as we did in Chapter 5 for $\lambda\phi^n$ theories, is not at all convenient in a gauge theory.

[1] As in Section 5.6, this does not mean that the other Green's functions have no divergences, but that their divergencies are automatically cured by the renormalization of Green's functions with a smaller number of external legs.

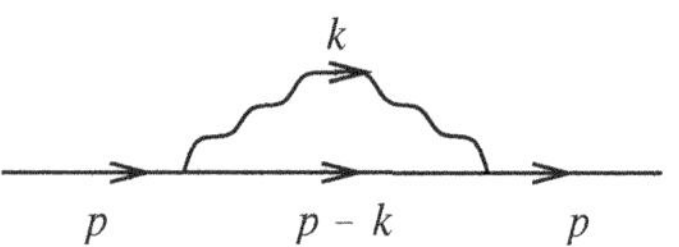

Fig. 7.1 The one-loop electron self-energy.

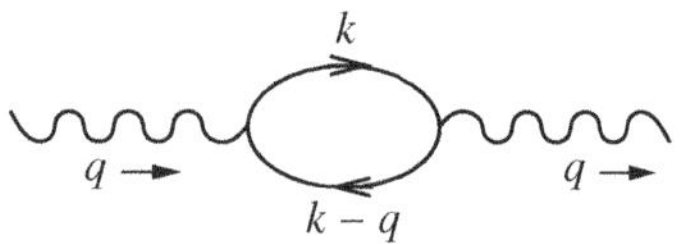

Fig. 7.2 The one-loop photon self-energy.

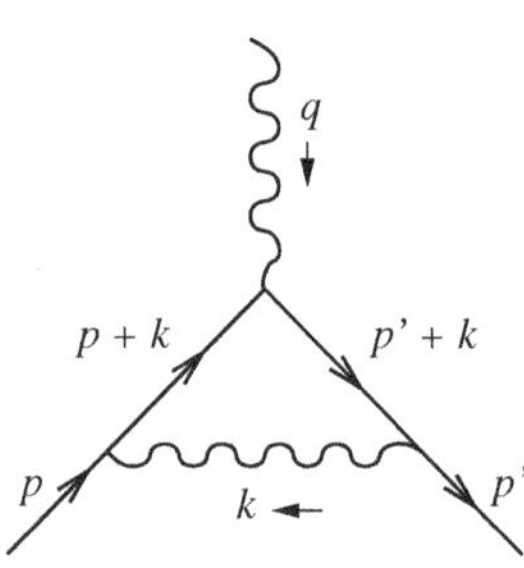

Fig. 7.3 The one-loop vertex correction.

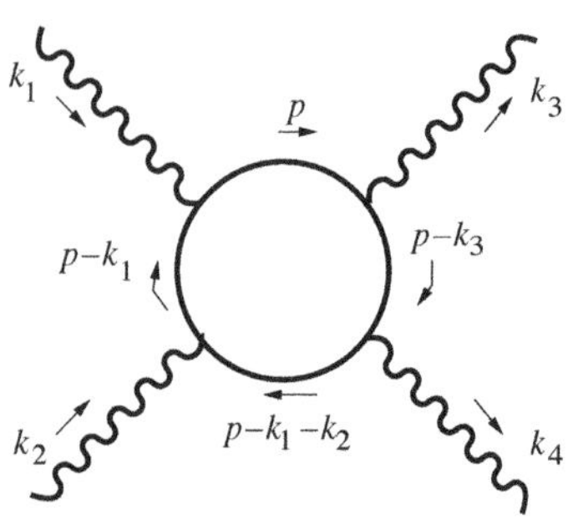

Fig. 7.4 The one-loop light–light scattering amplitude.

The problem is that putting such a cutoff means that we are setting to zero all momentum modes of the fields with $k > \Lambda$. However, even if we set to zero all Fourier modes of the gauge field $\tilde{A}_\mu(k)$ with $k > \Lambda$, these modes are regenerated by a gauge transformation $A_\mu(x) \to A_\mu(x) - \partial_\mu \theta$, with θ generic. In other words, a cutoff in momentum space is not compatible with gauge invariance. In general, it is very dangerous to break a symmetry of the theory by the regularization. One would naively expect that the symmetry is recovered when we remove the cutoff, but this turns out to be not at all automatic (if the symmetry is not recovered, one says that there is an *anomaly* in the theory). If the gauge symmetry became anomalous it would be a disaster. We saw in Section 4.3 that gauge invariance is crucial in order to eliminate the spurious degrees of freedom from the gauge field A_μ and remain with a massless particle with two helicity states $h = \pm 1$, as is the photon. It is therefore much more convenient to regularize the theory maintaining gauge invariance explicitly. The two most useful gauge invariant regularizations are *dimensional regularization* and *Pauli–Villars regularization.* The former is based on the following idea. Consider for example

$$I_d \equiv \int \frac{d^d k}{(2\pi)^d} \frac{1}{(k^2 + \Delta)^2}, \tag{7.23}$$

where we have already performed the Wick rotation, so k is now a Euclidean momentum, while Δ is some combination of external momenta and masses. We are interested in $d = 4$, but we keep for the moment d generic. For $d = 4$, this is one of the typical divergent parts of the diagrams written above. Now one observes that, if $d < 4$, the integral is convergent and the result can be written in terms of the Euler Γ function,

$$I_d = \frac{1}{(4\pi)^{d/2}} \frac{\Gamma(2 - \frac{d}{2})}{\Gamma(2)} \left(\frac{1}{\Delta}\right)^{2-\frac{d}{2}}. \tag{7.24}$$

The function $\Gamma(z)$ has isolated poles at $z = 0, -1, -2, \ldots$ and therefore the integral diverges in $d = 4, 6, 8, \ldots$, and is otherwise well defined, even for d non-integer. We can therefore take the right-hand side of eq. (7.24) as the *definition* of I_d for generic d, real or even complex. We can now study it in $d = 4 - \epsilon$ dimensions, and in the limit $\epsilon \to 0$ we recover our divergence; from the known behavior of the Γ function near the poles one finds that, as $\epsilon \to 0$,

$$I_{4-\epsilon} \to \frac{1}{(4\pi)^2} \left(\frac{2}{\epsilon} - \log\frac{\Delta}{4\pi} - \gamma + O(\epsilon)\right), \tag{7.25}$$

where $\gamma \simeq 0.5772\ldots$ is called the Euler–Mascheroni constant. We have therefore succeeded in writing the integral as a divergent part plus a finite term; ϵ plays the role of the cutoff.

The Pauli–Villars regularization is instead based on the idea of modifying the form of the propagator in the UV, so that it goes to zero more rapidly and helps the convergence of the loop integrals. For instance, the photon propagator is modified by the replacement

$$\frac{1}{k^2 - i\epsilon} \to \frac{1}{k^2 - i\epsilon} - \frac{1}{k^2 + \Lambda^2 - i\epsilon}. \tag{7.26}$$

In the limit $\Lambda \to \infty$ we recover the original propagator, but for finite Λ at large k^2 the propagator decreases as $1/k^4$ rather than $1/k^2$.

It can be shown that both dimensional and Pauli–Villars regularizations preserve gauge invariance. There is a well-developed technology for computing integrals and renormalizing the theory in these schemes, see, e.g. Peskin and Schroeder (1995) or Weinberg (1995).

Once we have regularized the integrals, respecting gauge invariance, we can adapt to QED the same reasoning explained in the case of $\lambda\phi^4$ theory; then, the graph in Fig. 7.1 can be treated exactly as we did in Section 5.5.2 for the scalar propagator, and the divergence is reabsorbed in a renormalization of the mass and of the wave function of the fermion. The graph in Fig. 7.2 instead gives a result of the form

$$\Pi_{\mu\nu}(q) = \left(\eta_{\mu\nu}q^2 - q_\mu q_\nu\right) \Pi(q^2) , \tag{7.27}$$

with $\Pi(q^2)$ divergent. This divergence is reabsorbed in a renormalization of the wave function of the photon. It is important that in $\Pi_{\mu\nu}(q)$ there is no term proportional to $\eta_{\mu\nu}$ times a constant (rather than $\eta_{\mu\nu}q^2$), since this would have provided a renormalization of the mass of the photon. However, a photon mass term $m^2 A_\mu A^\mu$ in the Lagrangian is forbidden by gauge invariance, and therefore such a term is not produced using a gauge-invariant regularization. Finally, the graph in Fig. 7.3 renormalizes the electric charge.

The graph in Fig. 7.4, in a naive power counting, seems logarithmically divergent. However the explicit computation shows that this graph is finite because the would-be divergent term inside the integral actually vanishes.

After reabsorbing the divergences into the renormalized fields, renormalized mass and renormalized charge, all other Green's function are one-loop finite. This turns out to hold at all loops, and QED is renormalizable.

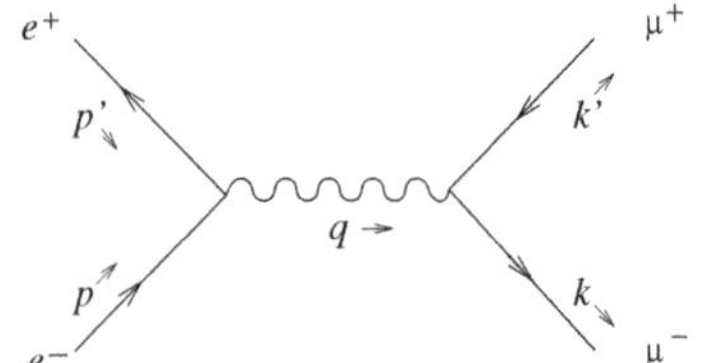

Fig. 7.5 The amplitude for $e^+e^- \to \mu^+\mu^-$ at order e^2; p, p' are the incoming momenta of the electron and positron, respectively, and k, k' the outgoing momenta of μ^- and μ^+.

7.3 Solved problems

Problem 7.1. $e^+e^- \to \gamma \to \mu^+\mu^-$

As a prototype of many similar computations, we evaluate the cross-section for the process $e^+e^- \to \mu^+\mu^-$ in QED. Remember, however, that when we approach the electroweak scale we cannot limit ourselves to QED and there is also a contribution to the amplitude from the Z^0, $e^+e^- \to Z^0 \to \mu^+\mu^-$, which becomes resonant at $E = m_Z \simeq 90$ GeV. As long as the CM energy E is much smaller that m_Z we can neglect it.

In QED at lowest order there is only one Feynman graph, shown in Fig. 7.5. The Feynman rules give

$$i\mathcal{M}_{fi} = \bar{v}^{s'}(p')(-ie\gamma^\mu)u^s(p)\,\frac{-i}{q^2}\left(\eta_{\mu\nu} - (1-\xi)\frac{q_\mu q_\nu}{q^2}\right)\bar{u}^r(k)(-ie\gamma^\nu)v^{r'}(k') , \tag{7.28}$$

where $q = p + p'$ and the assignment of the momenta to the various particles is as in Fig. 7.5. First of all, we observe that the term $\sim q_\mu q_\nu$ in the photon propagator gives zero. In fact, using $q = p + p'$,

$$q_\mu \left(\bar{v}^{s'}(p')\gamma^\mu u^s(p)\right) = \left(\bar{v}^{s'}(p')\not{p}'\right) u^s(p) + \bar{v}^{s'}(p')\left(\not{p} u^s(p)\right). \tag{7.29}$$

Using the Dirac equations for u and for $\bar{v}$, given in eqs. (3.100) and (3.115), we see that the right-hand side of eq. (7.29) is equal to

$$-m\bar{v}^{s'}(p')u^s(p) + m\bar{v}^{s'}(p')u^s(p) = 0. \tag{7.30}$$

Therefore the matrix element is independent of the gauge fixing parameter ξ, as we expected from gauge invariance. The origin of this result is the fact that in the interaction Lagrangian A_μ is coupled to the current $\bar{\Psi}\gamma^\mu\Psi$, which is conserved on the equations of motion since it is the Noether current of the $U(1)$ symmetry. We therefore recover, at the quantum level, a condition that we already found classically in Section 3.5.4: to preserve gauge-invariance, a gauge field must be coupled to a conserved current.

Let us now perform the computation of the scattering cross-section for $e^+e^- \to \mu^+\mu^-$. Using the notation $s = q^2$ for the square of the CM energy we have

$$\begin{aligned} |\mathcal{M}_{fi}|^2 = \frac{e^4}{s^2} & \left(\bar{u}(\mu^-)\gamma^\mu v(\mu^+)\right)\left(\bar{v}(\mu^+)\gamma^\nu u(\mu^-)\right) \\ & \times \left(\bar{v}(e^+)\gamma_\mu u(e^-)\right)\left(\bar{u}(e^-)\gamma_\nu v(e^+)\right). \end{aligned} \tag{7.31}$$

We have used the notation $u(e^-) = u^s(p)$, etc. in which instead of writing explicitly the momentum and spin we have written the particles to which they refer, and we used the identity $(\bar{u}\gamma^\mu v)^* = \bar{v}\gamma^\mu u$, which is easily derived using $(\gamma^\mu)^\dagger = \gamma^0\gamma^\mu\gamma^0$.

If we are interested in a process with a specific spin structure, i.e. if we know the spin of the initial particles and we are interested in the amplitude with a given value of the spin of the final particle, we can use the explicit expression for $u^s(p), v^s(p)$ given in Section 4.2. However, it is more common that we have an unpolarized beam, so we do not know the spin of the initial particle, and we accept in the detector all final particles, without measuring their spin state. In this case we must *average* the cross-section over the initial spin state and *sum* it over the final spin state. To understand how to perform the sum over spins it can be convenient, even if a bit tedious, to write out explicitly the Dirac indices and rewrite the above expression as

$$\begin{aligned} |\mathcal{M}_{fi}|^2 &= \frac{e^4}{s^2}\,\bar{u}_a(\mu^-)\gamma^\mu_{ab}v_b(\mu^+)\bar{v}_c(\mu^+)\gamma^\nu_{cd}u_d(\mu^-) \\ &\quad \times \bar{v}_{a'}(e^+)(\gamma_\mu)_{a'b'}u_{b'}(e^-)\bar{u}_{c'}(e^-)(\gamma_\nu)_{c'd'}v_{d'}(e^+) \\ &= \frac{e^4}{s^2}\left[u(\mu^-)\bar{u}(\mu^-)\right]_{da}\gamma^\mu_{ab}\left[v(\mu^+)\bar{v}(\mu^+)\right]_{bc}\gamma^\nu_{cd} \\ &\quad \times \left[v(e^+)\bar{v}(e^+)\right]_{d'a'}(\gamma_\mu)_{a'b'}\left[u(e^-)\bar{u}(e^-)\right]_{b'c'}(\gamma_\nu)_{c'd'}. \end{aligned} \tag{7.32}$$

Recalling now eqs. (3.112) and (3.113) we see that, summing over the spin states, $[v(\mu^+)\bar{v}(\mu^+)]_{bc}$ can be replaced by $(\not{k}' - m_\mu)_{bc}$, and $[u(\mu^-)\bar{u}(\mu^-)]_{da}$ by $(\not{k} + m_\mu)_{da}$. Similarly for the initial electrons; here however we have to average, rather than to sum, over the two spin states, so $u(e^-)\bar{u}(e^-)$ is replaced by $(1/2)(\not{p}+m_e)$ and $v(e^+)\bar{v}(e^+)$ by $(1/2)(\not{p}' - m_e)$. Looking at the structure of the Dirac indices, we see that they are cyclic and can be rewritten in matrix form as traces. Therefore

$$\frac{1}{4}\sum_{\text{spin}} |\mathcal{M}_{fi}|^2 = \frac{e^4}{4s^2} \operatorname{Tr}\left[(\not{k}+m_\mu)\gamma^\mu(\not{k}'-m_\mu)\gamma^\nu\right] \operatorname{Tr}\left[(\not{p}+m_e)\gamma_\nu(\not{p}'-m_e)\gamma_\mu\right] . \tag{7.33}$$

The traces can be performed using the identities

$$\operatorname{Tr}(\gamma^\mu\gamma^\nu) = 4\eta^{\mu\nu} , \tag{7.34}$$

$$\operatorname{Tr}(\gamma^\mu\gamma^\nu\gamma^\rho\gamma^\sigma) = 4(\eta^{\mu\nu}\eta^{\rho\sigma} - \eta^{\mu\rho}\eta^{\nu\sigma} + \eta^{\mu\sigma}\eta^{\nu\rho}) , \tag{7.35}$$

while the trace of an odd number of γ matrices vanishes (for other useful identities, see, e.g. Peskin and Schroeder (1995), page 133). The factors $\eta^{\mu\nu}$ can then be used to contract the various momenta between them. The resulting scalar products are most easily computed in the CM frame. In this frame $p = (E, \mathbf{p}), p' = (E, -\mathbf{p})$ with $(2E)^2 = s$ and $\mathbf{p}^2 = E^2 - m_e^2$, while $k = (E, \mathbf{k}), k' = (E, -\mathbf{k})$ with $\mathbf{k}^2 = E^2 - m_\mu^2$. Denoting by θ the angle between $\mathbf{p}$ and $\mathbf{k}$, the result is

$$\frac{1}{4}\sum_{\text{spin}} |\mathcal{M}_{fi}|^2 = e^4 \left[1 + 4\,\frac{m_e^2+m_\mu^2}{s} + (1-\frac{4m_e^2}{s})(1-\frac{4m_\mu^2}{s})\cos^2\theta\right] . \tag{7.36}$$

To compute the cross-section we use eq. (6.43), with

$$|\mathcal{M}_{fi}|^2 \to \frac{1}{4}\sum_{\text{spin}} |\mathcal{M}_{fi}|^2 \tag{7.37}$$

and with $|\mathbf{p}| = \sqrt{E^2 - m_e^2}$, $|\mathbf{p}'| = \sqrt{E^2 - m_\mu^2}$. Introducing $\alpha = e^2/(4\pi)$ we get

$$\frac{d\sigma}{d\Omega} = \frac{\alpha^2}{4s}\left(\frac{1-(4m_\mu^2/s)}{1-(4m_e^2/s)}\right)^{1/2} \left[1 + 4\,\frac{m_e^2+m_\mu^2}{s} + (1-\frac{4m_e^2}{s})(1-\frac{4m_\mu^2}{s})\cos^2\theta\right] . \tag{7.38}$$

In the large energy limit, $m_e, m_\mu \ll \sqrt{s}$ (but still $\sqrt{s} \ll m_Z$ otherwise QED is not the correct theory to use, and we must resort to the Standard Model) we have

$$\frac{d\sigma}{d\Omega} \simeq \frac{\alpha^2}{4s}\,(1+\cos^2\theta) \tag{7.39}$$

and, performing the angular integration, the total cross-section in this limit is

$$\sigma \simeq \frac{4\pi\alpha^2}{3s} . \tag{7.40}$$

Problem 7.2. Electromagnetic form factors

In this problem we study the most general form of the radiative corrections to the electron–photon vertex, and we will show that the effect of loop corrections, to all orders in α, is contained in two form factors, describing the electric charge density and the magnetic dipole density.

Consider first of all the graph in Fig. 7.6, where the initial and final electrons are on-shell, i.e. $p_1^2 = p_2^2 = m_e^2$, while the photon line with momentum q^μ can be an internal line of a more general graph, and therefore q^2 is generic.

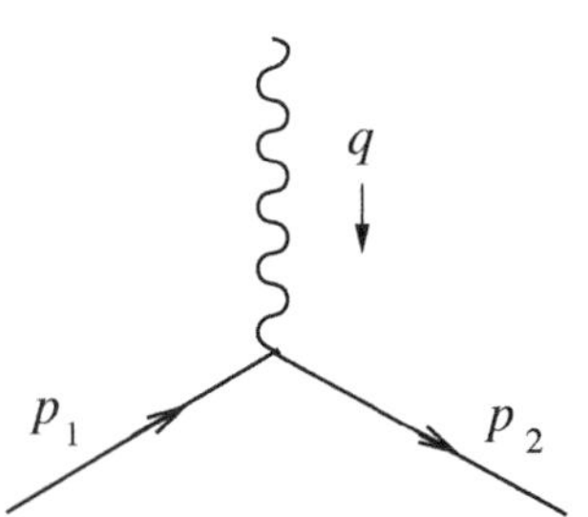

Fig. 7.6 The tree-level vertex. The photon is off-shell.

Consider the electromagnetic current operator

$$J^\mu_{\text{em}}(x) = \bar{\Psi}(x)\gamma^\mu\Psi(x) , \tag{7.41}$$

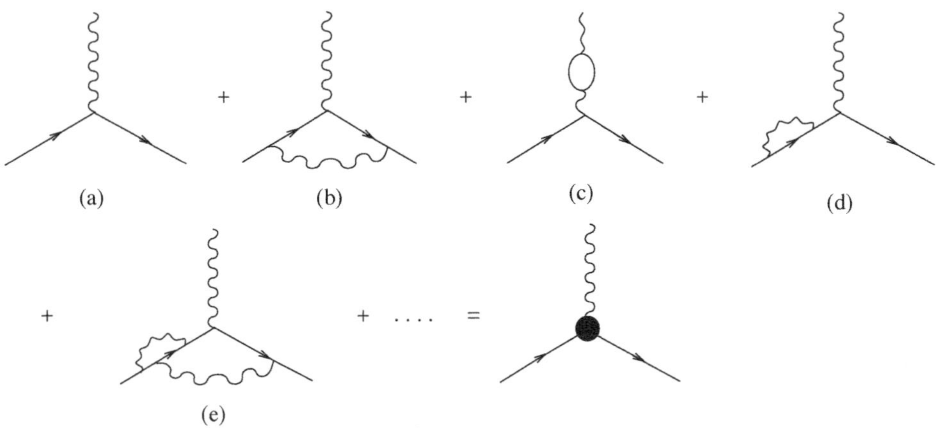

Fig. 7.7 Loop corrections to the vertex. The graph (a) is the tree-level contribution. The one-loop contributions are given by the graphs (b), (c), (d) and by a graph like (d), but with the photon line on the other electron line. The graph (e) is an example of a two-loop graph. The sum over all possible graphs is indicated by the blob.

where Ψ is the full quantum field, rather than the free field in the interaction picture. The information on the electron–photon vertex, to all orders in α, is contained in the matrix element of this current between the initial and final electron states,

$$\langle p_2|J^\mu_{\rm em}(x)|p_1\rangle \,. \tag{7.42}$$

(For notational simplicity, we suppress the spin labels in the initial and final state.) At tree level, we just substitute Ψ with the free field (4.32) and we compute the matrix element explicitly,

$$\langle p_2|J^\mu_{\rm em}(x)|p_1\rangle_{|{\rm tree}} = \bar{u}(p_2)\gamma^\mu u(p_1)\, e^{-i(p_1-p_2)x} \,. \tag{7.43}$$

This is the contribution to the matrix element of the iT operator. The factors $e^{-i(p_1-p_2)x}$, together with similar factors from the other external lines of the complete Feynman diagram, contribute to the overall Dirac delta expressing energy–momentum conservation, which is extracted from the definition of $\mathcal{M}_{fi}$, see eq. (5.115). These exponential factors can be extracted to all orders in perturbation theory, simply observing that, if $\hat{P}$ is the momentum operator, we have

$$J^\mu_{\rm em}(x) = e^{i\hat{P}x} J^\mu_{\rm em}(0) e^{-i\hat{P}x} \tag{7.44}$$

and therefore

$$\langle p_2|J^\mu_{\rm em}(x)|p_1\rangle = \langle p_2|e^{i\hat{P}x} J^\mu_{\rm em}(0)e^{-i\hat{P}x}|p_1\rangle = e^{-i(p_1-p_2)x}\langle p_2|J^\mu_{\rm em}(0)|p_1\rangle \,. \tag{7.45}$$

So, the object which enters in $\mathcal{M}_{fi}$ is $\langle p_2|J^\mu_{\rm em}(0)|p_1\rangle$ and at tree level

$$\langle p_2|J^\mu_{\rm em}(0)|p_1\rangle_{|{\rm tree}} = \bar{u}(p_2)\gamma^\mu u(p_1) \,. \tag{7.46}$$

The one-loop corrections and an example of a two-loop graph are shown in Fig. 7.7, and we have generically denoted by a blob the sum of all possible radiative corrections, to all orders in α. Of course, it is not possible to compute this sum explicitly. However, the result is constrained by Lorentz invariance, parity and gauge invariance. Lorentz invariance implies that the result must be a four-vector, so we should ask what four-vectors can be constructed with the spinors $\bar{u}(p_2)$ and $u(p_1)$

The most general fermion bilinears have been classified in eq. (3.119). Using the variables

$$p^\mu = p_1^\mu + p_2^\mu\,, \qquad q^\mu = p_2^\mu - p_1^\mu\,, \tag{7.47}$$

the most general four-vectors that we can construct with the spinors $\bar{u}_2 = \bar{u}(p_2)$ and $u_1 = u(p_1)$ is a linear combination of

$$\bar{u}_2 p^\mu u_1\,, \quad \bar{u}_2 q^\mu u_1\,, \quad \bar{u}_2\gamma^\mu u_1\,, \quad \bar{u}_2\sigma^{\mu\nu}p_\nu u_1\,, \quad \bar{u}_2\sigma^{\mu\nu}q_\nu u_1\,, \tag{7.48}$$

which are true four-vectors, plus the corresponding pseudovectors obtained inserting γ^5. Therefore $\langle p_2|J^\mu_{\rm em}(0)|p_1\rangle$ must be a combination of these quantities.

Under parity $J^\mu_{\rm em}(x)$ is a true four-vector, so it transforms as $J^0_{\rm em}(t,\mathbf{x}) \to J^0_{\rm em}(t,-\mathbf{x})$ and $J^i_{\rm em}(t,\mathbf{x}) \to -J^i_{\rm em}(t,-\mathbf{x})$ or, more compactly, as $J^\mu_{\rm em}(t,\mathbf{x}) \to \eta^{\mu\mu}J^\mu_{\rm em}(t,-\mathbf{x})$, with no sum over the μ index. At the same time, under parity the state $|p_1\rangle \to \eta_e|p_1'\rangle$, where η_e is the intrinsic parity of the electron and p_1' is the parity-reversed momentum, $p_1' = (p_1^0, -p_1^i)$; similarly $\langle p_2| \to \langle p_2'|\eta_e^*$, with the same η_e since we have an electron both in the initial and in the final state. The term $\eta_e^*\eta_e = 1$ therefore cancels and the matrix element $\langle p_2|J^\mu_{\rm em}(0)|p_1\rangle$, under parity, picks a factor $\eta^{\mu\mu}$ while the momenta p_1, p_2 become parity-reversed. This is the same transformation properties of the five terms displayed in eq. (7.48), while the corresponding quantities constructed with γ^5 pick an overall $-\eta^{\mu\mu}$ factor. Since parity is a symmetry of QED, only the terms in eq. (7.48) can enter in the parametrization of the matrix element that we are considering, while the terms with γ^5 are absent.

Another simplification comes from the fact that, when $u_1, \bar{u}_2$ are solutions of the Dirac equation with masses m_1, m_2 respectively (in our case $m_1 = m_2 = m_e$), there is an algebraic identity, known as the Gordon identity: from the definition of $\sigma^{\mu\nu}$,

$$\bar{u}_2\sigma_{\mu\nu}q^\nu u_1 = \frac{i}{2}\bar{u}_2[\gamma_\mu,\gamma_\nu](p_2^\nu - p_1^\nu)u_1 = \frac{i}{2}\bar{u}_2\left[\gamma_\mu \not{p}_2 - \gamma_\mu\not{p}_1 - \not{p}_2\gamma_\mu + \not{p}_1\gamma_\mu\right]u_1\,. \tag{7.49}$$

In the first and in the fourth term we anticommute $\not{p}$ with γ^μ, so that $\not{p}_1$ is next to u_1 and $\not{p}_2$ is next to $\bar{u}_2$, using

$$\gamma_\mu\not{p}_2 = (\{\gamma_\mu,\gamma_\nu\} - \gamma_\nu\gamma_\mu)\,p_2^\nu = 2p_{2,\mu} - \not{p}_2\gamma_\mu\,, \tag{7.50}$$

and similarly for $\not{p}_1\gamma_\mu$. Then we use $\not{p}_1 u_1 = m_1 u_1$ and $\bar{u}_2\not{p}_2 = m_2\bar{u}_2$, see eqs. (3.100) and (3.114), and we find

$$\bar{u}_2\sigma_{\mu\nu}q^\nu u_1 = i\bar{u}_2\left[p_\mu - (m_1+m_2)\gamma_\mu\right]u_1\,. \tag{7.51}$$

Similarly, considering $\bar{u}_2\sigma_{\mu\nu}p^\nu u_1$, we find

$$\bar{u}_2\sigma_{\mu\nu}p^\nu u_1 = i\bar{u}_2\left[q^\mu + (m_1 - m_2)\gamma^\mu\right]u_1\,. \tag{7.52}$$

We can use these identities to eliminate $\bar{u}_2\sigma^{\mu\nu}p_\nu u_1$ and $\bar{u}_2 p^\mu u_1$ from the list of independent bilinears. Therefore we find that $\langle p_2|J^\mu_{\rm em}(0)|p_1\rangle$ is at most a linear combination of $\bar{u}(p_2)\gamma^\mu u(p_1), \bar{u}(p_2)\sigma^{\mu\nu}q_\nu u(p_1)$ and $\bar{u}(p_2)q^\mu u(p_1)$. The coefficients must be Lorentz invariant functions. With p^μ and q^μ we can construct the invariants q^2, p^2 and qp. However, $qp = p_2^2 - p_1^2 = 0$, while q^2 and p^2 are not independent since $q^2 = p_2^2 + p_1^2 - 2p_2p_1 = 2m_e^2 - 2p_2p_1$ and

$p^2 = p_2^2 + p_1^2 + 2p_2p_1 = 2m_e^2 + 2p_2p_1$, so that $q^2 + p^2 = 4m_e^2$. We choose q^2 as the independent variable. Then

$$\langle p_2|J^\mu_{\rm em}(0)|p_1\rangle = f_1(q^2)\,\bar{u}_2\gamma^\mu u_1 + f_2(q^2)\frac{i}{2m_e}\,\bar{u}_2\sigma^{\mu\nu}q_\nu u_1 + f_3(q^2)\,\bar{u}_2 q^\mu u_1\,. \tag{7.53}$$

The factor i in $i\sigma^{\mu\nu}q_\nu$ is chosen so that $f_2(q^2)$ is real, and the factor $1/2m_e$ is a convenient normalization, so that $f_2(q^2)$ is dimensionless, as $f_1(q^2)$. We now make use of the fact that, as a consequence of gauge invariance, the electromagnetic current is conserved, $\partial_\mu J^\mu_{\rm em}(x) = 0$. Using eq. (7.45) we see that this means that

$$q_\mu\langle p_2|J^\mu_{\rm em}(0)|p_1\rangle = 0\,. \tag{7.54}$$

The term $\bar{u}_2\gamma^\mu u_1$ satisfies this condition, since $q_\mu\bar{u}_2\gamma^\mu u_1 = \bar{u}_2(\not{p}_2 - \not{p}_1)u_1$ vanishes because u_1 and u_2 are solutions of the equations of motion. Also the term $\bar{u}_2\sigma^{\mu\nu}q_\nu u_1$ satisfies the condition: $\sigma^{\mu\nu}q_\nu q_\mu$ vanishes identically because $\sigma^{\mu\nu}$ is antisymmetric. Instead $q_\mu\bar{u}_2 q^\mu u_1 = q^2\bar{u}_2 u_1$ does not vanish, since the photon is in general off-shell, and $q^2 \neq 0$. Current conservation is exact to all orders in α, since it is a consequence of gauge invariance, which means that the function $f_3(q^2)$ must be identically zero.

In conclusion, the most general parametrization of the matrix element of the electromagnetic current, compatible with Lorentz invariance, parity and current conservation (i.e. with gauge invariance) is

$$\boxed{\langle p_2|J^\mu_{\rm em}(0)|p_1\rangle = f_1(q^2)\,\bar{u}_2\gamma^\mu u_1 + f_2(q^2)\frac{i}{2m_e}q_\nu\,\bar{u}_2\sigma^{\mu\nu}u_1\,.} \tag{7.55}$$

The functions $f_1(q^2)$ and $f_2(q^2)$ are called form factors. Comparison with eq. (7.46) shows that at tree level $f_1(q^2) = 1$ and $f_2(q^2) = 0$. We now want to understand their physical meaning. This can be obtained considering their effect on the scattering amplitude. The meaning of f_1 can be understood considering its effect on the scattering of the electron on a static source, such as a heavy atom. We computed it at lowest order in Section 6.6. Including f_1, eq. (6.77) becomes

$$M_{fi}(q) = \frac{Ze^2}{\mathbf{q}^2}f_1(\mathbf{q})\,, \tag{7.56}$$

where we have taken into account that, for elastic scattering, $q^\mu = (0,\mathbf{q})$ and we denoted by $f_1(\mathbf{q})$ the function $f_1(q^2)$ evaluated at $q^\mu = (0,\mathbf{q})$. The interaction potential (6.78) then becomes

$$V(\mathbf{x}) = -Ze^2\int\frac{d^3q}{(2\pi)^3}\frac{f_1(\mathbf{q})}{\mathbf{q}^2}e^{i\mathbf{q}\cdot\mathbf{x}}\,. \tag{7.57}$$

We denote by $\rho(\mathbf{x})$ the inverse Fourier transform of $f_1(\mathbf{q})$,

$$f_1(\mathbf{q}) = \int d^3x'\rho(\mathbf{x}')e^{-i\mathbf{q}\cdot\mathbf{x}'}\,. \tag{7.58}$$

Then

$$\begin{aligned} V(\mathbf{x}) &= -Ze^2\int d^3x'\rho(\mathbf{x}')\int\frac{d^3q}{(2\pi)^3}\frac{1}{\mathbf{q}^2}e^{i\mathbf{q}\cdot(\mathbf{x}-\mathbf{x}')} \\ &= -\frac{Ze^2}{4\pi}\int d^3x'\frac{1}{|\mathbf{x}-\mathbf{x}'|}\rho(\mathbf{x}')\,. \end{aligned} \tag{7.59}$$

This is the Coulomb potential generated by a charge distribution $\rho(\mathbf{x})$, and therefore the form factor $f_1(\mathbf{q})$ is the Fourier transform of the charge distribution. We see that the effect of loop corrections is to delocalize the charge distribution of the electron, which is a Dirac delta at tree level and becomes $\rho(\mathbf{x})$ after the electron is "dressed" by the radiative corrections. We also see that, to all orders in α, $f_1(0) = 1$, since it is just the total electron charge, in units of e.

The meaning of the second form factor is more easily understood looking rather at the scattering in an external magnetic field. Instead of specifying the structure of the source, we can more simply write the interaction with an external field $A_\mu^{\rm ext}$ in the form

$$\mathcal{L}_{\rm ext} = -eA_\mu^{\rm ext} J^\mu_{\rm em}\,. \tag{7.60}$$

We consider again a static external field, so $q^0 = 0$, but now we take $A_0^{\rm ext} = 0$ and $\nabla \times \mathbf{A} = \mathbf{B}$. The amplitude is

$$\begin{aligned}\mathcal{M}_{fi} &= -ef_1(\mathbf{q})\tilde{A}_\mu^{\rm ext}(q)\bar{u}(p')\gamma^\mu u(p) - ef_2(\mathbf{q})\frac{i}{2m_e}\tilde{A}_\mu^{\rm ext}(q)q_\nu\bar{u}(p')\sigma^{\mu\nu}u(p)\\ &= ef_1(\mathbf{q})\tilde{A}^i_{\rm ext}(q)\bar{u}(p')\gamma^i u(p) - i\frac{e}{2m_e}f_2(\mathbf{q})\tilde{A}^i_{\rm ext}(q)q^j\bar{u}(p')\sigma^{ij}u(p).\end{aligned} \tag{7.61}$$

We consider a slowly varying field, so we take $\mathbf{q} \to 0$ and we keep only the first non-vanishing contribution. The computation is performed in detail in Peskin and Schroeder (1995), Section 6.2, so we simply quote the result. The expansion in powers of $\mathbf{q}$ starts from a constant spin-independent term proportional to $\mathbf{p} + \mathbf{p}'$. This is the contribution of the operator $\mathbf{p}\cdot\mathbf{A} + \mathbf{A}\cdot\mathbf{p}$ from the non-relativistic Hamiltonian $(\mathbf{p} - e\mathbf{A})^2/2m_e$. We are more interested in the next term, which is linear in $\mathbf{q}$ and depends on the spin, and gets a contribution both from f_1 and from f_2. The result, limiting ourselves to processes with the same spin state for the initial and final electron, is

$$M_{fi} \simeq \frac{e}{2m_e}[f_1(0) + f_2(0)]\tilde{B}^i_{\mathbf{q}}\xi^\dagger\sigma^i\xi\,. \tag{7.62}$$

where $M_{fi} = \mathcal{M}_{fi}/(2m_e)$ is the matrix element with a non-relativistic normalization for the electron, and is related to the scattering potential as in eq. (6.71); $\tilde{B}^i_{\mathbf{q}}$ is the Fourier component of the magnetic field. Using the results of Section 6.6, we can see that this is the scattering amplitude that would be generated, in the non-relativistic theory, by a potential

$$V(\mathbf{x}) = -\boldsymbol{\mu}\cdot\mathbf{B}(\mathbf{x}) \tag{7.63}$$

with

$$\boldsymbol{\mu} = \frac{e}{m_e}[f_1(0) + f_2(0)]\left(\xi^\dagger\frac{\boldsymbol{\sigma}}{2}\xi\right)\,. \tag{7.64}$$

From (7.63) we see that $\boldsymbol{\mu}$ is a magnetic dipole moment. Since $f_1(0) = 1$ to all orders in perturbation theory, the magnetic moment $\boldsymbol{\mu}$ is related to the expectation value of the spin operator,

$$\mathbf{S} = \xi^\dagger\frac{\boldsymbol{\sigma}}{2}\xi\,, \tag{7.65}$$

by

$$\boldsymbol{\mu} = g\frac{e}{2m_e}\mathbf{S}\,, \tag{7.66}$$

where $g = 2 + 2f_2(0)$, or

$$\frac{g-2}{2} = f_2(0)\,. \tag{7.67}$$

The form factor $f_2(q^2)$ therefore gives a correction to the magnetic dipole moment. At tree level, $f_2(0) = 0$ and we recover the result that we found from the Dirac equation, see eq. (3.186). Deviations from $g = 2$ therefore come from the loop corrections to $f_2(0)$. The one-loop result, first derived by Schwinger in 1948, turns out to be $f_2(0) = \alpha/(2\pi)$.

Summary of chapter

- QED describes the interactions of spin 1/2 charged particles with photons. The Feynman rules are summarized in Fig. 7.8. The wave functions associated to the external legs were given on page 135.
- QED is renormalizable. The one-loop divergences are studied in Section 7.2 and are reabsorbed into the renormalization of the fermion mass, fermion wave function, photon wave function and electric charge. The photon mass is protected against loop corrections by gauge invariance.
- Invariance principles constrain the form of loop corrections to all orders. In particular, the vertex is parametrized by two form factors, representing the charge density and the magnetic dipole density.

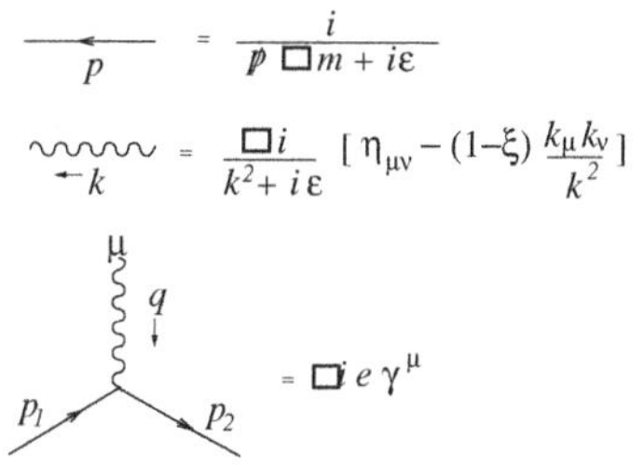

p $= \frac{i}{\not p \,\Box m + i\varepsilon}$

k $= \frac{\Box i}{k^2 + i\varepsilon}\,[\eta_{\mu\nu} - (1-\xi)\frac{k_\mu k_\nu}{k^2}]$

μ, q, p_1, p_2 $= \Box i e \gamma^\mu$

Fig. 7.8 The Feynman rules for QED.

Further reading

- QED is discussed in great detail in many books. See, in particular, Itzykson and Zuber (1980), Peskin and Schroeder (1995), Weinberg (1995) and Landau and Lifshitz, vol. IV (1982).
- Many useful theoretical and experimental results are collected in *Quantum Electrodynamics*, T. Kinoshita ed., World Scientific 1990. This includes reviews on high-precision tests of QED, a description of measurements and of calculations of the magnetic moment of electrons and muons, hydrogenic bound states, Lamb shift experiments, hyperfine structure experiments, precision measurements in positronium, etc.
- For an explicit calculation of the $g - 2$ of the electron at $O(\alpha)$ and of the Lamb shift see, e.g. Mandl and Shaw (1984), Section 9.6.

Exercises

(7.1) (i) Write the Feynman diagrams for the annihilation $e^+e^- \to 2\gamma$, to lowest perturbative order.

(ii) Write the modulus squared of the amplitude as a trace, averaging over the initial spins and summing over the final helicities. Evaluate the trace in the limit $\mathbf{p} \to 0$, where $\mathbf{p}$ is the electron momentum in the CM.

(iii) Compute the cross-section $\bar{\sigma}$ of the process (averaged and summed over the initial and final spins) in the limit $\mathbf{p} \to 0$. Show that the flux factor I can be written as

$$I = E_1 E_2 v\,, \tag{7.68}$$

where E_1, E_2 are the energies of the two incoming particles in the CM and v their relative velocity. Using this, and the value of $|\mathcal{M}|^2$ computed above, verify that in the limit $\mathbf{p} \to 0$,

$$\bar{\sigma}_{e^+e^-\to 2\gamma} \to \frac{\pi r_B^2}{v}\,, \tag{7.69}$$

with $r_B = \alpha/m_e$.

(7.2) In this exercise we compute the decay rate of positronium (the hydrogenoid bound state of e^+ and e^-) into two photons, restricting to positronium states with orbital angular momentum $L = 0$.

(i) Using the results of Exercise 4.1, show that, if $L = 0$, the annihilation can take place only when the e^+e^- pair has total angular momentum $J = 0$, and that the cross-section $\bar{\sigma}$, averaged over initial spin and summed over the final helicities, is given by

$$\bar{\sigma} = \frac{1}{4}\sigma^{(J=0)}\,, \tag{7.70}$$

where $\sigma^{(J=0)}$ is the cross-section with the e^+e^- pair in the state with $J = 0$.

(ii) Consider first the cross-section for the annihilation $e^+e^- \to 2\gamma$,

$$\sigma = \frac{1}{4I}\int |\mathcal{M}_{e^+e^-\to 2\gamma}|^2 d\Phi^{(2)} \tag{7.71}$$

(we include the 1/2! for the identical photons in $d\Phi^{(2)}$). Using eq. (7.68) show that

$$\sigma v = \int |M_{e^+e^-\to 2\gamma}|^2 d\Phi^{(2)}\,, \tag{7.72}$$

where M_{fi} is the matrix element with the normalization of one particle per unit volume. Verify also that

$$\Gamma_{\rm Pos\to 2\gamma} = \int |M_{\rm Pos\to 2\gamma}|^2 d\Phi^{(2)}\,, \tag{7.73}$$

where in $M_{\rm Pos\to 2\gamma}$ the positronium state (labeled "Pos") is normalized as one particle per unit volume.

(iii) Write $M_{e^+e^-\to 2\gamma} = \langle 2\gamma|\mathbf{p}\,, -\mathbf{p}\,\rangle$ where $\mathbf{p}$ is the electron momentum in the CM, and $M_{\rm Pos\to 2\gamma} = \langle 2\gamma|{\rm Pos}\rangle$, where the positronium is at rest. Show that

$$M_{\rm Pos\to 2\gamma} = \int \frac{d^3p}{(2\pi)^3}\langle 2\gamma|\mathbf{p}\,, -\mathbf{p}\,\rangle\tilde{\psi}(\mathbf{p}\,)\,, \tag{7.74}$$

where $\tilde{\psi}(\mathbf{p}\,)$ is the positronium wave function in momentum space.

(iv) Justify the fact that $\tilde{\psi}(\mathbf{p}\,)$ is peaked at small values of the momentum, $|\mathbf{p}| \sim (1/2)m_e\alpha$, and from this derive that, at lowest order in α,

$$M_{\rm Pos\to 2\gamma} \simeq \psi(0)\lim_{\mathbf{p}\to 0}\langle 2\gamma|\mathbf{p}\,, -\mathbf{p}\,\rangle\,, \tag{7.75}$$

where $\psi(x)$ is the wave function in position space. Using eqs. (7.72) and (7.73) derive the relation

$$\Gamma_{\rm Pos\to 2\gamma} = |\psi(0)|^2 \lim_{\mathbf{p}\to 0}\sigma^{(J=0)}v\,. \tag{7.76}$$

(v) The wave functions of positronium are the same as the hydrogen atom, with the replacement of the reduced mass, which is approximately m_e in the hydrogen atom, with $m_e/2$ for the e^+e^- system. Then, in the state with quantum numbers $n = 0, L = 0$,

$$\psi(r) = \frac{1}{\sqrt{\pi a^3}}e^{-r/a}\,, \tag{7.77}$$

with $a = 2/(m_e\alpha)$. Using the cross-section (7.69) and eqs. (7.76) and (7.70) show that, for the state $n = L = 0$,

$$\Gamma_{\rm Pos\to 2\gamma} = \frac{1}{2}m_e\alpha^5\,. \tag{7.78}$$

Compare this result with the experimental value of the lifetime, $\Gamma = 7.994(11)\ {\rm ns}^{-1}$.

The low-energy limit of the electroweak theory

8

In this section we begin our study of weak interactions. The electroweak theory is described by the Standard Model (SM). A systematic explanation of the SM is beyond the scope of this course, but in the next chapters we will introduce two of the most important theoretical tools for its construction, i.e. non-abelian gauge fields and the Higgs mechanism.

However, the full structure of the SM is only revealed at energies comparable to the masses of the bosons $W^{\pm}$ and Z^0 that, together with the photon, mediate the electroweak interaction. Since $m_W = 80.425(38)$ GeV and $m_Z = 91.1876(21)$ GeV, the weak decays of particles with masses between a few hundred MeV and a few GeV, as for instance the muons, the pions, the kaons, the neutron, charmed mesons like the D^0, etc., can be studied in a low-energy approximation to the SM. For instance in the β-decay of the free neutron, $n \rightarrow pe^-\bar{\nu}_e$, we have a mass difference $m_n - m_p \simeq 1.29$ MeV. Therefore, even if at the fundamental level the decay is mediated by the W-boson, the fact that the maximum momentum transfer is much smaller than m_W allows us to use a low-energy effective theory. The same approximation holds for nuclear β-decays.

For the same reason, we can use the low-energy theory when we study a scattering process mediated by weak interactions, as for instance $e^-\nu_e \rightarrow e^-\nu_e$, at center-of mass energies well below m_W.

In this chapter we introduce this low-energy approximation, which is given by a four-fermion model, and we will understand how it can be obtained "integrating out" some heavy gauge bosons. We will then illustrate in detail in the Solved Problems section how to use it to compute explicitly many weak decays.

8.1 A four-fermion model

As a preliminary exercise, we consider a theory with a single Dirac fermion Ψ, and a Lagrangian

$$\mathcal{L} = \bar{\Psi}\,(i\partial\!\!\!/ - m)\,\Psi + G(\bar{\Psi}\Psi)^2\,. \tag{8.1}$$

The interaction is given by a four-fermion term, and in the Feynman diagrams we have vertices where four fermionic lines meet at a point,

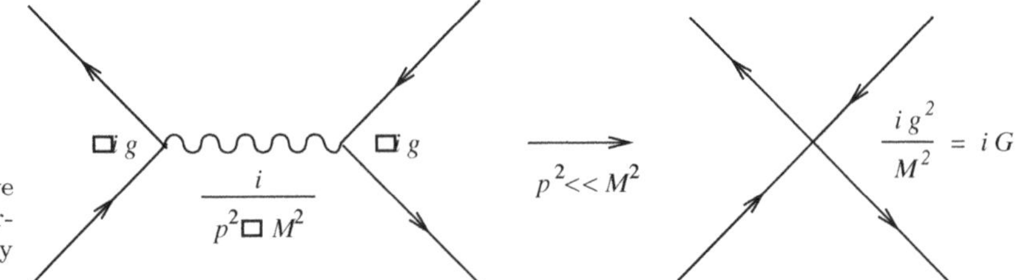

Fig. 8.1 The exchange of a massive intermediate boson reduces to a four-fermion interaction in the low-energy limit.

similarly to the $\lambda\phi^4$ theory. The difference, however, is that the coupling constant G is not dimensionless. In fact, since $\int d^4x\,\mathcal{L}$ is dimensionless, from the kinetic term we see that Ψ has dimensions of $(\text{mass})^{3/2}$, and therefore $G \sim (\text{mass})^{-2}$. As we saw in Section 5.6, a negative mass dimension for the coupling means that the theory is not renormalizable.

We denote the fermion by f and the antifermion by $\bar{f}$. If we compute a scattering amplitude $f\bar{f} \to f\bar{f}$ at tree level in this theory, we find

$$i\mathcal{M}_{f\bar{f}\to f\bar{f}} = iG\,[\bar{v}(p_1)u(p_2)\bar{u}(p_2')v(p_1')]\,, \qquad (8.2)$$

where u, v are the wave functions, p_1, p_2 the initial momenta and p_1', p_2' the final momenta. We want to compare this four-fermion theory with a theory where the fermion is coupled not directly to itself, but rather to a massive real scalar field with mass M, so we consider a theory with action

$$S = \int d^4x\,[\frac{1}{2}\partial^\mu\phi\partial_\mu\phi - \frac{1}{2}M^2\phi^2 + \bar{\Psi}\,(i\partial\!\!\!/ - m)\,\Psi - g\phi\bar{\Psi}\Psi]\,, \qquad (8.3)$$

where g is a coupling constant which, by dimensional analysis, is immediately seen to be dimensionless. The $f\bar{f} \to f\bar{f}$ scattering amplitude at tree level in this theory is given by the Feynman diagram on the left of Fig. 8.1 and the Feynman rules give

$$i\mathcal{M}_{f\bar{f}\to f\bar{f}} = (-ig)^2\,\frac{i}{p^2 - M^2}\,[\bar{v}(p_1)u(p_2)\bar{u}(p_2')v(p_1')]\,, \qquad (8.4)$$

where $p^2 = (p_1 + p_2)^2 = E_{\text{cm}}^2$, and E_{cm} is the energy in the center of mass. Therefore

$$i\mathcal{M}_{f\bar{f}\to f\bar{f}} = i\frac{g^2}{M^2 - E_{\text{cm}}^2}\,[\bar{v}(p_1)u(p_2)\bar{u}(p_2')v(p_1')]\,. \qquad (8.5)$$

We see that at low energies, $E_{\text{cm}} \ll M$, the amplitudes (8.5) and (8.2) coincide if we make the identification

$$G = \frac{g^2}{M^2}\,. \qquad (8.6)$$

More generally, consider an arbitrary amplitude, with all external momenta small compared to M and regularize the theory putting a cutoff in momentum space $\Lambda \lesssim M$, so that also all the momenta circulating in the loops are smaller than M. Then, in all internal lines the scalar propagator can be approximated as

$$\frac{i}{p^2 - M^2} \to -\frac{i}{M^2} \tag{8.7}$$

and, with the identification (8.6), the Feynman rules of this scalar-fermion theory become identical to those of the four-fermion theory (8.1). Therefore the four-fermion theory (8.1) can be seen as the low-energy limit of the theory (8.3), which has only dimensionless couplings and indeed is renormalizable. We see that, even if the four-fermion theory is not renormalizable, it can nevertheless be a useful low-energy approximation to a renormalizable theory. The presence of a coupling G with dimensions $(\text{mass})^{-2}$ is now seen to be a signal of the existence of a particle with mass M and coupling g, with G, M and g related by eq. (8.6).

The situation for the low-energy limit of the electroweak theory is conceptually similar, but with more complicated four-fermion terms compared to this toy model. In the next section we will first of all introduce the fermionic fields that appear in the electroweak theory, and we will then present their interactions, as we know them nowadays from the Standard Model. We will finally see that in the low-energy limit weak interactions are described by a four-fermion theory, which is the Fermi theory of weak interactions, with the so-called $V - A$ structure of the currents proposed by Feynman and Gell-Mann.

8.2 Charged and neutral currents in the Standard Model

The fermions that appear in the Standard Model are the leptons and the quarks. The leptons are organized into three families: the electron e and its neutrino ν_e, the muon μ and its neutrino ν_μ, and the τ and its neutrino ν_τ. Similarly the quarks are organized into three families: u and d, c and s, t and b. In units of $|e|$, the quarks u, c, t have electric charge $+2/3$ while d, s, b have $-1/3$. For the antiparticles the charge is reversed, and $\bar{u}, \bar{c}, \bar{t}$ have charge $-2/3$ while $\bar{d}, \bar{s}, \bar{b}$ have charge $+1/3$. We denote by e the Dirac spinor describing the electron and the positron, by μ the Dirac spinor describing $\mu^\pm$, etc. We denote by e_L, e_R the Dirac spinors

$$e_L = \frac{1-\gamma^5}{2}\, e\,, \qquad e_R = \frac{1+\gamma^5}{2}\, e\,, \tag{8.8}$$

and similarly for all other particles. Therefore $e_{L,R}$ are the Weyl spinors written in a four-component Dirac notation. Neglecting the possibility of neutrino masses, the neutrinos have only the left-handed component ν_L, and we will use a Dirac notation for them, with a Dirac spinor ν that satisfies $(1/2)(1-\gamma^5)\nu = \nu$, see eq. (3.93).

Nowadays we understand the Fermi theory as a low-energy limit of the Standard Model, so we will give the full structure of the interaction in terms of leptons and quarks that derives from the Standard Model.

The *charged* leptonic current is defined as

$$J_l^{-,\mu} = \frac{1}{\sqrt{2}} \left(\bar{e}_L \gamma^\mu \nu_{e,L} + \bar{\mu}_L \gamma^\mu \nu_{\mu,L} + \bar{\tau}_L \gamma^\mu \nu_{\tau,L} \right) . \tag{8.9}$$

The operator $\bar{e}$ creates an electron or destroys a positron, so it lowers the charge by one unit, while ν_e destroys a neutrino or creates an antineutrino. Since the electric charge of the neutrino is zero, $J_l^{-,\mu}$ is an operator that lowers the electric charge by one unit, and is therefore called a charged current. Its hermitian conjugate is

$$J_l^{+,\mu} = \frac{1}{\sqrt{2}} \left(\bar{\nu}_{e,L} \gamma^\mu e_L + \bar{\nu}_{\mu,L} \gamma^\mu \mu_L + \bar{\nu}_{\tau,L} \gamma^\mu \tau_L \right) , \tag{8.10}$$

and raises the charge by one unit. Observe that, for any spinor Ψ, the definition $\Psi_L = (1/2)(1-\gamma^5)\Psi$ implies $\Psi_L^\dagger = \Psi^\dagger (1/2)(1-\gamma^5)$ since γ^5 is hermitian, and therefore $\bar{\Psi}_L = \bar{\Psi}(1/2)(1+\gamma^5)$, since γ^0 anticommutes with γ^5. Furthermore from $(\gamma^5)^2 = 1$, it follows that

$$\left(\frac{1-\gamma^5}{2} \right)^2 = \left(\frac{1-\gamma^5}{2} \right) . \tag{8.11}$$

Therefore we can write

$$\bar{\nu}_{e,L} \gamma^\mu e_L = \bar{\nu}_e \frac{1+\gamma^5}{2} \gamma^\mu \frac{1-\gamma^5}{2} e = \bar{\nu}_e \gamma^\mu (\frac{1-\gamma^5}{2})^2 e = \frac{1}{2} \bar{\nu}_e \gamma^\mu (1-\gamma^5) e , \tag{8.12}$$

and similarly for all other terms. The charged current is therefore proportional to the vector current minus the axial current.[1] This is referred to as the "$V - A$" structure of the charged weak currents.

Similarly, there is a quark charged current, which mediates transitions between hadronic states. Its explicit form in terms of quarks is quite similar[2] to the form of the leptonic current,

$$J_h^{-,\mu} = \frac{1}{\sqrt{2}} \left(\bar{d}'_L \gamma^\mu u_L + \bar{s}'_L \gamma^\mu c_L + \bar{b}'_L \gamma^\mu t_L \right) . \tag{8.13}$$

Here (d', s', b') are linear combinations of (d, s, b). In a first approximation $b' \simeq b$, while

$$d' = d \cos\theta_C + s \sin\theta_C , \tag{8.14}$$

$$s' = -d \sin\theta_C + s \cos\theta_C , \tag{8.15}$$

and θ_C is known as the Cabibbo angle. Numerically, $\sin\theta_C = 0.220(3)$. If we include also the (small) mixing of the b quark, the 3×3 matrix relating (d', s', b') to (d, s, b) is called the Cabibbo–Kobayashi–Maskawa (CKM) matrix.

The total charged current is therefore

$$J^{-,\mu} = J_l^{-,\mu} + J_h^{-,\mu} , \tag{8.16}$$

and its hermitian conjugate is denoted $J^{+,\mu}$. Even if eqs. (8.9) and (8.13) are formally similar, there is an important difference. When we

[1] The vector and axial current, and their relations with vector and axial $U(1)$ symmetries, were discussed in Section 3.4.3.

[2] Quark fields also have a color index, as we will see in Chapter 10. All the currents that we will consider in this chapter are color singlets, i.e. the color index in the bilinears is simply summed over, and we do not write it explicitly.

compute weak decays the initial and final states will be leptons and hadrons, rather then leptons and quarks. Hadrons are bound states of quarks, held together by complicated non-perturbative QCD effects. Therefore, while the matrix element of the leptonic current between leptonic states can be computed explicitly, we are not able to compute the matrix element of the quark current between hadronic states because we are not able to write a wave function of the hadronic states in terms of the constituent quarks. The best we can do (unless we resort to non-perturbative techniques beyond the scope of this book) is to parametrize them using symmetry principles. We will give many explicit examples of these computations in the Solved Problems section.

Beside charged currents, there are also neutral currents. Restricting to the first family (the other families just give a replication), the neutral lepton current is

$$J_l^{0,\mu} = a_1\, \bar{\nu}_{e,L}\gamma^\mu \nu_{e,L} + a_2\, \bar{e}_L\gamma^\mu e_L + a_3\, \bar{e}_R\gamma^\mu e_R\,, \tag{8.17}$$

and the neutral quark current is

$$J_h^{0,\mu} = b_1\, \bar{u}_L\gamma^\mu u_L + b_2\, \bar{u}_R\gamma^\mu u_R + b_3\, \bar{d}_L\gamma^\mu d_L + b_4\, \bar{d}_R\gamma^\mu d_R\,. \tag{8.18}$$

The Standard Model predicts the coefficients a_i, b_i in terms of a single parameter θ_W called the Weinberg angle,

$$a_1 = \frac{1}{2}\,, \qquad a_2 = -\frac{1}{2} + \sin^2\theta_W\,, \qquad a_3 = \sin^2\theta_W\,, \tag{8.19}$$

and

$$\begin{aligned} b_1 &= \frac{1}{2} - \frac{2}{3}\sin^2\theta_W & b_2 &= -\frac{2}{3}\sin^2\theta_W \\ b_3 &= -\frac{1}{2} + \frac{1}{3}\sin^2\theta_W & b_4 &= \frac{1}{3}\sin^2\theta_W\,. \end{aligned} \tag{8.20}$$

The experimental value[3] of the Weinberg angle is $\sin^2\theta_W = 0.23120(15)$. The total neutral current is

$$J^{0,\mu} = J_l^{0,\mu} + J_h^{0,\mu}\,. \tag{8.21}$$

In the Standard Model the charged and neutral currents are coupled to the vector fields $W_\mu^\pm$ and Z_μ^0 respectively, with an interaction Lagrangian given by

$$\mathcal{L}_{\text{int}} = g(W_\mu^+ J^{+,\mu} + W_\mu^- J^{-,\mu}) + \bar{g}\, Z_\mu^0 J^{0,\mu}\,. \tag{8.22}$$

Observe that the left- and right-handed spinors enter the theory in an asymmetric way. The charged currents, in particular, depend only on the left-handed spinors. In the neutral currents we also have the right-handed electron e_R but its coupling is different from the coupling of e_L, see eqs. (8.17) and (8.19). This means that parity is broken by weak interactions, both in processes mediated by charged currents and

[3] Actually, the precise definition of θ_W depends on the renormalization scheme. This is the value in the so-called $\overline{\text{MS}}$ scheme.

in processes mediated by neutral currents. The two gauge couplings g and $\bar{g}$ are related by

$$\bar{g} = \frac{g}{\cos\theta_W}, \tag{8.23}$$

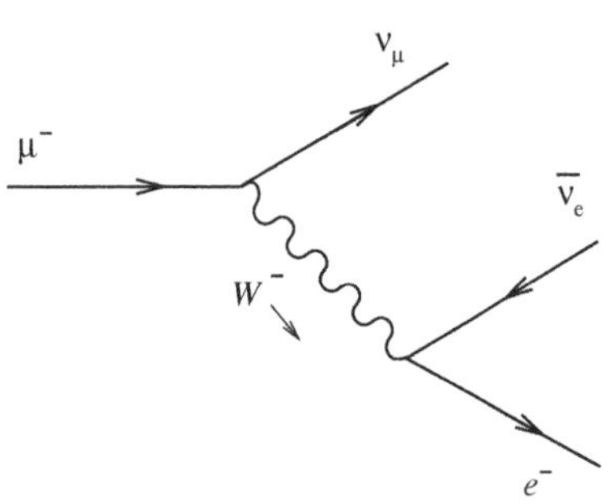

Fig. 8.2 The decay $\mu^- \to e^-\bar{\nu}_e\nu_\mu$.

and g is related to the electric charge by $|e| = g\sin\theta_W$. The bosons $W^\pm$ and Z^0 are massive: $m_W = 80.425(38)$ GeV and $m_Z = 91.1876(21)$ GeV. We will study in Section 11.4 the mechanism that generates the masses for these gauge bosons. In any case, when we study the decay of particles like the muon (with $m_\mu \simeq 105$ MeV) or the pions (with $m_{\pi^\pm} \simeq 140$ MeV), etc., the masses m_W, m_Z are much larger than the energy scales involved, and therefore we are in a situation similar to that examined in Section 8.1: the interaction, that at a fundamental level is described by the exchange of the $W^\pm, Z^0$ bosons, is described at an effective level by a four-fermion Lagrangian.

To be specific, let us study the muon decay, $\mu^- \to e^-\bar{\nu}_e\nu_\mu$. At the fundamental level, the interaction is mediated by the W^- boson as in Fig. 8.2. From the form (8.10) of the current and (8.22) of the Lagrangian we see that at the first vertex the relevant term in the interaction Lagrangian is

$$\frac{g}{\sqrt{2}}\left(\bar{\nu}_{\mu,L}\gamma^\mu\mu_L\right)W_\mu^+, \tag{8.24}$$

which destroys the incoming μ^-, creates a ν_μ and creates a W^- (the operator W^- can destroy a boson W^- or create a boson W^+, while the operator W^+, which is its hermitian conjugate, can destroy a boson W^+ or create a boson W^-). Similarly, at the second vertex the relevant term in the interaction Lagrangian is

$$\frac{g}{\sqrt{2}}\left(\bar{e}_L\gamma^\mu\nu_{e,L}\right)W_\mu^-. \tag{8.25}$$

The propagator of the W boson turns out to be

$$\tilde{D}_{\mu\nu}(q) = \frac{-i}{q^2 - m_W^2 + i\epsilon}\left(\eta_{\mu\nu} - \frac{q_\mu q_\nu}{m_W^2}\right). \tag{8.26}$$

For our present purposes the derivation of this result is not necessary, and the only important point is that, when all masses of the initial and final particles are much smaller than m_W, this reduces to

$$\tilde{D}_{\mu\nu} \simeq \frac{i}{m_W^2}\eta_{\mu\nu}. \tag{8.27}$$

This result is easily understood by comparison with the propagator of a massive scalar particle, $i/(q^2 - M^2) \to -i/M^2$. The spatial components W_μ with $\mu = 1, 2, 3$, at low energies, have the same propagator as a scalar field, while $W_{\mu=0}$, as usual, has the "wrong" sign.

We can now compute the amplitude for the muon decay. Since $m_W \gg m_\mu, m_e$,

$$\begin{aligned} i\mathcal{M}_{fi} &\simeq \left(-i\frac{g}{\sqrt{2}}\right)^2 \left[\bar{u}(\nu_\mu)\gamma^\mu\frac{1-\gamma^5}{2}u(\mu)\right] \frac{i}{m_W^2}\eta_{\mu\nu} \left[\bar{u}(e)\gamma^\nu\frac{1-\gamma^5}{2}v(\nu_e)\right] \\ &= -i\frac{g^2}{8m_W^2}\left[\bar{u}(\nu_\mu)\gamma^\mu(1-\gamma^5)u(\mu)\right]\left[\bar{u}(e)\gamma_\mu(1-\gamma^5)v(\nu_e)\right], \end{aligned} \tag{8.28}$$

where $u, v, \bar{u}$ are the appropriate wave functions, and we have written the current in the $V - A$ form using eq. (8.12). The *Fermi constant* G_F is defined by

$$\frac{G_F}{\sqrt{2}} = \frac{g^2}{8m_W^2} \,. \tag{8.29}$$

We see that the amplitude $\mathcal{M}_{fi}$ in eq. (8.28) can be derived directly from the effective Lagrangian

$$\mathcal{L} = -\frac{G_F}{\sqrt{2}} \left[\bar{\nu}_\mu \gamma^\mu (1-\gamma^5)\mu\right] \left[\bar{e}\gamma_\mu (1-\gamma^5)\nu_e\right] . \tag{8.30}$$

The same reasoning can be made for all other processes mediated by the W boson. Therefore, the total effective Lagrangian for processes involving charged currents is

$$\mathcal{L}_F = -\frac{G_F}{\sqrt{2}} j_\mu^\dagger j^\mu \,, \tag{8.31}$$

where $j^\mu = j_l^\mu + j_h^\mu$,

$$j_l^\mu = \sum_{l=e,\mu,\tau} \bar{l}\gamma^\mu (1-\gamma^5)\nu_l \,, \tag{8.32}$$

and

$$j_h^\mu = \bar{d}'\gamma^\mu (1-\gamma^5)u + \bar{s}'\gamma^\mu (1-\gamma^5)c + \bar{b}'\gamma^\mu (1-\gamma^5)t \,. \tag{8.33}$$

(Observe that j^μ differ by J^μ, defined in eqs. (8.10) and (8.13), by a normalization factor $2\sqrt{2}$.) Equation (8.31) is known as the Fermi Lagrangian.[4]

The same analysis can be repeated for the neutral currents. In this case the coupling g is replaced by $\bar{g}$, see eq. (8.22), and m_W by m_Z. The masses m_W and m_Z are related by $m_W = m_Z \cos\theta_W$ (see Section 11.4). Using eq. (8.23), we see that $\bar{g}^2/m_Z^2 = g^2/m_W^2$, and we find

$$\mathcal{L}_{\text{neutral}} = -\frac{4G_F}{\sqrt{2}} J_\mu^0 J^{0,\mu} \,, \tag{8.34}$$

with $J^{0,\mu}$ given in eqs. (8.17), (8.18) and (8.21).[5] It can be convenient to change the normalization also for the neutral currents. We then write

$$j_l^{0,\mu} = a_1\, \bar{\nu}_e \gamma^\mu (1-\gamma^5)\nu_e + a_2\, \bar{e}\gamma^\mu (1-\gamma^5)e + a_3\, \bar{e}\gamma^\mu (1+\gamma^5)e \,, \tag{8.35}$$

and

$$\begin{aligned} j_h^{0,\mu} = {} & b_1\, \bar{u}\gamma^\mu (1-\gamma^5)u + b_2\, \bar{u}\gamma^\mu (1+\gamma^5)u \\ & + b_3\, \bar{d}\gamma^\mu (1-\gamma^5)d + b_4\, \bar{d}\gamma^\mu (1+\gamma^5)d \,, \end{aligned} \tag{8.36}$$

with a_i, b_i still given by eqs. (8.19) and (8.20). Then $j_\mu^0 = 2J_\mu^0$ and the Lagrangian for neutral currents reads

$$\mathcal{L}_{\text{neutral}} = -\frac{G_F}{\sqrt{2}} j_\mu^0 j^{0,\mu} \,. \tag{8.37}$$

In the next section we will use these effective Lagrangians to compute explicitly a number of weak decays.

[4] The original Fermi theory, proposed in the 1930s to explain beta decay, contained only vector currents and obviously it was formulated in terms of proton and neutron fields rather than quarks. Here we have presented the full structure, as we understand it today from the Standard Model. See, e.g. (Weinberg), vol. II, Chapter 21 for references on the discovery of the Standard Model.

[5] Writing the Lagrangian (8.34) we have taken into account that each given neutral current process gets two contributions from $J_\mu^0 J^{0,\mu}$. E.g. in the process $e_L\bar{e}_L \to \nu_L\bar{\nu}_L$ we can pick the term $e_L\bar{e}_L$ from the first factor J_μ^0 and $\nu_L\bar{\nu}_L$ from the second factor $J^{0,\mu}$, or vice versa. This does not happen in the charged current Lagrangian, where we have a current times its hermitian conjugate, rather than a current times itself as in eq. (8.34).

8.3 Solved problems: weak decays

Problem 8.1. $\mu^- \to e^- \bar{\nu}_e \nu_\mu$

This exercise is a prototype of many similar calculations and we perform it in great detail. We already computed the matrix element for the decay in eq. (8.28). At lowest order in the Fermi coupling,

$$\mathcal{M}_{fi} = -\frac{G_F}{\sqrt{2}} \left[\bar{\nu}_\mu \gamma^\mu (1-\gamma^5)\mu\right] \left[\bar{e}\gamma_\mu (1-\gamma^5)\nu_e\right] . \tag{8.38}$$

We denote the wave functions of the initial and final particles simply by the name of the particle. This notation does not displays explicitly the information of whether we had a particle or an antiparticle (i.e. whether the wave function is u or v), but has the advantage of being lighter.

The complex conjugate of the matrix element is computed using $(\gamma^\mu)^\dagger = \gamma^0\gamma^\mu\gamma^0$ and $(\gamma^5)^\dagger = \gamma^5$. For any two wave functions Ψ_1, Ψ_2

$$\begin{aligned}(\bar{\Psi}_1\gamma^\mu(1-\gamma^5)\Psi_2)^* &= \left(\Psi_1^\dagger\gamma^0\gamma^\mu(1-\gamma^5)\Psi_2\right)^\dagger = \Psi_2^\dagger(1-\gamma^5)(\gamma^0\gamma^\mu\gamma^0)\gamma^0\Psi_1 \\ &= \bar{\Psi}_2\gamma^\mu(1-\gamma^5)\Psi_1 .\end{aligned} \tag{8.39}$$

Then

$$\mathcal{M}_{fi}^* = -\frac{G_F}{\sqrt{2}} \left[\bar{\mu}\gamma^\mu(1-\gamma^5)\nu_\mu\right] \left[\bar{\nu}_e\gamma_\mu(1-\gamma^5)e\right] \tag{8.40}$$

and

$$|\mathcal{M}_{fi}|^2 = \frac{G_F^2}{2} \left[\bar{\nu}_\mu\gamma^\mu(1-\gamma^5)\mu\right]\left[\bar{\mu}\gamma^\nu(1-\gamma^5)\nu_\mu\right]\left[\bar{e}\gamma_\mu(1-\gamma^5)\nu_e\right]\left[\bar{\nu}_e\gamma_\nu(1-\gamma^5)e\right] . \tag{8.41}$$

We denote by k the four-momentum of the electron, p of the muon, q_1 of $\bar{\nu}_e$ and q_2 of ν_μ. We average over the polarization of the initial muon and we sum over the final electron polarization, using eqs. (3.112) and (3.113). This means that (writing explicitly the Dirac indices a, b) we replace

$$e_a\bar{e}_b \to (\not{k} + m_e)_{ab} , \tag{8.42}$$

$$\mu_a\bar{\mu}_b \to \frac{1}{2}(\not{p} + m_\mu)_{ab} , \tag{8.43}$$

$$(\nu_e)_a(\bar{\nu}_e)_b = (\not{q}_1)_{ab} , \tag{8.44}$$

$$(\nu_\mu)_a(\bar{\nu}_\mu)_b = (\not{q}_2)_{ab} . \tag{8.45}$$

The last two equations follow automatically from the fact that the massless neutrinos, written in a Dirac notation, have only left-handed components. Substituting into eq. (8.41) we see that, in the product of the first two brackets, the structure of Dirac indices becomes cyclic,

$$\begin{aligned}&\left[\bar{\nu}_\mu\gamma^\mu(1-\gamma^5)\mu\right]\left[\bar{\mu}\gamma^\nu(1-\gamma^5)\nu_\mu\right] \\ &= (\bar{\nu}_\mu)_a\left(\gamma^\mu(1-\gamma^5)\right)_{ab}\mu_b\bar{\mu}_c\left(\gamma^\nu(1-\gamma^5)\right)_{cd}(\nu_\mu)_d \\ &= \left[(\nu_\mu)_d(\bar{\nu}_\mu)_a\right]\left(\gamma^\mu(1-\gamma^5)\right)_{ab}\left[\mu_b\bar{\mu}_c\right]\left(\gamma^\nu(1-\gamma^5)\right)_{cd} \\ &\to (\not{q}_2)_{da}\left(\gamma^\mu(1-\gamma^5)\right)_{ab}\frac{1}{2}(\not{p}+m_\mu)_{bc}\left(\gamma^\nu(1-\gamma^5)\right)_{cd} \\ &= \frac{1}{2}\mathrm{Tr}\left[\not{q}_2\gamma^\mu(1-\gamma^5)(\not{p}+m_\mu)\gamma^\nu(1-\gamma^5)\right] .\end{aligned} \tag{8.46}$$

The γ^5 factors can be simplified observing that

$$
\begin{aligned}
(1-\gamma^5)(\not p+m_\mu)\gamma^\nu(1-\gamma^5) &= \not p(1+\gamma^5)\gamma^\nu(1-\gamma^5)+m_\mu(1-\gamma^5)\gamma^\nu(1-\gamma^5) \\
&= \not p\gamma^\nu(1-\gamma^5)^2+m_\mu\gamma^\nu(1+\gamma^5)(1-\gamma^5) \\
&= 2\not p\gamma^\nu(1-\gamma^5)\,, \qquad (8.47)
\end{aligned}
$$

where we used the fact that $(1-\gamma^5)^2=2(1-\gamma^5)$ while $(1+\gamma^5)(1-\gamma^5)=0$. The product of the last two brackets in eq. (8.41) is treated similarly, and we find

$$
|\mathcal{M}_{fi}|^2 = G_F^2 \mathrm{Tr}\left[\not q_2\gamma^\mu\not p\gamma^\nu(1-\gamma^5)\right]\mathrm{Tr}\left[\not q_1\gamma_\nu\not k\gamma_\mu(1-\gamma^5)\right]. \qquad (8.48)
$$

The traces are computed using the identities given in the Notation section, that we recall here

$$
\begin{aligned}
\mathrm{Tr}(\gamma^\mu\gamma^\nu) &= 4\,\eta^{\mu\nu}\,, && (8.49)\\
\mathrm{Tr}(\gamma^\mu\gamma^\nu\gamma^\rho\gamma^\sigma) &= 4\,(\eta^{\mu\nu}\eta^{\rho\sigma}-\eta^{\mu\rho}\eta^{\nu\sigma}+\eta^{\mu\sigma}\eta^{\nu\rho})\,, && (8.50)\\
\mathrm{Tr}(\gamma^5\gamma^\mu\gamma^\nu\gamma^\rho\gamma^\sigma) &= -4i\epsilon^{\mu\nu\rho\sigma}\,. && (8.51)
\end{aligned}
$$

As a result

$$
\begin{aligned}
|\mathcal{M}_{fi}|^2 = 16G_F^2\, & p_\beta q_{2,\alpha}k^\rho q_1^\sigma\,(\eta^{\alpha\mu}\eta^{\beta\nu}-\eta^{\alpha\beta}\eta^{\mu\nu}+\eta^{\alpha\nu}\eta^{\beta\mu}+i\epsilon^{\alpha\mu\beta\nu}) \\
& \times(\eta_{\rho\mu}\eta_{\sigma\nu}-\eta_{\rho\sigma}\eta_{\mu\nu}+\eta_{\rho\nu}\eta_{\mu\sigma}+i\epsilon_{\rho\mu\sigma\nu})\,. \qquad (8.52)
\end{aligned}
$$

Performing the contractions, and using the identity

$$
\epsilon^{\alpha\mu\beta\nu}\epsilon_{\rho\mu\sigma\nu} = -2(\delta^\alpha_\rho\delta^\beta_\sigma-\delta^\alpha_\sigma\delta^\beta_\rho)\,, \qquad (8.53)
$$

we finally find

$$
|\mathcal{M}_{fi}|^2 = 64G_F^2\,(p_\mu q_1^\mu)\,(k_\nu q_2^\nu)\,. \qquad (8.54)
$$

We can now find the decay rate using eq. (6.20). We work in the rest frame of the muon. Then

$$
\begin{aligned}
d\Gamma &= \frac{1}{2m_\mu}\,|\mathcal{M}_{fi}|^2\,d\Phi^{(3)} \qquad (8.55)\\
&= \frac{32G_F^2}{m_\mu}\frac{d^3k}{(2\pi)^3 2E_{\mathbf{k}}}\frac{d^3q_1}{(2\pi)^3 2E_1}\frac{d^3q_2}{(2\pi)^3 2E_2}(2\pi)^4\delta^{(4)}(p-k-q_1-q_2)p_\mu q_1^\mu\,k_\nu q_2^\nu\,,
\end{aligned}
$$

where we denote $q^0_{1,2}=E_{1,2}$. Experimentally, the neutrinos are of course much more difficult to observe than the electron, so it is more common that one is interested in the decay rate as a function of just the electron energy. Then we integrate first over q_1, q_2. Let us define $q=p-k$. We must compute

$$
I^{\mu\nu}(q) \equiv \int \frac{d^3q_1}{E_1}\frac{d^3q_2}{E_2}\,\delta^{(4)}(q-q_1-q_2)q_1^\mu q_2^\nu\,. \qquad (8.56)
$$

It is not necessary to compute all the components of $I^{\mu\nu}(q)$ separately. Lorentz invariance dictates that the most general form of $I^{\mu\nu}(q)$ is

$$
I^{\mu\nu}(q) = A(q^2)\eta^{\mu\nu}+B(q^2)\frac{q^\mu q^\nu}{q^2}\,. \qquad (8.57)
$$

Taking the trace, $I^\mu_\mu=4A+B$, while $q_\mu q_\nu I^{\mu\nu}=q^2(A+B)$, so it suffices to compute I^μ_μ and $q_\mu q_\nu I^{\mu\nu}$ to have A, B and therefore the full tensor $I^{\mu\nu}(q)$. We compute first

$$
I^\mu_\mu = \int \frac{d^3q_1}{E_1}\frac{d^3q_2}{E_2}\,\delta^{(4)}(q-q_1-q_2)(q_1q_2)\,, \qquad (8.58)
$$

where $(q_1 q_2)$ denotes the scalar product $q_{1,\mu} q_2^\mu$. The Dirac delta gives $q = q_1 + q_2$ and therefore $q^2 = q_1^2 + q_2^2 + 2(q_1 q_2) = 2(q_1 q_2)$, since for massless neutrinos $q_1^2 = q_2^2 = 0$. Then

$$I_\mu^\mu = \frac{q^2}{2} \int \frac{d^3q_1}{E_1} \frac{d^3q_2}{E_2} \delta^{(4)}(q - q_1 - q_2) = \frac{q^2}{2} \int \frac{d^3q_1}{E_1 E_2} \delta(q^0 - E_1 - E_2) . \quad (8.59)$$

Since I_μ^μ is Lorentz invariant, we can compute it in the frame that we prefer; in particular, in the center-of-mass frame of the two neutrinos, we have $E_1 = E_2$ and therefore

$$I_\mu^\mu = \frac{q^2}{2} \int \frac{E_1^2 dE_1 d\Omega}{E_1^2} \delta(q^0 - 2E_1) = \frac{q^2}{2} 4\pi \int dE_1 \frac{1}{2} \delta(E_1 - \frac{q^0}{2}) = \pi q^2 . \quad (8.60)$$

To compute $q_\mu q_\nu I^{\mu\nu}$ we again use the rest frame of the two neutrinos, so that $q = (q^0, 0)$ (and therefore $(q^0)^2 = q^2$) and $(q_1 q) = E_1 q^0, (q_2 q) = E_2 q^0$. Then

$$\begin{aligned} q_\mu q_\nu I^{\mu\nu} &= \int \frac{d^3q_1}{E_1} \frac{d^3q_2}{E_2} \delta^{(4)}(q - q_1 - q_2)(q_1 q)(q_2 q) \\ &= q_0^2 \int d^3q_1 d^3q_2 \, \delta^{(4)}(q - q_1 - q_2) \\ &= q^2 \int d^3q_1 \, \delta(q^0 - 2E_1) \\ &= q^2 \int E_1^2 dE_1 d\Omega \frac{1}{2} \delta(E_1 - \frac{q^0}{2}) \\ &= \frac{\pi}{2} q^4 , \end{aligned} \quad (8.61)$$

where of course q^4 is a notation for $(q^2)^2$. We therefore find A, B and

$$I^{\mu\nu}(q) = \frac{\pi}{6} (q^2 \eta^{\mu\nu} + 2q^\mu q^\nu) . \quad (8.62)$$

We insert this into eq. (8.55) and, since $m_e \ll m_\mu$, we set for simplicity the electron mass to zero. Then we find, in the muon rest frame,

$$d\Gamma = \frac{G_F^2}{48\pi^4 m_\mu} \frac{E_{\mathbf{k}}^2 dE_{\mathbf{k}} d\Omega}{E_{\mathbf{k}}} \left[q^2(pk) + 2(qp)(qk) \right] . \quad (8.63)$$

In the muon rest frame $p = (m_\mu, 0)$, so $pk = m_\mu E_{\mathbf{k}}$; since $q = p - k$, we have: $qp = p^2 - pk = m_\mu^2 - m_\mu E_{\mathbf{k}}$; $q^2 = (p-k)^2 = p^2 + k^2 - 2pk = m_\mu^2 - 2m_\mu E_{\mathbf{k}}$; and finally, $qk = pk - k^2 = m_\mu E_{\mathbf{k}}$. We insert these expressions into eq. (8.63). The integration over $d\Omega$ simply gives a factor 4π, since the integrand is independent of the angles, and we finally find

$$\frac{d\Gamma}{dE} = \frac{G_F^2}{12\pi^3} E^2 (3m_\mu^2 - 4m_\mu E) , \quad (8.64)$$

where $E = E_{\mathbf{k}}$ is the electron energy. The minimum value of E is equal to $m_e \simeq 0$, and corresponds to the kinematical configuration where the electron is at rest and the two neutrinos have equal and opposite momenta. The maximum value is obtained instead when the electron goes in one direction carrying away half of the energy, and the two neutrinos are collinear, in the opposite direction (see the discussion of three-body kinematics in Problem 6.1). Then $E_{\min} = m_e \simeq 0$ and $E_{\max} = m_\mu/2$. Then, integrating eq. (8.64),

$$\Gamma \simeq \frac{G_F^2 m_\mu^5}{192\pi^3} . \quad (8.65)$$

Comparison with the muon lifetime allows us to determine the Fermi constant, and one finds (after a more accurate theoretical computation that includes electromagnetic loop corrections in which a photon is exchanged between the electron and the muon)

$$G_F = 1.16637(1) \times 10^{-5}\,\text{GeV}^{-2}\,. \tag{8.66}$$

Problem 8.2. $\pi^+ \to l^+\nu_l$

We consider the decay of the charged pion π^+ into a lepton l^+ and its neutrino ν_l, with $l = e, \mu$ (the τ lepton has $m_\tau \simeq 1777$ MeV and therefore is heavier than the pion). In terms of quarks, $\pi^+ = u\bar{d}$. Using eqs. (8.31), (8.32) and (8.33) the relevant interaction term is

$$\mathcal{L}_F = -\frac{G_F}{\sqrt{2}}\left(\bar{\nu}_l\gamma_\mu(1-\gamma^5)l\right)\left(\bar{d}'\gamma^\mu(1-\gamma^5)u\right)\,. \tag{8.67}$$

Furthermore, from eq. (8.14), we have $d' = d\cos\theta_C + s\sin\theta_C$ and only the term $d\cos\theta_C$ is relevant for this process. Therefore we use

$$\mathcal{L}_F = -\frac{G_F\cos\theta_C}{\sqrt{2}}\left(\bar{\nu}_l\gamma_\mu(1-\gamma^5)l\right)\left(\bar{d}\gamma^\mu(1-\gamma^5)u\right)\,. \tag{8.68}$$

In Chapter 5 we derived the LSZ formula, which gives the matrix element of the operator iT (related to the S matrix by $S = 1+iT$), in the case where the fields that appear in the Lagrangian describe the particles that we see in the initial and final states. We then saw that a field $\phi(x)$ describing the annihilation of an initial particle with four-momentum k_i brings a factor e^{-ik_ix}, while a field $\phi(x)$ describing the creation of a final particle with four-momentum p_j brings a factor e^{+ip_jx} (together with the appropriate spinor wave functions for spin-1/2 particles, or polarization vectors for spin-1 particles, as discussed in Section 5.5.4). Upon integration over d^4x, these exponential factors gave $(2\pi)^4\delta^{(4)}(P_i - P_f)$, which expresses energy–momentum conservation. The matrix element $i\mathcal{M}_{fi}$ was defined extracting this Dirac delta from the matrix element of iT, see eq. (5.98).

Here the situation is slightly different, because in the Lagrangian we have the quark fields but in the initial state we instead have the pion. To extract energy–momentum conservation we must come back to our derivation and be more general. In the interaction picture, the evolution operator is $\exp\{-i\int d^4x\mathcal{H}_{\text{int}}(x)\}$ and therefore we are interested in the matrix element of this operator between the initial and final state. To first order we are interested in

$$-i\int d^4x\,\langle f|\mathcal{H}_{\text{int}}(x)|i\rangle \tag{8.69}$$

where we denote by $|i\rangle$ and $|f\rangle$ the initial and final states, respectively. We now use the fact that, if $\hat{P}^\mu$ is the space-time translation operator, we have

$$\mathcal{H}_{\text{int}}(x) = e^{i\hat{P}x}\mathcal{H}_{\text{int}}(0)e^{-i\hat{P}x}\,. \tag{8.70}$$

Inserting this into eq. (8.69), we find

$$\begin{aligned}-i\int d^4x\,\langle f|e^{i\hat{P}x}\mathcal{H}_{\text{int}}(0)e^{-i\hat{P}x}|i\rangle &= -i\int d^4x\,e^{i(P_f-P_i)x}\langle f|\mathcal{H}_{\text{int}}(0)|i\rangle \\ &= -i(2\pi)^4\delta^{(4)}(P_f-P_i)\langle f|\mathcal{H}_{\text{int}}(0)|i\rangle\,. \end{aligned} \tag{8.71}$$

We see that energy–momentum conservation is a general consequence of space-time translation invariance, as it should be. Of course, the same manipulations can be performed on arbitrary powers of $\mathcal{H}_{\rm int}(x)$ and therefore the Dirac delta is extracted in this way to all orders in perturbation theory. The factor $(2\pi)^4\delta^{(4)}(P_f - P_i)$ is reabsorbed in the definition of $\mathcal{M}_{fi}$, and we see that $\mathcal{M}_{fi}$ is determined by the matrix element of the interaction Hamiltonian density evaluated at $x = 0$.

Coming back to our problem (and observing that in Fermi theory in the interaction Lagrangian there are no derivatives, and therefore $\mathcal{L}_{\rm int} = -\mathcal{H}_{\rm int}$), the matrix element between the initial pion state and the final $l\nu_l$ state is

$$\begin{aligned}\mathcal{M}_{fi} &= -\frac{G_F \cos\theta_C}{\sqrt{2}} \langle l^+\nu_l| \left[\bar{\nu}_l\gamma_\mu(1-\gamma^5)l\right](0) \left[\bar{d}\gamma^\mu(1-\gamma^5)u\right](0)|\pi^+\rangle \qquad (8.72)\\ &= -\frac{G_F \cos\theta_C}{\sqrt{2}} \langle l^+\nu_l| \left[\bar{\nu}_l\gamma_\mu(1-\gamma^5)l\right](0)|0\rangle\langle 0| \left[\bar{d}\gamma^\mu(1-\gamma^5)u\right](0)|\pi^+\rangle\,,\end{aligned}$$

with the currents evaluated at $x = 0$. The leptonic matrix element is easily computed,

$$\langle l^+\nu_l| \left[\bar{\nu}_l\gamma_\mu(1-\gamma^5)l\right](0)|0\rangle = \bar{u}(\nu_l)\gamma_\mu(1-\gamma^5)v(l)\,, \qquad (8.73)$$

where, as usual, $\bar{u}$ and v are the spinor wave functions. However, the matrix element of the quark current between the pion state and the vacuum cannot be computed like this. First of all, the quarks inside the pions of course are not free particles. The situation is further complicated by the fact that the interaction that confines the quarks inside the pions cannot be treated perturbatively. This can be understood comparing the bound state of quarks with the simplest example of bound state that we know, the hydrogen atom. The hydrogen atom is a bound state of an electron and a proton, and its total mass is $m_p + m_e -$ (binding energy). The binding energy in the ground state is the Rydberg, $(1/2)m_e\alpha^2$, and since $\alpha \ll 1$ it is a small correction compared to m_e and therefore to the mass $m_p + m_e$ of the free system. This is what we expect in a system in which the interaction term can be treated perturbatively. If this were the case also in QCD, we should expect that m_π is equal to m_u plus m_d minus a small binding energy. However, this expectation is completely wrong. The u and d quarks have masses[6] of the order of a few MeV, while $m_{\pi^+} \simeq 140$ MeV. Therefore the contribution of the masses of the constituent quarks to the total pion mass is completely negligible! Almost the totality of the pion mass comes from the energy of the gluon field (the QCD analog of the photon field) created by the u and d quarks, as well as from vacuum fluctuations involving creation of quark–antiquark pairs. These are complicated non-perturbative effects and, at the level of this book, we are unable to compute them.[7]

[6] Since the quarks do not appear as free particles, there are subtleties in how their masses are defined. We will not enter into these issues.

[7] In principle, matrix elements of the electroweak currents (often called simply "weak matrix elements") can be computed with numerical simulations in lattice gauge theory. In practice, despite much progress, there are still technical difficulties and one is limited by computer power.

The fact that we are unable to compute the matrix element of the hadronic current does not mean that we cannot proceed further. In fact, similarly to the electromagnetic form factors discussed in Problem 7.2, Lorentz covariance dictates the most general form of the matrix element, so we can at least parametrize it, in such a way that all our ignorance is hidden in a few quantities. In the case of the pion decay the parametrization is rather simple. The matrix element $\langle 0|\bar{d}\gamma^\mu(1-\gamma^5)u|\pi^+\rangle$ is a Lorentz four-vector. The pion state is described by its four-momentum p_π^μ and by nothing else, since the pion has spin zero. Therefore p_π^μ is the only four-vector on which this matrix element depends and, no matter what complicated computation we perform, the result must be proportional to p_π^μ. Therefore we can write

$$\langle 0| \left[\bar{d}\gamma^\mu(1-\gamma^5)u\right](0)|\pi^+\rangle = -f_\pi p_\pi^\mu \,, \qquad (8.74)$$

where f_π is a proportionality constant, and the minus sign is a convention in the definition of f_π. In general f_π will be a function of all the Lorentz invariant quantities on which the matrix element depends. But again, the only four-vector on which the matrix element depends is p_π^μ, and having only p_π^μ at our disposal the only Lorentz invariant quantity that we can construct is p_π^2. However $p_\pi^2 = m_\pi^2$ is a constant and therefore also f_π is a constant. It is called the pion decay constant, and all our ignorance on the inner structure of the pion is hidden in it.[8]

[8] The definition of f_π in the literature can sometime differ by a factor of 2 or of $\sqrt{2}$ from the one that we have adopted.

It is also instructive to look separately at the contributions of the vector and axial current in eq. (8.74). The pion is a pseudoscalar, i.e. it has intrinsic parity -1. Consider first the matrix element of the vector current, $\langle 0|\bar{d}\gamma^\mu u|\pi^+\rangle$. Under a parity transformation $\bar{d}\gamma^\mu u$ transforms as a true four-vector, but the matrix element picks an extra minus sign due to the intrinsic parity of the pion. Therefore, overall, the matrix element of the vector current between a pion state and the vacuum is a pseudo-four-vector, and conversely the matrix element of the axial current is a true four-vector. The value of the matrix element is determined by the strong interaction, which conserves parity. This means that the parametrization of the matrix element must hold also in the parity-transformed frame, and therefore a true vector must be equated to a true vector and a pseudovector to a pseudovector. Since p_π^μ is a true four-vector, and there is no pseudovector at our disposal, in eq. (8.74) the two separate contributions are

$$\langle 0| \left(\bar{d}\gamma^\mu u\right)(0)|\pi^+\rangle = 0\,, \qquad \langle 0| \left(\bar{d}\gamma^\mu \gamma^5 u\right)(0)|\pi^+\rangle = f_\pi p_\pi^\mu \,, \qquad (8.75)$$

and therefore all the contribution comes from the axial current.

The rest of the computation is straightforward. Plugging eqs. (8.73) and (8.74) into eq. (8.72) we find

$$\mathcal{M}_{fi} = \frac{G_F f_\pi \cos\theta_C}{\sqrt{2}}\, \bar{u}(\nu_l) \not{p}_\pi (1-\gamma^5) v(l)\,. \qquad (8.76)$$

We write $p_\pi = p_l + p_\nu$, where p_l and p_ν are the four-momenta of the lepton and of the neutrino, respectively. Then

$$\bar{u}(\nu_l)\not{p}_\pi(1-\gamma^5)v(l) = \bar{u}(\nu_l)\not{p}_\nu(1-\gamma^5)v(l) + \bar{u}(\nu_l)(1+\gamma^5)\not{p}_l v(l) \qquad (8.77)$$

and we use fact that $\bar{u}, v$ satisfy the Dirac equations (3.114) and (3.101). Therefore

$$\mathcal{M}_{fi} = -\frac{G_F f_\pi \cos\theta_C}{\sqrt{2}}\, m_l\, \bar{u}(\nu_l)(1+\gamma^5)v(l)\,. \qquad (8.78)$$

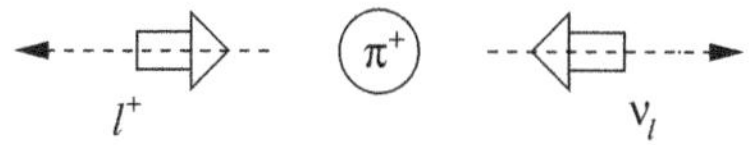

Fig. 8.3 The decay $\pi^+ \to l^+\nu_l$. The momenta of the particles are denoted by dashed lines and the spin by the large arrows.

Observe that the amplitude vanishes if $m_l = 0$. This can be understood observing that the charged weak currents depend only on the left-handed spinors, as in $\bar{\nu}_{l,L}\gamma^\mu l_L$. As discussed in Section 4.2.2, in the massless limit a left-handed operator describes a particle with $h = -1/2$ and its antiparticle with $h = +1/2$. Therefore the neutrino ν_l in the final state of the π^+ decay is left-handed, $h = -1/2$. Since the pion has spin zero, conservation of momentum and of angular momentum requires that also the antilepton l^+ has negative helicity $h = -1/2$, see Fig. 8.3. However, the antilepton l^+ is created by the left-handed Weyl field that appears in the weak charged current and therefore, according to the discussion in Section 4.2.2, in the massless limit it is a pure $h = +1/2$ state. Therefore in the massless limit the process is forbidden because it would violate angular momentum conservation.

Even if the contribution comes only from the spin configuration with negative helicities for both the neutrino and the antilepton it is still convenient, in the computation of $|\mathcal{M}_{fi}|^2$, to formally sum over all spins with the usual rules given in eqs. (3.112) and (3.113), $u\bar{u} \to \not{p}_\nu, v\bar{v} \to (\not{p}_l - m_l)$, since these rules are completely general. To compute $\mathcal{M}_{fi}^*$ we use

$$\left[\bar{u}(1+\gamma^5)v\right]^\dagger = \bar{v}(1-\gamma^5)u\,. \tag{8.79}$$

Then

$$\begin{aligned}|\mathcal{M}_{fi}|^2 &= \frac{G_F^2 f_\pi^2 \cos^2\theta_C}{2} m_l^2 \text{Tr}\left[(\not{p}_l - m_l)(1-\gamma^5)\not{p}_\nu(1+\gamma^5)\right] \\ &= G_F^2 f_\pi^2 \cos^2\theta_C\, m_l^2 \text{Tr}\left[\not{p}_l \not{p}_\nu\right] = 4G_F^2 f_\pi^2 \cos^2\theta_C\, m_l^2 (p_l p_\nu)\,. \end{aligned} \tag{8.80}$$

We have used $(1-\gamma^5)\not{p}_\nu(1+\gamma^5) = \not{p}_\nu(1+\gamma^5)^2 = 2\not{p}_\nu(1+\gamma^5)$ and the identity

$$\text{Tr}\left[\gamma^\mu\gamma^\nu\gamma^5\right] = 0\,. \tag{8.81}$$

(The first non-vanishing trace involving the matrix γ^5 is the trace of the product of γ^5 with four matrices γ^μ.) In the pion rest frame, $p_\pi = (m_\pi, 0)$; using $p_l = p_\pi - p_\nu$, we have $p_l p_\nu = p_\pi p_\nu - p_\nu^2 = m_\pi E_\nu$. Therefore

$$|\mathcal{M}_{fi}|^2 = 4G_F^2 f_\pi^2 \cos^2\theta_C\, m_l^2 m_\pi E_\nu \tag{8.82}$$

and, in the pion rest frame,

$$\begin{aligned} d\Gamma &= \frac{1}{2m_\pi} 4G_F^2 f_\pi^2 \cos^2\theta_C\, m_l^2 m_\pi E_\nu\, d\Phi^{(2)} \\ &= G_F^2 f_\pi^2 \cos^2\theta_C\, m_l^2 \frac{d^3p_l}{(2\pi)^3 2E_l}\frac{d^3p_\nu}{(2\pi)^3}(2\pi)^4\delta^{(4)}(p_\pi - p_l - p_\nu) \\ &= \frac{G_F^2 f_\pi^2 \cos^2\theta_C\, m_l^2}{8\pi^2}\frac{d^3p_l}{E_l}\delta(m_\pi - E_l - E_\nu)\,. \end{aligned} \tag{8.83}$$

Using $p_l dp_l = E_l dE_l$, we write $d^3p_l = p_l^2 dp_l d\Omega = p_l E_l dE_l d\Omega = (E_l^2 - m_l^2)^{1/2} E_l dE_l d\Omega$. Integrating over the solid angle, we find

$$d\Gamma = \frac{G_F^2 f_\pi^2 \cos^2\theta_C\, m_l^2}{2\pi}\sqrt{E_l^2 - m_l^2} dE_l\, \delta\left(E_l + \sqrt{E_l^2 - m_l^2} - m_\pi\right)\,. \tag{8.84}$$

The Dirac delta has only one zero, at $E_l = (m_\pi^2 + m_l^2)/(2m_\pi) \equiv E_0$. Using the identity $\delta(f(E)) = (1/|f'(E_0)|)\delta(E - E_0)$, we finally find

$$\Gamma(\pi^+ \to l^+\nu_l) = \frac{G_F^2 f_\pi^2 \cos^2\theta_C}{8\pi} m_l^2 m_\pi \left(1 - \frac{m_l^2}{m_\pi^2}\right)^2\,. \tag{8.85}$$

The parameter f_π can be fixed comparing the rate $\Gamma(\pi^+ \to \mu^+\nu_\mu)$ with the experimental value. Taking into account that $\cos\theta_C \simeq 0.97$, one finds $f_\pi \simeq 130$ MeV.

If we look at the branching ratio $\Gamma(\pi^+ \to e^+\nu_e)/\Gamma(\pi^+ \to \mu^+\nu_\mu)$ the parameter f_π cancels and we get the prediction

$$\frac{\Gamma(\pi^+ \to e^+\nu_e)}{\Gamma(\pi^+ \to \mu^+\nu_\mu)} = \frac{m_e^2}{m_\mu^2}\left(\frac{1 - m_e^2/m_\pi^2}{1 - m_\mu^2/m_\pi^2}\right)^2 \simeq 1.28 \times 10^{-4}\,. \tag{8.86}$$

This can be compared to the experimental value, which is $1.230(4) \times 10^{-4}$. We see that this lowest-order calculation already gives agreement at a level of a few per cent. This is the best that we could expect; for instance, electromagnetic corrections due to the exchange of a photon between the pion and the final lepton modify the matrix element by a factor $1 + O(\alpha)$ and therefore give a correction of the order of a few per cent.

Complement: Isospin and flavor $SU(3)$

Here we examine some symmetries of strong interactions, and their effects on the vector and axial quark currents. We will then use these results in the next problem, when we will need the quark currents in the computation of a weak matrix element involving hadrons. Consider the currents

$$V^\mu_{ab} = \bar{q}_a \gamma^\mu q_b \,, \qquad A^\mu_{ab} = \bar{q}_a \gamma^\mu \gamma^5 q_b \tag{8.87}$$

where q is the Dirac field describing a quark and the indices a, b label the type of quark $(u, d, s, \ldots)$ and is called a flavor index. Consider at first *free* quarks, so that they satisfy the free Dirac equation, $i\partial\!\!\!/ q_a = m_a q_a$, and its hermitian conjugate $i\partial_\mu \bar{q}_a \gamma^\mu = -m_a \bar{q}_a$. Then

$$\partial_\mu V^\mu_{ab} = i(m_a - m_b)\bar{q}_a q_b \,, \qquad \partial_\mu A^\mu_{ab} = i(m_a + m_b)\bar{q}_a \gamma^5 q_b \,. \tag{8.88}$$

We see that in the limit in which all quark masses are zero, both the axial and vector currents are conserved; if instead the masses are non-zero but two or more of them are degenerate, still the corresponding vector currents are conserved, at the level of the free theory. We can trace this to the existence of a symmetry in the free Lagrangian for N quark flavors,

$$\mathcal{L}_{\rm free} = \sum_{a=1}^{N} \left[i\bar{q}_a \partial\!\!\!/ q_a - m_a \bar{q}_a q_a \right] . \tag{8.89}$$

Consider in fact the global transformation

$$q_a \to U_{ab} q_b \tag{8.90}$$

where U_{ab} is a $U(N)$ matrix in flavor space. Since U acts only on flavor space and is a global transformation (i.e. it is independent of x) it commutes both with the γ matrices and with the derivatives. Therefore, under (8.90), $\bar{q}\partial\!\!\!/ q \to \bar{q}U^\dagger U \partial\!\!\!/ q = \bar{q}\partial\!\!\!/ q$, so the kinetic term is invariant. Instead, a generic mass term does not respect this symmetry. In fact, if M is the matrix in flavor space with matrix elements $M_{ab} = m_a \delta_{ab}$, the mass term in the Lagrangian can be written in matrix form as $\bar{q}Mq$, and under eq. (8.90) it becomes $\bar{q}U^\dagger M U q$. A generic $U(N)$ matrix U does not commute with M, unless M is proportional to the identity matrix. Therefore we have a $U(N)$ symmetry only when N masses are equal. The corresponding conserved currents are just the vector currents V^μ_{ab}. Consider now the transformation

$$q_a \to \left(e^{i\gamma^5 \beta} \right)_{ab} q_b \tag{8.91}$$

where $\beta = \beta_c T^c$, T^c are the $U(N)$ generators and β_c the $U(N)$ parameters. In other words, the transformation matrix is $U_L = \exp\{-i\beta\}$ on the left-handed quarks and $U_R = U_L^\dagger$ on the right-handed quarks. Equations (8.90) and (8.91) are the generalization to N flavors of the vector and chiral transformations discussed in Section 3.4.3. The chiral transformation (8.91) is still a symmetry of the kinetic term. In fact, as in Section 3.4.3, we use $\gamma^\mu e^{i\gamma^5\beta} = -e^{i\gamma^5\beta}\gamma^\mu$, which follows from $\{\gamma^5, \gamma^\mu\} = 0$. Then, performing the transformation on the kinetic terms, in order to bring $U^\dagger$ near U and use $U^\dagger U = 1$ we must anticommute it twice with γ matrices, once with the γ^0 implicit in $\bar{q}$ and the second time with γ^μ. Therefore we pick a minus sign twice, and the kinetic term is invariant. In the mass term, instead, even if all the masses are equal, so that we have the structure $\bar{q}q$, we must anticommute only once, with the γ^0 implicit in $\bar{q}$. Therefore the mass term picks a minus sign, and it is not invariant. So, the axial symmetry (8.91) is a symmetry only if all masses are zero, and in this case their conserved currents are the axial currents A^μ_{ab} given in eq. (8.87). We therefore understand why, when the masses are non-zero, the divergence of the vector currents is proportional to the mass difference while for the axial currents it is proportional to their sum.

Until now we have discussed these symmetries for the free quark Lagrangian. However, we will see in Section 10.3 that the interaction term in QCD respects these symmetries so, in strong interactions, the only non-invariant term in the Lagrangian comes from the quark masses. In particular, the masses of the u and d quarks are almost degenerate, $m_u \simeq m_d$. Therefore we have an approximated global $U(2) = U(1) \times SU(2)$ vector symmetry, to which correspond four conserved currents. The $SU(2)$ part of this symmetry is called isospin. The u, d quarks are an isospin doublet; u has $(I = 1/2, I_z = +1/2)$ while d has $(I = 1/2, I_z = -1/2)$. All other quarks have isospin zero. The three currents associated to isospin can be written compactly introducing the notation

$$Q = \begin{pmatrix} u \\ d \end{pmatrix} \tag{8.92}$$

for the (u, d) doublet. As we recalled in Section 2.5, the generators of an $SU(2)$ group on a doublet are represented by one half times the Pauli matrices. These, when they act in isospin space, are conventionally denoted by τ^a rather than σ^a, so the isospin generators on the (u, d) doublet are written $T^a = \tau^a/2$. The $SU(2)$ conserved currents can be written in the form

$$V^a_\mu = \bar{Q}\gamma_\mu \frac{\tau^a}{2} Q\,, \tag{8.93}$$

which shows that these currents are an $SU(2)$ triplet. Introducing the linear combinations

$$\tau^+ = \frac{1}{2}(\tau^1 + i\tau^2) = \begin{pmatrix} 0 & 1 \\ 0 & 0 \end{pmatrix}, \qquad \tau^- = \frac{1}{2}(\tau^1 - i\tau^2) = \begin{pmatrix} 0 & 0 \\ 1 & 0 \end{pmatrix}, \tag{8.94}$$

and $V^\pm_\mu = V^1_\mu \pm iV^2_\mu$, we see that

$$V^+_\mu = \bar{u}\gamma_\mu d\,, \qquad V^-_\mu = \bar{d}\gamma_\mu u\,, \qquad V^3_\mu = \frac{1}{2}(\bar{u}\gamma_\mu u - \bar{d}\gamma_\mu d)\,. \tag{8.95}$$

The $U(1)$ current is instead

$$j^\mu_{U(1)} = \bar{Q}\gamma^\mu Q = \bar{u}\gamma_\mu u + \bar{d}\gamma_\mu d\,. \tag{8.96}$$

The electromagnetic current involving the u and d quarks is

$$j^\mu_{\rm em} = \frac{2}{3}\bar{u}\gamma_\mu u - \frac{1}{3}\bar{d}\gamma_\mu d\,, \tag{8.97}$$

since the u quark has a charge $(2/3)|e|$ and the d quark has charge $-(1/3)|e|$; $j^\mu_{\rm em}$ can be written as a linear combination of $j^\mu_{U(1)}$ and of V^3_μ,

$$j^\mu_{\rm em} = \frac{1}{6}\bar{Q}\gamma^\mu Q + \bar{Q}\gamma^\mu \frac{\tau^3}{2} Q\,. \tag{8.98}$$

This shows that the electromagnetic interaction does not conserve isospin, since this current is the sum of the isospin singlet $\bar{Q}\gamma^\mu Q$, which induces transitions with $\Delta I = 0$ and therefore conserves isospin, and of a term $\bar{Q}\gamma^\mu \tau^3 Q$ which instead induces transitions with $|\Delta I| = 1$ and $\Delta I_z = 0$. An example of the former is the decay $\rho^0 \to \pi^0\gamma$, since both the ρ^0 and the π^0 mesons have isospin $I = 1, I_z = 0$. An example of the latter is the decay $\rho^0 \to \eta^0\gamma$, since the meson η^0 has isospin $I = 0$, or $\Sigma^0 \to \Lambda^0\gamma$, where Σ^0 is a baryon with $I = 1, I_z = 0$ and Λ^0 is a baryon with $I = 0$.

From the form of the weak currents discussed in Section 8.2, we see that isospin is violated also by weak interactions. We leave it as an instructive exercise to classify the possible isospin violations in weak decays of hadrons, both in semileptonic decays (i.e. when in the final state there is another hadron plus a lepton and its neutrino) and in non-leptonic decays, i.e. when there is no lepton in the final state.

When we have a symmetry, the corresponding charge is conserved. In non-relativistic quantum mechanics, this means that the charge commutes with the Hamiltonian; therefore we can diagonalize simultaneously the charge and the Hamiltonian, and the energy eigenstates can be labeled by the value of the charge. For instance, in a system invariant under spatial rotations, the angular momentum is conserved and the states can be labeled by their energy E, angular momentum J and by J_z. At each energy level we have $2J + 1$ degenerate states, corresponding to the $2J + 1$ possible values of J_z.

In QFT the situation is more complicated because it also depends on the behavior of the vacuum state under the symmetry transformation. There are two possibilities: if the vacuum state is unique then it must be invariant under the symmetry. If however it is degenerate, it is possible that a symmetry transformation sends it into a new vacuum state. The latter possibility corresponds to spontaneous symmetry breaking, and will be discussed in Chapter 11, where we will also explain why this possibility cannot take place in non-relativistic quantum mechanics.

If the vacuum is non-degenerate, the situation is completely analogous to non-relativistic quantum mechanics and the eigenstates of the Hamiltonian (i.e. the particles) are labeled by the quantum numbers of the conserved charge. In the case of isospin, each strongly interacting particle is labeled by its isospin I and by I_z, and at each mass level we have $2I_z + 1$ degenerate hadronic states. Isospin is only an approximate symmetry of strong interactions because m_u is not exactly equal to m_d, and is also violated by weak and electromagnetic interactions, therefore the states are only approximately degenerate. A comparison with the mass spectrum of hadrons made of u and d quarks shows that the option chosen by Nature is indeed that the vacuum is invariant under isospin and hadrons are organized in isospin multiplets with approximately degenerate masses. For examples the three pions have $m_{\pi^0} \simeq 135$ MeV and $m_{\pi^\pm} \simeq 140$ MeV. The equality of the π^+ and π^- masses follows from the CPT theorem, as we saw in Section 4.2.3. However, the approximate equality with the π^0 mass is a consequence of isospin symmetry, and we conclude that the three pions are an isospin triplet, i.e. they have $I = 1$. Similarly the approximate equality of the proton and neutron masses, $m_p \simeq 938.2$ MeV, $m_n \simeq 939.5$ MeV, is due to the fact that they are an isospin doublet, i.e. they have $I = 1/2$.

The fact that isospin is an approximate symmetry, and therefore the vector current is approximately conserved, is sometimes referred to as the CVC (conserved vector current) approximation. In this approximation, the isospin multiplets are exactly degenerate, so when one uses the CVC approximation one must, by consistency, set $m_{\pi^0} = m_{\pi^\pm}$, $m_p = m_n$, etc. in the calculation of the amplitudes.

The approximation where m_u and m_d are both set to zero is of course less accurate, and it is known as PCAC (partially conserved axial current) since in this limit also the axial currents are conserved. If the vacuum were invariant under this approximate symmetry we should see a parity doubling of the hadron spectrum. Since this is not the case, we must conclude that either the approximate symmetry is not a good approximation at all, or it is spontaneously broken. The latter option turns out to be the correct one.

Coming back to the vector symmetry, one can enlarge it to the next lightest quark, which is the s quark, assuming that $m_u \simeq m_d \simeq m_s$. In this limit we have an approximate $SU(3)$ symmetry and particles are organized according to the $SU(3)$ representations. For example, the light pseudoscalar mesons are in an octet, formed by the three pions, the four kaons $K^\pm, K^0, \bar{K}^0$ and the η meson. This symmetry is more approximate than isospin, as we see, e.g. from the fact that the mass of $K^\pm$ is $m_K \simeq 494$ MeV, rather larger than the pion masses. Still, the mass differences are smaller than the typical masses of hadrons. Furthermore, a number of techniques exist for computing deviations from the exact symmetry. The interested reader can see for instance Georgi (1984).

Problem 8.3. $K^0 \to \pi^- l^+ \nu_l$

We now consider the decay of a spin-0 hadron into another spin-0 hadron, plus a lepton $l^+ = e^+, \mu^+$ and its neutrino. We will consider for definiteness the decay of the neutral kaon, $K^0 \to \pi^- l^+ \nu_l$, but the same type of computation can be performed for, say, $\pi^+ \to \pi^0 e^+ \nu_e$ or for the beta decay of a spin-0 nucleus into another spin-0 nucleus. Decays where in the final state there is both a lepton–neutrino pair and a hadron are called semileptonic.

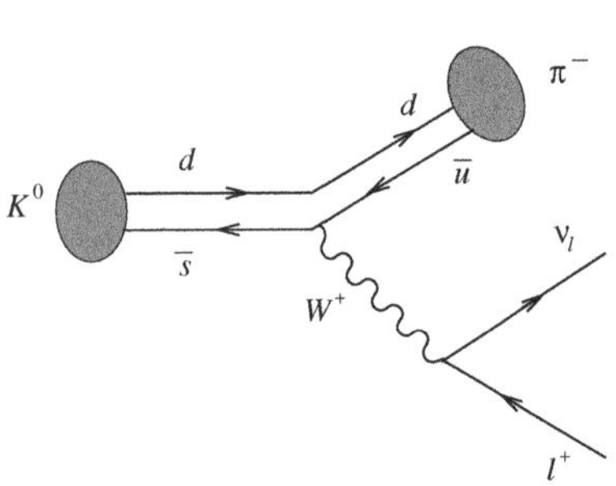

Fig. 8.4 The decay $K^0 \to \pi^- l^+ \nu_l$. The shaded blobs indicate that the hadrons are complicated bound states of quarks.

In terms of quarks, $K^0 = d\bar{s}$ and $\pi^- = d\bar{u}$. At the quark level the transition proceeds as shown in Fig. 8.4. The d quark is a "spectator", while the quark $\bar{s}$, with electric charge $+1/3$ (in units of $|e|$) is transformed into a $\bar{u}$ quark, with charge $-2/3$, by the emission of a (virtual) W^+ boson which then decays into $l^+\nu_l$. The shaded blobs in the graph indicate that $d\bar{s}$ and $d\bar{u}$ are complicated bound states, rather than free quarks. The relevant hadronic current is, using eqs. (8.33) and (8.14),

$$j_h^\mu = \sin\theta_C \ \bar{s}\gamma^\mu(1-\gamma^5)u \tag{8.99}$$

and the relevant term in the Fermi Lagrangian is

$$\mathcal{L}_F = -\frac{G_F \sin\theta_C}{\sqrt{2}} \left(\bar{\nu}_l \gamma_\mu (1-\gamma^5) l\right) \left(\bar{s}\gamma^\mu(1-\gamma^5)u\right) . \tag{8.100}$$

With the same steps discussed in Problem 8.2 we find the matrix element

$$\mathcal{M}_{fi} = -\frac{G_F \sin\theta_C}{\sqrt{2}} \bar{u}(\nu_l)\gamma_\mu(1-\gamma^5)v(l) \langle \pi^- | \bar{s}\gamma^\mu(1-\gamma^5)u | K^0 \rangle , \tag{8.101}$$

where it is understood that the current $\bar{s}\gamma^\mu(1-\gamma^5)u$ is evaluated at $x=0$. As in Problem 8.2, we parametrize the hadronic matrix element in the most general form compatible with Lorentz invariance. Since both the pion and the kaon have spin zero, the matrix element can depend only on their momenta, that we denote p^μ_π and p^μ_K, respectively. We introduce the combinations

$$p^\mu = p^\mu_K + p^\mu_\pi\,, \qquad q^\mu = p^\mu_K - p^\mu_\pi\,. \tag{8.102}$$

There is only one independent Lorentz scalar that we can construct with p^μ and q^μ, which we take to be q^2. In fact, $qp = p_K^2 - p_\pi^2 = m_K^2 - m_\pi^2$ is a constant while, from their definitions, $p^2 = m_K^2 + m_\pi^2 + 2p_K p_\pi$ and $q^2 = m_K^2 + m_\pi^2 - 2p_K p_\pi$. Therefore $p^2 + q^2 = 2(m_K^2 + m_\pi^2)$ and they are not independent. So the most general parametrization of the hadronic matrix element consistent with Lorentz invariance is

$$\langle \pi^- |\bar{s}\gamma^\mu(1-\gamma^5)u|K^0\rangle = f_1(q^2)p^\mu + f_2(q^2)q^\mu\,, \tag{8.103}$$

with f_1, f_2 functions of the only independent Lorentz scalar at our disposal. They are the form factors for the process in question. All our ignorance of the internal structure of the pion and of the kaon is hidden in these two functions. A number of considerations allow us to simplify the problem further:

(i). The internal structure of the pion and of the kaon is determined by the strong interactions, and strong interactions conserve time-reversal invariance. We now show that invariance under time reversal implies that $f_1(q^2)$ and $f_2(q^2)$ are relatively real.

The proof is as follows. We saw in Section 4.2.3 that time reversal is implemented by an anti-unitary operator T, i.e. by an operator that, given two states $|a\rangle$ and $|b\rangle$, satisfies $\langle Ta|Tb\rangle = \langle a|b\rangle^*$. Let $|a\rangle = |\pi^-\rangle$ and $|b\rangle = j^\mu(0)|K^0\rangle$, where $j^\mu(x) = \left[\bar{s}\gamma^\mu(1-\gamma^5)u\right](x)$. Since $|\pi^-\rangle$ is fully characterized by the pion momentum p^μ_π, we can rather use the notation $|\pi^-\rangle = |p_\pi\rangle$. The time-reversed state $|T\pi^-\rangle$ describes a pion with time-reversed four-momentum $p'_\pi = (p^0_\pi, -\mathbf{p}_\pi)$, i.e. $p'^\mu_\pi = \eta^{\mu\mu}p^\mu_\pi$ (with no sum over the μ index). Since at the quantum level we also have the possibility of a phase, we write

$$|Tp_\pi\rangle = \eta_\pi|p'_\pi\rangle\,. \tag{8.104}$$

Similarly, the time reversal of the state $j^\mu(0)|K^0\rangle$ is obtained changing the kaon momentum into $p'^\mu_K = \eta^{\mu\mu}p^\mu_K$ and allowing for a phase η_K. At the same time, under time reversal a current transforms as $j^\mu(t,\mathbf{x}) \to \eta^{\mu\mu}j^\mu(-t,\mathbf{x})$, therefore $j^\mu(x=0) \to \eta^{\mu\mu}j^\mu(x=0)$. In conclusion, the matrix element that we denoted $\langle Ta|Tb\rangle$ is obtained from the matrix element $\langle a|b\rangle$ replacing $p^\mu \to \eta^{\mu\mu}p^\mu$, $q^\mu \to \eta^{\mu\mu}q^\mu$ and $j^\mu(0) \to \eta^{\mu\mu}j^\mu(0)$, and multiplying by the relative phase $\eta_{\rm rel} = \eta_\pi^*\eta_K$.

Since in our case T is a good symmetry, if the parametrization (8.103) holds in a frame it must hold also in the time-reversed frame. Therefore the equation $\langle Ta|Tb\rangle = \langle a|b\rangle^*$ translates into

$$\eta_{\rm rel}\,\eta^{\mu\mu}\left(f_1(q^2)\eta^{\mu\mu}p^\mu + f_2(q^2)\eta^{\mu\mu}q^\mu\right) = \left(f_1(q^2)p^\mu + f_2(q^2)q^\mu\right)^*\,, \tag{8.105}$$

where on the left-hand side the overall factor $\eta^{\mu\mu}$ comes from the transformation of the current, and the factors $\eta^{\mu\mu}$ inside the parentheses come from the transformation of the momenta, and there is no sum over μ. Since, for each μ, we have $(\eta^{\mu\mu})^2 = 1$, we find

$$\eta_{\rm rel}\left(f_1(q^2)p^\mu + f_2(q^2)q^\mu\right) = f_1^*(q^2)p^\mu + f_2^*(q^2)q^\mu\,. \tag{8.106}$$

This means that $f_1^*(q^2) = \eta_{\rm rel}f_1(q^2)$ and $f_2^*(q^2) = \eta_{\rm rel}f_2(q^2)$, i.e. $f_1(q^2)$ and $f_2(q^2)$ pick the same phase $\eta_{\rm rel}$ (independent of q^2) under complex conjugation. Apart from an overall phase, that is irrelevant when we take the modulus squared of the matrix element, we can then take $f_1(q^2)$ and $f_2(q^2)$ to be simultaneously real.

(ii) Use of CVC. Since the pion and the kaon have the same intrinsic parity (they are both pseudoscalars) only the vector current contributes to the matrix element,

$$\langle\pi^-|[\bar{s}\gamma^\mu\gamma^5 u](0)|K^0\rangle = 0\,, \tag{8.107}$$

$$\langle\pi^-|\,[\bar{s}\gamma^\mu u]\,(0)|K^0\rangle = f_1(q^2)p^\mu + f_2(q^2)q^\mu\,. \tag{8.108}$$

(We have written explicitly that the current in the matrix element is evaluated at $x = 0$.) Let us now use the CVC approximation discussed in the Complement on page 209. In this context, this means that we assume an approximate $SU(3)$ flavor symmetry, since it is only in this case that the vector current involving the u and s quarks is conserved (of course isospin only provides the conservation of the currents involving the u and d quarks). Denoting $[\bar{s}\gamma^\mu u]\,(x) = V^\mu(x)$, this means that we are assuming that $\partial_\mu\langle\pi^-|V^\mu(x)|K^0\rangle = 0$. Using the fact that $V^\mu(x) = e^{i\hat{P}x}V^\mu(0)e^{-i\hat{P}x}$, we have

$$0 = \partial_\mu\langle\pi^-|e^{i\hat{P}x}V^\mu(0)e^{-i\hat{P}x}|K^0\rangle = \partial_\mu e^{-i(p_K-p_\pi)x}\langle\pi^-|V^\mu(0)|K^0\rangle\,. \tag{8.109}$$

Since $p_K - p_\pi = q$ we find that, in the CVC approximation,

$$q_\mu\langle\pi^-|V^\mu(0)|K^0\rangle = 0\,. \tag{8.110}$$

Inserting the parametrization (8.108), this gives

$$f_1(q^2)pq + f_2(q^2)q^2 = 0\,. \tag{8.111}$$

The scalar product $pq = m_K^2 - m_\pi^2$ is automatically zero within the CVC approximation since, as we explained in the previous section, when we assume an $SU(3)$ flavor symmetry we must set for consistency $m_K = m_\pi$.[9] The term q^2 is instead non-vanishing. Therefore, if CVC were an exact symmetry, we should conclude that $f_2(q^2) = 0$. In reality, $m_K \neq m_\pi$. However, flavor $SU(3)$ is still an approximate symmetry, which means that $f_2(q^2)$ is small with respect to $f_1(q^2)$. In a first approximation, we can then neglect it.

[9] Of course, this approximation is made only in the calculation of the matrix element, and gives the zeroth-order term in an expansion of the matrix element in powers of $m_K - m_\pi$. We do not set $m_K = m_\pi$ in the expression for the phase space, since otherwise the decay would be forbidden!

(iii) There is a further reason to neglect the term $f_2(q^2)q^\mu$. The four-momentum $q = p_K - p_\pi$ can be rewritten in terms of the lepton and neutrino momenta, p_l and p_{ν_l} using energy–momentum conservation, as $q = p_l + p_{\nu_l}$. When we multiply it by the leptonic matrix element in eq. (8.101) we find, using the Dirac equation,

$$\begin{aligned} q^\mu\bar{u}(\nu_l)\gamma_\mu(1-\gamma^5)v(l) &= \bar{u}(\nu_l)\not{p}_{\nu_l}(1-\gamma^5)v(l) + \bar{u}(\nu_l)(1+\gamma^5)\not{p}_l v(l) \\ &= -m_l\,\bar{u}(\nu_l)(1+\gamma^5)v(l)\,. \end{aligned} \tag{8.112}$$

The lepton mass m_l, especially when the lepton is the electron, is small compared with the kaon mass. Therefore the term proportional to $f_2(q^2)q^\mu$ in the matrix element gives a small contribution both because f_2 is small and because it is suppressed by m_l/m_K with respect to the leading term. In the following we will neglect it (of course, it would not take great effort to carry out the computation more generally, keeping also f_2, but it is important to understand where the dominant contribution comes from).

(iv) Our final approximation is to observe that, in the spirit of CVC, the momentum transfer q^2 is small, since we consider the mass difference between m_K and m_π as small, and we therefore approximate $f_1(q^2) \simeq f_1(0)$.

In principle all these approximations can be improved systematically, order by order in the deviations from $SU(3)$ symmetry. Furthermore, using $SU(3)$ symmetry, $f_1(0)$ can be related to a similar quantity for pion decay, and it can be shown that $f_1(0) = 1$ in the limit of exact symmetry. The interested reader is referred to Okun (1982), Sections 4.3 and 6.4.

In conclusion, we write

$$\langle \pi^- | \bar{s}\gamma^\mu (1-\gamma^5) u | K^0 \rangle \simeq p^\mu \,. \tag{8.113}$$

The rest of the computation is long but straightforward. The matrix element is

$$\mathcal{M}_{fi} = -\frac{G_F \sin\theta_C}{\sqrt{2}}\, p^\mu \bar{u}(\nu_l)\gamma_\mu (1-\gamma^5) v(l) \,. \tag{8.114}$$

Summing over the polarization of the final lepton,

$$|\mathcal{M}_{fi}|^2 = \frac{G_F^2 \sin^2\theta_C}{2} p^\mu p^\nu \, \mathrm{Tr}\left[\not{p}_{\nu_l}\gamma_\mu (1-\gamma^5)(\not{p}_l + m_l)(1+\gamma^5)\gamma_\nu\right] \,. \tag{8.115}$$

The calculation of the traces and the subsequent contraction of Lorentz indices gives

$$|\mathcal{M}_{fi}|^2 = 4G_F^2 \sin^2\theta_C \left[2(pp_{\nu_l})(pp_l) - p^2 (p_l p_{\nu_l})\right] \,. \tag{8.116}$$

We compute the scalar products in the kaon rest frame, which gives

$$p^2 = m_K^2 + m_\pi^2 + 2m_K E_\pi \,, \tag{8.117}$$

$$(pp_{\nu_l}) = \frac{1}{2}(3m_K^2 - m_\pi^2 + m_l^2) - m_K(2E_l + E_\pi) \,, \tag{8.118}$$

$$(pp_l) = -\frac{1}{2}(m_K^2 + m_\pi^2 + m_l^2) + m_K(2E_l + E_\pi) \,, \tag{8.119}$$

$$(p_l p_{\nu_l}) = \frac{1}{2}(m_K^2 + m_\pi^2 - m_l^2) - m_K E_\pi \,. \tag{8.120}$$

The quickest way to obtain eq. (8.120) is to write energy–momentum conservation in the form $p_K - p_\pi = p_l + p_{\nu_l}$. Taking the squared modulus of both sides, we get $m_K^2 + m_\pi^2 - 2p_K p_\pi = m_l^2 + 2p_l p_{\nu_l}$. In the kaon rest frame, $p_K p_\pi = m_K E_\pi$, and we obtain eq. (8.120). The other relations are obtained similarly.

Using eqs. (6.20) and (6.89) and choosing the lepton and pion energies as independent variables, the decay width is

$$d\Gamma = \frac{1}{2m_K}|\mathcal{M}_{fi}|^2 \, d\Phi^{(3)} \,, \qquad d\Phi^{(3)} = \frac{1}{32\pi^3} dE_l dE_\pi \,. \tag{8.121}$$

Neglecting for simplicity m_l, we find after some algebra

$$d\Gamma = \frac{G_F^2 \sin^2\theta_C m_K}{8\pi^3}\left[E_\pi^2 - m_\pi^2 - (m_K - E_\pi - 2E_l)^2\right] dE_l dE_\pi \,. \tag{8.122}$$

We can now integrate over the lepton energy, at fixed E_π. Using the results of Problem 6.1, the kinematical limits, when $m_l = 0$, are

$$E_l^{\min} = \frac{m_K - E_\pi - |\mathbf{p}_\pi|}{2} \,, \qquad E_l^{\max} = \frac{m_K - E_\pi + |\mathbf{p}_\pi|}{2} \,. \tag{8.123}$$

This gives

$$d\Gamma = \frac{G_F^2 \sin^2\theta_C m_K}{12\pi^3}\left(E_\pi^2 - m_\pi^2\right)^{3/2} dE_\pi \,. \tag{8.124}$$

We finally integrate over dE_π, with the limits

$$E_\pi^{\min} = m_\pi \,, \qquad E_\pi^{\max} = \frac{m_K^2 + m_\pi^2}{2m_K} \,. \tag{8.125}$$

In principle the integration can be performed exactly, but the result is slightly complicated. If, for simplicity, we limit ourselves to the leading and next-to-leading expansion in m_π/m_K we can expand the integrand in eq. (8.124) in powers of m_π and we find

$$\Gamma \simeq \frac{G_F^2 \sin^2\theta_C m_K^5}{768\pi^3}\left(1 - \frac{8m_\pi^2}{m_K^2}\right). \tag{8.126}$$

Summary of chapter

- Weak decays in which the momentum transfer is small compared to the masses of the $W^\pm$ and Z^0 boson, and scattering processes with $E_{\rm CM} \ll m_W$, can be computed using a low-energy approximation to the Standard Model.
- The low-energy interaction Lagrangian is a four-fermion theory, obtained as the product of two fermionic currents involving leptons or quarks. There are two distinct types of processes: those involving neutral currents (which at the fundamental level correspond to Feynman graphs involving the Z^0 boson) and those involving charged currents (which are mediated by the $W^\pm$ bosons).
- The leptonic charged current is given in eq. (8.32) and the hadronic charged current is given in eq. (8.33). The effective Lagrangian for processes involving charged currents is given in eq. (8.31).
- The leptonic neutral current is given in eq. (8.35) and the hadronic neutral current is given in eq. (8.36). The effective Lagrangian for processes involving neutral currents is given in eq. (8.37).
- Left-handed and right-handed fields enter the theory in a different way. In particular, charged currents are built only with left-handed fields. Neutral currents contain also right-handed fields, but with couplings that differ from the left-handed ones. Therefore weak interactions violate parity.
- While the calculation of the matrix elements of the leptonic currents is straightforward, the computation of the matrix elements of the hadronic currents is more difficult, because the current is written in terms of quark fields while the initial and final states are hadrons. In general, we are not able to compute the hadronic matrix elements explicitly, but we can parametrize them in such a way that our ignorance of the internal hadronic structure is hidden in a few form factors. Some detailed examples have been given in the Solved Problem section.

Further reading

- Two excellent books on the phenomenology of weak interactions at low energies are Okun (1982), where the reader can find a large number of explicit calculations of weak decays, and Georgi (1984). Note that Okun defines $\gamma^5 = -i\gamma^0\gamma^1\gamma^2\gamma^3\gamma^4$ while we define $\gamma^5 = +i\gamma^0\gamma^1\gamma^2\gamma^3\gamma^4$.
- Many explicit calculations of weak processes can be found in the collection of solved problems by Di Giacomo, Paffuti and Rossi (1994).
- For particle physics phenomenology, with emphasis on the experimental aspects, an excellent book is Perkins (2000), especially the updated 4th edition.

Exercises

(8.1) Consider the neutron β-decay, $n \to pe^-\bar{\nu}_e$.

(i) Show that the matrix element of the hadronic current between the neutron state $|n\rangle$ and the proton states $|p\rangle$ is $\langle p|j^h_\mu|n\rangle = \langle p|V^h_\mu|n\rangle - \langle p|A^h_\mu|n\rangle$, with

$$\langle p|V^h_\mu|n\rangle = f_1(q^2)\bar{u}_p\gamma_\mu u_n + f_2(q^2)\bar{u}_p\sigma_{\mu\nu}q^\nu u_n + f_3(q^2)q_\mu\bar{u}_p u_n\,, \tag{8.127}$$

$$\langle p|A^h_\mu|n\rangle = g_1(q^2)\bar{u}_p\gamma_\mu\gamma^5 u_n + g_2(q^2)\bar{u}_p\sigma_{\mu\nu}q^\nu\gamma^5 u_n + g_3(q^2)q_\mu\bar{u}_p\gamma^5 u_n\,, \tag{8.128}$$

where u_n and u_p are the neutron and proton wave functions, respectively, and $q^\mu = p^\mu_n - p^\mu_p$ is the difference between the neutron and the proton four-momentum.

(ii) Show that the form factors can be approximated by their values at $q^2 = 0$.

(iii) Using CVC and symmetry arguments, it can be shown that the main contribution comes from $f_1(0) \equiv g_V$ and $g_1(0) \equiv g_A$ and furthermore $g_V \simeq 1$ (see Okun (1982), Sections 5.5 and 5.6). Therefore

$$\langle p|j^h_\mu|n\rangle \simeq \bar{u}_p\gamma_\mu(1 - g_A\gamma^5)u_n\,. \tag{8.129}$$

Using this approximation for the matrix element of the hadronic current, compute the decay width as a function of the electron energy E_e and find the Fermi spectrum of beta decay,

$$\frac{d\Gamma}{dE_e} = \frac{G_F^2\cos^2\theta_C}{2\pi^3}(1+3g_A^2)E_e(\Delta-E_e)^2\sqrt{E_e^2 - m_e^2}\,, \tag{8.130}$$

where $\Delta = m_n - m_p$. Experimentally, g_A and g_V are relatively real (which follows from time-reversal invariance as in Solved Problem 8.3) and $g_A/g_V = 1.261(4)$.

(iv) Plot the result, eq. (8.130), against E_e, and investigate how the end-point of the spectrum changes if the neutrino mass is non-zero.

(v) Compute the total decay rate and compare with the neutron lifetime, $\tau = 1/\Gamma = 885.7 \pm 0.8$ s.

(8.2) Experimentally, the decay $\mu^- \to e^-\gamma$ is not observed and the present limit is

$$\frac{\Gamma(\mu^- \to e^-\gamma)}{\Gamma_{\text{tot}}} < 1.2 \times 10^{-11}\,. \tag{8.131}$$

In the Standard Model, the decay $\mu^- \to e^-\gamma$ is indeed forbidden.

(i) Consider a hypothetical interaction term in the Lagrangian of the form $\mathcal{L} = j_\mu A^\mu$, where j_μ is a conserved vector current with a non-vanishing matrix element between the electron and the muon. Show that, for a process with an on-shell photon, so that $q^2 = 0$, where q^μ the photon four-momentum,

$$\langle e^-|j^\mu(0)|\mu^-\rangle = f_2(0)\bar{u}_e\sigma^{\mu\nu}q_\nu u_\mu + f_3(0)q^\mu\bar{u}_e u_\mu\,, \tag{8.132}$$

where u_e, u_μ are the electron and muon wave functions, and the argument of the form factors is $q^2 = 0$.

(ii) Show that the corresponding matrix element for $\mu^- \to e^-\gamma$ is

$$\mathcal{M}_{fi} = f_2(0)\epsilon^*_\mu q_\nu\bar{u}_e\sigma^{\mu\nu}u_\mu\,. \tag{8.133}$$

Verify that, dimensionally, $f_2(0)$ is the inverse of a mass. From the discussion in Problem 7.2 it follows that this is a magnetic dipole transition. This motivates the definition of the quantity F_2 as $f_2(0) = F_2\, e/(2m_\mu)$.

(iii) Show that the decay rate is

$$\Gamma(\mu^- \to e^-\gamma) = \frac{e^2|F_2|^2}{32\pi}\, m_\mu \left(1 - \frac{m_e^2}{m_\mu^2}\right)^3 . \tag{8.134}$$

To perform the sum over the polarizations of the photon in a covariant way make the replacement

$$\epsilon_\mu^* \epsilon_\nu \to -\eta_{\mu\nu} \tag{8.135}$$

(see Peskin and Schroeder (1995), pages 159-160 for the proof). Compute the experimental bound on $|F_2|$ that follows from eq. (8.131).

(8.3) (i) Estimate the cross-section for neutrino–electron scattering at $s \ll m_W^2$ (with s the square of the CM energy) using only the general form of the Fermi Lagrangian and dimensional considerations.

(ii) Using the explicit form of the currents in the Fermi Lagrangian, write the amplitude for the process $e^-\nu_e \to e^-\nu_e$. Realize that there is both a contribution from neutral currents and one from charged currents. Perform the calculation of the cross-section.

Hint: when summing the contribution to the amplitude from neutral currents and from charged currents, use in the charged current term the *Fierz identity*, valid for any four Dirac spinors wavefunctions $u_1, \ldots, u_4$

$$[\bar{u}_1\gamma^\mu(1-\gamma^5)u_2]\,[\bar{u}_3\gamma_\mu(1-\gamma^5)u_4]$$
$$= [\bar{u}_1\gamma^\mu(1-\gamma^5)u_4]\,[\bar{u}_3\gamma_\mu(1-\gamma^5)u_2]\,. \tag{8.136}$$

This and similar identities between products of fermion bilinears are proved, e.g. in Okun (1982), Section 29.3.4. (Observe that, if instead of using the wave functions $u_1, \ldots, u_4$, we write the identity in terms of the quantized Dirac fields, there is a further minus sign from the anticommutation of the creation and annihilation operators.)

(8.4) (i) Show that, in a medium with target density n (number of particles per unit volume), the interaction rate, that is, the number of scattering events per unit time performed by a given particle, is

$$\Gamma = n\sigma v\,, \tag{8.137}$$

where σ is the cross-section and v the average relative velocity.

(ii) In the early Universe, a particle is in equilibrium with the primordial plasma as long as the interaction rate Γ of the processes that maintains equilibrium is larger than the expansion rate of the Universe, which is given by the Hubble parameter H. In the radiation-dominated era, the Hubble parameter is related to the temperature of the Universe by $H \sim T^2/M_{\rm Pl}$, where $M_{\rm Pl} \sim 10^{19}$ GeV is the Planck mass.

Show that, at a temperature T, for particles with mass $m \ll T$, $n \sim T^3$ and, using the cross-section found in Exercise 8.3, give an estimate of the temperature at which neutrinos decoupled from the rest of the primordial plasma.

Path integral quantization

9

Until now we have used canonical quantization, that is, we have promoted the classical fields to quantum operators. In this chapter we present an alternative formulation of quantum mechanics and of QFT, due to Feynman (who elaborated on work of Dirac) and known as path integral quantization, in which the coordinates in quantum mechanics (QM), and the fields in QFT, remain ordinary functions rather than operators.

Canonical quantization, with its creation and annihilation operators, gives a more immediate understanding of the notion of particle in QFT (or of excitations in condensed matter). However, the path integral technique has a number of other advantages. Namely,

- We have seen that the basic object in the computation of amplitudes in QFT is the vacuum expectation value of the T-products of fields. In the canonical quantization this is expressed in terms of the exponential of the interaction Hamiltonian, see eq. (5.67), and to compute this quantity we expand the exponential order by order in the coupling constant. Indeed, the exponential of an operator is *defined* by its Taylor expansion. This definition of QFT is therefore intrinsically perturbative. Non-perturbative terms, that is, terms non-analytic in the coupling constant g, like $\exp\{-O(1/g^2)\}$, cannot be computed, and in fact they are not even well defined, within canonical quantization.[1]
  The path integral formulation provides instead a definition of the theory which is in principle non-perturbative, and to which it is possible to apply a number of methods, from numerical simulations to semiclassical approximations, that allow us to compute non-perturbative effects. This is especially important in theories, like QCD, where non-perturbative effects can be crucial.
- The path integral formulation shows the existence of deep relations between quantum field theory, statistical mechanics and critical phenomena. This has produced, especially in the 1970s–1980s, a flow of ideas in both directions.
- Even from a computational point of view, there are situations (e.g. in string theory) where the actual computations based on the path integral can be simpler than in the operator formalism.

Therefore, canonical and path integral quantization complement each other, and are both basic tools of QFT. In this chapter we will illustrate this technique and some of its applications even if, within the scope of this course, we will only be able to describe the most elementary aspects.

[1] In particular cases, typically in some models living in 1+1 space-time dimensions, one can obtain non-perturbative results with the operator formalism, using special techniques beyond the scope of this book, but these models are exceptions rather than the rule.

9.1 Path integral formulation of quantum mechanics

The basic idea can be understood directly in non-relativistic quantum mechanics. In this section we denote the operators by a hat, so the position and momentum operators are $\hat{q}$ and $\hat{p}$, respectively, and the Hamiltonian is $\hat{H}(\hat{q},\hat{p})$. For simplicity, we consider a one-dimensional system. As we already recalled in Chapter 5, in the Schrödinger representation the states evolve with time,

$$|\psi_S\rangle(t) = e^{-i\hat{H}t}|\psi_S\rangle(0)\,, \tag{9.1}$$

while the operators are time independent. In the Heisenberg representation the states

$$|\psi_H\rangle \equiv e^{i\hat{H}t}|\psi_S\rangle(t) \tag{9.2}$$

are by construction time independent, and the operators evolve as

$$\hat{A}_H(t) = e^{i\hat{H}t}\hat{A}_S e^{-i\hat{H}t}\,. \tag{9.3}$$

In the Heisenberg representation we denote by $|q,t_i\rangle$ the vector which is an eigenstate of the operator $\hat{q}(t_i)$, with eigenvalue q,

$$\hat{q}(t_i)|q,t_i\rangle = q|q,t_i\rangle\,. \tag{9.4}$$

In the Schrödinger representation, the amplitude for evolving between the state $|q_i\rangle$ at time T_i and the state $|q_f\rangle$ at time T_f is

$$\mathcal{A} = \langle q_f|e^{-i\hat{H}(T_f-T_i)}|q_i\rangle\,. \tag{9.5}$$

In the Heisenberg representation we have

$$e^{i\hat{H}T_i}|q_i\rangle \equiv |q_i,T_i\rangle\,, \qquad e^{i\hat{H}T_f}|q_f\rangle \equiv |q_f,T_f\rangle\,, \tag{9.6}$$

and therefore

$$\mathcal{A} = \langle q_f,T_f|q_i,T_i\rangle\,. \tag{9.7}$$

At any fixed t, the states $|q,t\rangle$ form a complete set,

$$1 = \int dq\,|q,t\rangle\langle q,t|\,, \tag{9.8}$$

where here and in the following it is understood that the integral runs between $-\infty$ and ∞. We choose a set of intermediate values of time $t_0, t_1, \ldots t_N$ with $T_i = t_0 < t_1 < t_2 < \ldots < t_N = T_f$; we take for simplicity the t_m equally spaced, $t_m = t_0 + m\epsilon$, with $N\epsilon = T_f - T_i$. Then we can write

$$\begin{aligned}\mathcal{A} &= \langle q_f,T_f|q_i,T_i\rangle \\ &= \int dq_1\,\langle q_f,T_f|q_1,t_1\rangle\langle q_1,t_1|q_i,T_i\rangle \\ &= \int dq_1 dq_2\,\langle q_f,T_f|q_2,t_2\rangle\langle q_2,t_2|q_1,t_1\rangle\langle q_1,t_1|q_i,T_i\rangle \\ &= \int dq_1 dq_2 \ldots dq_{N-1} \prod_{m=0}^{N-1}\langle q_{m+1},t_{m+1}|q_m,t_m\rangle\,, \end{aligned} \tag{9.9}$$

where in the last line we used the notation $q_i = q_0, q_f = q_N$. Now we evaluate $\langle q_{m+1}, t_{m+1} | q_m, t_m \rangle$ going to the Schrödinger representation,

$$\begin{aligned}\langle q_{m+1}, t_{m+1} | q_m, t_m \rangle &= \langle q_{m+1} | e^{-i\hat{H}t_{m+1}} e^{i\hat{H}t_m} | q_m \rangle \\ &= \langle q_{m+1} | e^{-i\hat{H}\epsilon} | q_m \rangle \\ &= \int dp_m \, \langle q_{m+1} | p_m \rangle \langle p_m | e^{-i\hat{H}\epsilon} | q_m \rangle . \end{aligned} \tag{9.10}$$

We define $\hat{H}(\hat{p}, \hat{q})$ so that, if in the classical Hamiltonian there are terms with products of p and q, in $\hat{H}$ we order $\hat{p}$ on the left of $\hat{q}$. We also normalize the plane wave as $\langle q|p \rangle = e^{iqp}$, which corresponds to the normalization (6.1) in a unit volume. Then

$$\begin{aligned}&\langle q_{m+1}, t_{m+1} | q_m, t_m \rangle \qquad (9.11)\\ &\quad = \int dp_m \, e^{iq_{m+1}p_m} \langle p_m | 1 - i\hat{H}(\hat{p},\hat{q})\epsilon + O(\epsilon^2) | q_m \rangle \\ &\quad = \int dp_m \, e^{iq_{m+1}p_m} \left[1 - iH(p_m, q_m)\epsilon + O(\epsilon^2) \right] \langle p_m | q_m \rangle \\ &\quad = \int dp_m \, e^{i(q_{m+1}-q_m)p_m} \left[e^{-iH(p_m,q_m)\epsilon} + O(\epsilon^2) \right] \\ &\quad = \int dp_m \, \exp\left\{ -i\epsilon \left[H(p_m, q_m) - p_m \frac{q_{m+1} - q_m}{\epsilon} \right] \right\} + O(\epsilon^2) . \end{aligned}$$

Inserting this result into eq. (9.9) we find

$$\begin{aligned}\langle q_f, T_f | q_i, T_i \rangle = &\int dp_0 (dq_1 dp_1) \ldots (dq_{N-1} dp_{N-1}) \qquad (9.12)\\ &\times \exp\left\{ -i\epsilon \sum_{m=0}^{N-1} \left[H(p_m, q_m) - p_m \frac{q_{m+1} - q_m}{\epsilon} \right] \right\} + O(\epsilon^2) . \end{aligned}$$

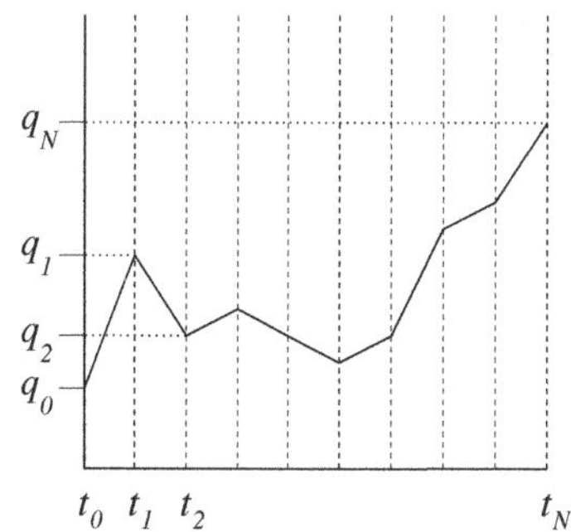

Fig. 9.1 The function $q(t)$ defined by an interpolation of the values $q(t_0) = q_0$, $q(t_1) = q_1, \ldots, q(t_N) = q_N$.

Consider the function $q(t)$ such that $q(t_0) = q_0$, $q(t_1) = q_1, \ldots, q(t_N) = q_N$, and defined for instance by a linear interpolation for intermediate values of t, as in Fig. 9.1. Since in eq. (9.12) we integrate over all possible values of $q_1, \ldots, q_{N-1}$, we are actually integrating over a very large class of functions with fixed boundary values.

Now we want to take the limit $\epsilon \to 0$. In this limit the integral in eq. (9.12) runs over an infinite number of integration variables, and defines a *functional integral*, i.e. an integral over all possible functions $q(t), p(t)$, with the boundary conditions $q(T_i) = q_i, q(T_f) = q_f$. Considering for instance the free-particle case $H = p^2/(2m)$, we see that, after performing the integral over the momenta, we remain with an integral over the dq_m of $\exp\{i(m/2\epsilon)(q_{m+1} - q_m)^2\}$. Therefore the integral over each slice dq_m produces a factor $\sim \epsilon^{1/2}$ and, to obtain a finite limit as $\epsilon \to 0$, in the correct discretization we must associate to each dq_i a factor $\sim 1/\epsilon^{1/2}$ that compensate for this, and also a numerical coefficient that reproduces the normalization factor for the amplitude which is obtained in the operator formalism. Actually, even if it is possible to track in detail the normalization factors (and in non-relativistic quantum mechanics they are important for reproducing the correct amplitudes), we are not be interested in them, since, as we will see below, in QFT they cancel anyhow between numerator and denominator in expressions such as eq. (5.67). We then indicate with $[dpdq]$ the formal limit of the integration measure, including some appropriate normalization factor $\mathcal{N}(\epsilon)$ that we will not need explicitly,

$$[dpdq] = \lim_{\epsilon \to 0} \mathcal{N}(\epsilon)\, dp_0 (dq_1 dp_1) \ldots (dq_{N-1} dp_{N-1}) \,. \tag{9.13}$$

To discuss the non-relativistic limit, it is convenient to reinstate explicitly $\hbar$. Then, letting $\epsilon \to 0$ in eq. (9.12), we get

$$\langle q_f, T_f | q_i, T_i \rangle = \int_{q(T_i)=q_i}^{q(T_f)=q_f} [dpdq] \exp\left\{ \frac{i}{\hbar} \int_{T_i}^{T_f} dt\, [p\dot{q} - H(p,q)] \right\} . \tag{9.14}$$

If the Hamiltonian has the simple form $H(p,q) = p^2/2 + V(q)$ the integral over p is Gaussian and can be performed explicitly. The result (again reabsorbing into the measure $[dq]$ some overall factor) is

$$\boxed{\langle q_f, T_f | q_i, T_i \rangle = \int_{q(T_i)=q_i}^{q(T_f)=q_f} [dq] \exp\left\{ \frac{i}{\hbar} S \right\} ,} \tag{9.15}$$

where S is the classical action,

$$S = \int_{T_i}^{T_f} dt\, L(q, \dot{q}) \,, \qquad L = \frac{1}{2}\dot{q}^2 - V(q) \,. \tag{9.16}$$

The transition amplitude is therefore written as a sum over all possible paths $q(t)$ which satisfy the boundary conditions $q(T_i) = q_i$ and $q(T_f) = q_f$, and each path is weighted by the exponential of i times its action, in units of $\hbar$. A first interesting consequence of this expression is a novel understanding of the classical limit. The classical limit corresponds formally to $\hbar \to 0$, and in this limit the integral can be computed in the stationary phase approximation: the configurations which are far from the extrema of S give contributions which oscillate wildly under small deformations of the path, $q(t) \to q(t) + \delta q(t)$ and therefore these contributions integrate to zero. Then, in the limit $\hbar \to 0$, the only non-vanishing contributions to the integral come from the extrema of the action,

$$\frac{\delta}{\delta q} S = 0 \,. \tag{9.17}$$

This is the equation that defines the classical trajectory in classical mechanics. We have therefore recovered the classical limit: as $\hbar \to 0$, only the trajectories that satisfy the classical equations of motion contribute. We then find a beautiful physical picture, due to Feynman, of quantum mechanics: a particle "explores" all possible paths simultaneously, and the quantum amplitude is obtained by summing over *all* trajectories which interpolate between the initial and the final point. In the classical limit this sum is dominated by the configurations which are solutions of the equations of motion, but when $\hbar$ is not small compared to the scales in question, many other trajectories contribute.

Having understood how to recover the classical limit, we now set again $\hbar = 1$. Consider the insertion into the path integral of a factor $q(t)$, with t given and such that $T_i < t < T_f$,

$$\int_{q(T_i)=q_i}^{q(T_f)=q_f} [dq]\; q(t) e^{iS} . \tag{9.18}$$

Consider a generic path which satisfies the boundary conditions $q(T_i) = q_i, q(T_f) = q_f$. At the intermediate value t it will have a generic value $\bar{q}$. We can write

$$\int_{q(T_i)=q_i}^{q(T_f)=q_f} [dq] = \int_{-\infty}^{\infty} d\bar{q} \int_{q(T_i)=q_i}^{q(t)=\bar{q}} [dq] \int_{q(t)=\bar{q}}^{q(T_f)=q_f} [dq] \;, \tag{9.19}$$

while at the same time the action $\int_{T_i}^{T_f} dt\, L$ can be split as $\int_{T_i}^{t} dt\, L + \int_{t}^{T_f} dt\, L$. Therefore the expression in eq. (9.18) becomes

$$\int_{-\infty}^{\infty} d\bar{q} \left(\int_{q(T_i)=q_i}^{q(t)=\bar{q}} [dq] \exp\left\{ i \int_{T_i}^{t} dt\, L \right\} \right) \bar{q}$$
$$\times \left(\int_{q(t)=\bar{q}}^{q(T_f)=q_f} [dq] \exp\left\{ i \int_{t}^{T_f} dt\, L \right\} \right) . \tag{9.20}$$

However, the first term in parentheses is just $\langle \bar{q}, t | q_i, T_i \rangle$ and the second is $\langle q_f, T_f | \bar{q}, t \rangle$. Therefore

$$\begin{aligned} \int_{q(T_i)=q_i}^{q(T_f)=q_f} [dq]\; q(t)\, e^{iS} &= \int_{-\infty}^{\infty} d\bar{q}\, \langle q_f, T_f | \bar{q}, t \rangle\, \bar{q}\, \langle \bar{q}, t | q_i, T_i \rangle \\ &= \int_{-\infty}^{\infty} d\bar{q}\, \langle q_f, T_f | \hat{q}(t) | \bar{q}, t \rangle \langle \bar{q}, t | q_i, T_i \rangle . \end{aligned} \tag{9.21}$$

Using the completeness relation $\int_{-\infty}^{\infty} d\bar{q} |\bar{q}, t\rangle \langle \bar{q}, t| = 1$ we finally obtain

$$\int_{q(T_i)=q_i}^{q(T_f)=q_f} [dq]\; q(t)\, e^{iS} = \langle q_f, T_f | \hat{q}(t) | q_i, T_i \rangle . \tag{9.22}$$

Therefore the matrix element of the Heisenberg *operator* $\hat{q}(t)$ is obtained computing the path integral of the *function* $q(t)$, with the weight e^{iS}. Consider now

$$\int_{q(T_i)=q_i}^{q(T_f)=q_f} [dq]\; q(t_1) q(t_2) \ldots q(t_n)\, e^{iS} . \tag{9.23}$$

In the path integral the factors $q(t_i)$ are just functions, i.e. c-numbers, and therefore they commute, so we can order them as we prefer. In particular, we can decide to write them as $q(\bar{t}_1) q(\bar{t}_2) \ldots q(\bar{t}_n)$, where $\bar{t}_i$ is the permutation of the t_i such that $\bar{t}_1 > \bar{t}_2 > \ldots > \bar{t}_n$. At this point we can repeat the above argument, splitting the integration over paths from $t = T_i$ to $t = T_f$ into an integration over paths from T_i to $\bar{t}_n$, from $\bar{t}_n$ to $\bar{t}_{n-1}$, etc. and, repeating the same steps as above, we find

$$\int_{q(T_i)=q_i}^{q(T_f)=q_f} [dq]\; q(t_1)q(t_2)\ldots q(t_n)e^{iS} = \langle q_f, T_f|\hat{q}(\bar{t}_1)\ldots\hat{q}(\bar{t}_n)|q_i, T_i\rangle\,. \tag{9.24}$$

However, $\hat{q}(\bar{t}_1)\ldots\hat{q}(\bar{t}_n) = T\{\hat{q}(t_1)\ldots\hat{q}(t_n)\}$ so, finally,

$$\boxed{\int_{q(T_i)=q_i}^{q(T_f)=q_f} [dq]\; q(t_1)\ldots q(t_n)\, e^{iS} = \langle q_f, T_f|T\{\hat{q}(t_1)\ldots\hat{q}(t_n)\}|q_i, T_i\rangle\,.} \tag{9.25}$$

Therefore the integration of a product of functions in the path integral gives automatically the expectation value of the time-ordered product of the corresponding operators.

9.2 Path integral quantization of scalar fields

The extension from quantum mechanics to scalar field theory, at least at a formal level, is rather straightforward. One simply observes that, if we have a multi-dimensional space with coordinates q_j, in eq. (9.15) we must replace $\int [dq]$ with $\int \prod_j [dq_j]$. For a scalar field theory the discrete label j is replaced by the continuous spatial coordinate $\mathbf{x}$, and the role of the $q_j(t)$ is taken by $\phi(\mathbf{x}, t)$. Furthermore the Lagrangian L can be written as $\int d^3x\, \mathcal{L}$. Therefore the amplitude for a transition between a field configuration $\phi_i(\mathbf{x})$ at time T_i and a field configuration $\phi_f(\mathbf{x})$ at time T_f is

$$\langle \phi_f(\mathbf{x}), T_f|\phi_i(\mathbf{x}), T_i\rangle = \int_{\phi(\mathbf{x},T_i)=\phi_i(\mathbf{x})}^{\phi(\mathbf{x},T_f)=\phi_f(\mathbf{x})} \mathcal{D}\phi\, \exp\left\{ i\int_{T_i}^{T_f} d^4x\, \mathcal{L} \right\}, \tag{9.26}$$

where $\mathcal{D}\phi$ is the integration over all field configurations $\phi(\mathbf{x}, t)$ with the given boundary conditions, and $\int_{T_i}^{T_f} d^4x = \int_{T_i}^{T_f} dt \int d^3x$. Observe that $\phi_{i,f}(\mathbf{x})$ are the initial and final values of a field ϕ which evolves with the full equations of motion, rather than with the free KG equations (as the coordinate $q(t)$ of the previous section was a solution of the equations of motion of the complete Hamiltonian). The corresponding Heisenberg operator $\hat{\phi}(x)$ is therefore the full quantum field that appears, for instance, on the left-hand side of eq. (5.67), rather than the interaction-picture field, which instead evolves with the free KG equation.

We now choose as initial and final field configurations the vacuum, $\phi_i(\mathbf{x}) = \phi_f(\mathbf{x}) = 0$ and we send $T_i \to -\infty$, $T_f \to +\infty$. Therefore we find the vacuum-to-vacuum transition amplitude

$$\langle 0, t = +\infty | 0, t = -\infty \rangle = \int \mathcal{D}\phi \, e^{iS} \,, \tag{9.27}$$

where the integral is performed over all field configurations that vanish at $t \to \pm\infty$. Similarly eq. (9.25) gives

$$\langle 0, t = +\infty | T\{\hat{\phi}(x_1) \ldots \hat{\phi}(x_n)\} | 0, t = -\infty \rangle = \int \mathcal{D}\phi \, \phi(x_1) \ldots \phi(x_n) \, e^{iS} \,, \tag{9.28}$$

with $\hat{\phi}$ the full quantum field in the Heisenberg representation. As in Chapter 5, we denote $|0, t = -\infty\rangle$ simply by $|0\rangle$. Then eq. (5.67) becomes

$$\langle 0 | T\{\hat{\phi}(x_1) \ldots \hat{\phi}(x_n)\} | 0 \rangle = \frac{\int \mathcal{D}\phi \; \phi(x_1) \ldots \phi(x_n) \, e^{iS}}{\int \mathcal{D}\phi \; e^{iS}} \,. \tag{9.29}$$

This is the basic formula which connects the operator formalism to the path integral formalism. The left-hand side of eq. (9.29) is the quantity that appears in the LSZ formula (5.46), and in Chapter 5 we understood how to evaluate it in terms of fields in the interaction picture, see eq. (5.67). Now we see that it can also be evaluated computing the functional integral on the right-hand side of eq. (9.29). This means first of all that it should be possible to rederive from the path integral representation the Feynman rules that we obtained using the operator language. We will see in the next section how this can be done. However, the right-hand side of eq. (9.29) in principle is defined also beyond perturbation theory, and in Section 9.4 we will discuss some non-perturbative techniques.

Observe also that until now we have not been careful in collecting the overall numerical factors that define the measure $\mathcal{D}\phi$. However, as we anticipated, these factors cancel between the numerator and the denominator in the right-hand side of eq. (9.29).

9.3 Perturbative evaluation of the path integral

We now want to evaluate perturbatively the right-hand side of eq. (9.29), and verify that we reproduce the same results obtained in the operator formalism. To begin, we consider the numerator in eq. (9.29) for an n-point function in a *free* scalar theory,

$$G(x_1, \ldots, x_n) \equiv \int \mathcal{D}\phi \; \phi(x_1) \ldots \phi(x_n) \, e^{iS} \,, \tag{9.30}$$

with

$$S = \frac{1}{2} \int d^4x \, \left(\partial^\mu \phi \partial_\mu \phi - m^2 \phi^2 \right) \,. \tag{9.31}$$

We first observe that it is not at all evident that the integral so defined makes sense, because the oscillating factor e^{iS} is not necessarily sufficient to provide the convergence of the integration over large fluctuations, i.e. over the field configurations with a large value of the action. One way to solve this problem is to add a small convergence factor $\exp\{-\epsilon \int d^4x\, \phi^2/2\}$ and to take the limit $\epsilon \to 0^+$ later. We therefore consider

$$\int \mathcal{D}\phi\, \phi(x_1)\ldots\phi(x_n)\, \exp\left\{\frac{i}{2}\int d^4x\, \left[\partial^\mu\phi\partial_\mu\phi - (m^2 - i\epsilon)\phi^2\right]\right\} \tag{9.32}$$

and the convergence factor just amounts to the replacement $m^2 \to m^2 - i\epsilon$, exactly as in the Feynman prescription for the propagator; compare with eq. (5.78). We next introduce

$$W[J] \equiv \int \mathcal{D}\phi\, \exp\left\{\frac{i}{2}\int d^4x\, \left[\partial^\mu\phi\partial_\mu\phi - (m^2 - i\epsilon)\phi^2\right] + \int d^4x\, \phi(x)J(x)\right\}. \tag{9.33}$$

$W[J]$ is a functional of the field $J(x)$, and its usefulness comes from the fact that, performing functional differentiations with respect to J, we obtain the Green's functions $G(x_1,\ldots,x_n)$. The *functional derivative* is a formal manipulation defined by the rule

$$\frac{\delta}{\delta J(y)} J(x) = \delta^{(4)}(x-y)\,, \tag{9.34}$$

plus the standard rules for the derivative of composite functions, and by the rule that we can carry it inside the integral sign; this means that

$$\frac{\delta}{\delta J(y)} \int d^4x\, J(x)\phi(x) = \phi(y)\,, \tag{9.35}$$

and therefore

$$G(x_1,\ldots,x_n) = \left(\frac{\delta}{\delta J(x_1)}\cdots\frac{\delta}{\delta J(x_n)} W[J]\right)_{|J=0}. \tag{9.36}$$

For this reason, $W[J]$ is called the generating functional of the Green's functions. To compute $W[J]$ explicitly we proceed as follows. First of all, we write it in momentum space, using

$$\begin{aligned}&\frac{i}{2}\int d^4x\, \left[\partial^\mu\phi\partial_\mu\phi - (m^2 - i\epsilon)\phi^2\right] + \int d^4x\, \phi(x)J(x)\\ &= \frac{1}{2}\int \frac{d^4p}{(2\pi)^4}\left[\tilde\phi(-p)i(p^2 - m^2 + i\epsilon)\tilde\phi(p) + \tilde J(-p)\tilde\phi(p)\right].\end{aligned} \tag{9.37}$$

It can be convenient to work with a finite space-time volume, so the four-momenta are discrete and the integrals are replaced by sums. The integration over all possible functions $\phi(x)$ can be written as an integration over all possible Fourier modes $\tilde\phi(p)$,

$$\mathcal{D}\phi = \prod_p d\tilde\phi(p)\,. \tag{9.38}$$

Proportionality factors in this equation are irrelevant because they cancel in ratios like eq. (9.29), and we can take eq. (9.38) as a definition of the integration measure. The integral can now be performed with the help of a very useful identity valid for Gaussian integrals,

$$\int_{-\infty}^{+\infty} \prod_{i=1}^{N} dy_i \exp\left[-\frac{1}{2}\sum_{i,j=1}^{N} y_i A_{ij} y_j + \sum_{i=1}^{N} y_i z_i\right]$$
$$= (2\pi)^{N/2} (\det A)^{-1/2} \exp\left[+\frac{1}{2}\sum_{i,j=1}^{N} z_i (A^{-1})_{ij} z_j\right], \qquad (9.39)$$

where A_{ij} is an invertible matrix. Coming back to a continuum notation, we find

$$W[J] = W[0] \exp\left\{\frac{1}{2}\int \frac{d^4p}{(2\pi)^4} \tilde{J}(-p)\tilde{D}(p)\tilde{J}(p)\right\}, \qquad (9.40)$$

where

$$W[0] = \int \mathcal{D}\phi \; e^{iS}, \qquad (9.41)$$

$$\tilde{D}(p) = \frac{i}{p^2 - m^2 + i\epsilon}. \qquad (9.42)$$

We recognize that $\tilde{D}(p)$ is the Feynman propagator in momentum space. From the path integral formulation it becomes clear that the propagator is simply the inverse of the operator which appears in the kinetic term. Going back into coordinate space, eq. (9.40) becomes

$$W[J] = W[0] \exp\left\{\frac{1}{2}\int d^4x d^4y \, J(x) D(x-y) J(y)\right\}. \qquad (9.43)$$

Taking the functional derivatives we therefore find all the Green's functions of the free theory. For the two-point function we see that

$$\int \mathcal{D}\phi \; \phi(x_1)\phi(x_2) e^{iS} = \left(\frac{\delta}{\delta J(x_1)}\frac{\delta}{\delta J(x_2)} W[J]\right)_{|J=0} = W[0]\, D(x_1 - x_2) \qquad (9.44)$$

and therefore

$$\frac{\int \mathcal{D}\phi \; \phi(x_1)\phi(x_2) e^{iS}}{\int \mathcal{D}\phi \; e^{iS}} = D(x_1 - x_2), \qquad (9.45)$$

as we expected from eq. (9.29). Taking multiple derivatives we obtain all higher-point Green's functions; for instance,

$$\left(\int \mathcal{D}\phi \; \phi(x_1)\ldots\phi(x_4) e^{iS}\right) \Big/ \left(\int \mathcal{D}\phi \; e^{iS}\right)$$
$$= \left[\frac{\delta^4}{\delta J(x_1)\ldots\delta J(x_4)} \exp\left\{\frac{1}{2}\int d^4x d^4y \, J(x) D(x-y) J(y)\right\}\right]_{J=0}$$
$$= D_{12}D_{34} + D_{13}D_{24} + D_{14}D_{23}, \qquad (9.46)$$

with the usual notation $D_{ij} = D(x_i - x_j)$. We see that we generate all possible connected and disconnected Green's functions, with the same combinatorics of the Wick theorem. Observe also that if we introduce

$$Z[J] \equiv \log W[J] \tag{9.47}$$

in the free theory we have

$$Z[J] = Z[0] + \frac{1}{2}\int d^4x d^4y\, J(x)D(x-y)J(y) \tag{9.48}$$

and therefore $Z[J]$ is the generating functional of the *connected* Green's functions, which in the free case is of course only the two-point function.

Having understood how to compute the path integral in the free theory, it is not difficult to introduce perturbatively the interaction, for instance in the form of a term $V(\phi) = \lambda\phi^4/4!$. The path integral becomes

$$\int \mathcal{D}\phi\; \phi(x_1)\dots\phi(x_n)$$
$$\times \exp\left\{i\int d^4x\left[\frac{1}{2}\left(\partial^\mu\phi\partial_\mu\phi - (m^2 - i\epsilon)\phi^2\right) - \frac{\lambda}{4!}\phi^4\right]\right\} \tag{9.49}$$

and is now non-Gaussian, and cannot be performed exactly. However, we can expand the exponential perturbatively in λ. For example, the contribution to the two-point function to order λ is

$$-i\frac{\lambda}{4!}\int d^4x \int \mathcal{D}\phi\; \phi(x_1)\phi(x_2)\phi^4(x)\, e^{\frac{i}{2}\int d^4x\left[\partial^\mu\phi\partial_\mu\phi - (m^2-i\epsilon)\phi^2\right]}$$
$$= -i\frac{\lambda}{4!}\int d^4x\left(\frac{\delta^6}{\delta^4 J(x)\delta J(x_1)\delta J(x_2)}W[J]\right)_{|J=0}, \tag{9.50}$$

where $W[J]$ is the generating functional of the *free* theory computed above. One can then check that we recover the Wick theorem and the same results that we obtained from the operator formalism. For a detailed discussion, see Ramond (1990), Chapter 4. As in the free theory, using $Z[J]$ instead of $W[J]$ we generate directly the *connected* Green's function.

9.4 Euclidean formulation

We have seen that the perturbative expansion of the path integral reproduces the expansion in terms of Feynman graphs. The greatest virtue of the path integral approach, however, is that it allows us to give a non-perturbative *definition* of a field theory, and it actually makes possible the computation of non-perturbative effects.

To this end, it is convenient to ensure the convergence of the path integral in a different way. Rather than adding a convergence factor $\exp\{-\epsilon\int d^4x\, \phi^2/2\}$, which, as we have seen, corresponds to the Feynman prescription for the propagator, we rotate the theory in Euclidean space. That is, we define the Euclidean time t_E as

$$t_E = it\,. \tag{9.51}$$

Therefore $d^4x = dtd^3x = -idt_E d^3x \equiv -i(d^4x)_E$ and

$$\frac{\partial}{\partial t} = i\frac{\partial}{\partial t_E}\,. \tag{9.52}$$

The action of a scalar field, with a generic potential $V(\phi)$, becomes

$$\begin{aligned}
S &= \int d^4x \left[\frac{1}{2}\left(\partial^\mu\phi\partial_\mu\phi - m^2\phi^2\right) - V(\phi)\right] \\
&= \int d^4x \left[\frac{1}{2}(\partial_t\phi)^2 - \frac{1}{2}(\partial_i\phi)^2 - \frac{1}{2}m^2\phi^2 - V(\phi)\right] \\
&= -i\int (d^4x)_E \left[-\frac{1}{2}(\partial_{t_E}\phi)^2 - \frac{1}{2}(\partial_i\phi)^2 - \frac{1}{2}m^2\phi^2 - V(\phi)\right] \\
&\equiv iS_E\,,
\end{aligned} \tag{9.53}$$

where we have defined the *Euclidean action* S_E,

$$\begin{aligned}
S_E &= \int (d^4x)_E \left[\frac{1}{2}(\partial_{t_E}\phi)^2 + \frac{1}{2}(\partial_i\phi)^2 + \frac{1}{2}m^2\phi^2 + V(\phi)\right] \\
&= \int d^4x \left[\frac{1}{2}\partial_\mu\phi\partial_\mu\phi + \frac{1}{2}m^2\phi^2 + V(\phi)\right].
\end{aligned} \tag{9.54}$$

In the last line we suppressed the subscript E from Euclidean quantities since in this section, from now on, only Euclidean quantities will appear, and we have used the convention that repeated Lorentz indices, both lower or both upper, are summed with the Euclidean metric $\eta^E_{\mu\nu} = (+,+,+,+)$, e.g.

$$\partial_\mu\phi\partial_\mu\phi \equiv (\partial_{t_E}\phi)^2 + (\partial_i\phi)^2\,. \tag{9.55}$$

The potential $V(\phi)$ must be bounded from below, otherwise the theory does not have a stable ground state. Without loss of generality, we can shift the potential by an arbitrary constant so that $V(\phi) \geqslant 0$. Then we see that the Euclidean action is positive definite and the factor

$$e^{iS} = e^{-S_E} \tag{9.56}$$

ensures the convergence of the integration over large fluctuations, since paths which have a very large action are exponentially suppressed. Therefore the path integral over Euclidean field configurations $\phi(t_E, \mathbf{x})$ is well defined, and we can compute the Green's function in Euclidean space. The Green's functions in Minkowski space can then be reconstructed by analytic continuation.[2] So the basic quantity that we want to compute is

$$\boxed{\int \mathcal{D}\phi\; \phi(x_1)\ldots\phi(x_n)\, e^{-S}\,,} \tag{9.57}$$

[2] It can be shown that analytic continuation is possible to all orders in perturbation theory.

where $x_1, \dots x_n$ are points in four-dimensional Euclidean space. Of course, this path integral can be computed first of all perturbatively, expanding in powers of the coupling constants. This gives a set of Feynman rules in Euclidean space. For instance, repeating the derivation of the previous section, we find that the Euclidean propagator in momentum space is

$$\tilde{D}_E(p) = \frac{1}{p^2 + m^2}, \tag{9.58}$$

where $p^2 = p_0^2 + \mathbf{p}^2$. Similarly, a vertex in $\lambda\phi^4$ theory carries a factor $-\lambda$ instead of $-i\lambda$. These Feynman rules allow us to compute perturbatively the Euclidean Green's functions which then, upon analytic continuation, reproduce the Green's functions obtained with the Minkowski-space Feynman rules.

The perturbative expansion corresponds to studying the contributions coming from the "small" field configurations, i.e. from the fields which do not differ much from the perturbative vacuum $\phi(x) = 0$. However, the Euclidean path integral formulation allows us to go further, and to study the contribution of field configurations far from the perturbative vacuum.

One possible approach is to discretize space-time and study the Euclidean path integral on a space-time lattice, using furthermore a finite volume. Then we have a finite number of lattice sites labeled by x, and a field configuration is specified by a finite collection of variables, i.e. by the value ϕ_i that the field takes at the point x_i, where the index i labels the lattice sites. The integration measure can be taken to be

$$\mathcal{D}\phi = \prod_i d\phi_i \,. \tag{9.59}$$

The functional integral is then reduced to an ordinary multiple integral, which in principle can be studied numerically. Actually, a direct numerical evaluation of an integral over a very large number of integration variables is out of the question, but there is a technique, known as Monte Carlo simulation, which consists in generating a set of field configurations with a probability distribution e^{-S}, and the functional integral is replaced by the sum over this set of configurations. This approach is extremely useful for gauge theories, since the theory on the lattice can be defined so that it preserves gauge invariance.

There is then the problem of extrapolating the results to the continuum limit (and to take the infinite-volume limit). This is done using the renormalization group techniques discussed in Section 5.9 and in the next section. This approach has the advantage of providing a method for computing the exact, non-perturbative, Green's functions in a way that in principle is not subject to any perturbative approximation. The actual limit, however, comes from computer capabilities, especially in the presence of fermions.[3]

Another possible approach is based on a semiclassical expansion, and we discuss it in the Solved Problems section.

[3]The path integral over fermionic fields is defined using anticommuting variables, rather than ordinary c-numbers. We do not expand on this, but the interested reader can see e.g. Ramond (1990), Chapter 5.

9.5 QFT and critical phenomena

Another virtue of the path integral formulation is to show the existence of deep connections between QFT and critical phenomena, as was understood in particular by K. Wilson. This connection emerges from the Euclidean formulation of the path integral. Performing the rotation to Euclidean space, we have seen that the right-hand side of eq. (9.29) becomes

$$\frac{\int \mathcal{D}\phi\, \phi(x_1)\dots\phi(x_n)\, e^{-S}}{\int \mathcal{D}\phi\, e^{-S}}\,, \tag{9.60}$$

where S is the Euclidean action and the integral runs over all configurations that go to zero as the Euclidean time $t_E \to \pm\infty$.[4] Formally, eq. (9.60) is identical to a statistical average of a classical system living in four *spatial* dimensions, in which each configuration is weighted with e^{-S}. In the language of statistical mechanics, this is called a *correlation function.* Therefore the (Minkowskian) Green's function of quantum field theory in three spatial and one time dimensions (or more generally in d spatial dimensions plus time), $\langle 0|T\{\hat{\phi}(x_1)\dots\hat{\phi}(x_n)\}|0\rangle$, can be obtained by computing the correlation functions of a classical statistical system living in four spatial dimensions (or more generally in $d+1$ spatial dimensions), and performing the analytic continuation back to Minkowski space.

[4]Observe that, if the vacuum is non-degenerate, i.e. if in the action there is a potential $V(\phi)$ with just one minimum (that we can always set to be at $\phi = 0$ with a shift of ϕ, and to have the value $V(0) = 0$ with a shift in V) then only the fields that satisfy $\phi(t, \mathbf{x}) \to 0$ as $|\mathbf{x}| \to \infty$ have a finite action, otherwise $\int d^3x V$ diverges at large $|\mathbf{x}|$. Then the only non-vanishing contribution to the functional integral comes from field configurations that vanish at infinity in all four Euclidean directions.

We want to investigate what kind of statistical system can reproduce the properties of a QFT. As discussed in the previous section, to compute the path integral in eq. (9.60) it is convenient to put the system in a finite volume and to discretize the four-dimensional Euclidean space, using for instance a four-dimensional lattice with lattice spacing a. Then the system has a finite number of variables $\phi_i = \phi(x_i)$, corresponding to the lattice sites, and the integration measure in the path integral is given by eq. (9.59). When the number of lattice sites is finite, the path integral is a well-defined statistical sum. The question is how to take the continuum limit so that a non-trivial QFT emerges.

To answer this question consider the two-point function in the massive theory. The Euclidean propagator, in momentum space, is $1/(p^2+m^2)$, and therefore the correlation function of the corresponding statistical system is

$$\langle\phi(x)\phi(0)\rangle = \int \frac{d^4p}{(2\pi)^4}\frac{e^{ipx}}{p^2+m^2}\,, \tag{9.61}$$

where $|p|^2 = p_\mu p_\mu$ and $px = p_\mu x_\mu$, with all contractions performed with the Euclidean metric $\delta_{\mu\nu}$. The exact computation of the integral is involved, but for $m|x| \gg 1$ the result is proportional to $(1/|x|^2)e^{-m|x|}$. Consider a point x separated from the origin by n lattice sites, say along the first axis. Then $|x| = na$ and for $n \gg 1$, neglecting prefactors,

$$\langle\phi_n\phi_0\rangle \sim e^{-amn}\,. \tag{9.62}$$

In statistical mechanics, the behavior $\langle\phi_n\phi_0\rangle \sim e^{-n/\xi}$ defines the (dimensionless) correlation length ξ, so in our case

$$\xi = \frac{1}{am} \,. \tag{9.63}$$

We see that, if we want to take the continuum limit $a \to 0$ while keeping the physical mass m fixed, the correlation length ξ must go to infinity.

The correlation length is a function of the couplings of the statistical system. This means that, in order to obtain a continuum QFT in which the physical masses, such as m in eq. (9.63) (and similarly the physical couplings) are finite, we must tune the parameters of the corresponding statistical system so that the correlation length diverges. Here we are rephrasing, in the language of statistical mechanics, the renormalization procedure introduced in Chapter 5: the dependence of the bare couplings on the cutoff (here the lattice spacing a) is tuned so to obtain finite values for the renormalized masses and couplings.

A divergent correlation length is characteristic of critical systems, and therefore removing the cutoff in a QFT is analogous to tuning a statistical system toward a critical point. Considering quantum field theories as critical statistical systems allows us to use a whole body of physical intuition stemming from statistical mechanics. In particular, we will discuss in this section how to obtain a very physical understanding of the renormalization group (RG) equations, like the Callan–Symanzik equation, that we derived with rather formal arguments in Section 5.9. To make the connection, we first briefly recall some basic facts about critical phenomena.

A classic example is the two-dimensional Ising model, defined placing a "spin" variable $s_i = \pm 1$ on each site i of a two-dimensional lattice, and taking as Hamiltonian

$$H = -J \sum_{i,j} s_i s_j \,, \tag{9.64}$$

where the sum runs only over nearest-neighbor pairs (i,j). We take $J > 0$, so the interaction tends to align the spins, i.e. it is a ferromagnetic coupling. The statistical mechanics of the Ising model is obtained from the partition function $Z = e^{-\beta F} = \mathrm{tr}\, e^{-\beta H}$, with $\beta = 1/k_B T$, and therefore is governed by the dimensionless parameter

$$K \equiv \frac{J}{k_B T} \,. \tag{9.65}$$

As with any statistical system, at fixed temperature T and fixed number of particles the equilibrium state is given by the minimum of the free energy $F = E - TS$, and will be the result of the competition between the tendency to minimize the energy E, so in our case to align the spins, and the tendency to maximize the entropy S, so to disorder the system. At small T the tendency to minimize energy is more important while increasing T the tendency toward maximization of the entropy becomes progressively more relevant. In the Ising model this competition leads to a phase transition: below a critical temperature T_c (or, more precisely below a critical value of $k_B T/J$) the system develops a spontaneous magnetization, that is $M \equiv \langle s_i \rangle \neq 0$, while above T_c the magnetization M vanishes (see Fig. 9.2). As $T \to T_c$ from below, the magnetization goes to zero as

$$M \sim (T_c - T)^{\beta}, \tag{9.66}$$

and β is known as a critical index. In the two-dimensional Ising model $\beta = 1/8$ but this kind of behavior is quite general for critical phenomena.

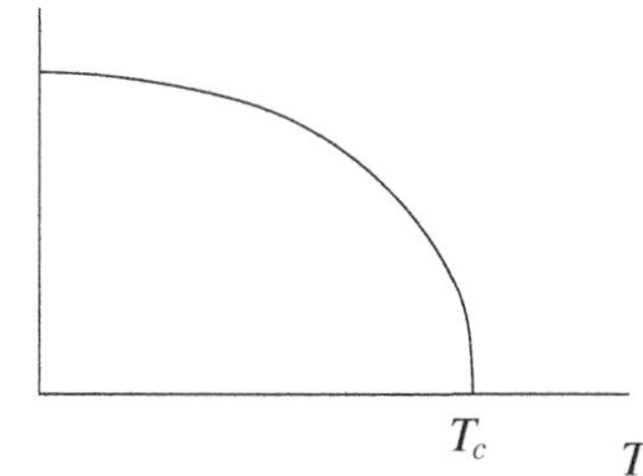

Fig. 9.2 The magnetization as a function of temperature in the Ising model.

In the magnetized phase there is long-range order. Since the spins tend to be aligned, if the spin s_0 at the site 0 has the value $s_0 = +1$ then, even if the site n is very far from the site 0, the probability of finding $s_n = +1$ is higher than the probability of having $s_n = -1$. Therefore there is no exponential decay of the correlation function (there will be, rather, a power-law fall off) or, in other words, the correlation length is infinite. In the disordered phase $T > T_c$, instead, ξ is finite. As $T \to T_c$ from above, ξ diverges as

$$\xi \sim \frac{1}{(T - T_c)^{\nu}}, \tag{9.67}$$

with ν another critical index ($\nu = 1$ in the two-dimensional Ising model). This behavior is illustrated in Fig. 9.3.

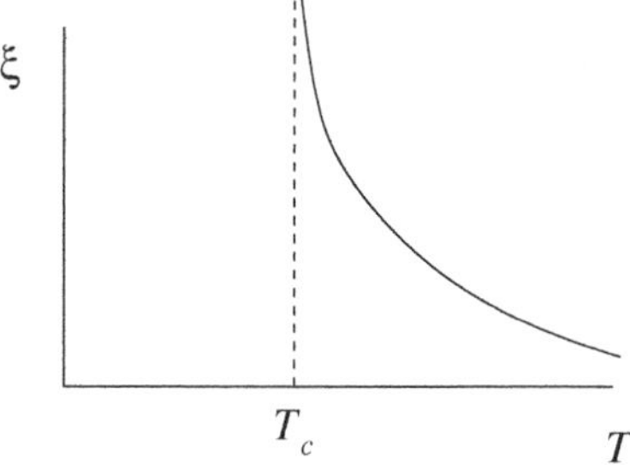

Fig. 9.3 The correlation length as a function of temperature in the Ising model.

We are now ready to discuss the renormalization group (RG) equations in the language of critical phenomena. First of all, observe that in QFT the lattice spacing (as any other UV cutoff) is not a physical quantity, and we are only interested in the limit $a \to 0$. For this reason, it is difficult to have an intuitive picture of how the bare theory should depend on the cutoff. In statistical mechanics, instead, the lattice spacing is a physical quantity, for instance of order 10^{-8} cm in a typical condensed matter system. The correspondence between QFT and statistical mechanics provided by the path integral is, more precisely, a correspondence between the *bare* Green's functions of QFT (which are objects about which it is difficult to form an intuitive physical picture) and the *physical* correlation functions of the statistical system. It is therefore not surprising that in statistical systems the RG equations have a more intuitive interpretation.

Consider for example the Ising model; the Hamiltonian (9.64) gives a *microscopic* description of the system in terms of interactions between nearest-neighbor spins separated by a distance a. However, we are usually not interested in the details of the interaction on a microscopic scale. We are more interested in understanding what happens at the macroscopic scale. To go from a microscopic to a macroscopic description is a non-trivial problem, since cooperative effects between the spins can take place. This is especially important when the correlation length ξ is large because, even if the fundamental interaction involves only nearest-neighbor pairs, the influence of each spin on the others is propagated, via this nearest-neighbor interaction, across a distance of order ξ.

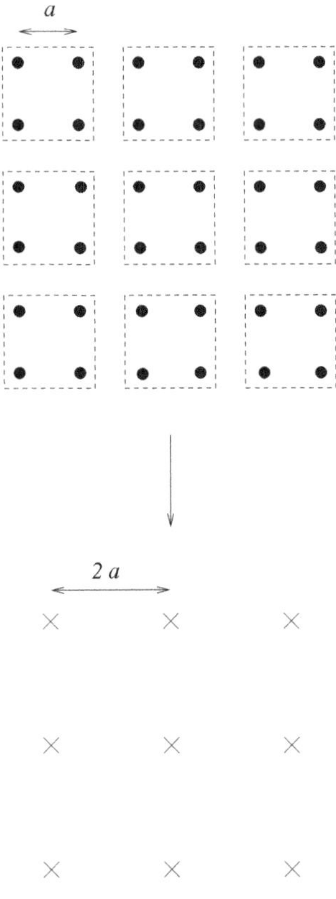

Fig. 9.4 The block spin transformation.

[5]Actually, there are better formulations of the RG transformation where the calculations can be done more explicitly and systematically. For instance, for a scalar field $\phi_i = \phi(x_i)$ one can integrate over the original variables in the path integral, with a Dirac delta of the form $\delta(\Phi_n - \sum_{i\in n}\phi_i)$, which forces the four variables ϕ_i in the n-th square to have a given value Φ_n, so we obtain an effective theory in terms of the "block fields" Φ_n. Alternatively, we can work in momentum space, using the momentum modes ϕ_k, regularizing the theory with a cutoff Λ which restricts the Euclidean momenta to $|k| < \Lambda$. Then, in the partition function Z, we integrate explicitly over the modes ϕ_k in a shell $\Lambda - \delta\Lambda < |k| < \Lambda$, remaining with an effective action for the low-momentum modes. We do not enter into these technical aspects, and we limit ourselves to illustrate the main ideas using Kadanoff's block spin transformation, which is rather intuitive.

To construct an effective Hamiltonian which describes the physics at the macroscopic scale we can proceed in steps, "integrating out" the details of the model at short scales. A possible way to do it, in the Ising model, is via a "block spin" transformation, first introduced by Kadanoff. Regroup the lattice sites into squares made of four adjacent spins, as in the upper part of Fig. 9.4. If $\xi \gg 1$, nearest-neighbor spins must be very strongly correlated. This means that, defining the variables $S_n = (1/4)\sum_{i\in n} s_i$, where the sum is over the four adjacent spins in the n-th square, each S_n will most of the time take the values ± 1, as the original spin variables s_i, and therefore we can consider them as effective spin variables, living on the centers of their respective squares, indicated by a cross in the lower part of Fig. 9.4.[5] Since the original spins s_i had only nearest-neighbor interaction, in a first approximation also these "block spin" variables will interact among themselves only with a nearest-neighbor interaction, but their effective coupling constant in general will not be the same as the coupling K which appears in eqs. (9.64) and (9.65). Rather, the coupling will have a value $K_2 \equiv f(K)$, for some function $f(K)$, determined by the condition that the partition function is preserved,

$$\mathrm{Tr}_{\{S_n\}} e^{-\beta H'[S_n]} = \mathrm{Tr}_{\{s_i\}} e^{-\beta H[s_i]}, \tag{9.68}$$

so that the theory obtained with a block spin transformation describes the same macroscopic physics. In general, $H'[S_n]$, as determined by the above equation, will contain also interaction terms that are not present in H, so the evolution of the couplings should really be followed in a multi-parameter space of coupling constants, rather than being restricted to the coupling K. For the moment, we neglect this aspect to simplify the presentation of the main ideas.

If K describes the interaction on a microscopic scale a, K_2 describes the effective interaction on the scale $2a$. We can then iterate the procedure, performing this block spin transformation over the S_n variables, again regrouping four of them together; since the Hamiltonian still has the same form as in eq. (9.64), with just the replacement $K \to f(K)$, the effective coupling at the scale $4a$ will be

$$K_3 \equiv f(K_2) = f(f(K)). \tag{9.69}$$

We can iterate the procedure n times, until $2^n a$ becomes almost of order $a\xi$. What we gain by this procedure is that at each stage the number of effective spin variables which are strongly correlated among themselves has been thinned. In fact, in terms of the variable S_n, the new (dimensionless) correlation length is $\xi_1 = \xi_0/2$ (where ξ_0 is the correlation length of the original variables s_i), since the spacing between the spins S_n is $2a$. After n steps the correlation length is $\xi_n = \xi_0/2^n$. When this number becomes of order one, collective effects do not play any role, and the physics of the system can simply be read off the effective Hamiltonian. Therefore the coupling K_n obtained iterating n times the block spin transformation, i.e. the coupling defined by

$$K_n = f(K_{n-1}), \tag{9.70}$$

with the initial condition $K_1 = K$, is the effective coupling which describes the physics at the length-scale $l = 2^n a$.

The relation with the renormalization group equations discussed in Section 5.9 now emerges. What we find, either using the QFT language in Section 5.9 or the language of statistical mechanics in this section, is that the behavior of the theory is described by effective coupling constants that depend on a length-scale l (or equivalently on an energy-scale $1/l$). In statistical mechanics l represents the scale over which microscopic fluctuations have been averaged out while, in QFT, $1/l$ is the cutoff in momentum space. All the necessary information is encoded in the functions that describe how the couplings change as we change the scale: the beta function in Section 5.9 or the function $f(K)$ in this section.

We see that the problem of taking into account the collective action of many degrees of freedom when the correlation length is large has been translated into the problem of computing the function $f(K)$. This can be a very difficult task, so it might seem that we have simply translated a difficult problem into another difficult problem. However, the power of the method emerges in connection with the notion of fixed point. A fixed point K_c is defined as a solution of the equation

$$K_c = f(K_c)\,. \tag{9.71}$$

Observe that ξ_n is a function of K_n, i.e. $\xi_n = \xi(K_n)$. Since $\xi_{n+1} = \xi_n/2$, the function $\xi(K)$ must satisfy

$$\xi[f(K)] = \frac{1}{2}\xi(K)\,. \tag{9.72}$$

At the fixed point $K_c = f(K_c)$ so $\xi(K_c) = (1/2)\xi(K_c)$, which has two solutions, $\xi(K_c) = \infty$ and $\xi(K_c) = 0$. We see from eq. (9.63) that the latter case is of no interest for constructing an interacting continuum QFT, and is called a "trivial" fixed point. The fixed points corresponding to $\xi(K_c) = \infty$ are instead called "critical".

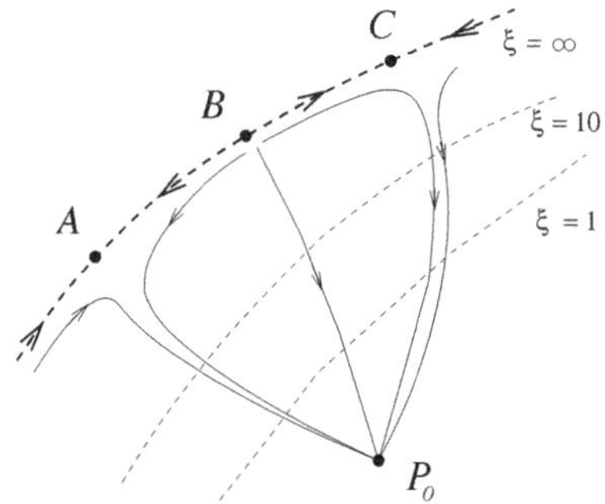

Fig. 9.5 An example of RG flow with three fixed points A, B, C on the critical surface and a fixed point P_0 at $\xi = 0$.

In a general system, the space of coupling constants will be multidimensional, and one must also take into account that, even if a term is not originally present in the microscopic Hamiltonian, it will be generated by the RG transformation, unless it is protected by some symmetry. Therefore we will have a RG flow in this multidimensional (actually, infinite-dimensional) space. However, most of the couplings will be simply driven to zero by the RG transformation. The operators with whom they are associated in the Lagrangian are termed *irrelevant*.

In the space of the remaining couplings, the RG flow will be basically determined by the fixed-point structure. Barring more exotic possibilities, like chaotic behavior, the RG trajectory will flow either to infinity or toward the fixed points.

A hypothetical example of a flow in a two-dimensional parameter space is shown in Fig. 9.5. In this figure we have drawn the critical surface, that is the surface in the parameter space where $\xi = \infty$, and other surfaces at constant ξ. We have considered a situation in which on the critical surface there are three fixed points, labeled A, B and C. Furthermore, there is a trivial fixed point P_0 with $\xi = 0$. Each point on this graph corresponds to a set of values for the couplings, i.e. a physical Hamiltonian in the language of statistical mechanics, or a bare QFT in field-theoretical language. Since under a RG step $\xi \to \xi/2$, if we apply the transformation to a point on the critical surface we remain on the critical surface, while if we start at ξ finite we will be eventually driven to the fixed point at $\xi = 0$.

A very important property of a fixed point is its stability. In Fig. 9.5 P_0 is a stable fixed point: the flow is such that, if we start infinitesimally close to P_0 in any direction, the RG transformation will bring us toward P_0. All fixed points on the critical surface have instead at least one unstable direction, which is the direction orthogonal to the critical surface itself: if we start close to the critical surface, but not exactly on it, the RG transformation decreases ξ and drives us further away from the critical surface. The fixed points A and C have one stable direction along the critical surface, while B has instead a second unstable direction.

Points that are on the same RG trajectory describe the same macroscopic physics, by construction of the RG transformation, see eq. (9.68). In principle, each point on the critical surface is suitable for removing the cutoff and defining a renormalized QFT, since there $\xi = \infty$ and $a = 0$. However, these theories are not all physically different. Since a RG trajectory connects equivalent theories, all the points on the critical surface that lie in the attraction basin of the same fixed point correspond to equivalent theories. In other words, the space of possible renormalized QFT is split into equivalence classes, known as *universality classes*, in one-to-one correspondence with the fixed points on the critical surface.

Universality is a powerful concept to explain why statistical systems with very different microscopic Hamiltonians turn out to have the same critical behavior, and in particular the same critical indices. For instance, for a generic system with one coupling constant K, we have seen in eq. (9.67) that near the critical coupling K_c

$$\xi(K) \simeq \frac{c}{(K-K_c)^{\nu}}\,, \tag{9.73}$$

with c a constant. Combining this with eq. (9.72), we have

$$\lim_{K\to K_c}\left(\frac{f(K)-K_c}{K-K_c}\right)^{\nu} = 2\,. \tag{9.74}$$

Close to the critical point, $f(K) \simeq K_c + f'(K_c)(K - K_c)$ and, substituting into eq. (9.74), we get $[f'(K_c)]^{\nu} = 2$, or

$$\nu = \frac{\log 2}{\log f'(K_c)}\,. \tag{9.75}$$

This shows that the critical index ν depends only on the form of the function $f(K)$ near the critical point, and not on the microscopic details of the Hamiltonian. In particular, all Hamiltonians that lie on the same RG trajectory describe systems that, even if apparently very different on the microscopic scale, have the same value of ν. The above results also nicely illustrate how the non-analytic behavior (9.73) emerges as a result of collective behavior, from a regular Hamiltonian and an analytic function $f(K)$.

In the context of QFT, universality means first of all that (at least within a universality class) the renormalized theory does not depend on the details of the regularization. Universality however is important also for a different reason. In Section 5.9 we saw that we can write the RG equations in two different forms:

(1) We can write them as equations that govern the dependence of the bare coupling constants on the cutoff, as in eq. (5.164). With a lattice cutoff $\Lambda = 1/a$, eq. (5.164) reads

$$a\frac{dg_0}{da} = -\beta(g_0(a)), \tag{9.76}$$

where, in the general case of many coupling constants, g_0 and β are vectors in the coupling constant space.

(2) We can write the RG equations in the form of equations that govern the dependence of the renormalized coupling constants on the energy. From eq. (5.178), setting $E = u\mu$ (where μ is the reference scale),

$$E\frac{dg_{\text{eff}}}{dE} = \beta(g_{\text{eff}}(E)). \tag{9.77}$$

Both eqs. (9.76) and (9.77) originated from eq. (5.161), in the former case taking the derivative with respect to the cutoff and in the latter with respect to the renormalization point μ.

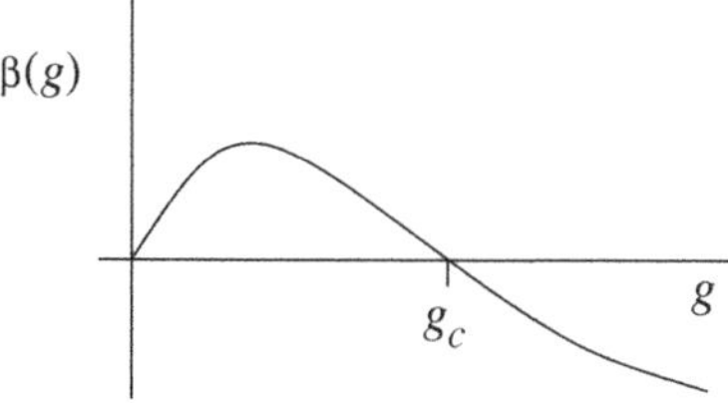

Fig. 9.6 An example of a beta function with two zeros, such that $g = 0$ is an IR fixed point and $g = g_c$ is a UV fixed point.

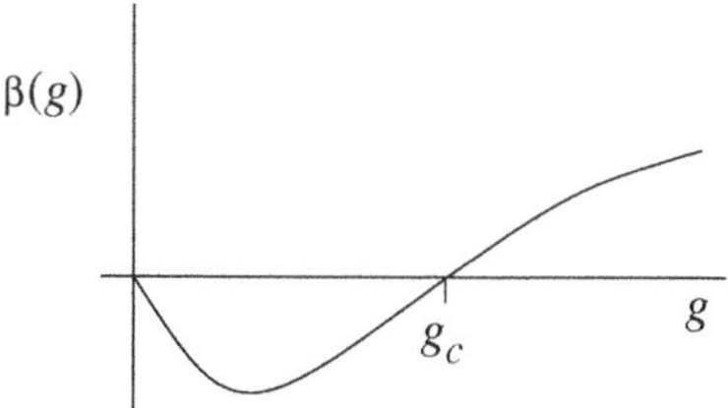

Fig. 9.7 Reversing the sign of the beta function, $g = 0$ becomes a UV fixed point and $g = g_c$ an IR fixed point.

Equation (9.76), as we have seen, can be understood in the language of critical phenomena, because the cutoff a can be interpreted as the physical lattice spacing of a condensed matter system, and a change in cutoff as the result of taking into account collective effects. However, eq. (9.77) has the same form as eq. (9.76), so we can again apply the notions of fixed points, universality, etc. While in eq. (9.76) the fixed points are at $\xi = 0, \infty$ (i.e. when the momentum space cutoff $1/a$ is 0 or ∞), in eq. (9.77) they will be at $E = 0$ or at $E = \infty$. In eq. (9.76) only the fixed points at $\xi = \infty$ were interesting in the QFT context. In eq. (9.77), instead, both types of fixed points are very interesting; a fixed point at $E = 0$ (called an infrared, or simply IR, fixed point) can govern the behavior of a theory at low energies while a fixed point at $E = \infty$ (called a UV fixed point) will be relevant at high energies.

This can be understood from Figs. 9.6 and 9.7, where for simplicity we take the space of coupling constants to be one-dimensional. We consider a theory with a beta function which vanishes at $g_{\text{eff}} = 0$ (the perturbative fixed point, which always exists) and which furthermore has a zero at a value $g_{\text{eff}} = g_c$. In Fig. 9.6 the beta function is positive for $0 < g_{\text{eff}} < g_c$ and negative for $g_{\text{eff}} > g_c$. From this, it follows that a solution of eq. (9.77), as $E \to \infty$, will always be attracted toward $g_{\text{eff}} = g_c$, independently of the initial value $g(\mu)$. In fact, if $0 < g(\mu) < g_c$, $\beta(g_{\text{eff}}) > 0$, therefore $dg_{\text{eff}}/dE > 0$ and $g_{\text{eff}}(E)$ increases asymptotically toward g_c. Conversely, if the initial value $g(\mu) > g_c$, the beta function is negative and $g_{\rm eff}(E)$ decreases toward g_c. Therefore in this example g_c is a UV fixed point. Universality manifests itself in the fact that the large-energy properties of the theory are governed by this value of the coupling, irrespectively of the initial value of $g(\mu)$.

To study the limit $E \to 0$, instead, we can run the RG flow backward and, again using Fig. 9.6, we see that, if $0 < g(\mu) < g_c$, $g_{\rm eff}(E)$ runs toward zero as $E \to 0$ while, if $g(\mu) > g_c$, $g_{\rm eff}(E)$ runs toward infinity. Therefore $g = 0$ is an IR fixed point (with attraction domain $0 < g(\mu) < g_c$), and all theories in this universality class become free at low energies.

In Fig. 9.7 we have reversed the sign of the beta function, and the same analysis as above shows that now $g = 0$ is a UV fixed point with attraction domain $0 < g(\mu) < g_c$, while in the infrared the theory flows at g_c.

In QCD, the beta function has only the perturbative zero at $g = 0$, and is negative for $g > 0$. There is no other fixed point. Therefore the theory in the UV flows to $g = 0$, which is the property of asymptotic freedom that we already mentioned in Section 5.9. In the IR limit the coupling grows large and enters in the strong coupling domain, where perturbative methods cannot be applied.

9.6 QFT at finite temperature

Another interesting application of the path integral technique is to QFT at finite temperature. To understand this relation we consider first quantum mechanics and we start from eq. (9.15), written using the Schrödinger representation,

$$\langle q_f | e^{-i\hat{H}(T_f - T_i)} | q_i \rangle = \int_{q(T_i)=q_i}^{q(T_f)=q_f} [dq] \, e^{iS} \,. \tag{9.78}$$

We now rotate from Minkowski to Euclidean space, $t \to -it$, so that $\exp\{-iHt\}$ becomes $\exp\{-Ht\}$, while $e^{iS/\hbar}$ becomes $e^{-S/\hbar}$, as discussed in the previous section. In this section, the notation t will hereafter denote Euclidean time. Then eq. (9.78) becomes

$$\langle q_f | e^{-\beta\hat{H}} | q_i \rangle = \int_{q(T_i)=q_i}^{q(T_f)=q_f} [dq] \, e^{-S} \,, \tag{9.79}$$

where $\beta \equiv T_f - T_i$ and T_i, T_f are the minimum and maximum values of Euclidean time; S is the Euclidean action so, for a particle in a potential V,

$$S = \frac{1}{2} m\dot{q}^2 + V(q) \,. \tag{9.80}$$

We now take $q_i = q_f \equiv q$ and we sum over all possible values of q,

$$\sum_q \langle q | \, e^{-\beta\hat{H}} | q \rangle = \int_{q(t)=q(t+\beta)} [dq] \, e^{-S} \,, \tag{9.81}$$

where the sum is over all possible configurations with $q(T_i) = q(T_f)$ or, in other words, over all periodic configurations with period $\beta = T_f - T_i$. Since the states $|q\rangle$ form a complete set, the left-hand side is the trace of $e^{-\beta\hat{H}}$ over the Hilbert space, so we find the relation

$$\mathrm{tr}\, e^{-\beta\hat{H}} = \int_{q(t)=q(t+\beta)} [dq]\; e^{-S} . \tag{9.82}$$

For a scalar field theory we can perform the same steps with just a change in notation, and we get

$$\mathrm{tr}\, e^{-\beta\hat{H}} = \int_{\phi(\mathbf{x},t)=\phi(\mathbf{x},t+\beta)} \mathcal{D}\phi\; e^{-S} , \tag{9.83}$$

with S the Euclidean action given in eq. (9.54) and H the field theory Hamiltonian. The left-hand side of eq. (9.82) or eq. (9.83) is the thermal partition function of a system with Hamiltonian H, at a temperature given by $k_B T = 1/\beta$. This means that the thermal averages of the system can be computed using the path integral, restricting to paths periodic in Euclidean time.

From a practical point of view, in quantum mechanics it is simpler to compute directly the trace on the left-hand side of eq. (9.82) using the operator formalism. In QFT, instead, it is usually much simpler to evaluate the path integral in eq. (9.83).

9.7 Solved problems

Problem 9.1. Instantons and tunneling

It happens in many theories that, beside the trivial solution $\phi = 0$, there are other, non-trivial, solutions of the Euclidean equations of motion, which vanish at infinity. A typical situation is given by solutions which describe tunneling phenomena between different vacua. We consider the action of a scalar field theory in D Euclidean dimensions,

$$S_E = \int d^D x \left[\frac{1}{2} \partial_\mu \phi \partial_\mu \phi + V(\phi) \right] , \tag{9.84}$$

and we choose

$$V(\phi) = \frac{m^2}{2} \phi^2 \left(1 - \frac{\phi}{\eta} \right)^2 , \tag{9.85}$$

so we have a mass term, a cubic and a quartic coupling. As a first step, we restrict to $D = 1$, so the action is

$$S_E = \int dt \left[\frac{1}{2} (\partial_t \phi)^2 + V(\phi) \right] . \tag{9.86}$$

This is Euclidean quantum mechanics, with $\phi(t)$ playing the role of the position $q(t)$. However, we will still use the notation ϕ and the typical field theory language, to emphasize that many general ideas carry through (with some qualifications to be discussed below) in field theory in D Euclidean dimensions.

This potential has two degenerate minima: one is at $\phi = 0$ and, in the field theory language, it is the perturbative vacuum; the other is located at $\phi = \eta$. Of course, with a redefinition of the field of the form $\phi \to \phi - \eta$ we could shift the rightmost minimum at $\phi = 0$ and call this the perturbative vacuum; in any case, the point is that there are two minima, and perturbation theory is defined as an expansion around one of them. So we choose $\phi = 0$ as the perturbative vacuum, and we study the path integral with the boundary conditions that the field goes to zero as $t \to \pm\infty$. We now ask whether there are classical solutions of the equations of motion with these boundary conditions. One is obviously the perturbative vacuum, $\phi(t) = 0$ for all t. To search for other solutions, observe that formally eq. (9.86) is the same as the action of a particle with coordinate $\phi(t)$ and unit mass, moving in *Minkowski* space in a potential $-V(\phi)$, shown in Fig. 9.8. From this formal analogy we immediately understand that there is a solution starting at $t = -\infty$ at $\phi = 0$ with a "speed" $\dot{\phi} \to 0^+$, which approaches the point $\phi = \eta$ at $t = +\infty$. In the original Minkowskian field theory, $\phi = 0$ and $\phi = \eta$ are the two degenerate vacua, separated by a potential barrier, and therefore this Euclidean solution, from the point of view of Minkowski field theory, represents a tunneling process between the two vacua. We will call this Euclidean solution an "instanton".

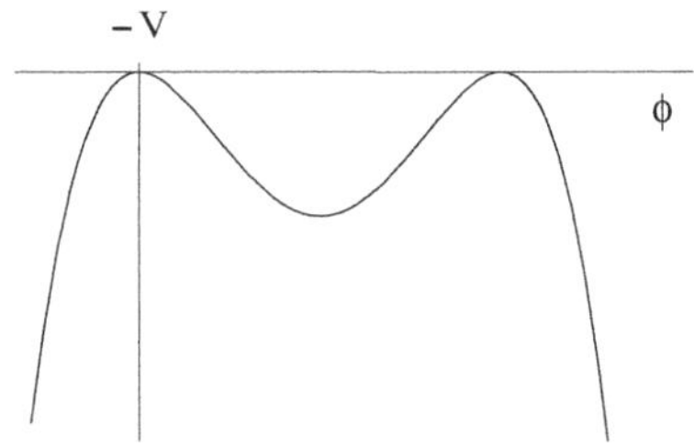

Fig. 9.8 The function $-V(\phi)$ against ϕ.

Actually, this solution does not satisfy our boundary conditions, since $\phi \neq 0$ at $t \to +\infty$. However, there is also a solution, that we call an anti-instanton, which goes from $\phi = \eta$ to $\phi = 0$, and we can combine the two solutions to obtain a path that starts and ends at $\phi = 0$; for instance we can consider a path that starts at $\phi = \epsilon > 0$. This will reach an inversion point close to $\phi = \eta$ and then will go back to $\phi = \epsilon$. In the limit $\epsilon \to 0^+$ we get the desired solution, which is called an instanton–anti-instanton ($I\bar{I}$) pair.

The path integral can therefore be approximated as a sum of two terms: the contributions of the small fluctuations around the perturbative vacuum $\phi = 0$, which reproduces the perturbative expansion, and the contribution of the fluctuations around the $I\bar{I}$ pair. The latter gives an example of a non-perturbative contribution. Let $\phi_0(t)$ be the $I\bar{I}$ classical solution. We can write a small fluctuation over ϕ_0 in the form

$$\phi = \phi_0 + \varphi\,, \tag{9.87}$$

with φ small. The action can be expanded in powers of φ as

$$S[\phi] = S[\phi_0] + S_2[\varphi] + S_3[\varphi] + \dots\,, \tag{9.88}$$

where S_2 is quadratic in φ, S_3 is cubic, etc. It is important that there is no term linear in φ, since ϕ_0 is a classical solution and therefore is an extremum of the action.

Then the contribution to the path integral coming from the fluctuations over ϕ_0 is

$$e^{-S[\phi_0]} \int \mathcal{D}\varphi\, e^{-(S_2[\varphi]+S_3[\varphi]+\dots)}\,. \tag{9.89}$$

We see that this contribution is proportional to $e^{-S[\phi_0]}$, where $S[\phi_0] \equiv S_{I\bar{I}}$ is the action of the $I\bar{I}$ configuration, or more generally of the classical configuration over which we are expanding. In this example, we have $S_{I\bar{I}} = S_I + S_{\bar{I}} = 2S_I$ so to compute $S_{I\bar{I}}$ we must find explicitly the instanton solution and its action S_I. In this model this is easily done: the equation of motion is

$$\ddot{\phi} = m^2\phi\left(1 - \frac{\phi}{\eta}\right)\left(1 - 2\frac{\phi}{\eta}\right)\,, \tag{9.90}$$

and the instanton solution, which interpolates between $\phi = 0$ at $t = -\infty$ and $\phi = \eta$ at $t = +\infty$, is

$$\phi(t) = \frac{\eta}{1 + \exp\{-mt\}} . \tag{9.91}$$

Inserting this into eq. (9.86) and setting $\tau = mt/2$ we find

$$S_I = \frac{m\eta^2}{8} \int_{-\infty}^{\infty} d\tau \, \frac{1}{\cosh^4 \tau} = \frac{m\eta^2}{6} . \tag{9.92}$$

Observe that the coefficient of the ϕ^4 term in the action (9.84, 9.85) is $m^2/(2\eta^2)$. We then define the quartic self-coupling λ from $\lambda/4! = m^2/(2\eta^2)$, and we see that the contribution to the path integral of the small fluctuations over this classical configurations is proportional to

$$e^{-2S_I} = \exp\left\{-\frac{c}{\lambda}\right\} , \tag{9.93}$$

with $c = 4m^3$. The fact that c is not adimensional is due to the fact that in $D = 1$ the coupling λ is not adimensional. The important point is that this contribution is *non-analytic* in the coupling λ and can never be seen in a perturbative expansion around $\lambda = 0$; the Taylor expansion of the function $f(x) = \exp\{-1/x\}$ around $x = 0$ is in fact identically zero. Instantons therefore provide an example of non-perturbative contributions that cannot be computed in a perturbative expansion in terms of Feynman graphs. Parametrically, as $\lambda \to 0$, $\exp\{-c/\lambda\}$ is smaller than any power of λ; therefore, in a theory in which the renormalized coupling is sufficiently small, like in QED, effects of this type are negligible. However, in a theory like QCD where the coupling is strong, non-perturbative effects are important.

The generalization to $D > 1$ Euclidean dimensions is not completely straightforward. For instance, if we start with eq. (9.84), we might look for a classical solution $\phi(t, \mathbf{x})$ which is independent of $\mathbf{x}$. Then we would just find again the solution $\phi(t)$ discussed above. However now its action contains also a divergent volume factor coming from the integration over the spatial coordinates, and therefore the contribution to the path integral of these configurations is $\sim e^{-\infty} = 0$. Thus we must look for solutions whose action is localized not only in time, as in eq. (9.92), but also in space. Solutions of this type turn out to exist in a number of interesting theories, including QCD, and are generically called instantons.

Summary of chapter

- In the path integral formulation of quantum mechanics the position and momentum (and any other variable) remain c-numbers rather than being promoted to operators. The amplitudes are computed summing over all possible trajectories with the given boundary conditions, weighting each trajectory with e^{iS} (or e^{-S} in Euclidean space). The connection between the operator formalism and the path integral quantization is provided by eqs. (9.15) and (9.25).
- The vacuum expectation value of a T-product of fields can be computed performing a path integral over all field configurations that go to zero at infinity, eq. (9.29). The perturbative expansion is recovered writing the action as $S = S_2 + S_{\rm int}$, where S_2 is the

part quadratic in the fields and S_{int} the interaction term. The inverse of the operator which appears in S_2 gives the propagator, and the perturbative expansion is reproduced expanding e^{iS} (or e^{-S} in Euclidean space) in powers of S_{int}.

- An important aspect of the path integral formulation is that it allows us to define the theory non-perturbatively, and it makes it possible to compute (at least in principle, or with suitable numerical or semiclassical techniques) terms non-analytic in the coupling constant.
- The Euclidean formulation of the path integral reveals deep connections between QFT and critical phenomena. The bare Green's functions of a QFT in d spatial dimensions are equivalent to the physical correlation functions of a statistical system living in $d+1$ spatial dimensions. The dependence of the bare coupling constants on the cutoff in QFT can then be understood, in the language of statistical mechanics, as a result of the collective interaction between many degrees of freedom when the correlation length is large, and can be computed "integrating out" the small scale fluctuations. This generates the RG transformation. The existence of fixed points of the RG transformation leads to the notion of universality.

Further reading

- A clear and concise review of the path integral technique can be found in Appendix A of Polchinski (1998). For more details on the path integral see Peskin and Schroeder (1995), Chapter 9. For the path integral quantization of gauge theories see also Ramond (1990).
- For the connection between QFT and critical phenomena, the original review papers are still excellent readings; a classical reference is K. G. Wilson and J. B. Kogut, Phys. Rept. **12** (1974) 75. See also J.B. Kogut, Rev. Mod. Phys. 51 (1979) 659. For a historical perspective of the developments leading to renormalization group and the relation between QFT and critical phenomena, see the Nobel lecture of K. Wilson, Rev. Mod. Phys. 55 (1983) 583.
- Two advanced books on the relations between field theory and critical phenomena are Parisi (1988) and Zinn-Justin (2002).
- For instantons, a classical reference is Coleman (1985). See also Chapter 39 of Zinn-Justin (2002).

Non-abelian gauge theories

10

In this chapter we introduce non-abelian gauge theories, or Yang–Mills theories. Their importance stems from the fact that strong interactions are described by a non-abelian gauge theory with gauge group $SU(3)$, known as quantum chromodynamics or QCD, while the electromagnetic and weak interactions are unified in a gauge theory with gauge group $SU(2) \times U(1)$, the electroweak theory. Together, QCD and the electroweak theory form the Standard Model, which to date reproduces all known experimental results of particle physics, up to energies of the order of a few hundred GeV. To have an idea of the degree of accuracy of these measurements, we mention that the mass of the Z^0 boson is measured with a precision of almost two parts in 10^5, $m_Z = 91.1876(21)$ GeV, its width is $\Gamma_Z = 2.4952(23)$ GeV, and many other observables are known with a precision of the order of a few parts in 10^3.

A full presentation of the Standard Model is beyond the scope of this course. In this and the next chapter we will however introduce two of its main ingredients, namely Yang–Mills theories and the Higgs mechanism.

Non-abelian gauge theories, beside having an extraordinary experimental success, have also a very rich theoretical structure, at the classical and especially at the quantum level. Within the scope of this course, we can only limit ourselves to just a few elementary aspects; in particular, we will discuss how to generalize gauge transformations to non-abelian groups and how to write the corresponding invariant Lagrangians.

10.1 Non-abelian gauge transformations

As a first step, it is useful to rewrite the abelian gauge transformation of electrodynamics in a form more suitable for generalization. We saw in Chapter 3 that electrodynamics has a local $U(1)$ gauge invariance. We write a generic x-dependent $U(1)$ element as

$$U(x) = e^{i\theta(x)}\,, \tag{10.1}$$

with $0 \leqslant \theta(x) \leqslant 2\pi$. A field Ψ with charge q transforms as

$$\Psi(x) \to U_q(x)\Psi(x) \tag{10.2}$$

where $U_q(x) = e^{iq\theta(x)}$ is a representation of the $U(1)$ transformation, labeled by the parameter q. The transformation of the gauge field instead is

$$A_\mu(x) \to A_\mu(x) - \partial_\mu\theta\,. \tag{10.3}$$

In terms of the gauge group element $U(x)$, this can be rewritten as

$$A_\mu(x) \to A_\mu + i(\partial_\mu U)U^\dagger\,. \qquad (10.4)$$

We make the (obvious) observation that the transformation law of A_μ is an intrinsic property of the gauge field, and does not know anything about q, which instead is a parameter that labels the representation to which the matter field Ψ belongs. The coupling between A_μ and Ψ is obtained using the covariant derivative, eq. (3.167), which depends on q, i.e. on the representation to which Ψ belongs,

$$D_\mu\Psi = (\partial_\mu + iqA_\mu)\Psi\,. \qquad (10.5)$$

The important property of the covariant derivative is that, even under x-dependent transformations, it transforms in the same way as Ψ,

$$D_\mu\Psi \to U_q(x)D_\mu\Psi\,, \qquad (10.6)$$

as we already saw in eq. (3.168). As discussed in Section 3.5.4, using covariant derivatives there is a very simple way to construct a theory with local $U(1)$ invariance: we start from a theory with global $U(1)$ invariance and we just replace all the ordinary derivatives with covariant derivatives. This method of coupling matter to the electromagnetic field is known as the minimal coupling. We have also seen, again in Section 3.5.4, that non-minimal couplings are also possible, but they are characterized by coupling constants with dimensions of inverse powers of mass. From the discussion in Section 5.8 and the explicit example of the Fermi theory presented in Section 8.1, we understand that couplings with inverse mass dimensions are less fundamental than dimensionless couplings, and emerge as the low-energy limit of some more fundamental dimensionless coupling. Therefore, it is the minimal coupling that we want to generalize.

We find it convenient to redefine $\theta(x) \to e\theta(x)$, where $e < 0$ is the electron charge. Therefore we write

$$U(x) = e^{ie\theta(x)}\,, \qquad (10.7)$$

where now $0 \leqslant \theta(x) \leqslant 2\pi/|e|$, and the gauge transformation becomes

$$\Psi(x) \to e^{iqe\theta(x)}\Psi(x)\,, \qquad (10.8)$$

$$A_\mu(x) \to A_\mu + \frac{i}{e}(\partial_\mu U)U^\dagger\,. \qquad (10.9)$$

We want to generalize the above transformations to the case where $U(x)$ belongs to a non-abelian group G, rather than just to $U(1)$, and we want to construct a Lagrangian invariant under such local transformations. We will limit ourselves to the the case $G = SU(N)$, although the construction is very general; G is called the *gauge group.*[1]

We start by generalizing eq. (10.8). We consider a set of fields $\Psi^\alpha(x)$ transforming in a given representation R of the gauge group. The fields

[1] We recall a few basic facts about the group $SU(N)$. It has N^2-1 generators T^a, which are hermitian and satisfy $\mathrm{Tr}\,T^a = 0$. By definition they obey the Lie algebra $[T^a, T^b] = if^{abc}T^c$, where f^{abc} are the structure constants of $SU(N)$, which are completely antisymmetric and real. For example, for $SU(2)$, $f^{abc} = \epsilon^{abc}$. If T^a_R is a representation of the algebra and V a unitary matrix of the same dimension as T^a_R, then $VT^a_RV^\dagger$ is still a solution of the Lie algebra and therefore provides an equivalent representation. We can fix V requiring that it diagonalizes the matrix $D^{ab}(R) \equiv \mathrm{Tr}\,(T^a_RT^b_R)$, so that $\mathrm{Tr}\,(T^a_RT^b_R) = C(R)\delta^{ab}$. The normalization factor $C(R)$ is fixed (since the Lie algebra is not invariant under a rescaling of T^a) and it depends on the representation R. For $SU(N)$, it turns out that $C(R) = 1/2$ for the fundamental representation and $C(R) = N$ for the adjoint representation; the reader can check it for $SU(2)$, using $T^i = \sigma^i/2$ for the fundamental and $(T^i)^{jk} = -i\epsilon^{ijk}$ for the adjoint, as we discussed in eq. (2.37). By definition for $SU(N)$ we raise or lower the index a with δ^{ab}, so we will conventionally always write it as an upper index, and repeated upper indices are summed over.

are then labeled by an index $\alpha = 1, \ldots, \dim(R)$. For definiteness we take Ψ^α to be Dirac fermions, but all the subsequent considerations are very general and apply to any matter fields, e.g. to bosonic fields or to Weyl fermions.

The fact that Ψ transforms in the representation R means that, under a gauge transformation,

$$\boxed{\Psi \to U_R \Psi\,,} \tag{10.10}$$

or, in components, $\Psi^\alpha(x) \to (U_R)^\alpha{}_\beta(x)\Psi^\beta(x)$. In eq. (10.10),

$$U_R(x) = \exp\{ig\theta^a(x)T_R^a\}\,, \tag{10.11}$$

where T_R^a are the generators of the gauge group in the representation R and $\theta^a(x)$ are the parameters of the transformation. We have redefined the parameters $\theta^a(x) \to g\theta^a(x)$, where g is a constant. We will see below that g will be the coupling constant of the theory.

The free Dirac Lagrangian,

$$\mathcal{L}_{\text{free}} = i\bar{\Psi}^\alpha \gamma^\mu \partial_\mu \Psi^\alpha\,, \tag{10.12}$$

(with the sum over the index α understood) is invariant under global $SU(N)$ transformations, since if $\Psi \to U_R\Psi$ then $\bar{\Psi} \to U_R^\dagger$ and, if U_R is independent of x, it goes through ∂_μ and cancels against $U_R^\dagger$. However, if U_R depends on x, performing the transformation we also get a term proportional to $\partial_\mu U$ and this Lagrangian is no longer invariant.

To construct an invariant Lagrangian, we introduce a set of gauge fields A_μ^a labeled by an index a, with one gauge field for each generator of the gauge group; the A_μ^a are called *non-abelian gauge fields.* In particular, $SU(N)$ has $N^2 - 1$ generators, so we have three gauge fields for $SU(2)$ and eight gauge fields for $SU(3)$. We introduce the matrix field

$$A_\mu(x) = A_\mu^a(x)T^a\,. \tag{10.13}$$

Of course A_μ^a does not depend on the representation (just as in electromagnetism the gauge field, and therefore its transformation properties, does not know anything about the parameter q that labels the matter representation), while the generators T^a, and therefore the matrix A_μ, have an explicit form which depends on the representation R. We define the gauge transformation of A_μ as

$$\boxed{A_\mu \to UA_\mu U^\dagger - \frac{i}{g}(\partial_\mu U)U^\dagger\,,} \tag{10.14}$$

where $A_\mu = A_\mu^a(x)T_R^a$ and $U(x) = \exp\{ig\theta^a(x)T_R^a\}$ are in the same representation R. This definition is consistent because the transformation that it induces on A_μ^a is independent of R, as it should be. This can be shown considering first an infinitesimal transformation,

$$U(x) = 1 + ig\theta^a(x)T_R^a + O(\theta^2)\,. \tag{10.15}$$

Then eq. (10.14) becomes

$$A^a_\mu T^a_R \to (1 + ig\theta^a T^a_R) A^b_\mu T^b_R (1 - ig\theta^c T^c_R) - \frac{i}{g}(igT^a_R \partial_\mu \theta^a) + O(\theta^2)$$
$$= A^a_\mu T^a_R + ig\theta^a A^b_\mu [T^a_R, T^b_R] + T^a_R \partial_\mu \theta^a + O(\theta^2)\,. \qquad (10.16)$$

Therefore

$$A^a_\mu \to A^a_\mu + \partial_\mu \theta^a - g f^{abc} \theta^b A^c_\mu + O(\theta^2)\,, \qquad (10.17)$$

and no dependence on the representation R appears. For Lie groups the infinitesimal transformation fixes uniquely also the finite transformation, and therefore even the finite transformation of A^a_μ is independent of R. Equation (10.14) generalizes eq. (10.9) to non-abelian groups. The constant g will play the role of the gauge coupling, as we will see below.[2]

[2] We take $g > 0$, while the electron charge is $e < 0$. This is the origin of some apparent sign differences in the definitions for the $U(1)$ and the non-abelian case.

In particular, under a global gauge transformation, $A_\mu \to U A_\mu U^\dagger$. It is interesting to ask what this transformation property means in terms of the $N^2 - 1$ fields A^a_μ, and we will see in Section 10.4 that it means that, under global gauge transformations, $A^a_\mu \to (U_{\rm adj})^a{}_b A^b_\mu$, where $U_{\rm adj}$ is the adjoint representation of the gauge group.

We now define the covariant derivative on the field Ψ as

$$\boxed{D_\mu \Psi = \left(\partial_\mu - ig\, A^a_\mu T^a_R\right) \Psi\,,} \qquad (10.18)$$

where T^a_R are the generators in the same representation R as the field Ψ. Using eqs. (10.10) and (10.14) we see that

$$D_\mu \Psi \to \partial_\mu (U_R \Psi) - ig \left(U_R A_\mu U_R^\dagger - \frac{i}{g} (\partial_\mu U_R) U_R^\dagger \right) U_R \Psi = U_R D_\mu \Psi\,, \qquad (10.19)$$

where we used the fact that $A^a_\mu T^a_R$ transforms with the same matrix U_R which appears in the transformation of Ψ. Therefore $D_\mu \Psi$ transforms in the same way as Ψ, even under local transformations.

10.2 Yang–Mills theory

Using the covariant derivative, it is now easy to write a Lagrangian with local non-abelian gauge invariance. We just replace $\partial_\mu \to D_\mu$ in the free theory, that is, we write

$$\mathcal{L} = \sum_\alpha \bar{\Psi}^\alpha \left[i\gamma^\mu (D_\mu \Psi)^\alpha - m \Psi^\alpha \right]. \qquad (10.20)$$

This Lagrangian contains the kinetic term of the fermionic field and its interaction with the gauge fields. The interaction term, which is hidden in the covariant derivative, is

$$\mathcal{L}_{\rm int} = g A^a_\mu\, \bar{\Psi}^\alpha \gamma^\mu (T^a_R)_{\alpha\beta} \Psi^\beta\,, \qquad (10.21)$$

and we see that g is a coupling constant. We also need a kinetic term for the gauge fields. One might try to define the field strength tensor of each of the gauge fields A^a_μ as $F^a_{\mu\nu} = \partial_\mu A^a_\nu - \partial_\nu A^a_\mu$, but it is immediate to verify that this quantity does not have any simple transformation property under (10.14). Instead, a straightforward computation (using the identity $0 = \partial_\mu(UU^\dagger) = (\partial_\mu U)U^\dagger + U(\partial_\mu U^\dagger)$ and therefore $\partial_\mu U^\dagger = -U^\dagger(\partial_\mu U)U^\dagger$) shows that the quantity

$$\boxed{F_{\mu\nu} = \partial_\mu A_\nu - \partial_\nu A_\mu - ig[A_\mu, A_\nu]} \tag{10.22}$$

transforms as

$$F_{\mu\nu}(x) \to U(x)F_{\mu\nu}(x)U^\dagger(x)\,. \tag{10.23}$$

$F_{\mu\nu}$ is called the non-abelian field strength. From eqs. (10.22) and (10.13) we see that we can rewrite $F_{\mu\nu}$ as

$$F_{\mu\nu} = F^a_{\mu\nu}T^a \tag{10.24}$$

with

$$F^a_{\mu\nu} = \partial_\mu A^a_\nu - \partial_\nu A^a_\mu + gf^{abc}A^b_\mu A^c_\nu\,. \tag{10.25}$$

Now it is easy to construct a gauge-invariant kinetic term for the gauge field; it is given by

$$\mathcal{L}_{\text{gauge}} = -\frac{1}{2}\,\text{Tr}\,F_{\mu\nu}F^{\mu\nu} = -\frac{1}{4}\,F^a_{\mu\nu}F^{a\,\mu\nu}\,, \tag{10.26}$$

where $F_{\mu\nu}$ has been taken in the fundamental representation, and we used the fact that $\text{Tr}(T^a_F T^b_F) = (1/2)\delta^{ab}$. Under gauge transformations $\text{Tr}\,F_{\mu\nu}F^{\mu\nu} \to \text{Tr}\,(UF_{\mu\nu}F^{\mu\nu}U^\dagger) = \text{Tr}\,F_{\mu\nu}F^{\mu\nu}$ due to the cyclic property of the trace.

The complete Lagrangian of the $SU(N)$ Yang–Mills theory with Dirac fermions in the representation R is therefore

$$\mathcal{L}_{\text{YM}} = i\bar{\Psi}^\alpha \partial\!\!\!/ \Psi^\alpha - m\bar{\Psi}^\alpha\Psi^\alpha + gA^a_\mu\bar{\Psi}^\alpha\gamma^\mu(T^a_R)_{\alpha\beta}\Psi^\beta - \frac{1}{4}\,F^a_{\mu\nu}F^{a\,\mu\nu}\,, \tag{10.27}$$

or, in more compact form,

$$\boxed{\mathcal{L}_{\text{YM}} = \bar{\Psi}\,(iD\!\!\!\!/ - m)\,\Psi - \frac{1}{2}\,\text{Tr}\,F_{\mu\nu}F^{\mu\nu}\,.} \tag{10.28}$$

Observe, from eq. (10.25), that the term F^2 contains not only the standard kinetic term of the gauge fields, but also an interaction vertex with three gauge bosons, proportional to g, and a vertex with four gauge bosons, proportional to g^2, as shown in Fig. 10.1. Observe also that gauge invariance has fixed the three-boson, four-boson, and boson–fermion–fermion vertices in terms of a single parameter, the gauge coupling g.

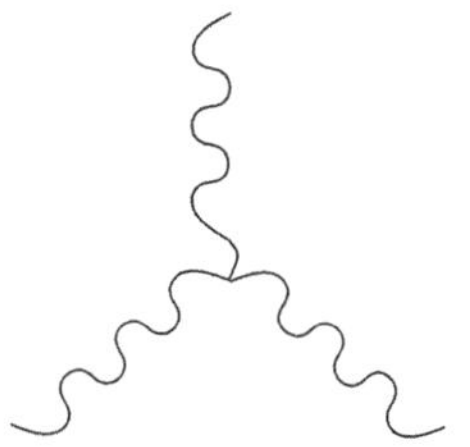

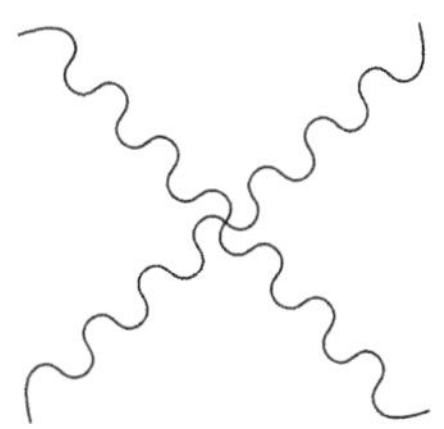

Fig. 10.1 The vertices with three and with four non-abelian gauge bosons.

10.3 QCD

Quantum chromodynamics (QCD) is a Yang–Mills theory with gauge group $SU(3)$. The matter fields are the *quarks*. They are in the fundamental representation of the gauge group and have spin 1/2. As we already discussed in Chapter 8, there are six type of quarks, denoted as u (up), d (down), c (charm), s (strange), t (top) and b (bottom). The type of quark is called the *flavor*, while the index of the gauge group is called the *color* index. Therefore a generic quark field has two indices, $\Psi^{\alpha,A}$ with $\alpha = 1,2,3$ the color index and $A = u,d,c,s,t,b$ the flavor index. Each quark flavor is described by a Lagrangian of the type (10.27), with a different mass for each flavor. The $3^2 - 1 = 8$ gauge bosons are called *gluons*. Therefore the QCD Lagrangian is

$$\begin{aligned}\mathcal{L}_{QCD} = & \, i\bar{\Psi}^{\alpha,A}\partial\!\!\!/\,\Psi^{\alpha,A} - m_A\bar{\Psi}^{\alpha,A}\Psi^{\alpha,A} - \frac{1}{4}F^a_{\mu\nu}F^{a\,\mu\nu} \\ & +gA^a_\mu\,\bar{\Psi}^{\alpha,A}\gamma^\mu T^a_{\alpha\beta}\Psi^{\beta,A}\,,\end{aligned} \tag{10.29}$$

where we sum over both the color indices α, β and the flavor index A, and T^a are the generators of $SU(3)$ in the fundamental representation.

QCD is the fundamental theory of strong interactions. A crucial property of QCD, that we already discussed in Sections 5.9 and 9.5, is asymptotic freedom, which means that the running coupling constant $g_{\text{eff}}(E)$ (defined in Section 5.9) is small at high energies and large at low energies. At small distances QCD is well described in terms of weakly interacting quarks and gluons, while at large distances, of the order of 1 fm, the theory becomes non-perturbative and quarks are *confined*. This means that quarks cannot be observed as free particles, but we can only observe color-singlet bound states of quark–antiquarks (mesons) or of three quarks or three antiquarks (baryons). Mesons and baryons are collectively denoted as hadrons and, being composed of quarks, are subject to strong interactions. The strong interactions generate dynamically a characteristic energy scale $\Lambda_{\text{QCD}} \sim (1\,\text{fm})^{-1} \simeq 200$ MeV. The lightest hadron is the pion, whose mass is in fact of this order of magnitude, $m_\pi \simeq 140$ MeV.

Besides the exact *local* $SU(3)$ color symmetry, QCD also has important approximate *global* symmetries, due to the possibility of performing a coordinate-independent rotation in flavor space. We saw in Section 3.4.3 that the free Lagrangian of a single *massless* Dirac fermion has a $U(1) \times U(1)$ symmetry, in which we rotate independently the left-handed and right-handed Weyl spinors,

$$\psi_L \to e^{i\theta_L}\psi_L\,, \qquad \psi_R \to e^{i\theta_R}\psi_R\,. \tag{10.30}$$

In terms of the Dirac spinor Ψ the two independent transformations with $\theta_R = \theta_L = \alpha$ and $\theta_R = -\theta_L = \beta$ have been written in eqs. (3.125) and (3.126), and we recall them here,

$$\Psi \to e^{i\alpha}\Psi\,, \qquad \Psi \to e^{i\beta\gamma^5}\Psi\,. \tag{10.31}$$

The transformation parametrized by α is called the vector $U(1)$, while the one parametrized by β is called the axial $U(1)$, or $U_A(1)$.

Consider now the QCD Lagrangian with N_f quark flavors $\Psi^1, \ldots, \Psi^{N_f}$ (i.e. Ψ^1 is the Dirac spinor describing the u-quark, Ψ^2 the d-quark, etc.). Denote by q_L a column vector with N_f components whose entries are the Weyl spinors $\psi_L^1, \ldots, \psi_L^{N_f}$ which describes the left-handed quarks,

$$q_L = \begin{pmatrix} \psi_L^1 \\ \cdot \\ \cdot \\ \psi_L^{N_f} \end{pmatrix}, \tag{10.32}$$

and similarly for q_R (the color index is not written explicitly). Recalling the relation between the Dirac Lagrangian written in terms of Dirac spinors and in terms of Weyl spinors, Sections 3.4.1 and 3.4.2, we can rewrite the quark part of the QCD Lagrangian (10.29) in the form

$$\mathcal{L}_{\text{quarks}} = iq_L^\dagger \bar{\sigma}^\mu D_\mu q_L + iq_R^\dagger \sigma^\mu D_\mu q_R - (q_L^\dagger M q_R + q_R^\dagger M q_L), \tag{10.33}$$

where M is a mass matrix, diagonal in flavor space

$$M_{AB} = m_A \delta_{AB}. \tag{10.34}$$

If we set the mass term to zero, in the above Lagrangian there is no coupling between left-handed and right-handed quarks, and we can perform a $SU(N_f)$ transformation independently on the left-handed and right-handed quarks,

$$q_L \to U_L q_L, \qquad q_R \to U_R q_R, \tag{10.35}$$

with U_L, U_R two independent $SU(N_f)$ matrices acting in flavor space. The operator D_μ acts on the coordinates, through ∂_μ, and in color space, because of $A_\mu^a T^a$; however, it knows nothing about flavor. Therefore, if the matrices $U_{L,R}$ do not depend on the coordinates x^μ, they commute with D_μ. Then under eq. (10.35)

$$q_L^\dagger \bar{\sigma}^\mu D_\mu q_L \to q_L^\dagger U_L^\dagger \bar{\sigma}^\mu D_\mu U_L q_L = q_L^\dagger U_L^\dagger U_L \bar{\sigma}^\mu D_\mu q_L, \tag{10.36}$$

so it is invariant, since $U_L^\dagger U_L = 1$, and similarly for q_R. This means that, in the limit in which we can neglect the masses of N_f quark flavors, QCD has an approximate global $SU_L(N_f) \times SU_R(N_f)$ invariance.[3]

We introduce the Dirac spinor Q, in the chiral representation of the γ-matrices,

$$Q = \begin{pmatrix} q_L \\ q_R \end{pmatrix}. \tag{10.37}$$

Then the symmetry $SU_L(N_f) \times SU_R(N_f)$ can be written as a product of a vector $SU(N_f)$ and an axial $SU(N_f)$, similarly to eq. (10.31)

$$Q \to e^{i\alpha} Q, \qquad Q \to e^{i\beta\gamma^5} Q, \tag{10.38}$$

[3] Actually, we could more generally consider a $U_L(N_f) \times U_R(N_f)$ transformation, so we also have a vector $U(1)$, which corresponds to baryon number, and an axial $U(1)$. The axial $U(1)$ symmetry is however spoiled by subtle quantum effects that we will not discuss.

with $\alpha = \alpha^a T^a$ and $\beta = \beta^a T^a$, where T^a are the generators of the flavor symmetry in the fundamental representation.

If the mass term is non-zero, but N_f masses are equal, so that $M = mI$ is a multiple of the identity matrix, we no longer have a $SU(N_f) \times SU(N_f)$ global symmetry, but we still have a $SU(N_f)$ global symmetry in which the left- and right-handed quarks are rotated in the same way, since in this case the mass term is

$$m(q_L^\dagger q_R + q_R^\dagger q_L) \tag{10.39}$$

and is invariant under eq. (10.35) with $U_L = U_R$.

The approximation of neglecting the quark masses or of neglecting their differences is useful only for the lightest quarks, u and d (and, to a lesser accuracy, s). In particular, if we take $m_u \simeq m_d$, we have an approximate $SU(2)$ global symmetry called isospin, while if we further assume $m_u \simeq m_d \simeq m_s$ we have an approximate $SU(3)$ flavor symmetry. We discussed these symmetries in the Complement on page 209, where we also explained how to use them to extract information on hadronic matrix elements. We will further examine the axial $SU(N_f)$ symmetry in Section 11.2, when we discuss spontaneous symmetry breaking and Goldstone bosons.

10.4 Fields in the adjoint representation

We have seen that the form of the covariant derivative depends on the transformation property of the object on which it acts, since in eq. (10.18) the generators are in the same representation R as the field Ψ. Apart from fields transforming in the fundamental representation of $SU(N)$, another typical case that one encounters is that of fields in the adjoint representation. Let us consider for definiteness a real scalar field. As we saw in Section 2.4, the adjoint representation exists for any group and has the same dimension as the number of generators, i.e. $N^2 - 1$ for $SU(N)$. A scalar field in the adjoint can be written as $\phi^a(x), a = 1, \ldots, N^2 - 1$ (while for a field in the fundamental, as in the previous section, we use the notation ϕ^α with $\alpha = 1, \ldots, N$), and the indices a, b are of the same type as the indices labeling the generators. Under a gauge transformation, a field in the adjoint of $SU(N)$ transforms by definition as

$$\phi \to \left(e^{ig\theta^a(x)T^a_{\text{adj}}}\right)\phi\,, \tag{10.40}$$

where ϕ is the vector column with components ϕ^a. Using the fact that we have as many fields as generators, we can form the matrix field

$$\Phi(x) = \phi^a(x) T^a \tag{10.41}$$

(with the generators T^a in any representation that we wish to use, not necessarily in the adjoint). We now show that, in terms of Φ, eq. (10.40) becomes

$$\Phi(x) \to U(x)\Phi(x)U^\dagger(x)\,. \tag{10.42}$$

Here $U(x) = \exp\{ig\theta^a(x)T^a\}$, where the generators T^a are in the same representation that we used in the definition of Φ, eq. (10.41).[4] To prove this assertion, it is sufficient to consider an infinitesimal transformation. The explicit form of the generators in the adjoint of $SU(N)$ is $(T^a_{\rm adj})^{bc} = -if^{abc}$. Writing all indices as upper indices, raised with δ^{ab}, under an infinitesimal transformation we have

$$\delta\phi^c = ig\theta^a (T^a_{\rm adj})^{cb}\phi^b = -gf^{abc}\theta^a\phi^b\,, \tag{10.43}$$

which implies that

$$\delta\Phi = \delta\phi^c T^c = -gf^{abc}\theta^a\phi^b T^c\,. \tag{10.44}$$

On the other hand, the infinitesimal form of eq. (10.42) is

$$\delta\Phi = ig[\theta^a T^a, \Phi] = ig\theta^a\phi^b[T^a, T^b] = -gf^{abc}\theta^a\phi^b T^c\,, \tag{10.45}$$

which agrees with eq. (10.44).[5] Of course, nothing here depends on the Lorentz indices of the field, so we see that $F_{\mu\nu}$ is an example of a field transforming in the adjoint representation under local gauge transformation; compare with eq. (10.23). Instead A_μ transforms in the adjoint only under global transformations, while for local transformations it acquires also the inhomogeneous term $\sim \partial_\mu U$. Observe also that if we choose ϕ^a real then Φ is hermitian, and the gauge transformation is compatible with the hermiticity condition.

The covariant derivative of a field in the adjoint is

$$(D_\mu\phi)^a = \partial_\mu\phi^a - ig\,A^c_\mu (T^c_{\rm adj})^a{}_b\phi^b\,. \tag{10.46}$$

Using $(T^a_{\rm adj})_{bc} = -if^a{}_{bc}$ we have

$$(D_\mu\phi)^a = \partial_\mu\phi^a - gf^{abc}\phi^b A^c_\mu\,. \tag{10.47}$$

By definition $(D_\mu\phi)^a$ transforms as ϕ^a under local gauge transformations. We can also write the covariant derivative in terms of Φ; defining $D_\mu\Phi = (D_\mu\phi)^a T^a$, eq. (10.47) gives

$$D_\mu\Phi = \partial_\mu\Phi - ig[A_\mu, \Phi]\,. \tag{10.48}$$

Using eqs. (10.14) and (10.42), we easily check that under gauge transformations

$$D_\mu\Phi \to U(x)(D_\mu\Phi)U^\dagger(x)\,, \tag{10.49}$$

confirming that $D_\mu\Phi$ transforms as Φ. An invariant Lagrangian is

$$\mathcal{L} = {\rm Tr}\, D^\mu\Phi D_\mu\Phi = \frac{1}{2}\left(\partial_\mu\phi^a - gf^{abc}\phi^b A^c_\mu\right)^2\,. \tag{10.50}$$

Here the generators which appear in eq. (10.41) have been chosen in the fundamental representation, so the trace gives a factor $1/2$ and we recover the standard normalization of the kinetic term. The gauge invariance of eq. (10.50) follows from the cyclicity of the trace. Again, we see that the requirement of gauge invariance fixes the interaction terms, and in eq. (10.50) we have a cubic interaction $-gf^{abc}(\partial_\mu\phi^a)\phi^b A^c_\mu$ and a quartic interaction $O(g^2\phi^2 A^2_\mu)$.

[4] In other words, eq. (10.42) holds at the abstract group level, without any reference to the representation.

[5] Actually, to prove eq. (10.42) it was not really necessary to perform an explicit computation. It suffices to realize that eq. (10.42) is the same transformation law obeyed by the tensor $T^{\alpha\beta} = \psi^\alpha\psi^{\dagger\,\beta}$ where ψ is in the fundamental representation N and $\psi^\dagger$ in the antifundamental $\bar{N}$. The product $N\otimes\bar{N}$ decomposes into $(N^2-1)\oplus 1$, i.e. in the adjoint plus the singlet. However, the singlet is absent in Φ because ${\rm Tr}\,T^a{=}0$, and therefore Φ is purely in the adjoint.

Summary of chapter

- Non-abelian gauge transformations generalize the local invariance of electrodynamics, with gauge group $U(1)$, to non-abelian gauge groups like $SU(N)$. Instead of a single gauge field, we now have a set of gauge fields A^a_μ, with one gauge field for each generator of the gauge group. Matter fields are in a representation R of the gauge group and therefore carry an internal index $\alpha = 1, \ldots, \dim\,(R)$. The transformation laws are given by eqs. (10.10) and (10.14).
- The Yang–Mills Lagrangian is given by eq. (10.28). Besides an interaction term between matter and gauge fields, dictated by the covariant derivative, there are also interaction vertices involving only three and four gauge bosons, fixed by the form of the non-abelian field strength. Therefore all these interaction terms are fixed by the requirement of gauge invariance.
- QCD is a Yang–Mills theory with gauge group $SU(3)$; the matter fields are the quarks and the gauge fields are the gluons. The Lagrangian is given in eq. (10.29).

Further reading

- Non-abelian gauge theories are the building blocks of modern particle physics. Given their extraordinary experimental success and their rich theoretical structure, the literature on them is vast. A detailed introduction is provided in Peskin and Schroeder (1995) and in Weinberg vol. II, (1996).
- A detailed survey of QCD is given by the three volumes of *At the frontier of Particle Physics–Handbook of QCD*, M. Shifman ed., World Scientific 2001.

Spontaneous symmetry breaking

11

In this chapter we present the phenomenon of spontaneous symmetry breaking (SSB). This is a mechanism of great importance both in particle physics and in condensed matter physics. Its generality and importance stem from the fact that it deals with how a symmetry of the action in QFT (or of the Hamiltonian in a statistical system) is reflected on the ground state of the system. As we will see in Section 11.1, SSB strictly speaking can only take place in a system with an infinite number of degrees of freedom. It is therefore a genuinely field-theoretical phenomenon, which does not appear in quantum mechanical systems with a finite number of variables.

We will examine the effect of SSB on different types of symmetries. In Section 11.2 we will discuss the SSB of global symmetries, and the emergence of Goldstone bosons. In Section 11.3 we will examine the SSB of local abelian symmetries, and we will see that it is a crucial element in the BCS theory of superconductivity, when the latter is formulated in field theoretical language. We will finally examine the SSB of non-abelian gauge symmetries, and we will see that in this case it gives rise to the masses of non-abelian gauge bosons, like the $W^{\pm}$ and Z^0 in the Standard Model.

11.1 Degenerate vacua in QM and QFT

Spontaneous symmetry breaking is a very general phenomenon characterized by the fact that the action has a symmetry (global or local) but the quantum theory, instead of having a unique vacuum state which respects this symmetry, has a family of degenerate vacua that transform into each other under the action of the symmetry group.

A simple example is given by a ferromagnet. The action governing its microscopic dynamics is invariant under spatial rotations. For instance, we can describe a ferromagnet by a generalization of the Ising Hamiltonian given in eq. (9.64), introducing a vector variable $\mathbf{s}_i$ associated to each site i,

$$H = -J\sum_{i,j} \mathbf{s}_i{\cdot}\mathbf{s}_j\,, \tag{11.1}$$

where $J > 0$ and the sum is restricted to nearest-neighbor pairs. As we discussed in Section 9.5, above a critical temperature a ferromagnet has a unique ground state, with zero magnetization. Of course this state respects the rotational invariance, since on it the expectation value of the magnetization $\mathbf{M} = \langle \mathbf{s}_i \rangle$ vanishes, and therefore no preferred direction is selected. Below a critical temperature instead it becomes thermodynamically favorable to develop a non-zero magnetization, and in this new vacuum $\mathbf{M} \neq 0$ and the full $SO(3)$ rotational symmetry is broken to the subgroup $SO(2)$ of rotations around the magnetization axis.

The original invariance of the Lagrangian is now reflected in the fact that, instead of a single vacuum state, there is a whole family of vacua related to each other by rotations, since the magnetization can in principle develop in any direction. However, the system will choose one of these states as its vacuum state. The symmetry is then said to be *spontaneously broken* by the choice of a vacuum.

SSB is a phenomenon that cannot take place in a quantum mechanical system with a finite number of degrees of freedom, since in this case, if we have a family of "vacua", the true vacuum state is a superposition of them which respects the original symmetry. To illustrate this point, we consider for instance the quantum mechanics of a particle, described by a coordinate $q(t)$, in a potential

$$V(q) = \frac{1}{2}\lambda^2(q^2(t) - \eta^2)^2 , \tag{11.2}$$

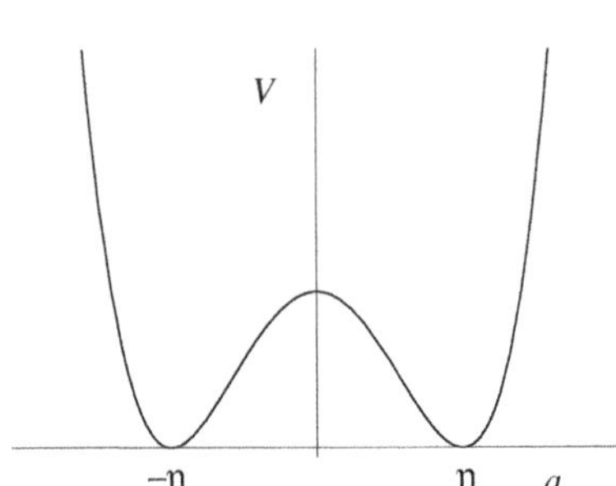

Fig. 11.1 A double-well potential

with λ, η parameters. This potential is shown in Fig. 11.1, and is called a double-well potential. The Lagrangian is

$$L = \frac{1}{2}m\dot{q}^2 - V(q) , \tag{11.3}$$

and is symmetric under the parity transformation $q(t) \to -q(t)$ (this is also called a Z_2 symmetry, where Z_2 is the finite group formed by 1 and -1 under multiplication). The potential has two minima, at $q = \pm\eta$. We can solve the Schrödinger equation expanding the potential around the minimum at $q = +\eta$, retaining only the quadratic term in the Taylor expansion of the potential around η (so that we have a harmonic oscillator), and treating in perturbation theory all higher powers of the expansion of the potential. We call $|+\rangle$ the ground state obtained in this way; more precisely, this is a *perturbative vacuum*. We can do the same expanding around $-\eta$, and we call $|-\rangle$ the corresponding perturbative vacuum. However, the true ground state of the theory is neither $|+\rangle$ nor $|-\rangle$. At the non-perturbative level there is a non-vanishing amplitude for the transition between these two states, due to the possibility of tunneling under the barrier which separates the two minima, and which can be computed in a WKB approximation (or using the instanton technique developed in Solved Problem 9.1). Because of the tunneling process, the Hamiltonian is not diagonal in the $|\pm\rangle$ basis. Rather, we will have

$$\begin{aligned} \langle +|H|+\rangle &= \langle -|H|-\rangle \equiv a \\ \langle +|H|-\rangle &= \langle -|H|+\rangle \equiv b , \end{aligned} \tag{11.4}$$

with $b \ll a$, since the tunneling amplitude is exponentially suppressed. Diagonalizing this Hamiltonian we immediately find that the eigenstates are the symmetric and antisymmetric combinations

$$|S\rangle = |+\rangle + |-\rangle\,, \qquad |A\rangle = |+\rangle - |-\rangle\,, \tag{11.5}$$

with energies $a \pm b$, respectively. Therefore the degeneracy between these states is lifted by the fact that $b \neq 0$, and the true ground state is the combination with energy $a - |b|$.

Under a parity transformation $q \to -q$, $|S\rangle$ is invariant while $|A\rangle$ picks a minus sign. Recalling that physical states are defined up to an overall phase, we see that the true ground state of the Hamiltonian goes into itself under parity, and there is no SSB of the Z_2 symmetry.

Consider now a real scalar field with Lagrangian

$$\mathcal{L} = \partial^\mu \phi \partial_\mu \phi - \frac{1}{2}\lambda^2(\phi^2 - \eta^2)^2\,. \tag{11.6}$$

Here again we have a Z_2 symmetry $\phi \to -\phi$. The crucial difference is that the tunneling amplitude in this case is proportional to $\exp\{-cV\}$ with c a constant and V the spatial volume. In fact, this tunneling amplitude can be evaluated as in the instanton computation that we have discussed in Solved Problem 9.1, with a classical configuration which is not localized in space, so its action is proportional to the volume, $S_{\rm cl} = cV$, and the tunneling amplitude is proportional to $\exp\{-S_{\rm cl}\} = \exp\{-cV\}$. This result can be understood physically by discretizing space, so that our field theory corresponds to a quantum mechanical system in which for each spatial point $\mathbf{x}$ we have a variable $q_{\mathbf{x}}(t) \equiv \phi(\mathbf{x}, t)$, and in order to tunnel into the other vacuum each of the $q_{\mathbf{x}}$ must tunnel. Let the tunneling amplitude for a single variable $q_{\mathbf{x}}$ be proportional to $e^{-c'}$, for some constant c'. The total amplitude is the product of the separate amplitudes so, if N is the number of lattice sites,

$$\text{tunneling amplitude} \sim \prod_{\mathbf{x}} e^{-c'} = e^{-c'N} = e^{-cV}\,. \tag{11.7}$$

In an infinite volume this amplitude vanishes and there is no mixing between the two vacua. In other words, the effective height of the barrier is infinite and therefore we truly have two distinct sectors of the theory, i.e. two different Hilbert spaces $\mathcal{H}_+, \mathcal{H}_-$ constructed above the two vacua $|\pm\rangle$ with the usual rules of second quantization. There is no possibility to restore the symmetry via tunneling, and all local operators have vanishing matrix elements between a state in $\mathcal{H}_+$ and a state in $\mathcal{H}_-$.

A characteristic of SSB is the existence of an order parameter which takes a non-zero expectation value on the chosen vacuum. In the example of the ferromagnet the order parameter is the magnetization, i.e. a spatial vector, while in the previous example it was an element of Z_2, $\langle\phi\rangle/\eta = \pm 1$. In the following we will be interested in situations where the order parameter is a scalar field ϕ, real or complex. In any case, the order parameter is a quantity which is *not* invariant under the symmetry in question, so that a non-vanishing expectation value means that the symmetry is broken.

For a Lie group, we can restate the condition of SSB in terms of the action of the generators on the vacuum state. We denote by $U = \exp\{i\theta^a T^a\}$ a generic element of the symmetry group in question, and by T^a the generators. If the vacuum state is invariant, then for any value of the parameters θ^a we have $U|0\rangle = |0\rangle$ and therefore all generators must annihilate the vacuum,[1] so $T^a|0\rangle = 0$ for each a. Instead, if the vacuum state is not invariant, there must be one or more generators T^a that do not give zero when acting on the vacuum state,

$$T^a|0\rangle \neq 0\,. \tag{11.8}$$

[1] More precisely, since vectors that differ by a phase describe the same physical state, we do not have SSB if $U|0\rangle = e^{i\alpha}|0\rangle$, for some constant phase α. Conversely, in order to have SSB, beside eq. (11.8) we must also require that $T^a|0\rangle$ is not proportional to $|0\rangle$ itself.

For example, for a ferromagnet in the ordered phase the $SO(3)$ rotation group is broken. The $SO(3)$ generators are the angular momentum operators J_x, J_y, J_z and if the magnetization is, say, along the z-axis, we have $J_z|0\rangle = 0$ (since rotations around the z-axis still leave the vacuum state invariant) but $J_x|0\rangle \neq 0, J_y|0\rangle \neq 0$. The full $SO(3)$ group is therefore broken to the $SO(2)$ subgroup generated by J_z.

11.2 SSB of global symmetries and Goldstone bosons

Consider the Lagrangian for a complex scalar field

$$\mathcal{L} = \partial_\mu\phi^*\partial^\mu\phi - V(|\phi|)\,, \tag{11.9}$$

with

$$V(|\phi|) = \frac{1}{2}\lambda^2\left(|\phi|^2 - \eta^2\right)^2\,. \tag{11.10}$$

This is a double-well potential for $|\phi|$ and therefore it has a *continuous* set of minima; writing $\phi = |\phi|e^{i\alpha}$, the vacua are characterized by $\langle|\phi|\rangle = \eta$, and $\langle\alpha\rangle$ arbitrary. The Lagrangian has a global $U(1)$ invariance

$$\phi \to e^{i\theta}\phi\,, \tag{11.11}$$

with θ an arbitrary constant. The scalar field will choose one of these vacua, so that $\langle\alpha\rangle = \alpha_0$, and the $U(1)$ symmetry is spontaneously broken. Without loss of generality we can redefine α so that $\alpha_0 = 0$, and therefore on the vacuum

$$\langle\phi\rangle = \eta\,. \tag{11.12}$$

We want to understand the spectrum of the theory after SSB. This can be done studying the small oscillations around the vacuum. We therefore write

$$\phi(x) = \eta + \frac{1}{\sqrt{2}}\left(\chi(x) + i\psi(x)\right) \tag{11.13}$$

where χ and ψ are real fields (the normalization $1/\sqrt{2}$ is chosen for later convenience). Observe that the set of vacua is a circle of radius η in the complex field plane, and since we are expanding around the point $(\mathrm{Re}\,\phi = \eta, \mathrm{Im}\,\phi = 0)$, χ is a fluctuation in the direction orthogonal to the manifold of vacua, while ψ is a fluctuation in the tangential direction, as shown in Fig. 11.2. In other words, $\eta + i\psi$, for ψ constant and infinitesimal, is another vacuum. A small displacement in the direction of ψ does not cost energy since we are moving along a flat direction of the potential (at least to lowest order, i.e. retaining terms quadratic in ψ in the Lagrangian and neglecting cubic and higher-order terms). Instead with a small displacement in the direction of χ we feel an approximately quadratic rise of the potential, so this fluctuation costs energy. It is therefore clear that, after quantization, ψ is associated to a massless mode, while χ is a massive mode. To check this formally, we insert eq. (11.13) into the Lagrangian (11.9), and we find

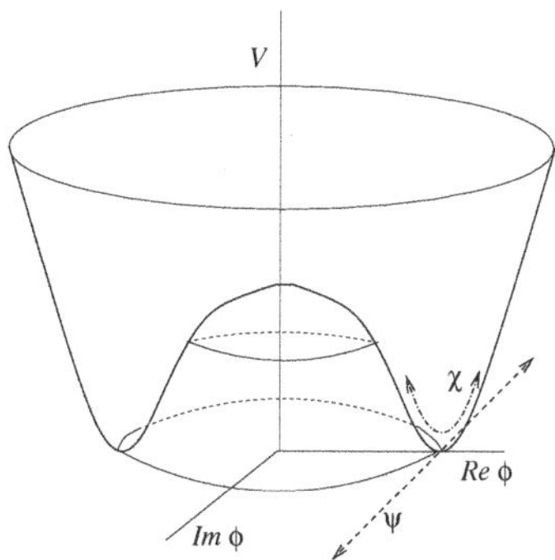

Fig. 11.2 The directions in field space parametrized by χ and ψ.

$$L = \frac{1}{2}\partial_\mu\chi\partial^\mu\chi + \frac{1}{2}\partial_\mu\psi\partial^\mu\psi - \frac{\lambda^2}{8}\left[(2\sqrt{2}\eta)\,\chi + \chi^2 + \psi^2\right]^2 . \qquad (11.14)$$

We see indeed that in this Lagrangian there is a mass term for χ,

$$\frac{1}{2}m_\chi^2 = \frac{\lambda^2}{8}(2\sqrt{2}\eta)^2 = \lambda^2\eta^2 , \qquad (11.15)$$

but there is no term of the form $(1/2)m_\psi^2\psi^2$, so ψ is massless. In conclusion, in this model the $U(1)$ symmetry (11.11) is spontaneously broken by the choice of vacuum, and at the same time a massless spin-0 boson appears in the spectrum.[2]

[2] Our discussion is oversimplified, because we assumed that the relevant quantity, for determining whether the vacuum is degenerate, is the *classical* potential $V(\phi)$. Quantum corrections in general can modify the form of the potential, and generate an effective potential (known as the Coleman–Weinberg effective potential), which is the quantity that really determines whether there is SSB or not. However, replacing $V(\phi)$ with this effective potential, our considerations are correct.

This is an example of a general theorem, the Goldstone theorem, which states that, given a field theory which is Lorentz invariant, local, and has a Hilbert space with a positive definite scalar product, if a continuous global symmetry is spontaneously broken, then in the expansion around the symmetry-breaking vacuum there appears a massless particle for each generator that breaks the symmetry. This particle is called a Goldstone (or Nambu–Goldstone) particle.

As in the above example, also in the general case the emergence of massless particles corresponds to the possibility of moving, in field space, in the direction of the manifold of vacua. The dimensionality of the manifold of vacua is equal to the number of generators which break the symmetry. In fact, setting the vacuum energy to zero, by definition we have $H|0\rangle = 0$. Since T^a is the generator of a symmetry transformation, it satisfies $[T^a, H] = 0$ and therefore

$$H(T^a|0\rangle) = T^a H|0\rangle = 0 . \qquad (11.16)$$

So, if $T^a|0\rangle \neq 0$ (and if it is not proportional to $|0\rangle$ itself, see note 1) we have found a new state with the minimum energy, i.e. another vacuum state. This is the origin of the fact that we have a Goldstone particle for each generator which breaks the symmetry.

The Goldstone theorem further states that the quantum numbers of the Goldstone particles are the same as the corresponding generator. In most cases, the global symmetry transformations are internal transformations in the field space which do not act on the Lorentz indices of the fields. For instance, in the above example the symmetry which is broken is $U(1)$, or, equivalently, an $O(2)$ rotation symmetry in the space $(\mathrm{Re}\,\phi, \mathrm{Im}\,\phi)$. These rotations do not touch Lorentz indices, and therefore the generators are Lorentz scalars. Correspondingly, the associated massless particle is a spin-0 boson.[3]

[3] However, in supersymmetry the generators exchange fermions with bosons and carry half integer spin. As a consequence, the Goldstone particles associated to global supersymmetry breaking are fermions.

In particle physics, an important example of Goldstone bosons is provided by the pions. From the discussion in Section 10.3 we know that, in the limit in which the masses of the up and down quarks can be neglected, QCD has an approximate global $SU(2) \times SU(2)$ symmetry. We can now ask how it is realized on the vacuum. If the vacuum is invariant, then the situation is completely analogous to ordinary quantum mechanics, and the spectrum of the system is organized in multiplets (degenerate in mass) of the symmetry group. On the contrary, if a generator fails to annihilate the vacuum, we have seen that there is a corresponding massless particle in the spectrum.

Therefore, if the $SU(2) \times SU(2)$ approximate global symmetry of QCD were unbroken, all strongly interacting particles should be approximately arranged in representations of $SU(2) \times SU(2)$. Since the two $SU(2)$ factors are obtained one from the other with a parity transformation, this means in particular that for each strongly interacting particle there should be a second one, approximately degenerate in mass, and with the opposite parity. Experimentally this is not the case. For instance, the three pions are pseudoscalars, and there exists no triplet of real scalars close in mass to the pions.

Rather, the experimental values of masses and quantum numbers of the strongly interacting particles point toward a different alternative: the vector $SU(2)$ is unbroken, and is in fact the isospin symmetry. Correspondingly, particles are organized in isospin multiplets almost degenerate in mass; the three pions form a triplet, the proton and neutron a doublet, etc. This explains why their mass differences, which are $O(1)$ MeV, are tiny compared to the strong interaction scale which is rather $O(100)$ MeV.

On the contrary the axial $SU(2)$ is spontaneously broken, and as a consequence we have three (because of the three generators of $SU(2)$) Goldstone bosons, which are pseudoscalar because of the γ^5 in the generators of the axial $SU(2)$ transformation, see eq. (10.38). More precisely, one uses the term quasi-Goldstone bosons to stress that these are particles which would be massless in the limit of exact symmetry; since instead this $SU(2) \times SU(2)$ symmetry is only approximate, these particles are light compared to the other hadrons. Indeed, the three pions fulfill these conditions. They are pseudoscalars, and they are the lightest strongly interacting particles, with masses $O(140)$ MeV rather than the values $O(1)$ GeV typical of the neutron and the proton.

Identifying the pions as the pseudo-Goldstone bosons of chiral symmetry allows us to write down effective Lagrangians which govern their dynamics, burying all our ignorance of QCD at large distances into a few phenomenological parameters. This is a more advanced subject, and we refer the reader to the Further Reading section.

11.3 Abelian gauge theories: SSB and superconductivity

To illustrate the effect of SSB on a theory with a local symmetry we start again from the Lagrangian (11.9), but now we gauge the $U(1)$ symmetry. Therefore we introduce a $U(1)$ gauge field A_μ and we take as Lagrangian

$$\mathcal{L} = (D_\mu\phi)^* D^\mu\phi - V(|\phi|) - \frac{1}{4}F_{\mu\nu}F^{\mu\nu}\,, \tag{11.17}$$

with

$$D_\mu\phi = (\partial_\mu + iqA_\mu)\,\phi\,. \tag{11.18}$$

As before,

$$V(|\phi|) = \frac{1}{2}\lambda^2\left(|\phi|^2 - \eta^2\right)^2. \tag{11.19}$$

To understand the physical content of the theory, it is convenient to write the complex field ϕ in terms of its modulus and a phase, and to expand the modulus around η,[4]

$$\phi(x) = |\phi(x)|\,e^{i\alpha(x)} = \left(\eta + \frac{1}{\sqrt{2}}\varphi(x)\right)e^{i\alpha(x)}\,. \tag{11.20}$$

Now observe that, since under the $U(1)$ local transformation ϕ transforms as

$$\phi(x) \to e^{iq\theta(x)}\phi(x) \tag{11.21}$$

with $\theta(x)$ the parameter of the gauge transformation, we can fix the gauge freedom setting $\alpha(x) = 0$ in eq. (11.20). In other words, we have used the gauge freedom to remove one degree of freedom from the complex field ϕ, so that we are left with just a single *real* scalar field φ. The phase $\alpha(x)$ parametrizes the manifold of vacua, so it is the field that, in the case of global symmetries, describes the Goldstone boson.[5] We see that when we break a local symmetry the Goldstone boson is eliminated from the physical spectrum by gauge invariance. After setting $\alpha(x) = 0$, using eqs. (11.18) and (11.20) we get

$$D_\mu\phi = \frac{1}{\sqrt{2}}\,(\partial_\mu\varphi) + iq\left(\eta + \frac{1}{\sqrt{2}}\varphi\right)A_\mu \tag{11.22}$$

and, substituting into eq. (11.17),

$$\begin{aligned}\mathcal{L} = {} & \frac{1}{2}\partial^\mu\varphi\partial_\mu\varphi - \lambda^2\left(\eta^2\varphi^2 + \frac{\eta\sqrt{2}}{2}\varphi^3 + \frac{1}{8}\varphi^4\right) \\ & + q^2\left(\eta + \frac{1}{\sqrt{2}}\varphi\right)^2 A_\mu A^\mu - \frac{1}{4}F_{\mu\nu}F^{\mu\nu}\,.\end{aligned} \tag{11.23}$$

We recognize a standard kinetic term for a real massive scalar field φ, with mass $m_\varphi^2 = 2\lambda^2\eta^2$. For the gauge field, the quadratic term is now

$$\mathcal{L}_A = -\frac{1}{4}F_{\mu\nu}F^{\mu\nu} + \frac{1}{2}m_A^2 A_\mu A^\mu\,, \tag{11.24}$$

[4] Polar coordinates become singular at the origin, therefore this parametrization is only useful when $\varphi \ll \eta$. However, to understand the particle content of the theory it is sufficient to limit ourselves to φ infinitesimal and therefore, as long as $\eta \neq 0$, we can use this parametrization without problems.

[5] In fact, α parametrizes exactly the direction in field space corresponding to the vacuum manifold, while the field ψ in the previous section parametrizes this direction only for infinitesimal displacements.

with

$$m_A^2 = 2q^2\eta^2 \,. \tag{11.25}$$

Taking the variation of eq. (11.24) we find the equation of motion

$$\partial_\mu F^{\mu\nu} + m_A^2 A^\nu = 0 \,. \tag{11.26}$$

Equation (11.26) is known as the Proca equation (we already met it in Exercise 4.4). Contracting it with ∂_ν we have $\partial_\mu\partial_\nu F^{\mu\nu} + m_A^2 \partial_\nu A^\nu = 0$; since $\partial_\mu\partial_\nu F^{\mu\nu} = 0$ automatically, and $m_A \neq 0$, we find

$$\partial_\nu A^\nu = 0 \,. \tag{11.27}$$

Using this condition, $\partial_\mu F^{\mu\nu} = \partial_\mu(\partial^\mu A^\nu - \partial^\nu A^\mu)$ becomes equal to $\Box A^\nu$ and eq. (11.26) gives

$$(\Box + m_A^2) A^\nu = 0 \,. \tag{11.28}$$

Therefore the Proca equation describes a massive gauge boson.

We saw in Section 2.4.1 that a vector field A^μ, from the point of view of spatial rotations, decomposes into $\mathbf{0} \oplus \mathbf{1}$. Expanding $A_\mu(x)$ in plane waves, the condition $\partial_\mu A^\mu = 0$ becomes $k_\mu \epsilon^\mu(k) = 0$ and eliminates, in a covariant way, the component with polarization vector $\epsilon_\mu(k) \sim k_\mu$, since for this polarization we have $k_\mu \epsilon^\mu(k) \sim k^2 = m_A^2 \neq 0$. In the rest frame of the particle (which exists, since $m_A \neq 0$), $k^\mu = m_A(1,0,0,0)$ and the polarization vector which has been eliminated is $\epsilon^\mu(k) = (1,0,0,0)$ which, from the point of view of spatial rotations, is a scalar. Therefore eq. (11.27) eliminates the spin-0 part and we are left with a pure spin 1. In conclusion, eq. (11.26) describe a massive spin-1 particle.

From this example we learn that the spontaneous breaking of a *local* symmetry does not produce Goldstone bosons, but instead the gauge field has acquired a mass proportional to the vacuum expectation value of the scalar field. In this context the scalar field ϕ is called a Higgs field, and the mechanism that produces a mass for the gauge boson is called the *Higgs mechanism.* It is interesting to compare the number of degrees of freedom with and without SSB. If in the potential we set $\eta = 0$, then there is no SSB; the scalar field has two real components. We cannot use the gauge invariance to eliminate the phase θ as before, because when $\eta = 0$ the decomposition (11.20) of ϕ in terms of two real fields φ, θ is not well defined: in fact in this case $\varphi = \sqrt{2}|\phi|$ and therefore $\varphi \geqslant 0$, so it is no longer a scalar field which can freely perform at least infinitesimal fluctuations around $\varphi = 0$. Rather, gauge invariance can be used to eliminate the longitudinal components of A_μ, as we studied when we quantized the free electromagnetic field in Section 4.3.2, and the remaining gauge field has two physical degrees of freedom, the two transverse polarizations. In total we have two physical degrees of freedom from the Higgs field and two from the gauge field. After SSB, the scalar field has just one real component, but the gauge field is massive, and a massive spin-1 particle has three degrees of freedom. In total, we have $1 + 3 = 4$ degrees of freedom. So, the Higgs mechanism implies a reshuffling of the degrees of freedom. The field that, in the case of a global symmetry, was a Goldstone boson, is turned into the third polarization state of a massive spin-1 particle.

One might ask what do we really gain by giving a mass to the gauge boson with the Higgs mechanism, rather than adding by hand a mass term $(1/2)m_A^2 A_\mu A^\mu$ to the Lagrangian (a mass term for the gauge field generated by SSB is called a *soft* mass term, in contrast to a term added by hand, which is called a *hard* mass term). The point is that, in the Higgs mechanism, the Lagrangian *is* gauge invariant, which is not the case if we instead add by hand a mass term. The breaking of the symmetry takes place at the level of the vacuum. It can be shown that such a spontaneous breaking preserves a number of good properties of the unbroken theory, and in particular the theory is still renormalizable. Intuitively, this comes from the fact that at very high energies $E \gg \eta$ we can neglect η and the UV properties of the theory are the same as in the case $\eta = 0$. If we instead break the gauge symmetry by hand the theory is not renormalizable.

Spontaneous breaking of the $U(1)$ gauge symmetry is realized in Nature in the phenomenon of *superconductivity*. Let us recall that the relation between the electric current $\mathbf{j}$ and an applied external electric field $\mathbf{E}$ is $\mathbf{j} = \sigma\mathbf{E}$, where the proportionality constant σ is called the conductivity. A superconductor is an object where $\sigma = \infty$. In a piece of material with finite volume we have a finite number of electrons, so we cannot have an infinite current, and therefore the electric field $\mathbf{E}$ is forced to be zero inside the superconductor, and the Maxwell equation $\dot{\mathbf{B}} = -\nabla \times \mathbf{E}$ states that the magnetic field $\mathbf{B}$ is constant in time. Therefore, if $\mathbf{B}$ was zero at some initial time, it will remain zero inside the superconductor even if we switch on an external magnetic field outside the superconductor. This means the field lines of the applied external magnetic field cannot penetrate inside the superconductor (*Meissner effect*). At the microscopic level, what happens is that the electrons in the superconductor form currents on the surface, which screen the external field.[6] There is therefore a characteristic screening length l, and inside the superconductor the external magnetic field drops exponentially,

$$B(x) = B(0)e^{-x/l}, \tag{11.29}$$

where $x = 0$ represents the interface between the superconductor (at $x > 0$) and the external space. The physical mechanism behind superconductivity is that, due basically to an interaction mediated by phonons, pairs of electrons bind together in a singlet state, forming the so-called Cooper pairs. This composite object is therefore described, at the level of effective theory, by a charged scalar field, with charge equal to twice the electron charge. The effective Lagrangian describing the interaction of this scalar field with the electromagnetic field is given by eq. (11.17) (with $q = 2e$). The result (11.29) is then understood in terms of the Higgs mechanism: the scalar field describing the Cooper pair plays the role of the Higgs field and develops a vacuum expectation value; as a consequence, the photon acquires a mass μ, and its wave equation becomes eq. (11.28). Then also the electric and magnetic fields

[6] Since a given piece of superconducting material, with a finite volume, has only a finite number of electrons, there is a maximum magnetic field B_c that can be screened. If the applied external field is higher than B_c, it turns out that the field penetrates in a non-homogeneous manner. For type II superconductors, the magnetic field penetrates in the form of narrow flux tubes.

satisfy a massive KG,

$$(\Box + \mu^2)\,\mathbf{E} = 0\,, \qquad (\Box + \mu^2)\,\mathbf{B} = 0\,. \tag{11.30}$$

When we switch on an external magnetic field, after a transient time we will have a static field configuration. Therefore the equation for $\mathbf{B}$ becomes $\nabla^2\mathbf{B} = 0$ at $x < 0$ and $(\nabla^2 - \mu^2)\mathbf{B} = 0$ at $x > 0$. The solution of this equation, at $x > 0$, is given by (11.29) with l identified with μ^{-1}. The penetration length is therefore the inverse of the mass that the photon has inside a superconductor.

11.4 Non-abelian gauge theories: the masses of $W^\pm$ and Z^0

We consider now an $SU(2)$ gauge theory with a doublet of complex scalar fields ϕ^α, with $\alpha = 1, 2$, transforming in the fundamental representation. We call ϕ the Higgs field. The covariant derivative is

$$(D_\mu\phi)^\alpha = \partial_\mu\phi^\alpha - igA_\mu^a(T^a)^\alpha{}_\beta\phi^\beta\,, \tag{11.31}$$

and the generators T^a in the fundamental representation, for $SU(2)$, are $T^a = \sigma^a/2$, where σ^a are the Pauli matrices.

Since $\phi^\dagger\phi$ is invariant under $\phi \to U\phi$ with U unitary, any function of $\phi^\dagger\phi$ is gauge invariant, and we can also write a gauge-invariant potential term $V(\phi^\dagger\phi)$. Therefore the Lagrangian is

$$\mathcal{L}_{SU(2)-\text{Higgs}} = (D^\mu\phi)^\dagger(D_\mu\phi) - V(\phi^\dagger\phi) - \frac{1}{4}F_{\mu\nu}^a F^{a\,\mu\nu}\,, \tag{11.32}$$

and we choose

$$V(\phi^\dagger\phi) = \frac{1}{2}\lambda^2\left(\phi^\dagger\phi - \eta^2\right)^2\,. \tag{11.33}$$

We have a degenerate family of vacua, at $\phi^\dagger\phi = \eta^2$. Following the same strategy used for the SSB of the $U(1)$ gauge invariance, we use the gauge freedom to eliminate some components of ϕ (similarly to the elimination of $\alpha(x)$ in the previous section). Here we have a field ϕ with two complex components, i.e. four real components, and an $SU(2)$ transformation, which has three parameters. We can then use the gauge freedom to eliminate three of the four components of ϕ, writing it as

$$\phi = \begin{pmatrix} 0 \\ \eta + \frac{1}{\sqrt{2}}\chi \end{pmatrix}\,, \tag{11.34}$$

where χ is a *real* scalar field. It is convenient to introduce the matrices

$$\sigma^+ = \frac{1}{\sqrt{2}}\left(\sigma^1 + i\sigma^2\right) = \sqrt{2}\begin{pmatrix} 0 & 1 \\ 0 & 0 \end{pmatrix} \tag{11.35}$$

$$\sigma^- = \frac{1}{\sqrt{2}}\left(\sigma^1 - i\sigma^2\right) = \sqrt{2}\begin{pmatrix} 0 & 0 \\ 1 & 0 \end{pmatrix} \tag{11.36}$$

and the fields

$$A_\mu^\pm = \frac{1}{\sqrt{2}}\left(A_\mu^1 \pm iA_\mu^2\right) \tag{11.37}$$

so that

$$\sigma^a A_\mu^a = \sigma^+ A_\mu^- + \sigma^- A_\mu^+ + \sigma^3 A_\mu^3 . \tag{11.38}$$

Then, the covariant derivative becomes

$$D_\mu \phi = \begin{pmatrix} 0 \\ \frac{1}{\sqrt{2}}\partial_\mu \chi \end{pmatrix} - i\frac{g}{2}(\eta + \frac{\chi}{\sqrt{2}}) \begin{pmatrix} \sqrt{2}A_\mu^- \\ -A_\mu^3 \end{pmatrix} . \tag{11.39}$$

Recalling from the definition (11.37) that $(A_\mu^-)^* = A_\mu^+$, we find

$$(D^\mu\phi)^\dagger(D_\mu\phi) = \frac{1}{2}\partial^\mu\chi\partial_\mu\chi + \frac{g^2}{4}\left(\eta + \frac{\chi}{\sqrt{2}}\right)^2 A^{3\,\mu}A_\mu^3$$
$$+\frac{g^2}{2}\left(\eta + \frac{\chi}{\sqrt{2}}\right)^2 A_\mu^+ A^{-\,\mu} . \tag{11.40}$$

Apart from the standard kinetic term of the χ field and from cubic and quartic couplings between χ and the gauge fields, we recognize a mass term for A_μ^3,

$$\frac{1}{2}m_A^2 = \frac{g^2}{4}\eta^2 \tag{11.41}$$

and, using

$$A_\mu^+ A^{-\,\mu} = \frac{1}{2}(A_\mu^1 A^{1\,\mu} + A_\mu^2 A^{2\,\mu}) , \tag{11.42}$$

we see that the term $(g^2/2)\eta^2 A_\mu^+ A^{-\,\mu}$ gives the same mass m_A to both A_μ^1 and A_μ^2 or, equivalently, to their linear combinations $A_\mu^\pm$. Therefore all three gauge bosons become massive, with a mass

$$m_A = \frac{g\eta}{\sqrt{2}} . \tag{11.43}$$

In the Standard Model the situation is similar, but the gauge group now is $SU(2) \times U(1)$. We have three gauge bosons A_μ^a associated with $SU(2)$ and one gauge boson B_μ associated to $U(1)$, and two different gauge couplings, g for $SU(2)$ and g' for $U(1)$. This means that on a field in a generic representation the covariant derivative is

$$D_\mu = \partial_\mu - igT^a A_\mu^a - ig'SB_\mu \tag{11.44}$$

where T^a are the $SU(2)$ generators in the representation of interest, and S is the charge of the particle in question relative to the $U(1)$ group, i.e the parameter that labels the $U(1)$ representation.

The Higgs boson ϕ is an $SU(2)$ doublet (so that on it $T^a = \sigma^a/2$) and is given the assignment $S = 1/2$.[7] Therefore

$$D_\mu\phi = \left(\partial_\mu - ig\frac{\sigma^a}{2}A_\mu^a - i\frac{g'}{2}B_\mu\right)\phi . \tag{11.45}$$

[7] Note that both components of the doublet have the same assignment of S. In general, on any $SU(2)$ multiplet, S is a constant times the unit matrix, which means that S commutes with the $SU(2)$ generators, as it should be, since the gauge group is the direct product of $SU(2)$ and $U(1)$.

The potential for the Higgs field is the same as in eq. (11.33) so that again we can choose a gauge such that

$$\phi = \begin{pmatrix} 0 \\ \eta + \frac{1}{\sqrt{2}}\chi \end{pmatrix} . \tag{11.46}$$

Computing $(D^\mu \phi)^\dagger D_\mu \phi$ using eq. (11.45) we find terms quadratic in the gauge fields, of the form

$$\frac{1}{4}\eta^2 (gA^3_\mu - g' B_\mu)(gA^{3\,\mu} - g' B^\mu) + \frac{1}{2} g^2 \eta^2 A^+_\mu A^{-\,\mu} . \tag{11.47}$$

It is convenient to introduce the notation

$$\bar{g} = \sqrt{g^2 + g'^2} , \qquad g/\bar{g} = \cos\theta_W , \qquad g'/\bar{g} = \sin\theta_W \tag{11.48}$$

where θ_W is the Weinberg angle. We also change notation, $W^\pm_\mu = A^\pm_\mu$, and we define

$$Z^0_\mu \equiv A^3_\mu \cos\theta_W - B_\mu \sin\theta_W . \tag{11.49}$$

Then we see from eq. (11.47) that the Z boson gets a mass

$$m_Z = \frac{1}{\sqrt{2}} \bar{g} \eta \tag{11.50}$$

while the W-bosons get a mass

$$m_W = \frac{1}{\sqrt{2}} g \eta . \tag{11.51}$$

The ratio of the W to Z mass is therefore given in terms of the Weinberg angle,

$$\frac{m_W}{m_Z} = \cos\theta_W . \tag{11.52}$$

Instead, the other orthogonal combination of A^3_μ and B_μ,

$$A_\mu \equiv A^3_\mu \sin\theta_W + B_\mu \cos\theta_W \tag{11.53}$$

remains massless and is therefore identified with the photon.

Summary of chapter

- SSB takes place when, rather than a single vacuum invariant under the symmetry in question, we have a family of vacua which transform among themselves under the action of the symmetry group. The system will eventually settle into one of these vacua, and the symmetry is spontaneously broken by this choice.
- If we have a quantum system with a finite number of degrees of freedom there is in general an exponentially small, but nevertheless finite, amplitude for tunneling between the different perturbative vacua. The true vacuum will be a superposition of the perturbative vacua which respects the symmetry, and therefore there is no SSB. However, in a system with an infinite number of degrees of freedom, as in QFT, the tunneling amplitude is zero because each degree of freedom should tunnel, and therefore SSB is possible.

- When a *global* symmetry is spontaneously broken, in the spectrum of the theory there is a massless particle for each broken symmetry generators. In particular, the pions are the Goldstone bosons associated to the SSB of the axial $SU(2)$ symmetry of QCD. They would be exactly massless if the symmetry were exact. Since it is only approximate, they are just lighter than the other hadrons.
- When a *local* symmetry is spontaneously broken, the gauge field becomes massive and the would-be Goldstone boson is turned into the third physical degree of freedom of the massive spin-1 gauge field. This mechanism gives an effective mass to the photon in a superconductor (which is at the origin of the Meissner effect) and gives a mass to the gauge bosons $W^{\pm}$ and Z^0 of the electroweak theory.

Further reading

- For spontaneous symmetry breaking in gauge theories, a clear discussion is given for instance in Okun (1982), Chapter 20 and in Coleman (1985), Chapter 5.
- An advanced discussion of SSB can be found in (Weinberg), vol II, Chapters 19 and 21.
- For a discussion of pion dynamics and chiral Lagrangians see Georgi (1984), Chapter 5, Coleman (1985), Chapter 2 and (Weinberg), vol II, Chapter 19.

12 Solutions to exercises

12.1 Chapter 1

(1.1) Since photons are massless the only energy scale is provided by $k_B T$. Dimensionally, in units $\hbar = c = 1$, an energy density is $(\text{mass})^4$, therefore the photon density must be $\rho_\gamma \sim (k_B T)^4$. This gives ρ_γ in units $(\text{eV})^4$. Transforming to GeV/cm^3 using $200\,\text{MeV fm} \simeq 1$ gives $\rho_\gamma/\rho_c \sim 5 \times 10^{-5}$. At the present epoch of the Universe the energy density in photons, or more generally in relativistic particles, is much smaller than in non-relativistic matter.

(1.2) A temperature $T \simeq 4.5 \times 10^6$ K corresponds to an energy $k_B T \simeq 388$ eV (using $k_B T \simeq 1/38.68$ eV at $T = 300$ K). For a relativistic particle at the equilibrium temperature T, the average energy is $E \simeq 3k_B T$ and therefore the average photon energy is $E_\gamma = O(1)$ keV. Since $E_\gamma \ll m_e \ll m_p$, we can use the Thompson formula (1.16) for the scattering on electrons and the same formula, with m_e replaced by m_p, for the scattering on protons. Therefore $\sigma(\gamma p \to \gamma p) \simeq 8\pi\alpha^2/(3m_p^2)$. This is smaller than the $\gamma e \to \gamma e$ cross-section by a factor m_e^2/m_p^2 and therefore the contribution of the protons to l is negligible. Because of electric charge neutrality, in our simplified model of the Sun the electron number density is equal to the proton number density and is $n = \rho/(m_e + m_p) \simeq \rho/m_p \simeq 0.8 \times 10^{24}\,\text{cm}^{-3}$. Inserting the numerical value for the Thompson cross-section, $\sigma(\gamma e \to \gamma e) \simeq 6.65 \times 10^{-25}\,\text{cm}^2$, we find $l \simeq 1.8$ cm. More accurate modeling of the Sun gives $l \simeq 0.5$ cm. The photons therefore perform a random walk of step l inside the Sun. For a random walk in one dimension, after N steps we have $\langle x^2 \rangle = Nl^2$. In three dimensions a radial distance $R_\odot$ is covered in N steps with $R_\odot^2 = (1/3)Nl^2$ because, if we denote by x the axis along which the photon finally escaped, not all steps have been performed along the x direction. Rather in each step $\langle x^2 + y^2 + z^2 \rangle$ increases by l^2 , so $\langle x^2 \rangle$ effectively performs a random walk of step $l^2/3$. Therefore we get an escape time $t = Nl/c = 3R_\odot^2/(lc) \simeq 3 \times 10^4$ yr.

(1.3) For slow particles the largest length-scale is the De Broglie wavelength $\lambda = 1/(mv)$. For the neutron $m \sim 939.56$ MeV, so $E = (1/2)mv^2 \sim 1$ MeV gives $v \sim 0.046$, $\lambda \sim 4.5$ fm and $\sigma \sim \pi\lambda^2 \sim 0.7$ barn.

12.2 Chapter 2

(2.1) In the rest frame of the particle, $E = m$ and $\mathbf{p} = 0$. Performing a boost along a direction, say the x-axis, E and $p \equiv p^x$ transform as t and x in eq. (2.18), so after the boost $E = m\cosh\eta$ and $p = m\sinh\eta$, and therefore $(E+p)/(E-p) = e^{2\eta}$.
Performing a further boost with rapidity η' in the same direction, $E \to E\cosh\eta' + p\sinh\eta'$ and $p \to E\sinh\eta' + p\cosh\eta'$, so

$$\begin{aligned} e^{2\eta} &\to \frac{(E\cosh\eta' + p\sinh\eta') + (E\sinh\eta' + p\cosh\eta')}{(E\cosh\eta' + p\sinh\eta') - (E\sinh\eta' + p\cosh\eta')} \\ &= e^{2\eta'}\,\frac{E+p}{E-p} = e^{2\eta+2\eta'}\,. \end{aligned} \tag{12.1}$$

(2.2) A generic tensor $T^{i_1\ldots i_N}$ without any symmetry properties, from the point of view of angular momenta is the direct product of N times the vector representation, $T^{i_1\ldots i_N} = \mathbf{1}\otimes\mathbf{1}\otimes\ldots\otimes\mathbf{1}$, so it contains spin up to $j = N$. Decomposing $T^{i_1\ldots i_N}$ in irreducible representations, we must remove the traces and each pair of indices must be symmetrized or antisymmetrized. When we remove a trace two indices are contracted and we are left with a tensor with two less indices, which can have only up to spin $N-2$. When we antisymmetrize over two indices (i,j) we can then contract with ϵ^{ijk}, so we obtain a tensor with one less index, and maximum spin $N-1$. Therefore the spin N in $T^{i_1\ldots i_N}$ can be neither in the traces nor in the tensors in which some indices have been antisymmetrized, and must be in the totally symmetric and traceless tensor.
Typical examples are for instance the quadrupole moment of a mass distribution $\rho(x)$ (or of a charge distribution),

$$Q^{ij} = \int d^3x\,\rho(x)(x^ix^j - \frac{1}{3}\delta^{ij}x^2)\,. \tag{12.2}$$

which is a spin-2 operator. A spin-3 operator is the octupole moment,

$$\mathcal{O}^{ijk} = M^{ijk} - \frac{1}{5}\left(\delta^{ij}M^{llk} + \delta^{ik}M^{ljl} + \delta^{jk}M^{ill}\right)\,, \tag{12.3}$$

where the index l is summed over and

$$M^{ijk} = \int d^3x\,\rho(x)x^ix^jx^k\,. \tag{12.4}$$

(2.3) Let $v^0 = \xi_R^\dagger\psi_R$ and $v^i = \xi_R^\dagger\sigma^i\psi_R$. We verify that under boosts v^0 and v^i transform as appropriate for a contravariant four-vector. We can always take the x-axis as the boost direction, and it is also sufficient to consider an infinitesimal boost. Using eq. (2.60),

$$\begin{aligned} \xi_R^\dagger\psi_R &\to \xi_R^\dagger e^{\eta\sigma^1}\psi_R \simeq \xi_R^\dagger\psi_R + \eta\,\xi_R^\dagger\sigma^1\psi_R\,, \\ \xi_R^\dagger\sigma^1\psi_R &\to \xi_R^\dagger e^{\eta\sigma^1}\sigma^1\psi_R \simeq \eta\,\xi_R^\dagger\psi_R + \xi_R^\dagger\sigma^1\psi_R\,. \end{aligned} \tag{12.5}$$

Therefore $v^0 \to v^0 + \eta v^1$ and $v^1 \to \eta v^0 + v^1$, which is the infinitesimal form of eq. (2.18). Observe that instead $\bar{v}^\mu \equiv \xi_R^\dagger \bar{\sigma}^\mu \psi_R$ is *not* a contravariant four-vector since, under the transformation (2.18), $\bar{v}^0 \to \bar{v}^0 - \eta\bar{v}^1$ and $\bar{v}^1 \to -\eta\bar{v}^0 + \bar{v}^1$, i.e. they mix with $-\eta$ rather than $+\eta$. For the left-handed spinors the transformation matrix is $\exp\{-\eta\,\sigma^1/2\}$ instead of $\exp\{+\eta\,\sigma^1/2\}$, and the situation is reversed: $\xi_L^\dagger \bar{\sigma}^\mu \psi_L$ is a contravariant four-vector while $\xi_L^\dagger \sigma^\mu \psi_L$ is not. We can also verify directly the transformation properties under *finite* Lorentz transformation, using the identity

$$e^{\eta\sigma^1} = \cosh\eta + \sigma^1 \sinh\eta\,, \tag{12.6}$$

which can be proved performing the Taylor expansion of the exponential and using the fact that $(\sigma^1)^2 = 1$.

(2.4) $F^{\mu\nu} \to \Lambda^\mu{}_\rho \Lambda^\nu{}_\sigma F^{\rho\sigma}$, where $\Lambda^\mu{}_\rho = \exp\{-(i/2)\omega_{\alpha\beta}(J^{\alpha\beta})^\mu{}_\rho\}$ and $(J^{\alpha\beta})^\mu{}_\rho$ is given by eq. (2.23), since the tensor representations are obtained iterating on each index the transformation matrix of the four-vector representation. Expanding to first order in $\omega_{\alpha\beta}$ and performing the contractions,

$$\delta F^{\mu\nu} = \omega^\mu{}_\rho F^{\rho\nu} - \omega^\nu{}_\rho F^{\rho\mu}\,. \tag{12.7}$$

In terms of **E** and **B**,

$$\begin{aligned} \delta\mathbf{E} &= -\boldsymbol{\eta}\times\mathbf{B} + \boldsymbol{\theta}\times\mathbf{E}\,, \\ \delta\mathbf{B} &= +\boldsymbol{\eta}\times\mathbf{E} + \boldsymbol{\theta}\times\mathbf{B}\,. \end{aligned} \tag{12.8}$$

(2.5) (i) Writing explicitly the six conditions $A^{\mu\nu} = (1/2)\epsilon^{\mu\nu\rho\sigma}A_{\rho\sigma}$ we find $A^{01} = A_{23}$, $A^{02} = -A_{13}$, $A^{03} = A_{12}$, $A^{12} = A_{03}$, $A^{13} = -A_{02}$ and $A^{23} = A_{01}$. With the Minkowski metric, the first condition $A^{01} = A_{23}$ becomes $A^{01} = A^{23}$ while the last conditions $A^{23} = A_{01}$ becomes $A^{23} = -A^{01}$, and together they give $A^{01} = A^{23} = 0$. Similarly for the other conditions, so in the Minkowski case we are left with $A^{\mu\nu} = 0$.
(ii) If instead we raise the indices with $\delta^{\mu\nu}$ the conditions $A^{01} = A_{23}$ and $A^{23} = A_{01}$ are identical, so in total we have only three independent conditions $A^{01} = A^{23}$, $A^{02} = -A^{13}$ and $A^{03} = A^{12}$. Similarly an anti-self-dual tensor satisfies $A^{01} = -A^{23}$, $A^{02} = A^{13}$ and $A^{03} = -A^{12}$.
For $SO(4)$, $\epsilon^{\mu\nu\rho\sigma}$ is an invariant tensor (as for $SO(3,1)$, it follows from the condition $\det\Lambda = 1$). Therefore, if the condition $A^{\mu\nu} = (1/2)\epsilon^{\mu\nu\rho\sigma}A_{\rho\sigma}$ holds in a frame, it holds in all Lorentz-transformed frames, so a self-dual tensor remains self-dual, and an anti-self-dual tensor remains anti-self-dual. This means that self-dual and anti-self-dual tensors are irreducible representations of $SO(4)$, and that in Euclidean space a six-dimensional real antisymmetric tensor $A^{\mu\nu}$ decomposes into its self-dual and anti-self-dual parts.

(iii) With the Minkowski metric the conditions $A^{01} = iA_{23}$ and $A^{23} = iA_{01}$ become $A^{01} = iA^{23}$ and $A^{23} = -iA^{01}$ and therefore are identical, and similarly for the other conditions, so we are left with three independent conditions, $A^{01} = iA^{23}, A^{02} = -iA^{13}, A^{03} = iA^{12}$. The duality conditions are Lorentz-invariant so self-dual and anti-self-dual tensors are irreducible representations of $SO(3,1)$. However, the Minkowskian duality conditions make sense only if $A^{\mu\nu}$ is complex, so it can be used only to decompose a tensor $A^{\mu\nu}$ with six independent *complex* components into its self-dual and anti-self-dual parts, each with three complex components, i.e. each with six real degrees of freedom. Since under parity $\epsilon^{\mu\nu\rho\sigma}$ is a pseudotensor, a parity transformation exchanges the self-dual and anti-self-dual parts. Comparison with the classification of Lorentz representations in terms of the (j_-, j_+) quantum numbers show that they are the $(\mathbf{0}, \mathbf{1})$ and $(\mathbf{1}, \mathbf{0})$ representations. Observe that these representations have *complex* dimension three.

(iv) In terms of the electric and magnetic fields E^i, B^i and of the variables $a^i_+ = (-1/2)(E^i + iB^i)$, $a^i_- = (-1/2)(E^i - iB^i)$ we can write $F^{\mu\nu} = F^{\mu\nu}_+ + F^{\mu\nu}_-$ with

$$
\begin{aligned}
F^{\mu\nu} &= \begin{pmatrix} 0 & -E^1 & -E^2 & -E^3 \\ E^1 & 0 & -B^3 & B^2 \\ E^2 & B^3 & 0 & -B^1 \\ E^3 & -B^2 & B^1 & 0 \end{pmatrix}, \\
F^{\mu\nu}_{\pm} &= \begin{pmatrix} 0 & a^1_{\pm} & a^2_{\pm} & a^3_{\pm} \\ -a^1_{\pm} & 0 & \mp i a^3_{\pm} & \pm i a^2_{\pm} \\ -a^2_{\pm} & \pm i a^3_{\pm} & 0 & \mp i a^1_{\pm} \\ -a^3_{\pm} & \mp i a^2_{\pm} & \pm i a^1_{\pm} & 0 \end{pmatrix}.
\end{aligned}
\tag{12.9}
$$

The six independent real components of $F^{\mu\nu}$ have been written in terms of the three complex components a^i_+ of the self-dual tensor $F^{\mu\nu}_+$, and of their complex conjugate a^i_- which are the components of the anti-self-dual tensor $F^{\mu\nu}_-$. This is *not* a decomposition into representations of smaller dimensions. We have just rewritten a six-dimensional real representation in terms of a three-dimensional complex representation. Under a general Lorentz transformations the three components of E^i and the three components of B^i mix between themselves, so a real antisymmetric tensor is an irreducible representation of real dimension six.

(2.6) (i) In the (x, y) plane, $\mathbf{e}^1 = (1, 0) \to (\cos\theta, \sin\theta)$, $\mathbf{e}^2 = (0, 1) \to (-\sin\theta, \cos\theta)$, $\mathbf{e}^{\pm} \to e^{\mp i\theta}\mathbf{e}^{\pm}$, so from eq. (2.131) $\mathbf{e}^+$ has helicity $h = +1$ and $\mathbf{e}^-$ has $h = -1$. According to the discussion in Section 2.7, this means that electromagnetic waves are made of massless spin-1 particles, the photons.

(ii) The transformation of the tensor h^{ij} under rotations in the

(x, y) plane is $h^{ij} \to R^{ik} R^{jl} h^{kl}$, i.e. $h \to RhR^T$, with

$$h = \begin{pmatrix} h_+ & h_\times \\ h_\times & -h_+ \end{pmatrix}, \qquad R = \begin{pmatrix} \cos\theta & -\sin\theta \\ \sin\theta & \cos\theta \end{pmatrix}. \tag{12.10}$$

Performing the matrix multiplication we find

$$\begin{aligned} h_\times &\to h_\times \cos 2\theta + h_+ \sin 2\theta \,, \\ h_+ &\to -h_\times \sin 2\theta + h_+ \cos 2\theta \end{aligned} \tag{12.11}$$

and therefore $(h_\times \pm ih_+) \to e^{\mp 2i\theta}(h_\times \pm ih_+)$ which, according to eq. (2.131), means that they have helicities ± 2.

12.3 Chapter 3

(3.1) The dimensions are read from the kinetic terms. For a scalar $(\partial_\mu \phi)^2$ must have dimensions $(\text{mass})^4$ to compensate the factor d^4x. Since $\partial_\mu \sim$ mass, it follows that ϕ has dimensions of mass. Similarly $A_\mu \sim$ (mass) and $\psi \sim (\text{mass})^{3/2}$. In d space-time dimensions $\phi \sim A_\mu \sim (\text{mass})^{(d/2)-1}$ while $\psi \sim (\text{mass})^{(d-1)/2}$.

(3.2) Consider first u_L. Under a boost of rapidity η along the z axis we have (see eq. (2.59)) $u_L \to \exp\{-\eta\sigma^3/2\} u_L$. Use the identity

$$\exp\left\{\boldsymbol{\eta} \cdot \frac{\boldsymbol{\sigma}}{2}\right\} = \cosh \frac{|\boldsymbol{\eta}|}{2} + \hat{\boldsymbol{\eta}} \cdot \boldsymbol{\sigma} \sinh \frac{|\boldsymbol{\eta}|}{2} \,. \tag{12.12}$$

Inverting $\tanh\eta = v$ we get $e^{2\eta} = (1+v)/(1-v)$. From this verify that $\cosh(\eta/2) = \left(\frac{E+m}{2m}\right)^{1/2}$ and therefore $\sinh(|\eta|/2) = \left(\frac{E-m}{2m}\right)^{1/2}$. Pay attention to the fact that, in order to transform a particle at rest into a particle moving with velocity $+v$, we must perform a boost with velocity $-v$. Then verify that in the boosted frame

$$u_L = \frac{1}{\sqrt{2}} \left[\sqrt{E+m} - \sigma^3 \sqrt{E-m}\right] \xi \,. \tag{12.13}$$

Finally verify that, for a particle moving along the z axis,

$$\frac{1}{\sqrt{2}} \sqrt{E \pm m} = \frac{1}{2} \left(\sqrt{E+p^3} \pm \sqrt{E-p^3}\right), \tag{12.14}$$

and therefore eq. (3.103) is recovered. For u_R, under boost $u_R \to \exp\{+\eta\sigma^3/2\} u_R$ and therefore the result is recovered with the replacement $p^3 \to -p^3$.

(3.3) (i) d_ϕ must be equal to 1, i.e. to the mass dimensions of ϕ. Then $\partial\phi/\partial x^\mu \to \partial\phi'/\partial x'^\mu = e^{-2\alpha} \partial\phi/\partial x^\mu$ and $(\partial\phi)^2$ cancels the factor $e^{4\alpha}$ coming from d^4x. The current is

$$j_D^\mu = (\phi + x^\nu \partial_\nu \phi)\partial^\mu \phi - \frac{1}{2} x^\mu \partial_\nu \phi \partial^\nu \phi \,. \tag{12.15}$$

(ii) $\phi^2 \to e^{-2\alpha}\phi^2$ so $d^4x\, \phi^2$ is not invariant, while $d^4x\, \phi^4$ is invariant. Dilatations are a classical symmetry when there is no intrinsic mass-scale, so they are broken by a mass term but not by a term $\lambda\phi^4$ since λ is dimensionless.

(3.4) (i) $d_A = 1, d_\psi = 3/2$. (ii) From the Noether theorem (and eliminating terms that vanish upon use of the equations of motion)

$$j_D^\mu = x^\nu(\delta_\nu^\mu \frac{1}{4}F^2 - F^{\mu\rho}\partial_\nu A_\rho) + x^\nu \bar{\psi} i\gamma^\mu \partial_\nu \psi + \frac{3}{2}\bar{\psi} i\gamma^\mu \psi - F^{\mu\rho}A_\rho\,. \tag{12.16}$$

After some algebra, this can be rewritten as

$$j_D^\mu = \frac{3}{2}j^\mu + x^\nu T^\mu{}_\nu - \partial_\rho(F^{\mu\rho}x^\nu A_\nu) \tag{12.17}$$

where

$$T^\mu{}_\nu = \delta_\nu^\mu \frac{1}{4}F^2 - F^{\mu\rho}F_{\nu\rho} + \bar{\psi}\gamma^\mu(i\partial_\nu - eA_\nu)\psi\,, \tag{12.18}$$

and $j^\mu = i\bar{\psi}\gamma^\mu\psi$ is the $U(1)$ current. Since j^μ is conserved by itself, we can redefine the dilatation current subtracting it. Furthermore, the term $\partial_\rho(F^{\mu\rho}x^\nu A_\nu)$ does not contribute to the charge since its $\mu = 0$ component is a total spatial derivative, and also it is separately conserved, so we subtract it, too, from the definition of j_D^μ. Then $j_D^\mu = x^\nu T^\mu{}_\nu$ and $\partial_\mu j_D^\mu = x^\nu \partial_\mu T^\mu{}_\nu + T^\mu{}_\mu$. The term $\partial_\mu T^\mu{}_\nu$ vanishes because the energy–momentum tensor is conserved, while, from the above equation, $T^\mu{}_\mu = \bar{\psi}\gamma^\mu(i\partial_\mu - eA_\mu)\psi = 0$ using the massless Dirac equation.
(iii) Upon use of the equations of motion of the massive theory, j_D^μ happens to have the same form as in the massless case. However, again using the equations of motion of the massive theory, now

$$\partial_\mu j_D^\mu = T^\mu{}_\mu = m\bar{\psi}\psi\,. \tag{12.19}$$

The invariance under dilatations is broken if the trace of the energy–momentum tensor is non-vanishing.

(3.5) The two Lagrangians differ by a total derivative,

$$\mathcal{L}' = \mathcal{L} - (i/2)\partial_\mu(\bar{\psi}\gamma^\mu\psi)\,. \tag{12.20}$$

With $\mathcal{L}$, we find $T^{\mu\nu} = i\bar{\psi}\gamma^\mu\partial^\nu\psi$. With $\mathcal{L}'$, we find $T'^{\mu\nu} = T^{\mu\nu} - (i/2)\partial^\nu j^\mu$ with $j^\mu = \bar{\psi}\gamma^\mu\psi$. The extra term $(-i/2)\partial^\nu j^\mu$ does not spoil $\partial_\mu T^{\mu\nu} = 0$ because $\partial_\mu j^\mu = 0$. The conserved charges P^ν differ by a term proportional to $\int d^3x\, \partial^\nu j^0$. However this is zero because, if ν is a spatial index, it is a spatial derivative and then the spatial integral vanishes, assuming as always a sufficiently fast decrease of the fields at infinity. If instead $\nu = 0$ we use $\partial_0 j^0 = -\partial_i j^i$ so we get again a spatial divergence. Therefore the four-momentum computed with $T^{\mu\nu}$ and with $T'^{\mu\nu}$ is the same.

(3.6) We denote $(t, \mathbf{x})$ by x. Then the five-dimensional field is $\phi(x, y)$. We impose the boundary condition that $\phi(x, \pm R/2) = 0$, corresponding to the fact that the field vanishes at the boundary of space-time. The mode expansion compatible with these boundary conditions is

$$\phi(x, y) = \sum_{n=1}^{\infty} \phi_n(x)\,\mathrm{cs}\left(\frac{n\pi y}{R}\right)\,, \tag{12.21}$$

where $\mathrm{cs}(n\pi y/R)$ is $\cos(n\pi y/R)$ if n is odd and $\sin(n\pi y/R)$ if n is even. We therefore have an infinite set of four-dimensional fields $\phi_n(x)$. The fact $\phi(x,y)$ satisfies $(\Box_5 + m^2)\phi = 0$ implies that the fields $\phi_n(x)$ satisfy

$$\left[\Box + m^2 + \left(\frac{n\pi}{R}\right)^2\right] \phi_n(x) = 0 \qquad (12.22)$$

and therefore each $\phi_n(x)$, with $n = 1, \ldots, \infty$, describes a four-dimensional particle with mass m_n given by $m_n^2 = m^2 + (n\pi/R)^2$. This set of particles is called a Kaluza–Klein (KK) tower. In particular, if the five-dimensional mass $m = 0$, then $m_n = n\pi/R$ and the KK modes are equally spaced. Therefore the existence of an extra dimension of size R should manifest itself with the presence of new particles at an energy scale $O(\pi/R)$. Since no such particle is observed up to present accelerator energies E of order of a few hundreds GeV, we conclude from this that $R < \pi/(500\mathrm{GeV}) \sim 10^{-16}$ cm.
There is however a subtle way out of this limit. It is in principle possible (and indeed it is suggested by some theoretical considerations based on string theory) that the extra dimensions are not accessible to particles with the usual weak, electromagnetic or strong interaction, and that only gravity can propagate in the extra dimensions. In this case we can have a large R. The resulting KK modes would be light, but they would not be observed at accelerators because they interact too weakly. A limit on R would come from modifications of Newton's law of gravitation. Newton's law is well verified experimentally only down to the millimeter scale (below it is difficult to measure the gravitational force between two objects, because it is overwhelmed by the van der Waals forces). Therefore, the bound on extra dimensions in which only gravity can propagate is of order $R < 1$ mm (see N. Arkani-Hamed, S. Dimopoulos and G. R. Dvali, Phys. Rev. D59 (1999) 086004).

12.4 Chapter 4

(4.1) (i) The exchange of coordinates gives a factor $(-1)^L$, while the relative intrinsic parity of a fermion and an antifermion is -1, so in total we have $(-1)^{L+1}$. (ii) Consider $e^{\pm}$ as two charge state of the same particle, exchanged by C. Because of Fermi–Dirac statistics, the exchange of two identical fermions gives a minus sign. On the other hand, this exchange is performed applying the charge conjugation operator (which gives a factor C), exchanging the coordinates (which gives $(-1)^L$) and exchanging the spin. The spin exchange gives $(-1)^{S+1}$, i.e. the singlet state $S = 0$ has an antisymmetric spin wave function, while $S = 1$ has a symmetric spin wave function. Therefore $C(-1)^L(-1)^{S+1} = -1$, and it follows that $C = (-1)^{L+S}$. (iii) The ground state of para-positronium has

$L = 0, S = 0$ and therefore $C = +1$. Since the photon has $C = -1$, and QED conserves C, it can only decay into an even number of photons.

(4.2) Perform a boost along the z axis. Since the transverse components $\mathbf{p}_\perp$ of the momentum are not affected, $\delta^{(2)}(\mathbf{p}_\perp - \mathbf{k}_\perp)$ is invariant and we must consider only $E_\mathbf{p}\,\delta(p_z - k_z)$. Use the form of the Lorentz transformation of $E_\mathbf{p}, p_z$ together with the property of the Dirac delta $\delta(f(x)) = \delta(x - x_0)/|f'(x_0)|$ (valid when x_0 is the only solution of $f(x) = 0$).

(4.3) Use the fact that Ψ and Ψ^* anticommute at equal time, and the fact that the transpose of γ^μ can be written as $(\gamma^\mu)^T = \gamma^0\gamma^\mu\gamma^0$, as one verifies from the explicit expression of the γ matrices.

(4.4) (i) The mass term breaks gauge-invariance. The Euler–Lagrange equation is $\partial_\mu F^{\mu\nu} + m^2 A^\nu = 0$. Acting with ∂_ν, using $\partial_\nu\partial_\mu F^{\mu\nu} = 0$ and $m \neq 0$, gives $\partial_\nu A^\nu = 0$. Using this condition, $\partial_\mu F^{\mu\nu} = \partial_\mu\partial^\mu A^\nu - \partial_\mu\partial_\nu A^\mu$ reduces to $\Box A^\nu$ and therefore $\partial_\mu F^{\mu\nu} + m^2 A^\nu = 0$ becomes $(\Box + m^2)A^\mu = 0$. (ii) The expansion of A_μ in plane waves is as in eq. (4.104). However now the condition $(\Box + m^2)A^\mu = 0$ imposes $p^2 = m^2$, while $\partial_\nu A^\nu = 0$ gives $\epsilon_\mu p^\mu = 0$. Therefore there are three independent solutions for the polarization vectors ϵ_μ. Since all our equations are explicitly Lorentz covariant, we can study the particle content of the theory in the frame that we prefer and, since $m \neq 0$, we can choose the rest frame of the particle. In this frame $p = (m, 0, 0, 0)$ and the three independent orthogonal polarization vectors are $\epsilon^1 = (0, 1, 0, 0), \epsilon^2 = (0, 0, 1, 0)$ and $\epsilon^3 = (0, 0, 0, 1)$; they describe the three spin degrees of freedom of a massive vector field.

(4.5) (i) Acting on a generic multiparticle state $|\mathbf{p}_1, \ldots, \mathbf{p}_n\rangle$ we have

$$\begin{aligned}(2E_\mathbf{p})^{1/2} e^{-\beta H} a^\dagger_\mathbf{p} |\mathbf{p}_1, \ldots, \mathbf{p}_n\rangle \qquad (12.23)\\ = e^{-\beta H}|\mathbf{p}, \mathbf{p}_1, \ldots, \mathbf{p}_n\rangle \\ = \exp\{-\beta(E_\mathbf{p} + E_{\mathbf{p}_1} + \ldots + E_{\mathbf{p}_n})\}|\mathbf{p}, \mathbf{p}_1, \ldots, \mathbf{p}_n\rangle\,.\end{aligned}$$

On the other hand,

$$\begin{aligned}(2E_\mathbf{p})^{1/2} a^\dagger_\mathbf{p} e^{-\beta(H + E_\mathbf{p})} |\mathbf{p}_1, \ldots, \mathbf{p}_n\rangle \qquad (12.24)\\ = (2E_\mathbf{p})^{1/2} a^\dagger_\mathbf{p} e^{-\beta(E_{\mathbf{p}_1} + \ldots + E_{\mathbf{p}_n} + E_\mathbf{p})}|\mathbf{p}_1, \ldots, \mathbf{p}_n\rangle \\ = \exp\{-\beta(E_\mathbf{p} + E_{\mathbf{p}_1} + \ldots + E_{\mathbf{p}_n})\}|\mathbf{p}, \mathbf{p}_1, \ldots, \mathbf{p}_n\rangle\,,\end{aligned}$$

so the two expressions coincide on the most general state of the Fock space. An alternative derivation is obtained defining

$$f(\beta) = e^{-\beta H} a^\dagger_\mathbf{p} - a^\dagger_\mathbf{p} e^{-\beta(H + E_\mathbf{p})}\,. \qquad (12.25)$$

Clearly, $f(0) = 0$. Show that $[H, a^\dagger_\mathbf{p}] = E_\mathbf{p}\,a^\dagger_\mathbf{p}$, and using this check that $f'(\beta) = -Hf(\beta)$. The solution of this equation, with the boundary condition $f(0) = 0$, is $f(\beta) = 0$.

(ii) Using the above result and the cyclic property of the trace,

$$\begin{aligned}\mathrm{Tr}\left(e^{-\beta H}a^\dagger_{\mathbf{p}}a_{\mathbf{q}}\right) &= \mathrm{Tr}\left(a^\dagger_{\mathbf{p}}e^{-\beta(H+E_{\mathbf{p}})}a_{\mathbf{q}}\right)\\ &= \mathrm{Tr}\left(e^{-\beta(H+E_{\mathbf{p}})}a_{\mathbf{q}}a^\dagger_{\mathbf{p}}\right)\\ &= \mathrm{Tr}\left(e^{-\beta(H+E_{\mathbf{p}})}(a^\dagger_{\mathbf{p}}a_{\mathbf{q}}+[a_{\mathbf{q}},a^\dagger_{\mathbf{p}}])\right).\end{aligned} \tag{12.26}$$

Dividing by $\mathrm{Tr}\, e^{-\beta H}$,

$$\langle a^\dagger_{\mathbf{p}}a_{\mathbf{q}}\rangle_\beta = e^{-\beta E_{\mathbf{p}}}\langle a^\dagger_{\mathbf{p}}a_{\mathbf{q}}\rangle_\beta + e^{-\beta E_{\mathbf{p}}}(2\pi)^3\delta^{(3)}(\mathbf{p}-\mathbf{q})\,. \tag{12.27}$$

Solving for $\langle a^\dagger_{\mathbf{p}}a_{\mathbf{q}}\rangle_\beta$ we get the desired result. When $\mathbf{p}=\mathbf{q}$, in a finite volume, use eq. (4.7).
(iii) If $a^\dagger_{\mathbf{p}}$ and $a_{\mathbf{q}}$ obey anticommutation relations, in the last passage in eq. (12.26) $a_{\mathbf{q}}a^\dagger_{\mathbf{p}}$ is replaced by $-a^\dagger_{\mathbf{p}}a_{\mathbf{q}}+\{a_{\mathbf{q}},a^\dagger_{\mathbf{p}}\} = -a^\dagger_{\mathbf{p}}a_{\mathbf{q}}+(2\pi)^3\delta^{(3)}(\mathbf{p}-\mathbf{q})$ and therefore

$$\langle a^\dagger_{\mathbf{p}}a_{\mathbf{q}}\rangle_\beta = -e^{-\beta E_{\mathbf{p}}}\langle a^\dagger_{\mathbf{p}}a_{\mathbf{q}}\rangle_\beta + e^{-\beta E_{\mathbf{p}}}(2\pi)^3\delta^{(3)}(\mathbf{p}-\mathbf{q})\,, \tag{12.28}$$

so

$$\langle a^\dagger_{\mathbf{p}}a_{\mathbf{q}}\rangle_\beta = \frac{V}{e^{\beta E_{\mathbf{p}}}+1}\,. \tag{12.29}$$

(4.6) (i) The volume of the phase space is $V(4/3)\pi p_F^3$. Each cell has a volume $h^3=(2\pi)^3$ (in our units $\hbar=1$) and in each cell, by the exclusion principle, we can accommodate two electrons, with spin up and spin down. (ii) When $|\mathbf{p}|<p_F$, $a_{\mathbf{p},s}$ destroys a particle which is present in $|0\rangle_F$, so in this case $a_{\mathbf{p},s}|0\rangle_F\neq 0$. The fact that $A_{\mathbf{p},s}$ and $A^\dagger_{\mathbf{p},s}$ satisfy the canonical anticommutation relations follows easily from the identities $\theta(x)\theta(x)=\theta(x)$, $\theta(x)\theta(-x)=0$ and $\theta(x)+\theta(-x)=1$ satisfied by the step function. The operator $A^\dagger_{\mathbf{p},s}$, acting on $|0\rangle_F$, creates an electron above the Fermi surface or destroys an electron in the "filled Fermi sea". The latter process can be described as the creation of a "hole" in the Fermi sea, and the excitation of an electron from a level below p_F to a level above p_F can be described as the creation of an electron–hole pair. (iii) For instance,

$$\begin{aligned}\{A_{\mathbf{p},s},A^\dagger_{\mathbf{q},r}\} &= \alpha_{\mathbf{p}}\alpha^*_{\mathbf{q}}\{a_{\mathbf{p},s},a^\dagger_{\mathbf{q},r}\}+\beta_{\mathbf{p}}\beta^*_{\mathbf{q}}\{a^\dagger_{-\mathbf{p},-s},a_{-\mathbf{q},-r}\}\\ &= (|\alpha_{\mathbf{p}}|^2+|\beta_{\mathbf{p}}|^2)(2\pi)^3\delta^{(3)}(\mathbf{p}-\mathbf{q})\delta_{rs}\,.\end{aligned} \tag{12.30}$$

All other relations are proved similarly. (iv)

$$A^\dagger_{\mathbf{p}}A_{\mathbf{p}} = |\alpha_{\mathbf{p}}|^2a^\dagger_{\mathbf{p}}a_{\mathbf{p}}+|\beta_{\mathbf{p}}|^2a_{\mathbf{p}}a^\dagger_{\mathbf{p}}-\alpha_{\mathbf{p}}\beta^*_{\mathbf{p}}a_{\mathbf{p}}a_{\mathbf{p}}-\alpha^*_{\mathbf{p}}\beta_{\mathbf{p}}a^\dagger_{\mathbf{p}}a^\dagger_{\mathbf{p}}\,. \tag{12.31}$$

The terms $a_{\mathbf{p}}a_{\mathbf{p}}$ and $a^\dagger_{\mathbf{p}}a^\dagger_{\mathbf{p}}$ have a vanishing diagonal matrix element. Use $|\alpha_{\mathbf{p}}|^2=1+|\beta_{\mathbf{p}}|^2$ (since we are now considering bosons) and, in a unit volume, $a_{\mathbf{p}}a^\dagger_{\mathbf{p}}=a^\dagger_{\mathbf{p}}a_{\mathbf{p}}+1$.

12.5 Chapter 5

(5.1) Using $(\Box_x+m^2)\phi(x)=0$ and the relations $\partial_t\theta(t)=\delta(t)$, $\partial_t\theta(-t)=-\delta(t)$ (and therefore $\partial_t^2\theta(t)=\delta'(t)$, $\partial_t^2\theta(-t)=-\delta'(t)$) we find

$$(\Box_x+m^2)\left[\theta(t)\langle 0|\phi(x)\phi(0)|0\rangle+\theta(-t)\langle 0|\phi(0)\phi(x)|0\rangle\right]$$
$$=\delta'(t)\langle 0|[\phi(x),\phi(0)]|0\rangle+2\delta(t)\langle 0|[\partial_t\phi(x),\phi(0)]|0\rangle\,. \quad (12.32)$$

By definition of distributions, $\delta'(t)$ is defined integrating by parts, so $\delta'(t)\phi(x)=-\delta(t)\partial_t\phi(x)$. Since $\delta(t)$ has support only at $t=0$, the commutator $[\partial_t\phi(x),\phi(0)]$ above must be computed at equal time, and then $[\partial_t\phi(x),\phi(0)]=-i\delta^{(3)}(\mathbf{x})$, so we get the desired result. The derivation with $\phi(y)$ (here we set $y=0$) replaced by $\phi(y_1)\ldots\phi(y_n)$ is obtained similarly, writing explicitly all theta functions.

In momentum space we find $(-p^2+m^2)\tilde{D}(p)=-i$. Formally this gives $\tilde{D}(p)=i/(p^2-m^2)$, and therefore

$$D(x)=\int\frac{d^4p}{(2\pi)^4}\frac{i}{p^2-m^2}\,e^{-ipx}\,. \quad (12.33)$$

However, the integrand has two poles at $p^0=\pm\sqrt{\mathbf{p}^2+m^2}$, and therefore we must also specify how to go around these poles in the complex p^0 plane. For each pole we can go above or below it. After specifying a prescription, we can then compute the integral over p^0,

$$\int dp^0\frac{i}{(p^0)^2-\mathbf{p}^2-m^2}\,e^{-ip^0t}\,. \quad (12.34)$$

If $t>0$ we can close the contour in the lower half plane since, when $p^0=-iu$ with $u>0$ then $e^{-ip^0t}=e^{-ut}$, which for $t>0$ provides a convergence factor in the integral. Conversely, when $t<0$ we can close the contour in the upper half plane. If we go around both poles from below, then when $t>0$ (i.e. when we close the contour in the lower half plane) we encircle no pole (see Fig. 12.1), so the integral vanishes. Therefore, with this prescription, $D(t,\mathbf{x})=0$ for $t>0$, and $D(x)$ is called an *advanced* Green's function. Conversely, if we go around both poles from above we find that $D(t,\mathbf{x})=0$ for $t<0$, and we have a *retarded* Green's function. The Feynman propagator corresponds to a mixed case, see Fig. 5.1.

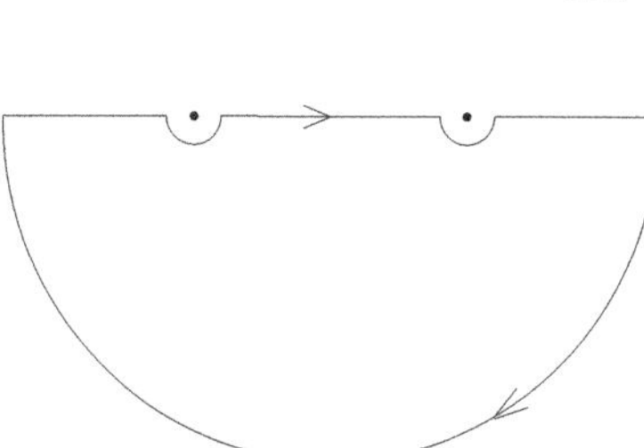

Fig. 12.1 The case when the poles in the complex p^0-plane are both encircled from below, corresponding to an advanced Green's function.

(5.2) $n(d-2)\leqslant 2d$. Observe that in $d=2$ the field ϕ is dimensionless and a term $\lambda\phi^n$ is renormalizable by power counting for every n, so we can take an arbitrary function $V(\phi)$ as the potential.

(5.3) The main point is to understand that, in the Wick theorem, we must omit the contractions between fields inside a normal ordered term. For instance, the $O(\lambda)$ contribution to the mass renormalization, in the theory with interaction $(\lambda/4!):\phi^4:$, is proportional to $\langle 0|T\{\phi(x_1)\phi(x_2):\phi^4(x):\}|0\rangle$. From the point of view of the combinatorics of the Wick theorem, $\varphi\equiv:\phi^4(x):$ can be treated just as a single field, so for instance

$$T\{\phi_1\phi_2\varphi(x)\} = :\phi_1\phi_2\varphi(x): + D_{12}\varphi(x) \\ +\langle 0|T\{\phi_1\varphi(x)\}|0\rangle\,\phi_2 + \langle 0|T\{\phi_2\varphi(x)\}|0\rangle\,\phi_1\,. \quad (12.35)$$

Using $\langle 0|T\{\phi_i\varphi(x)\}|0\rangle = 0$ (since it is odd under $\phi \to -\phi$) one finds

$$T\{\phi_1\phi_2 : \phi^4 :\} = :\phi_1\phi_2\phi^4: + D_{12} : \phi^4 :, \quad (12.36)$$

and therefore $\langle 0|T\{\phi(x_1)\phi(x_2) : \phi^4(x) :\}|0\rangle = 0$, so there is no mass renormalization at $O(\lambda)$. Alternatively, one can write $:\phi^4(x):$ $=: \phi(x_3)\phi(x_4)\phi(x_5)\phi(x_6):$ (letting $x_3 = x_4 = x_5 = x_6 \equiv x$ at the end of the calculation) and use eq. (5.85) to express $:\phi_3\phi_4\phi_5\phi_6:$ as $T\{\phi_3\phi_4\phi_5\phi_6\}$ minus the contraction terms, so in turn

$$T\{\phi_1\phi_2 : \phi_3\phi_4\phi_5\phi_6 :\} = T\{\phi_1\phi_2\phi_3\phi_4\phi_5\phi_6\} - (\text{contractions})\,. \quad (12.37)$$

One can now check explicitly that the "−(contractions)" term above cancel the terms in $T\{\phi_1\dots\phi_6\}$ where we have contractions between $\phi_i\phi_j$ with $i, j = 3, 4, 5, 6$.
In general, one can understand from this example that the introduction of the normal ordering in the interaction term eliminates the tadpole graphs.

(5.4) Setting $u = 1/\alpha_s$, eq. (5.194) becomes

$$\frac{du}{d\log E} = b_0 + \frac{b_1}{u}\,. \quad (12.38)$$

(i) Neglecting the term $\sim b_1$, the solution is $u(E) = u(\mu) + b_0\log(E/\mu)$. Substituting $\mu = \Lambda_{\rm QCD}\exp\{1/[b_0\alpha(\mu)]\}$, we find $u = b_0\log(E/\Lambda_{\rm QCD})$. (ii) We can solve perturbatively inserting the lowest-order solution into the term $\sim b_1$. The equation then becomes

$$\frac{du}{d\log E} = b_0 + \frac{b_1}{b_0\log(E/\Lambda_{\rm QCD})}\,. \quad (12.39)$$

The solution is $u(E) = b_0\log(E/\Lambda_{\rm QCD}) + (b_1/b_0)\log\log(E/\Lambda_{\rm QCD})$, where we have redefined $\Lambda_{\rm QCD}$ at two loops so that the integration constant vanishes.

12.6 Chapter 6

(6.1) Use $t = (p_1 - p_3)^2 = m_1^2 + m_3^2 - 2E_1E_3 + 2|\mathbf{p}_1||\mathbf{p}_3|\cos\theta$. Since $|\mathbf{p}_3|$ and E_3 are fixed by energy–momentum conservation, we have $dt = 2|\mathbf{p}_1||\mathbf{p}_3|d\cos\theta$. Inserting this into eq. (6.43) (with $\mathbf{p}_1 \equiv \mathbf{p}$, $\mathbf{p}_3 \equiv \mathbf{p}'$), integrating over $d\phi$ and using eq. (6.42) we get the desired result.

(6.2) Equation (6.132) is obtained performing a Lorentz boost with velocity $-v_2$. Since $E', \mathbf{p}'$ are fixed by energy–momentum conservation, only θ is a variable and eq. (6.133) follows. Then $d\Omega =$

$d\cos\theta d\phi = 2\pi dE_{\rm lab}/(\gamma_2 v_2|\mathbf{p}'|)$. Inserting this into eq. (6.41) gives the result. The kinematical limits are obtained setting $\cos\theta = \pm 1$ in eq. (6.132).

(6.3) (i) In eq. (6.41) set $\sqrt{s} \simeq M_A$ (since $M_A \gg \omega$) and, for the photon, $|\mathbf{p}'| = \omega$. (ii) Denoting by M_{fi} the matrix element with normalization for the atom equal to one particle in a volume $V = 1$, $\mathcal{M}_{fi} = \sqrt{2M_A}\sqrt{2M_{A^*}}M_{fi} \simeq 2M_A M_{fi}$ (since $M_{A^*} - M_A = \omega \ll M_A$). Using the phase space found above, eq. (6.20) gives

$$d\Gamma = \frac{1}{2M_A}(2M_A)^2|M_{fi}|^2 \frac{\omega}{16\pi^2 M_A} d\Omega \,. \tag{12.40}$$

M_A cancels and we get the desired result. (iii) Use eq. (6.43) with $s = M_A^2$ and $\mathcal{M}_{fi} = 2M_A M_{fi}$.

(6.4) Denoting by k_1, k_2, p the four-momenta in the CM of the photons and of the final atom, respectively, we have

$$d\Phi^{(3)} = \frac{1}{2!}\frac{1}{(2\pi)^5}\frac{d^3k_1}{2\omega_1}\frac{d^3k_2}{2\omega_2}\frac{d^3p}{2M_A}\delta^{(3)}(\mathbf{p}+\mathbf{k}_1+\mathbf{k}_2)\delta(\omega-\omega_1-\omega_2)\,. \tag{12.41}$$

The factor $1/2!$ takes into account the fact that the two photons are identical particles. Integrating over d^3p and using the notation $E_{A^*} - E_A = \omega$,

$$\begin{aligned} d\Phi^{(3)} &= \frac{\omega_1 d\omega_1 d\Omega_1 \omega_2 d\omega_2 d\Omega_2}{(2\pi)^5 16 M_A}\delta(\omega-\omega_1-\omega_2) \\ &= \frac{1}{(2\pi)^5 16 M_A}\omega_1(\omega-\omega_1)d\omega_1 d\Omega_1 d\Omega_2\,. \end{aligned} \tag{12.42}$$

Finally, to compute $d\Gamma$ use $\mathcal{M}_{fi} = 2M_A M_{fi}$, as in the previous exercise.

(6.5) Writing explicitly $d\Phi^{(j)}$ and $d\Phi^{(n-j+1)}$, the right-hand side of eq. (6.140) becomes

$$\begin{aligned} &\int_0^\infty \frac{d\mu^2}{2\pi}\left(\prod_{i=1}^{j}\frac{d^3p_i}{(2\pi)^3 2E_i}\right)(2\pi)^4\delta^{(4)}(p_1+\ldots+p_j-q) \\ &\times\left(\prod_{i=j+1}^{n}\frac{d^3p_i}{(2\pi)^3 2E_i}\right)\frac{d^3q}{(2\pi)^3 2q^0}(2\pi)^4\delta^{(4)}(p_{j+1}+\ldots+p_n+q-P)\,. \end{aligned} \tag{12.43}$$

The first Dirac delta forces $q = p_1 + \ldots + p_j$. Inserting this into the second Dirac delta, we can rewrite the above expression as

$$\begin{aligned} &\int_0^\infty \frac{d\mu^2}{2\pi}\left[\left(\prod_{i=1}^{n}\frac{d^3p_i}{(2\pi)^3 2E_i}\right)(2\pi)^4\delta^{(4)}(p_1+\ldots+p_n-P)\right] \\ &\times\frac{d^3q}{(2\pi)^3 2q^0}(2\pi)^4\delta^{(4)}(p_1+\ldots+p_j-q)\,. \end{aligned} \tag{12.44}$$

Now use the identity

$$\frac{d^3q}{2q^0} = d^4q\,\delta(q^2-\mu^2)\theta(q^0)\,, \tag{12.45}$$

which follows from the fact that, by definition, $\mu^2 = q_0^2 - \mathbf{q}^2$, and the θ function selects $q^0 = +\sqrt{\mathbf{q}^2 + \mu^2}$ as solutions of $q^2 - \mu^2 = 0$. Then the above expression becomes

$$\left(\prod_{i=1}^{n} \frac{d^3 p_i}{(2\pi)^3 2E_i}\right)(2\pi)^4 \delta^{(4)}(p_1 + \ldots + p_n - P) \tag{12.46}$$
$$\times \int_0^\infty d\mu^2 \int d^4 q\, \delta(q^2 - \mu^2))\theta(q^0)\delta^{(4)}(p_1 + \ldots + p_j - q)\,.$$

The last two integrals give

$$\int d^4 q\, \theta(q^0)\delta^{(4)}(p_1 + \ldots + p_j - q) \int_0^\infty d\mu^2 \delta(q^2 - \mu^2)$$
$$= \int d^4 q\, \theta(q^0)\delta^{(4)}(p_1 + \ldots + p_j - q) = 1\,, \tag{12.47}$$

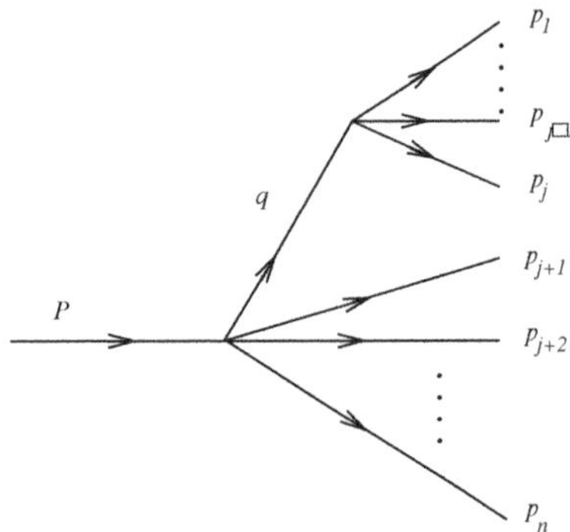

Fig. 12.2 A graphical representation of the decomposition of the phase space given in eq. (6.140).

and the desired result follows. Diagrammatically, we can represent eq. (6.140) as in Fig. 12.2, so this representation of the phase space is useful to discuss a process in which the n-body decay of the initial particle goes through a resonance of mass μ which later decays into j particles.

(6.6) (i) Denoting by p the external momentum and by q and $p - q$ the momenta in the loop, the graph gives[1]

$$i\mathcal{M} = (-ig)^2 \int \frac{d^4 q}{(2\pi)^4} \frac{i}{q^2 - m^2 + i\epsilon} \frac{i}{(q-p)^2 - m^2 + i\epsilon}\,. \tag{12.48}$$

In the rest frame of the initial particle, $p = (M_R, 0)$, where M_R is the (renormalized) mass of Φ. Then the poles in the integrand are at $q^0 = E_{\mathbf{q}} - i\epsilon$, $q^0 = -E_{\mathbf{q}} + i\epsilon$, $q^0 = M_R + E_{\mathbf{q}} - i\epsilon$ and $q^0 = M_R - E_{\mathbf{q}} + i\epsilon$, where $E_{\mathbf{q}} = +\sqrt{\mathbf{q}^2 + m^2}$. In the complex q^0-plane we can close the integration contour both in the lower or in the upper half-plane. Choosing for instance the lower-half plane, we pick the residues of the poles at $q^0 = E_{\mathbf{q}} - i\epsilon$ and at $q^0 = M_R + E_{\mathbf{q}} - i\epsilon$, and we get

$$\mathcal{M} = -g^2 \int \frac{d^3 q}{(2\pi)^3} \frac{1}{2M_R E_{\mathbf{q}}} \left(\frac{1}{M_R - 2E_{\mathbf{q}} + i\epsilon} + \frac{1}{M_R + 2E_{\mathbf{q}} - i\epsilon}\right). \tag{12.49}$$

In the second fraction we can set $\epsilon = 0$ since the denominator never vanishes. In the first we use the identity

$$\frac{1}{x \pm i\epsilon} = P\frac{1}{x} \mp i\pi\delta(x)\,, \tag{12.50}$$

where P denotes the principal part. Then we get an imaginary contribution to $\mathcal{M}$,

$$\mathrm{Im}\,\mathcal{M} = \pi g^2 \int \frac{d^3 q}{(2\pi)^3} \frac{1}{2M_R E_{\mathbf{q}}} \delta(M_R - 2E_{\mathbf{q}})\,. \tag{12.51}$$

[1] Observe that, if instead of an interaction term $g\phi_1\phi_2\Phi$ with two different fields ϕ_1, ϕ_2, we were to take a single ϕ field, with interaction Lagrangian $g\phi^2\Phi$, there would be an additional factor of two in the amplitude, because when we compute $\langle 0|T\{\phi^2(x)\Phi(x)\phi^2(y)\Phi(y)\}|0\rangle$ there are two possible contractions: we can contract the first $\phi(x)$ with the first $\phi(y)$ (and therefore the second $\phi(x)$ with the second $\phi(y)$) or the first $\phi(x)$ with the second $\phi(y)$. If instead we have $\langle 0|T\{\phi_1(x)\phi_2(x)\Phi(x)\phi_1(y)\phi_2(y)\Phi(y)\}|0\rangle$ and a Lagrangian whose kinetic term does not mix ϕ_1 and ϕ_2, we can only contract $\phi_1(x)$ with $\phi_1(y)$ and $\phi_2(x)$ with $\phi_2(y)$.

Since $E_{\mathbf{q}} = \sqrt{\mathbf{q}^2 + m^2} \geqslant m$, when $M_R < 2m$ the Dirac delta is never satisfied, and the imaginary part vanishes. Instead, when $M_R \geqslant 2m$, performing the integral with the help of the delta function gives

$$\mathrm{Im}\,\mathcal{M} = \frac{g^2}{16\pi}\sqrt{1 - \frac{4m^2}{M_R^2}}\,. \tag{12.52}$$

(ii) In Section 5.5.2 we have seen that the one-loop correction to the propagator produces a shift of the mass squared (see eqs. (5.108) and (5.112), and observe that the loop correction to the propagator, denoted here by $i\mathcal{M}$, is the quantity denoted as $-iB$ in eq. (5.108), so $B = -\mathcal{M}$), so

$$M^2 \to M^2 - \mathcal{M} = M^2 - \mathrm{Re}\mathcal{M} - i\,\mathrm{Im}\mathcal{M}\,. \tag{12.53}$$

The renormalized mass is given by $M_R^2 = M^2 - \mathrm{Re}\mathcal{M}$, and the quantity that appears in the denominator of the propagator, after inclusion of loop corrections, is therefore $M^2 = M_R^2 - i\mathrm{Im}\mathcal{M}$. On the other hand, from eq. (6.52), $M^2 = M_R^2 - iM_R\Gamma$. Therefore we expect that $M_R\Gamma = \mathrm{Im}\mathcal{M}$.
(iii) To verify this, we compute Γ explicitly. The amplitude for the process $\Phi \to \phi_1\phi_2$ is $(-g)$ and therefore, using the phase space (6.35),

$$\Gamma = \frac{1}{2M_R}(-g)^2\frac{4\pi}{32\pi^2}\sqrt{1 - \frac{4m^2}{M_R^2}} = \frac{g^2}{16\pi M_R}\sqrt{1 - \frac{4m^2}{M_R^2}}\,. \tag{12.54}$$

Comparing with eq. (12.52) we see that indeed $M_R\Gamma = \mathrm{Im}\mathcal{M}$. Observe that in this theory g has dimensions of mass, so Γ has the correct dimensions.

12.7 Chapter 7

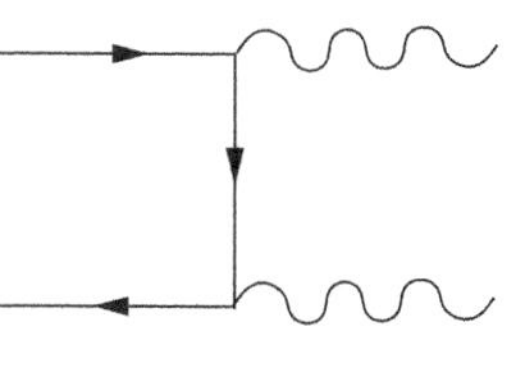

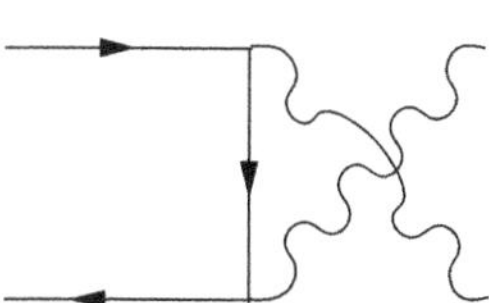

Fig. 12.3 The two Feynman diagrams for $e^+e^- \to \gamma\gamma$ to lowest order.

(7.1) (i) See Fig. 12.3. (ii) If s_1, s_2 are the initial spins and λ_1, λ_2 the final helicities,

$$\frac{1}{4}\sum_{s_1,s_2,\lambda_1,\lambda_2} |\mathcal{M}_{e^+e^-\to 2\gamma}|^2 = \frac{e^4}{4}\mathrm{Tr}\left[(\not{p}_1 + m_e)L^{\mu\nu}(\not{p}_2 - m_e)L^{\dagger}_{\mu\nu}\right] \tag{12.55}$$

where p_1 is the momentum of the electron, p_2 of the positron, k_1, k_2 of the two photons, and

$$L^{\mu\nu} = \gamma^\mu\frac{1}{\not{p}_1 - \not{k}_1 - m_e}\gamma^\nu + \gamma^\nu\frac{1}{\not{p}_1 - \not{k}_2 - m_e}\gamma^\mu\,. \tag{12.56}$$

In the limit $\mathbf{p} \to 0$, after long but straightforward algebra, the computation of the trace gives

$$\frac{1}{4}\sum_{s_1,s_2,\lambda_1,\lambda_2} |\mathcal{M}_{e^+e^-\to 2\gamma}|^2 = 32\pi^2\alpha^2\,. \tag{12.57}$$

To simplify the algebra, work directly in the CM, in the limit $\mathbf{p} \to 0$ (so that the photon energies are $\omega_1 = \omega_2 \simeq m_e$) and contract the γ matrices with repeated indices. For instance, using $\gamma_\mu \gamma^\mu = (\gamma_0)^2 - \sum_i (\gamma_i)^2 = 4$, one finds

$$\gamma_\mu \gamma^\nu \gamma^\mu = \gamma_\mu (\{\gamma^\nu, \gamma^\mu\} - \gamma^\mu \gamma^\nu) = \gamma_\mu 2\eta^{\mu\nu} - 4\gamma^\nu = -2\gamma^\nu \,. \quad (12.58)$$

In this way one can prove the useful identities

$$\gamma^\mu \not{A} \gamma_\mu = -2\not{A} \,, \quad \gamma^\mu \not{A} \not{B} \gamma_\mu = 4(AB) \,, \quad \gamma^\mu \not{A} \not{B} \not{C} \gamma_\mu = -2\not{C} \not{B} \not{A} \,. \quad (12.59)$$

Use also the cyclic property of the trace to bring closer γ matrices with repeated indices.
(iii) In the CM (considering for generality two particles with different masses m_1, m_2), $p_1 = (E_1, \mathbf{p})$, $p_2 = (E_2, -\mathbf{p})$, and

$$\begin{aligned} I^2 &= (p_1 p_2)^2 - m_1^2 m_2^2 \\ &= (E_1 E_2 + \mathbf{p}^2)^2 - (E_1^2 - \mathbf{p}^2)(E_2^2 - \mathbf{p}^2) \\ &= \mathbf{p}^2 (E_1 + E_2)^2 \,. \end{aligned} \quad (12.60)$$

Therefore

$$\begin{aligned} I &= |\mathbf{p}|(E_1 + E_2) = E_1 E_2 |\mathbf{p}| \left(\frac{1}{E_1} + \frac{1}{E_2} \right) \\ &= E_1 E_2 (|\mathbf{v}_1| + |\mathbf{v}_2|) \,, \end{aligned} \quad (12.61)$$

where $\mathbf{v}_1, \mathbf{v}_2$ are the respective velocities in the CM. The relative velocity has modulus $v = |\mathbf{v}_1| + |\mathbf{v}_2|$, so $I = E_1 E_2 v$. The result for σ then follows from the general formula (6.29), using eq. (12.57) and two-body phase space (6.35) with $m = 0$.

(7.2) (i) From Exercise 4.1 we know that an e^+e^- pair can annihilate into two photons only if it has $S = 0$; since we are considering a bound state with $L = 0$, then also $J = 0$. Alternatively, the result follows from the fact that a two-photon state cannot have $J = 1$, see Landau and Lifshitz, vol. IV (1982), Section 9 for the proof. Equation (7.70) then follows from

$$\bar{\sigma} = \frac{1}{4} \left[\sigma^{(J=0)} + \sum_{J_z = -1,0,1} \sigma^{(J=1)} \right] \,, \quad (12.62)$$

and $\sigma^{(J=1)} = 0$. (ii) Equations (7.72) and (7.73) follow immediately from eq. (6.8), with $V = 1$. (iii):

$$\langle 2\gamma | \text{Pos} \rangle = \int \frac{d^3 p_1}{(2\pi)^3} \frac{d^3 p_2}{(2\pi)^3} \langle 2\gamma | \mathbf{p}_1, \mathbf{p}_2 \rangle \langle \mathbf{p}_1, \mathbf{p}_2 | \text{Pos} \rangle \,. \quad (12.63)$$

In the CM,

$$\langle \mathbf{p}_1, \mathbf{p}_2 | \text{Pos} \rangle = (2\pi)^3 \delta^{(3)}(\mathbf{p}_1 + \mathbf{p}_2) \tilde{\psi}(\mathbf{p}_1) \quad (12.64)$$

where $\tilde{\psi}(\mathbf{p})$ is the wave function of positronium in momentum space, so

$$\langle 2\gamma|\text{Pos}\rangle = \int \frac{d^3p}{(2\pi)^3}\langle 2\gamma|\mathbf{p}\,,-\mathbf{p}\,\rangle\tilde{\psi}(\mathbf{p}\,)\,. \qquad (12.65)$$

(iv) From the order-of-magnitude estimates in the Introduction we know that in the hydrogen atom $v \sim \alpha$ and $|\mathbf{p}\,| \sim m_e\alpha$ (for positronium m_e becomes the reduced mass $m_e/2$). Then, to lowest order in α, in eq. (7.74) we can approximate $\langle 2\gamma|\mathbf{p}\,,-\mathbf{p}\,\rangle$ with its value at $\mathbf{p} \ll m_e$ and extract it from the integral. The remaining integral is $\psi(\mathbf{x}\,)$ at $\mathbf{x} = 0$. Equation (7.76) then follows from eqs. (7.72) and (7.73), recalling that only $J = 0$ contributes.
(v) The agreement is at the level of 0.5%. Including the first radiative correction, the theoretical prediction turns out to be

$$\Gamma = \frac{1}{2}m_e\alpha^5\left[1 - \frac{\alpha}{\pi}\left(5 - \frac{\pi^2}{4}\right)\right] \qquad (12.66)$$

and agrees with experiment, within the error (see D.W. Gidley et al., Phys. Rev. Lett. 49 (1982) 525).

12.8 Chapter 8

(8.1) (i) Compare with Solved Problem 7.2 on page 188. (ii) The maximum value of q^2 is $q^2_{\text{max}} = (m_n - m_p)^2 \simeq (1.3\,\text{MeV})^2$ (see e.g. Solved Problem 6.1). The typical scale of variation of the form factors is instead of order of the QCD scale, so $q_{\text{typical}} \sim$ a few hundred MeV. (iii):

$$\mathcal{M}_{fi} = -\frac{G_F\cos\theta_C}{\sqrt{2}}\bar{u}_e\gamma^\mu(1-\gamma^5)u_{\bar{\nu}}\bar{u}_p\gamma_\mu(1-g_A\gamma^5)u_n\,. \quad (12.67)$$

Averaging over the initial spin and summing over the final spins,

$$|\mathcal{M}_{fi}|^2 = \frac{G_F^2\cos^2\theta_C}{2}\text{Tr}[(\not{p}_e + m_e)\gamma^\mu(1-\gamma^5)\not{p}_{\bar{\nu}}(1+\gamma^5)\gamma^\nu]$$
$$\times\text{Tr}[(\not{p}_p + m_p)\gamma_\mu(1-g_A\gamma^5)\frac{\not{p}_n + m_n}{2}(1+g_A\gamma^5)\gamma^\nu]\,. \qquad (12.68)$$

Performing the traces,

$$|\mathcal{M}_{fi}|^2 = 16\,G_F^2\cos^2\theta_C[(1+g_A)^2(p_ep_p)(p_{\bar{\nu}}p_n) \qquad (12.69)$$
$$+(1-g_A)^2(p_ep_n)(p_{\bar{\nu}}p_p) - (1-g_A^2)m_pm_n(p_ep_{\bar{\nu}})]\,.$$

We next compute the scalar products in the neutron rest frame, $p_n = (m_n, 0)$. Observe that the maximum proton energy is

$$E_p^{\text{max}} = \frac{m_n^2 + m_p^2 - m_e^2}{2m_n} = m_p + \frac{\Delta^2 - m_e^2}{2m_n}\,, \qquad (12.70)$$

with $\Delta = m_n - m_p$. Since $(\Delta^2 - m_e^2)/(2m_n) \sim 10^{-4}$ MeV, we can neglect it with respect to $m_n \sim 10^3$ MeV and, in the scalar products, we can set the proton energy E_p to the fixed value $E_p \simeq m_p$. For the same reason, we can write $(p_e p_p) = E_e E_p - \mathbf{p}_e \cdot \mathbf{p}_p \simeq E_e m_p$, since $|\mathbf{p}_p| \ll m_p$. With this and similar approximations in the other scalar products,

$$|\mathcal{M}_{fi}|^2 = 16\, G_F^2 \cos^2\theta_C m_p m_n \\ \times[(1+3g_A^2)E_e E_{\bar{\nu}} + (1-g_A^2)\sqrt{E_e^2 - m_e^2}\, E_{\bar{\nu}} \cos\theta]\,, \tag{12.71}$$

where θ is the angle between the electron and the antineutrino. The width is given by

$$d\Gamma = \frac{1}{2m_n}|\mathcal{M}_{fi}|^2 \frac{d^3p_p}{(2\pi)^3 2m_p} \frac{d^3p_e}{(2\pi)^3 2E_e} \frac{d^3p_{\bar{\nu}}}{(2\pi)^3 2E_{\bar{\nu}}} \\ \times (2\pi)^4 \delta^{(3)}(\mathbf{p}_p + \mathbf{p}_e + \mathbf{p}_{\bar{\nu}})\delta(m_p + E_e + E_{\bar{\nu}} - m_n)\,. \tag{12.72}$$

Integrate first over d^3p_p with the help of the $\delta^{(3)}$. Write $d^3p_e = 4\pi p_e^2 dp_e$ and $d^3p_{\bar{\nu}} = 2\pi E_{\bar{\nu}}^2 dE_{\bar{\nu}} d\cos\theta$. Integrate over $dE_{\bar{\nu}}$ with the help of the remaining Dirac delta and finally perform the integration over $d\cos\theta$, between $\cos\theta = \pm 1$. The term linear in $\cos\theta$ in $|\mathcal{M}_{fi}|^2$ integrates to zero and the constant part gives the desired result.

(iv) The kinematical limits on E_e are $E_e^{\rm min} = m_e$ and

$$E_e^{\rm max} = \frac{m_n^2 - m_p^2 + m_e^2}{2m_n} = \Delta - \frac{\Delta^2 - m_e^2}{2m_n} \simeq \Delta\,. \tag{12.73}$$

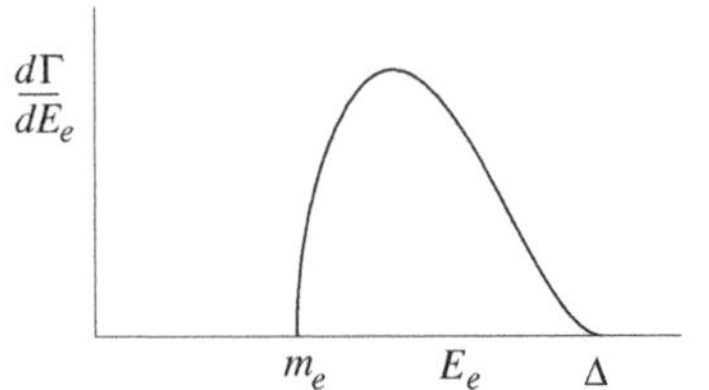

Fig. 12.4 The Fermi spectrum of β-decay.

The spectrum is shown in Fig. 12.4. When $m_\nu = 0$, at $E_e \simeq \Delta$ we have $d\Gamma/dE_e \sim (\Delta - E_e)^2$ and therefore the slope of the spectrum, $d^2\Gamma/dE_e^2$, goes to zero as $E_e \to \Delta$. For a small non-zero value of m_ν we can use the same expression for the matrix element and take into account $m_\nu \neq 0$ just in the phase space. The result is that now the slope diverges ($d^2\Gamma/dE_e^2 \to -\infty$) at the end-point of the spectrum.

(v) Integrating over E_e,

$$\Gamma = \frac{G_F^2 \Delta^5 \cos^2\theta_C}{2\pi^3}(1+3g_A^2)\int_{m_e/\Delta}^{1} dx\, x(1-x)^2\sqrt{x^2 - (m_e/\Delta)^2}\,. \tag{12.74}$$

If $m_e = 0$ the integral can be computed analytically, and is equal to $1/30$. For the physical values of m_e, Δ, numerical integration gives $0.472565/30$, and

$$\Gamma = 0.472565\,\frac{G_F^2 \Delta^5 \cos^2\theta_C}{60\pi^3}(1+3g_A^2)\,. \tag{12.75}$$

Inserting the numerical values, we get a result for $\tau = 1/\Gamma$ larger by about 8% than the experimental value. A more accurate computation requires us to keep all six form factors. Furthermore, an important correction comes from the exchange of photons between the electron and the proton; these Coulomb corrections between final states are large when, as in the present case, the relative speed of the final charged particles is small.

(8.2) (i) Using the same considerations as in Solved Problem 7.2, the most general parametrization of a vector current is

$$\langle e^-|j^\mu(0)|\mu^-\rangle = f_1(q^2)\bar{u}_e\gamma^\mu u_\mu + f_2(q^2)\bar{u}_e\sigma^{\mu\nu}q_\nu u_\mu + f_3(q^2)q^\mu\bar{u}_e u_\mu\,. \tag{12.76}$$

Imposing current conservation gives

$$q_\mu(f_1(q^2)\bar{u}_e\gamma^\mu u_\mu + f_2(q^2)\bar{u}_e\sigma^{\mu\nu}q_\nu u_\mu + f_3(q^2)q^\mu\bar{u}_e u_\mu) = 0\,. \tag{12.77}$$

In the first term we use the equations of motion together with $q = k - p$, where p and k are the electron and muon four-momenta, respectively. This gives $\bar{u}_e q_\mu\gamma^\mu u_\mu = (m_\mu - m_e)\bar{u}_e u_\mu$. In the second term $\sigma^{\mu\nu}q_\nu q_\mu = 0$ by symmetry, so we get

$$f_1(q^2)(m_\mu - m_e) + q^2 f_3(q^2) = 0\,. \tag{12.78}$$

Setting $q^2 = 0$ (which is the value of q^2 in which we are interested, since the photon is on-shell) we find that $f_1(0) = 0$.
(ii) The amplitude is obtained multiplying by ϵ^*_μ. Since $\epsilon^*_\mu(q)q^\mu = 0$, only the term $\sim f_2(0)$ survives.
(iii)

$$|\mathcal{M}_{fi}|^2 = \frac{e^2}{4m_\mu^2}|F_2|^2\,\epsilon^*_\mu\epsilon_\rho\,q_\nu q_\sigma(\bar{u}_e\sigma^{\mu\nu}u_\mu)\,(\bar{u}_\mu\sigma^{\rho\sigma}u_e)\,. \tag{12.79}$$

Perform the sum over the photon polarizations using $\epsilon^*_\mu\epsilon_\rho \to -\eta_{\mu\rho}$. To perform the sum over the spin of e^- and the average over the spin of μ^- one could replace here $u_e\bar{u}_e \to \not{p} + m_e$ and $u_\mu\bar{u}_\mu \to (\not{k} + m_\mu)/2$. The resulting trace apparently has up to six γ matrices but can be simplified using the γ matrix identities given in eq. (12.59). However, the calculation is much simpler if instead we eliminate immediately $\sigma^{\mu\nu}$ from $|\mathcal{M}_{fi}|^2$ using the Gordon identity, eq. (7.51); then we get

$$\begin{aligned}|\mathcal{M}_{fi}|^2 = &-\frac{e^2}{8m_\mu^2}|F_2|^2 \\ &\times \mathrm{Tr}\,\left[(\not{p} + m_e)(Q^\mu - m\gamma^\mu)(\not{k} + m_\mu)(Q_\mu - m\gamma_\mu)\right]\,,\end{aligned} \tag{12.80}$$

with $Q = p + k$ and $m = m_e + m_\mu$. Computing the trace and the resulting scalar products (we need only $2(pk) = m_e^2 + m_\mu^2$),

$$|\mathcal{M}_{fi}|^2 = \frac{e^2}{2m_\mu^2}|F_2|^2(m_\mu^2 - m_e^2)^2\,, \tag{12.81}$$

and the result for Γ follows immediately. The resulting bound on $|F_2|$ is $|F_2| < 1 \times 10^{-13}$.
Supersymmetric extensions of the SM predict a non-zero decay rate for $\mu^- \to e^-\gamma$, at a level not far from the experimental bound. Actually, in this case the effective current j_μ that mediates the transition has a structure $V - A$ rather than a pure vector current as we have taken in this exercise. The modification to the calculation amounts simply to the insertion of a projector $(1 - \gamma^5)/2$ between $\bar{u}_e$ and u_μ.

(8.3) (i) From the form of the Lagrangian, we see that the amplitude is proportional to G_F and therefore $\sigma \sim G_F^2$. Since G_F is the inverse of a mass squared, and the only other energy scale is the CM energy $\sqrt{s}$, for dimensional reasons we must have $\sigma \sim G_F^2 s$.
(ii) The amplitude is $\mathcal{M} = \mathcal{M}_W + \mathcal{M}_Z$ with

$$\mathcal{M}_W = -\frac{G_F}{\sqrt{2}}[\bar{\nu}_e\gamma_\mu(1-\gamma^5)e][\bar{e}\gamma^\mu(1-\gamma^5)\nu_e]\,, \tag{12.82}$$
$$\mathcal{M}_Z = -\frac{G_F}{\sqrt{2}}[\bar{\nu}_e\gamma_\mu(1-\gamma^5)\nu_e][a_2\,\bar{e}\gamma^\mu(1-\gamma^5)e + a_3\,\bar{e}\gamma^\mu(1+\gamma^5)e)]\,,$$

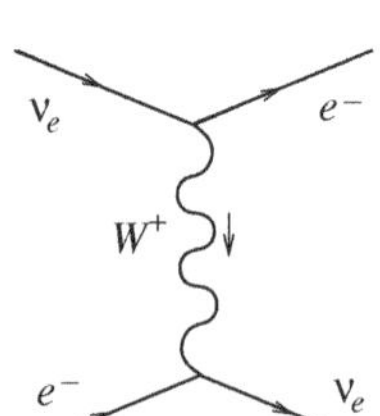

Fig. 12.5 The $e\nu_e$ scattering amplitude mediated by the W boson.

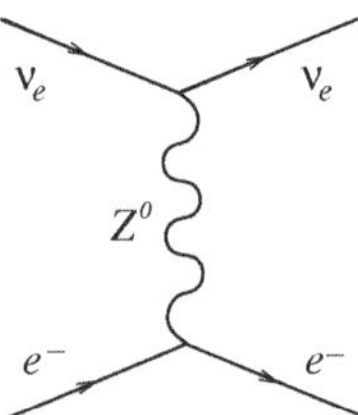

Fig. 12.6 The $e\nu_e$ scattering amplitude mediated by the Z boson.

with a_2, a_3 given in eq. (8.19); we already set $a_1 = 1/2$, and we took into account that the term $j^0_\mu j^{0,\mu}$ produces two equal contributions to the process, one in which the neutrino current is provided by the first j^0_μ (and therefore the electron current by the second) and one in which the neutrino current is provided by the second factor $j^{0,\mu}$. At a fundamental level, $\mathcal{M}_W$ and $\mathcal{M}_Z$ correspond to the graphs in Figs. 12.5 and 12.6.
Performing the Fierz rearrangement in $\mathcal{M}_W$, we get

$$\mathcal{M} = -\frac{G_F}{\sqrt{2}}[\bar{\nu}_e\gamma_\mu(1-\gamma^5)\nu_e] \times[(\frac{1}{2}+\sin^2\theta_W)\bar{e}\gamma^\mu(1-\gamma^5)e + \sin^2\theta_W\bar{e}\gamma^\mu(1+\gamma^5)e)]\,. \tag{12.83}$$

The computation of $|\mathcal{M}|^2$, with the usual average and sum over spins, and the subsequent computation of the scalar product in the CM frame is now rather straightforward, and the result is

$$\sigma(\nu_e e \to \nu_e e) = \frac{G_F^2 s}{\pi}\left[\left(\frac{1}{2}+\sin^2\theta_W\right)^2 + \frac{1}{3}\sin^4\theta_W\right] \simeq 0.176\, G_F^2 s\,. \tag{12.84}$$

(8.4) (i) Use eq. (6.21) with $n_2^0\, dV = 1$ (since we are considering a single target particle) and $\Gamma = dN/dt$.
(ii) $n \sim T^3$ follows from dimensional considerations if $m \ll T$, since then T is the only mass-scale and dimensionally $n = 1/\text{volume} = (\text{mass})^3$. Of course, it can also be obtained explicitly from the Boltzmann, Bose–Einstein or Fermi–Dirac distributions.
From the previous exercise, $\sigma \sim G_F^2 s$. At a temperature T much larger than all the masses in question, $s \sim T^2$. Furthermore we have seen that $n \sim T^3$, while for relativistic particles $v = 1$, so $\Gamma = n\sigma v \sim G_F^2 T^5$ and $\Gamma/H \sim (G_F^2 T^5)/(T^2/M_{\rm Pl}) \sim (T/1\,\text{MeV})^3$. Therefore for $T \gg$ MeV neutrino–electron scattering maintained the neutrinos in equilibrium, while when the temperature of the Universe dropped around $O(1)$ MeV the neutrinos decoupled. Observe that when $T \sim$ MeV the electron mass is not negligible compared to T, but $T \sim m_e$ so we still have only one mass-scale and the estimate $s \sim T^2$ is still correct.

Bibliography

The literature on quantum field theory and related topics is vast. Here we collect only a few general references that we find especially useful. More references on specific topics are given in the Further Reading sections, at the end of most chapters.

Coleman, S. (1985). *Aspects of Symmetry: Selected Erice Lectures.* Cambrigde University Press, Cambridge.

Di Giacomo, A., Paffuti, G. and Rossi, P. (1994). *Selected Problems in Theoretical Physics* (With Solutions). World Scientific, Singapore.

Georgi, H. (1984). *Weak Interactions and Modern Particle Theory.* Benjamin/Cummings, Menlo Park, CA.

Georgi, H. (1999). *Lie Algebras in Particle Physics*, 2nd edition. Perseus Books, Reading, MA.

Itzykson, C. and Zuber, J.-B. (1980). *Quantum Field Theory.* McGraw-Hill, Singapore.

Jackson, J. D. (1975). *Classical Electrodynamics.* Wiley, Chichester.

Kolb, E. W. and Turner, M. S. (1990). *The Early Universe.* Addison-Wesley, Reading, MA.

Landau, L. D. and Lifshitz, E. M. (1979). *Course of Theoretical Physics*, vol.II: *The Classical Theory of Fields.* Pergamon Press, Oxford.

Landau, L. D. and Lifshitz, E. M. (1977). *Course of Theoretical Physics*, vol.III: *Quantum Mechanics: Non-Relativistic Theory.* Pergamon Press, Oxford.

Landau, L. D. and Lifshitz, E. M. (1982). *Course of Theoretical Physics*, vol.IV (by Berestetskij, V. B., Lifshitz, E. M. and Pitaevskij, L. P.): *Quantum Electrodynamics.* Pergamon Press, Oxford.

Mandl, F. and Shaw, G. (1984). *Quantum Field Theory.* Wiley, Chichester.

Nakahara, M. (1990). *Geometry, Topology and Physics.* IOP Publishing, Bristol.

Okun, L. B. (1982). *Leptons and Quarks.* North-Holland, Amsterdam.

Parisi, G. (1988). *Statistical Field Theory.* Addison-Wesley, Redwood.

Perkins, D. H. (2000). *Introduction to High Energy Physics*, 4th edition. Cambridge University Press, Cambridge.

Peskin M. E. and Schroeder D. V. (1995). *An Introduction to Quantum Field Theory.* Perseus Books, Reading, MA.

Polchinski, J. (1998). *String Theory*, vol. I. Cambridge University Press, Cambridge.

Ramond, P. (1990). *Field Theory: A Modern Primer*, 2nd edition. Addison-Wesley, Redwood.

Weinberg, S. (1995). *The Quantum Theory of Fields.* vol. 1: *Foundations.* Cambridge University Press, Cambridge.

Weinberg, S. (1996). *The Quantum Theory of Fields.* vol. 2: *Modern Applications*, Cambridge University Press, Cambridge.

Zinn-Justin, J. (2002). *Quantum Field Theory and Critical Phenomena.* Oxford University Press, Oxford.

Index

Printed and bound by CPI Group (UK) Ltd, Croydon, CR0 4YY